Economic Commission for Europe
Geneva

ECONOMIC SURVEY OF EUROPE IN 1996-1997

Prepared by the
SECRETARIAT OF THE
ECONOMIC COMMISSION FOR EUROPE
GENEVA

UNITED NATIONS
New York and Geneva, 1997

NOTE

The designations employed and the presentation of the material in this publication do not imply the expression of any opinion whatsoever on the part of the Secretariat of the United Nations concerning the legal status of any country, territory, city or area, or of its authorities, or concerning the delimitation of its frontiers or boundaries.

UNITED NATIONS PUBLICATION
Sales No. E.97.II.E.1
ISBN 92-1-116666-7 ISSN 0070-8712

Copyright © United Nations, 1997
All rights reserved
Printed at United Nations, Geneva (Switzerland)

CONTENTS

	Page
Explanatory notes	viii
Abbreviations	ix
Preface	xi

			Page
Chapter 1	AN OVERVIEW OF RECENT DEVELOPMENTS AND SELECTED ISSUES IN THE ECE ECONOMIES		1
Chapter 2	THE WESTERN MARKET ECONOMIES		13
	2.1	The general context	13
	2.2	Output and demand	16
		(i) Recent developments	16
		(ii) Growth and volatility	25
		(iii) Germany's transition economy: development of the eastern *Länder*, 1990-1996	29
	2.3	Labour markets	34
		(i) Employment	34
		(ii) Unemployment	37
		(iii) Labour market policies	39
	2.4	Costs and prices	41
		(i) World commodity prices	41
		(ii) Domestic costs and prices	42
	2.5	External balances	46
	2.6	The stances of economic policy	47
		(i) Monetary policy	47
		(ii) Fiscal policy	52
	2.7	The short-run outlook	54
Chapter 3	THE TRANSITION ECONOMIES		59
	3.1	The general context	59
		(i) Expectations, outcomes and outlook	59
		(ii) Monetary policies	65
		(iii) The crisis in Bulgaria	75
	3.2	Output and demand	84
		(i) Output	84
		(ii) Demand	98
	3.3	Labour markets	110
		(i) Overview	110
		(ii) Employment	111
		(iii) Unemployment	114
	3.4	Costs and prices	119
		(i) Overview	119
		(ii) Consumer prices, performance and prospects	119
		(iii) Producer price inflation in industry	124
		(iv) Wages and unit labour costs in industry	125

			Page
3.5	Foreign trade		128
	(i)	Overall trade results	128
	(ii)	East European and Baltic countries	131
	(iii)	Russian Federation: trade with non-CIS countries	139
	(iv)	Other CIS countries: trade with non-CIS countries	142
	(v)	Intra-CIS trade	145
3.6	External financial developments in the transition economies		147
	(i)	Overview	147
	(ii)	Balance of payments, reserves and debts	148
	(iii)	External financing	162

Chapter 4 THE CENTRAL ASIAN ECONOMIES, 1991-1996 179

4.1	The impact of quitting a centripetal economy		179
	(i)	The management shock	181
	(ii)	The shock of price liberalization	182
4.2	Macroeconomic instabilities		182
	(i)	Policies to increase revenue	183
	(ii)	Policies to reduce expenditure	185
	(iii)	Overhaul of the public finance and statistical agencies	186
4.3	The post-independence recession		186
	(i)	Aggregate decline and upturn	186
	(ii)	Disinvestment and new capital formation	188
	(iii)	Labour supply and migration	189
	(iv)	The inactive population	191
	(v)	Levels of living	192
4.4	Institutions for a market economy		193
	(i)	Establishing new currencies	193
	(ii)	Privatization	197
	(iii)	Market-generating infrastructure	200
4.5	Diversification, competitiveness and sustainability		201
	(i)	Dependence on natural resources	201
	(ii)	Environmental damage and repair	203
	(iii)	New avenues of trade	204
	(iv)	Foreign investment and external support	207
4.6	Economic aims after five years of independent statehood		209

STATISTICAL APPENDIX 213

LIST OF TABLES

Table		Page
2.2.1	Quarterly changes in real GDP in the ECE market economies, 1995-1996	16
2.2.2	Annual changes in industrial production in the ECE market economies, 1993-1996	19
2.2.3	Annual changes in real GDP in the ECE market economies, 1994-1997	19
2.2.4	Annual changes in major domestic expenditure items in the ECE market economies, 1995-1996	22
2.2.5	Annual changes in government fixed investment in the ECE market economies, 1992-1996	23
2.2.6	Annual changes in the real foreign balance in the ECE market economies, 1995-1996	23
2.2.7	Contribution of major expenditure items to annual changes of GDP in the ECE market economies, 1995-1996	24
2.2.8	Average annual growth rates of real GDP in the ECE market economies, 1960-1996	25
2.2.9	Volatility of year-to-year changes in real GDP in the ECE market economies, 1961-1996	26
2.2.10	Convergence indicators for east Germany, 1991 and 1996	30
2.3.1	Employment in the ECE market economies, 1993-1996	35
2.3.2	Labour productivity in the ECE market economies, 1993-1996	36
2.3.3	Part-time employment in the ECE market economies, 1983, 1993 and 1995	37
2.3.4	Standardized unemployment rates in the ECE market economies, 1993-1996	38
2.3.5	Youth unemployment rates in the ECE market economies, 1994-1996	40
2.4.1	World commodity prices in dollars, 1993-1996	42
2.4.2	Unit labour costs in the ECE market economies, 1993-1996	43
2.4.3	Measures of inflation in the ECE market economies, 1993-1996	44
2.5.1	Changes in the volume of merchandise exports and imports in the ECE market economies, 1994-1996	46
2.5.2	Current account balances in the ECE market economies, 1994-1996	47
2.6.1	General government net lending in the ECE market economies, 1995-1997	53
3.1.1	Basic data on the transition economies, 1994-1996	60
3.1.2	GDP: expectations and outcomes in the transition economies, 1995-1997	61
3.1.3	Gross industrial output in the transition economies, 1995-1997	61
3.1.4	Foreign trade involvement of the transition economies, 1994-1996	62
3.1.5	Monetization in selected transition countries: share of monetary aggregates in GDP, 1994-1996	72
3.1.6	Constraints on monetary policy instruments under different exchange rate and monetary regimes	75
3.2.1	Share of major sectors in GDP in eastern Europe and the Baltic states, 1991-1996	89
3.2.2	Share of major sectors in GDP in the CIS countries, 1991-1996	90
3.2.3	Gross output of agriculture and construction in the transition countries, 1991-1996	91
3.2.4	Official estimates of the private sector's contribution to GDP in selected transition countries, 1991-1996	92
3.2.5	Measured labour productivity in the transition countries, 1991-1996	93
3.2.6	Growth of industrial output by branch in eastern Europe and the Baltic states, 1994-1996	94
3.2.7	Growth of industrial output by branch in the CIS, 1994-1996	95
3.2.8	Capacity utilization in manufacturing industry in selected transition countries, 1992-1996	96
3.2.9	Contribution of final demand components to real GDP growth in selected transition countries, 1992-1996	99
3.2.10	Composition of final demand in current prices in selected transition countries, 1994-1996	100
3.2.11	Real consumption in selected transition countries, 1993-1996	100
3.2.12	Retail trade and real net wages in the transition countries, 1993-1996	101
3.2.13	Investment in selected transition countries, 1993-1996	102
3.2.14	Machinery and equipment share in total investment of selected transition countries, 1991-1996	106
3.3.1	Total employment in the transition countries, 1990-1996	113
3.3.2	Registered unemployment in the transition countries, 1994-1996	115
3.3.3	Comparative measures of unemployment in selected transition countries, QI-QIII 1996	116
3.3.4	Changes in the labour force in selected transition countries, 1990-1995	117
3.3.5	Registered unemployment rates by sex in selected transition countries, 1995-1996	118
3.3.6	Share of women, youth and long-term unemployed in total unemployment in selected transition countries, 1995-1996	118
3.4.1	Consumer prices in the transition countries, 1995-1996	120
3.4.2	Producer prices in industry in the transition countries, 1995-1996	125
3.4.3	Wages and unit labour costs in industry in the transition economies, 1995-1996	127
3.5.1	Foreign trade of the European transition countries by direction, 1994-1996	129
3.5.2	Trade balances of the European transition countries, 1992-1996	130
3.5.3	Foreign trade of CEFTA countries by direction, 1994-1996	132
3.5.4	Foreign trade of south-east European countries by direction, 1994-1996	133
3.5.5	Foreign trade of the Baltic states by direction, 1994-1996	134
3.5.6	Indicators of foreign trade performance for selected east European and Baltic countries, 1995-1996	135
3.5.7	CIS countries' trade with non-CIS countries, 1994-1996	142

Table		Page
3.5.8	Foreign trade of selected CIS countries by direction, 1995-1996	143
3.5.9	CIS countries' trade within the CIS region, 1994-1996	146
3.6.1	Current account balances of eastern Europe, the Baltic countries and European members of the CIS, 1994-1997	149
3.6.2	Revisions to the Polish balance of payments, 1995-1996	151
3.6.3	Foreign trade in goods and non-factor services of eastern Europe, the Baltic countries and European members of the CIS, 1994-1996	152
3.6.4	Foreign trade in goods and non-factor services of eastern Europe, the Baltic countries and European members of the CIS, 1994-1996	153
3.6.5	Foreign exchange reserves of eastern Europe, the Baltic countries and European members of the CIS, 1990, 1993-1997	155
3.6.6	Total external debt of eastern Europe, the Baltic countries and European members of the CIS, 1990, 1993-1996	156
3.6.7	Measures of the debt burden of eastern Europe, the Baltic countries and European members of the CIS, 1990, 1994-1996	157
3.6.8	Czech Republic and Hungary: foreign liabilities by type of debtor, 1992-1996	158
3.6.9	Net capital flows into eastern Europe, the Baltic countries and the European members of the CIS countries, 1993-1996	162
3.6.10	Net capital flows into eastern Europe, the Baltic countries and the European members of the CIS, by type of capital, 1994-1996	163
3.6.11	Medium- and long-term funds raised on the international financial markets by eastern Europe, the Baltic countries and the CIS, 1992-1997	165
3.6.12	International credit ratings of eastern Europe, the Baltic countries and the CIS	166
3.6.13	International equity issues of eastern Europe, the Baltic countries and the CIS, 1993-1997	167
3.6.14	Sample of proposed international issues of eastern Europe, the Baltic countries and the CIS in 1997	168
3.6.15	Flows of net foreign direct investment into eastern Europe, the Baltic countries and the CIS, 1994-1996	169
3.6.16	Foreign direct investment of eastern Europe, the Baltic countries and the CIS, 1992, 1994-1996	170
3.6.17	Foreign direct investment inflows and proceeds from privatization, 1993-1997	171
3.6.18	Balance of payments of selected transition economies, 1994-1996	175
4.1.1	GDP growth in the central Asian economies, 1992-1996	181
4.1.2	Consumer price inflation in the central Asian economies, 1992-1996	182
4.2.1	Budget disequilibrium in relation to GDP and general government expenditure in the central Asian economies, 1992-1996	183
4.2.2	Government revenue and expenditure in relation to GDP in the central Asian economies, 1989-1996	184
4.3.1	Production aggregates in the central Asian economies, 1996	187
4.3.2	Projections of working age and total population in the central Asian economies, 1995 and 2010	190
4.3.3	Registered unemployment and not gainfully occupied population of working age in the central Asian economies, 1992, 1995 and 1996	192
4.3.4	Allocation of total household expenditure in the central Asian economies, 1991, 1993 and 1995-1996	193
4.4.1	Real exchange rates of new central Asian currencies against the dollar	195
4.4.2	Number of enterprises privatized in Kazakstan	197
4.4.3	Privatization plans in Kazakstan, 1994-1995 and 2000	198
4.4.4	Number of enterprises privatized in Kyrgyzstan, 1992-1996	198
4.4.5	Number of enterprises privatized in Tajikistan, 1991-1996	199
4.4.6	Privatization in Turkmenistan, end-1995	199
4.4.7	Privatization in Uzbekistan, 1992-1995	199
4.5.1	Annual water resources per person, average 1989-1990	204
4.5.2	Trade with CIS members and non-CIS countries, 1995	205
4.5.3	Trade with CIS members and non-CIS countries, 1996	205
4.5.4	Exports to and imports from non-CIS partners, 1994-1996	205
4.5.5	Progress in trilateral and quadrilateral integration, 1994-1996	206
4.5.6	Comparison of partner trade returns, 1995	206
4.5.7	Net inflows of foreign direct investment in the central Asian economies, 1991-1996	208
4.5.8	External indebtedness in the central Asian economies, 1993 and 1995	209

LIST OF CHARTS

Chart		Page
2.1.1	Consensus forecasts of annual changes in GDP in western Europe and the United States, 1996 and 1997	14
2.1.2	Comparison of actual outcomes and medium-term projections of GDP growth, 1988-2000	15
2.2.1	Quarterly changes in real GDP in the western market economies, 1993-1996	17
2.2.2	Industrial output in the western market economies, 1992-1996	17
2.2.3	Business cycle indicators for the European Union and the United States, January 1990-January 1997	18
2.2.4	Growth and volatility in the industrialized countries, 1960-1996	27
2.2.5	Volatility and growth in the industrialized countries, 1960-1996	29
2.2.6	Real GDP in east and west Germany, 1991-1996	30
2.3.1	Employment expectations in the European Union, 1991-1996	35
2.4.1	World commodity prices, 1994-1996	41
2.4.2	Selling-price expectations in manufacturing industry in the major west European countries, 1993-1996	45
2.6.1	Real short-term interest rates in the seven major economies, January 1995-January 1997	49
2.6.2	Nominal long-term interest rates in selected industrial countries, January 1995-January 1997	50
2.6.3	Average monthly bilateral exchange rates, January 1993-January 1997	51
2.6.4	Fiscal impulses, economic growth and output gaps in the European Union and the United States, 1989-1997	55
3.1.1	Indices of real money supply and composition of nominal broad money in selected transition economies, 1995-1996	67
3.1.2	Structure of domestic public debt in Bulgaria, 1993-1996	77
3.1.3	Budget expenditure and budget deficit in Bulgaria, 1992-1996	78
3.1.4	Exchange rate indices of Bulgaria, 1992-1996	79
3.1.5	Interest rate differential and forex stocks in Bulgaria, 1993-1996	80
3.2.1	Indices of gross industrial output in selected transition countries, 1995-1996	85
3.2.2	Change in income inequality during the transition and consumption in 1993-1995 for selected transition countries	108
3.2.3	Dollar exports and GDP in selected transition countries, 1993-1995	110
3.4.1	Consumer prices in the transition countries, 1995-1996	123
3.4.2	Consumer and industrial producer prices in the transition countries, 1995-1996	126
3.5.1	Specific western demand for selected transition countries' exports, 1992-1997	137

LIST OF BOXES

Box		Page
2.3.1	Standardized unemployment rates	39
3.1.1	Currency boards	70

EXPLANATORY NOTES

The following symbols have been used throughout this *Survey*:

    .. = not available or not pertinent

    − = nil or negligible

    * = estimate by the secretariat of the Economic Commission for Europe

    | = break in series

In referring to a combination of years, the use of an oblique stroke (e.g. 1995/96) signifies a twelve-month period (say, from 1 July 1995 to 30 June 1996). The use of a hyphen (e.g. 1995-1996) normally signifies either an average of, or a total for, the full period of calendar years covered (including the end-years indicated).

Unless the contrary is stated, the standard unit of weight used throughout is the metric ton. The definition of "billion" used throughout is a thousand million. The definition of "trillion" used throughout is a thousand billion. Minor discrepancies in totals and percentages are due to rounding.

References to dollars ($) are to United States dollars unless otherwise specified.

The membership of the United Nations Economic Commission for Europe (UN/ECE) consists of all the states of western Europe, eastern Europe and the territory of the former Soviet Union, North America and Israel.

The term *transition countries*, as used in the text and tables of this publication, refers to the formerly centrally planned economies of eastern Europe and the former Soviet Union. *Eastern Europe* refers to the economies of Albania, Bulgaria, Hungary, Poland, Romania, the successor states of the former Czechoslovakia (the Czech Republic and Slovakia) and the successor states of the former Socialist Federal Republic of Yugoslavia (Bosnia and Herzegovina, Croatia, Slovenia, The former Yugoslav Republic of Macedonia and Yugoslavia). Among the now independent republics of the former Soviet Union, distinction is made between the *Baltic states* (Estonia, Latvia and Lithuania) and the remaining republics which cooperate within the institutional framework of the Commonwealth of Independent States - the *CIS countries*.

The analysis in this *Survey* is based on data and information available to the secretariat of the Economic Commission for Europe up to mid-March 1997.

ABBREVIATIONS

ADB	Asian Development Bank
ADR	American depository receipt
BIS	Bank for International Settlements
CEFTA	Central European Free Trade Agreement
CETE	central European transition economies
c.i.f.	cost, insurance and freight
CIS	Commonwealth of Independent States
CMEA	Council for Mutual Economic Assistance
COMTRADE	Commodity Trade Database (of United Nations Statistical Office)
CPC	Caspian Pipeline Consortium
CPI	consumer price index
DIW	German Institute for Economic Research
EBRD	European Bank for Reconstruction and Development
ECB	European Central Bank
ECOFIN	Council of the Economic and Finance Ministers (of the European Union)
ECU	European currency unit
EFF	Extended Fund Facility (of IMF)
EMS	European monetary system
EMU	economic and monetary union
ERM	exchange rate mechanism
ESAF	Enhanced Structural Adjustment Facility (of IMF)
ESCAP	Economic and Social Commission for Asia and the Pacific
EU	European Union
FDI	foreign direct investment
FESAL	Financial and Enterprise Sector Adjustment Loan (of World Bank)
f.o.b.	free on board
G-7	Group of Seven
G-24	Group of Twenty-Four
GDP	gross domestic product
GDR	global depository receipt
GNP	gross national product
HS	Harmonized Commodity Description and Coding System
HWWA	Hamburg Institute for Economic Research
IDB	Islamic Development Bank
ILO	International Labour Office
IMF	International Monetary Fund
LFS	labour force survey
MFN	most favoured nation
NAIRU	non-accelerating inflation rate of unemployment
NIESR	National Institute of Economic and Social Research
NMP	net material product

OECD	Organization for Economic Co-operation and Development
OPT	outward processing trade
PMI	Purchasing Managers' Index
PPI	producer price index
R&D	research and development
SDR	special drawing rights
SETE	south European transition economies
SITC	Standard International Trade Classification
TACIS	Technical Assistance for the Commonwealth of Independent States
UNDP	United Nations Development Programme
UN/ECE	United Nations Economic Commission for Europe
UNESCO	United Nations Educational, Scientific and Cultural Organization
VAT	value added tax
WTO	World Trade Organization

PREFACE

The present *Survey* is the fiftieth in a series of reports prepared by the secretariat of the Economic Commission for Europe to serve the needs of the Commission and to help in reporting on world economic conditions.

The *Survey* is published on the responsibility of the secretariat, and the views expressed in it should not be attributed to the Commission or to its participating Governments.

The prepublication text of this *Survey* was completed in March 1997 as a document for the 52nd session of the Economic Commission for Europe. This text, finalized in May 1997, incorporates minor changes.

Chapter 1

AN OVERVIEW OF RECENT DEVELOPMENTS AND SELECTED ISSUES IN THE ECE ECONOMIES

(i) Introduction: the current outlook in the ECE region

Economic growth varied considerably last year throughout the ECE region of Europe, the CIS and North America. Among the market economies growth in western Europe (1.9 per cent) was fairly modest and less than in 1995, but it was much stronger in the United States (2.4 per cent), and unexpectedly, stronger than in 1995. In eastern Europe a fairly high average growth rate was maintained in the CEFTA countries of central Europe (5 per cent) and there was solid recovery in the Baltic states (3.4 per cent). The long decline in GDP continued in Russia (-6 per cent) but there were recoveries in many of the other members of the CIS.

Rates of inflation have continued to fall throughout the region. In western Europe the average rate of increase in consumer prices (2.4 per cent) was at its lowest in 30 years. In eastern Europe rates have also continued to decline although progress is still slow in several countries and there were major setbacks in south-east Europe. There were very large drops in the inflation rates of the CIS.

There were serious setbacks to the transition process in south-eastern Europe last year. The most dramatic were in Albania and Bulgaria but many of the factors behind these crises are present in other countries in the region and in the CIS, and many are rooted in history and the difficult initial conditions in which these countries began to construct market economies. (These issues are discussed below.)

The outlook for 1997 is not radically different from the outcome of last year. In western Europe GDP growth is expected to improve moderately (to perhaps just under 2.5 per cent), not enough to make much difference to current levels of unemployment. The outcome could possibly be weaker than forecast given the efforts being made to meet the Maastricht criteria and keep inflation at or below 2 per cent. After seven years of expansion, short-term indicators and forecasts in early 1997 were still pointing to a further slight acceleration in United States GDP growth to 2.5 per cent for the year as a whole. In eastern Europe, the average growth rate is expected to slow somewhat to 3½ per cent in 1997 (against 4 per cent last year). This slowdown mainly reflects developments in south-east Europe where tough stabilization programmes are expected to produce falls in GDP in Bulgaria and Romania. (Falling output in Albania also seems highly probable, with or without a stabilization programme.) Relatively high rates of growth (4-6 per cent) are expected to be maintained in Poland, Slovakia and Slovenia; and growth should pick up in Hungary after a successful stabilization programme. The recovery in the Baltic states is also expected to continue and strengthen with growth rates in the region of 3½ to 5 per cent.

There was a large increase in balance of payments deficits in eastern Europe and the Baltic states in 1996 and a further deterioration is expected in several of these countries in 1997. Many of them now have deficits over 5 per cent of GDP and, although for most of them finance has not been a problem, it is likely that some adjustments will eventually be made to check the growth of domestic demand.

Assessing the short-run economic outlook for Russia still remains extremely difficult. The government and some independent forecasters expect a modest (2 per cent) upturn in output in 1997 but there are still many Russian economists who think that even zero growth — i.e. bringing the seven-year decline to an end — might be the best to hope for. Much depends on whether the problem of arrears can be dealt with in a non-inflationary way. In Ukraine, the economy is still in deep depression but it could at least bottom out in 1997, although the government hopes for growth of 1.7 per cent. In the other CIS member countries the recovery seen in many of them last year is expected to continue, even if some governments' growth targets appear rather ambitious.

(ii) Eastern Europe, the Baltic states and the CIS

(a) Economic growth

In many respects 1996 was a disappointing year for the transition economies. Economic growth in eastern Europe averaged 4 per cent, less than in 1995 and generally less than expected; in Russia the slump in output deepened rather than bottoming out; the expected boost to activity in south-eastern Europe resulting from the ending of economic sanctions against Yugoslavia has been slow to materialize; there were major economic setbacks in Bulgaria and Romania; and in Albania an economic crisis developed into a state of political and social chaos in which the government lost control of large parts of the country.

But, as has been repeatedly emphasized in previous issues of this *Survey*, the transition economies are a very heterogenous group, and despite the many setbacks of the last year there was also notable progress in many areas. Growth remained relatively strong among some of the leading reformers and especially in Poland, where 6 per cent growth was more or less as forecast, and in Slovakia, where at nearly 7 per cent it was more than forecast. Growth was also sustained in the Czech economy although it was markedly lower than in Poland and Slovakia and a little below expectations. There was also positive growth in all three Baltic states (between 2.5 and 4 per cent); and, for the first time since the breakup of the Soviet Union, there was positive growth in a majority of the 12 members of the CIS.[1]

The slowdown in eastern Europe as a whole in 1996 was greater than expected at the start of the year. Some deceleration had been expected in those economies that had been growing relatively fast in 1995 (the Czech Republic, Poland, Romania and Slovakia) and this in fact occurred except in Slovakia; but it was also expected that this would be offset by faster growth in Bulgaria and the countries formerly constituting the SFR of Yugoslavia, and that growth, albeit fairly modest at 2 per cent, would be maintained in Hungary. In the event, a persistent failure to implement a coherent programme of structural reform finally came to a head in Bulgaria in the spring of 1996 with a massive financial and economic crisis which has still to be resolved; instead of growing by a forecast 3 per cent, Bulgarian GDP actually fell by 10 per cent.[2] This was one of the principal factors behind the slowdown in eastern Europe as a whole, but growth in the Hungarian economy was also weak, below the rate in 1995 and well below earlier forecasts. Although the measures introduced in 1995 to restore internal and external balance to the Hungarian economy have made considerable progress, their impact on domestic demand, and especially on household consumption, has been more severe and more long lasting than expected. Economic growth in Slovenia also continued to slow down in 1996 despite earlier forecasts of an acceleration: an important reason for this was the weakness of industrial production which was affected by the weaker demand abroad for exports.

In the other countries of south-east Europe (Albania apart) output increased (for the first time in The former Yugoslav Republic of Macedonia since 1989) but generally by much less than was forecast – or in some cases, hoped for – at the start of the year. The Croatian economy is still suffering from the economic effects of the war and the breakup of ties with the other countries of the former SFR of Yugoslavia; tight monetary and fiscal policies together with an appreciating exchange rate, all targeted at preventing any rekindling of inflation, had a restraining effect on the recovery of output until mid-year. The Yugoslav economy started to grow after economic sanctions had been lifted but, given the low level to which activity had fallen, the recovery has been very slow and far below the government's expectations. Sustained recovery of the Yugoslav economy will be highly dependent on imports but its export capacity has been considerably reduced as a result of the war and sanctions. Although a large trade deficit has been financed so far mainly by foreign currency accounts held abroad, these are unlikely to be able to sustain output for very long.

In most of the transition countries, economic growth in 1996 was sustained by domestic demand, both private consumption and fixed investment, with net exports making a negative contribution. A striking feature was the boom in fixed investment in most of the leading transition economies such as the Czech Republic, Poland, Slovakia and Slovenia. In the Czech Republic this was concentrated on the non-financial business sector, in Poland it was broadly based, while in Slovakia and Slovenia it has largely been made in infrastructure (roads, telecommunications, etc.). The revival of a sustained rate of fixed investment is one of the most positive features of recent developments in these countries as it demonstrates that domestic (and foreign) investors are now more firmly confident about the longer run economic prospects of these economies. There has also been a recovery, slight so far, in fixed investment in Estonia and Latvia; and it was beginning to pick up in Bulgaria before last year's crisis. In contrast, the rapid decline in fixed investment in Russia continued in 1996, a predictable consequence of the continuing uncertainty over the political and economic outlook.

(b) Employment and unemployment

One consequence of the slowdown in east European growth was that the hesitant improvement in the level of employment during 1995 came to a halt in the first three quarters of 1996. There were some increases in the numbers employed in some of the faster growing economies – the Czech Republic, Poland, Slovakia – but these were much smaller than the gains in 1995. There was, however, an increase in Lithuania (of 1.2 per cent), the first since independence and following three years of positive output growth. But everywhere else, and including most members of the CIS, total employment

---

[1] The fall in output ended in all the central Asian republics except Tajikistan. These republics are the subject of a special study in chap. 4 of this *Survey*.

[2] The Bulgarian crisis is discussed at length in chap. 3.1(iii) of this *Survey*.

continued to fall. In industry, where investment and productivity have been rising in most of eastern Europe and the Baltic states, the decline in employment continued in virtually every country for which data were available (for the first three quarters of 1996). Despite the rise in east European output since 1993 (GDP up 14 per cent and industrial output up by 23 per cent between 1993 and 1996), there was virtually no change in total employment. The recovery of output has in fact been accompanied by continued labour shedding in some countries and productivity in industry has grown rapidly: in addition to the gains from restructuring output and investing in new equipment, the recovery from low levels of capacity utilization greatly increased the productivity of those already employed. Also "excess employment" probably still exists in some of the remaining large, state owned enterprises in eastern Europe. The recovery in output will thus have to proceed much further before there is a significant net increase in the number of new jobs. (The problem of "excess employment" is probably very much greater in the Baltics and the CIS where the fall in aggregate output between 1989 and 1996 was three and four times larger than that in the level of employment.)

The slowdown in growth rates has also checked the decline in the number of people unemployed in eastern Europe. This had peaked at about 7.5 million people in early 1994 and declined slowly to 6 million in the third quarter of 1996; but in the last quarter there was an increase of nearly 110,000, more than two thirds of them in Bulgaria. Unemployment levels have continued to rise steadily in the CIS and at the end of 1996 the number registered as unemployed was just over 8 million. (This figure is almost certainly a considerable underestimate of the numbers unemployed according to the standard ILO definition.) Altogether, there were at least 14 million people unemployed in all the ECE transition economies at the end of 1996.

High levels and extended periods of unemployment tend to "discourage" some workers from even trying to find another job and they therefore tend to drop out of the labour force. This effect seems to be the main explanation for the continuing slight fall in unemployment *rates* in 1996. Nevertheless, the average rate in eastern Europe remained high, at 11.8 per cent at the end of last year, and for most individual countries it ranged between 10.5 per cent in Hungary and 16 per cent in Croatia. The rate remained exceptionally low in the Czech Republic, at 3.5 per cent at the end of 1996, although it had also risen from 2.7 per cent in the second quarter.

Unemployment rates in the Baltic states and in the CIS are generally much lower than those in eastern Europe, but this is largely a reflection of shortcomings in the coverage of the figures for registered unemployment. Estimates based on the more comprehensive labour force surveys, which are not yet being conducted in all countries, suggest that the true unemployment rates in these countries all tend to fall within the range quoted above for eastern Europe.

Apart from sharing similar high rates of unemployment, the composition of the unemployed in eastern Europe and the Baltic states is also similar in a number of respects to that in western Europe. First, unemployment rates for women are generally much higher than for men, the exception being Hungary, and women also tend to account for more than half of the number unemployed in most of the transition economies (including Russia). Secondly, young people (under the age of 25) also have much higher unemployment rates and generally account for between a quarter and one third of all persons unemployed (in Romania their share was nearly 50 per cent). And, thirdly, large numbers of people have now been unemployed for longer than one year: labour force survey data in 1996 indicate that the long-term unemployed account for some 30 to 60 per cent of all the unemployed. Given the characteristics of the relationship between economic recovery and job creation, described above, the prospects for a significant reduction of unemployment rates in the near future are not very good and, as the experience of western Europe suggests, that implies that the prospects for reducing the number of long-term unemployed are probably even worse. The presence of a large pool of long-term unemployed, many of them young people, is a potential source of social instability. In the short run, active labour market policies and reasonably effective social safety nets can provide some temporary relief, but effective solutions depend on a sustained revival of fixed investment; investment-promoting policies should therefore be central to efforts to deal with the problem of unemployment.

(c) Falling inflation rates and rising cost pressures

Rates of inflation generally continued to fall in 1996 with particularly marked improvements in the Baltic states, in Russia and all of the other members of the CIS except Kyrgyzstan.

In eastern Europe, however, progress was less striking and less general than in 1995. First, in Albania, Bulgaria and Romania there were major reversals, especially in Bulgaria where consumer price inflation rose 311 per cent in the 12 months to December 1996 (compared with 33 per cent in 1995). Although much lower than in Bulgaria, the inflation rate virtually tripled in Albania and more than doubled in Romania. Second, the reduction of inflation rates elsewhere in eastern Europe was quite small and often less than the targets that governments had set for 1996. In Hungary and Poland progress remains slow and inflation remains in the neighbourhood of 19-20 per cent, although their current budget programmes assume reductions in 1997 to 15 per cent and 13 per cent respectively.

In the Baltic states inflation over the year was in the range of 13-15 per cent, half the rates in 1995, while in the second half of 1996 the annualized rate was in single digits in Estonia and Lithuania and significantly lower than in most of the more advanced reforming countries of eastern Europe. Tight control of monetary and fiscal policies, together with more stable exchange rates, were the major factors behind this outcome.

There was also a large reduction in the rate of inflation in Russia and most of the other CIS countries: from typically four- and three-digit rates of increase in 1994 and 1995, the annual average rate in 1996 fell to within a range of 19 to 48 per cent in more than half of them. Most of the CIS members have been pursuing more effectual macro-policies, especially monetary, with the assistance of the international financial institutions. But the considerable deceleration in the Russian inflation rate has also been important for the other CIS members given their still high degree of dependence on the Russian economy. In Russia the rate of inflation over the twelve months to December 1996 was just under 22 per cent, compared with over 130 per cent in 1995. The maintenance of a tight monetary policy throughout the year and a relatively strong rouble in the first half were important factors behind this performance, but there was also an element of suppressed inflation as well: large arrears of payments for wages and pensions helped to dampen domestic demand, and increases in administered prices, not necessarily unwisely, were deliberately held below the average inflation rate. (Unlike most of the other CIS countries, the large drop in the inflation rate in Belarus was largely achieved through government price subsidies, cheap credits to agriculture and de facto price controls.)

One of the main factors behind the slow fall in rates of inflation in eastern Europe has been the rapid growth in real wages which has generally outpaced the growth in productivity, even in countries such as Poland where both output and productivity have been rising quite rapidly. The pressures here come from a variety of sources. In the first place there is probably a strong desire on the part of wage earners to make up for the considerable loss in real wages that they suffered in the early years of the transition, and the recovery of output gives them at last the opportunity to attempt to do so. Moreover the underlying trend in productivity is still uncertain in many countries: fixed investment in infrastructure as well as in machinery and equipment is clearly rising strongly in a number of countries (Poland, the Czech Republic, Slovakia and Slovenia) and this should lead to larger improvements in productivity over the medium run; but in many other countries, the rise in output per head is still heavily influenced by the recovery of output from very low levels of resource utilization. The process of restructuring large state owned (or formerly state owned) enterprises is still fairly slow and in several countries the privatization process has also been hesitant and lacking in direction: these factors tend to discourage fixed investment (foreign and domestic) and to hold back improvements in productivity. Thus, increases in unit labour costs are generally high and tending to accelerate in most of eastern Europe.

The pressure for high nominal wage increases is also influenced by two other factors. The first is that many governments are still in the process of liberalizing prices and this is still tending to give periodic shocks to the consumer price index (a 1 per cent rise in the CPI in the Czech Republic in July, more than a quarter of the annual increase in Slovenia). Price deregulation is also moving into a number of sensitive areas in many countries such as housing rents, utility prices (including public transport), and energy. How far these changes can be contained as one-off increases in the consumer price index will depend partly on careful timing (the increase in Czech energy prices and rents last summer was partly cushioned by the seasonal fall in food prices) but even more on whether they can be absorbed by real wage increases supported by productivity growth.

Another pressure for higher average nominal wages comes from the expansion of the service sector in most of the transition economies. In general, prices in the service sector have tended to rise faster than non-food prices and in 1996, as in previous years, they were the main source of the rise in consumer prices. This partly reflects the increasing concentration of administered price liberalization in this sector, but it also arises from the general tendency for service sector productivity to grow more slowly than in the rest of the economy and from its relatively greater insulation from foreign competition. In a market economy intersectoral or interindustry differences in wage increases tend to be much less than the variance in their rates of productivity growth. Consequently, prices will tend to rise more quickly in sectors with lower productivity growth and little competition from abroad. In this respect the east European economies are now little different from those in the west. But a crucial aspect of this so-called Scandinavian model of inflation is which sector sets the pace for the increase in wages. If the scope for wage increases is set by productivity growth in the tradeables sector (plus or minus any change in world market prices) price increases in the service sector will still outpace those in the rest of the economy, but the profitability and competitiveness of tradeables will not be eroded. If, however, the model goes into "reverse", as appears to have happened in several west European economies in the 1970s and 1980s, and wage increases in the non-tradeable sectors outpace those in the rest of the economy, the profitability of export and import-competing production will diminish.

It is difficult to say whether such a "reverse" mechanism is currently at work in the east European economies: part of the increase in relative service prices reflects the one-off adjustments of price liberalization and the resurgence of a sector artificially suppressed under the former central planning system, and there are still limitations to the availability of data by sector. But in most east European and Baltic economies, nominal wages and unit labour costs in industry have been rising faster than producer prices over the past year, implying a sharp squeeze on gross operating profits and profit margins. In general, producer prices rose much less than consumer prices last year, and their deceleration from 1995 was also much greater. To a large degree this probably reflects the effect of exchange rate stability, which not only helps to dampen the cost of imported inputs but also sharpens competition from imports at the factory gate. (The higher rate of consumer price increases reflects the role of services mentioned above, but relatively weak (uncompetitive) distribution systems probably also play a role in sustaining relatively larger increases, especially when consumer demand is rising strongly.) Too much reliance on the exchange rate as an anti-inflationary device, however, can be a dangerous strategy if pursued for too long. It can weaken the competitiveness not only of exporters, but of all producers in the tradeables sector.

Whether these pressures will actually result in higher rates of inflation will depend on the stance of macroeconomic policy and the effectiveness of the instruments of monetary and fiscal policy. The stance of macroeconomic policy in nearly all the transition economies has been clearly aimed at eliminating inflation and external imbalances, and especially attenuating the effects of price liberalization. Considerable progress has been made towards establishing macroeconomic equilibrium, but the experience of 1996 underlines many of the difficulties still faced by the governments of the transition economies in trying to pursue their goals with a range of policy instruments which is still limited by the standards of the western market economies and which do not always have the effects predicted by textbook models based on the experience and institutions of mature market economies. Thus, in the transition economies the demand for money appears to be unstable and there is only a weak correlation, if any, between changes in the money supply and subsequent changes in consumer prices.[3] This appears to be due to weak or still missing institutions and to the relative thinness or absence of a full range of financial markets. Nevertheless, a number of countries (Czech Republic, Hungary, Poland, Slovakia, Slovenia and Croatia) have achieved significant success in lowering and controlling their inflation rates; but one consequence of this success is that they have been attracting large and rapid inflows of foreign capital which threaten to destabilize domestic monetary control and to push up the exchange rate, thereby further weakening the competitiveness of the tradeables sector. Various measures have been introduced to deal with this dilemma, the most common being a widening of the bands in which the exchange rate is allowed to fluctuate, but they have also been accompanied by a tightening of monetary policy (in the Czech Republic, Hungary, Poland, Slovakia, for example). The monetary authorities in these countries are thus increasingly having to pursue a narrow path between trying to check surges in foreign capital inflows, on the one hand, and trying to avoid a tightening of policy to the point at which it would threaten the recovery of output and fixed investment, on the other.

Apart from the difficulties of implementing stabilization policies in a transition economy, another major factor in determining whether inflationary pressures actually translate into higher prices is whether governments can resist the pre-election temptation to loosen policies in an attempt to boost their share of the vote. In Albania and Romania an excessive loosening of monetary policy in the run-up to elections played a role in the deterioration of their economies in 1996, and similar developments occurred elsewhere (for example in Slovenia) but with less severe consequences. The temptation to try to boost household incomes before an election is particularly strong in countries where the long transition process has so far yielded meagre results for large numbers of voters. The recent increase in the number of strikes and street demonstrations against the effects of tough stabilization policies and restructuring programmes, not least in some of the more advanced reforming countries, will intensify the pressures on governments to soften their policy stance. Advice and warnings from the international financial institutions and others to resist seem to have little effect in this context.

(d) *Deteriorating external accounts*

In the 1980s most of the countries of eastern Europe and the former Soviet Union were in surplus for most of the time on their merchandise trade. But from 1990 more and more of them moved into deficit as trade liberalization and the transformation process got underway and by 1995 all the east European and Baltic states were in deficit. In contrast, about half of the CIS countries were still in surplus, the outstanding case being Russia with a large and rising surplus.

In 1996 trade deficits deteriorated quite considerably in most of eastern Europe and the Baltic states and in most cases at an accelerating rate in the second half of the year – they doubled in Poland and Yugoslavia, sextupled in Slovakia, and elsewhere increases of 50-70 per cent were common. The main exceptions to this development were a few countries

---

[3] The problems of implementing monetary policies are discussed at length in chap. 3.1(ii), below.

where domestic adjustments forced a reduction in imports (Bulgaria, where imports fell over 20 per cent, and to a lesser extent Romania and Slovenia). Altogether the aggregate of east European and Baltic trade deficits increased from $23.6 billion in 1995 to just over $37 billion in 1996.

Most of the CIS countries also moved into deficit although on a much smaller scale than eastern Europe: except for Ukraine, most of the deficits were less than $1 billion. But all of them were dwarfed by another increase in the Russian surplus, from $31 billion in 1995 to over $37 billion: or 9 per cent of GDP, in 1996. (Allowing for unrecorded trade, however, the Russian surplus is estimated at some $29 billion in 1996.)

The proximate reasons for the deterioration in eastern Europe and, to a lesser extent, in the Baltic states were a sharp deceleration in the growth of export values (from 25 per cent in 1995 to just over 1 per cent on average for eastern Europe, and from 36 to 16 per cent for the Baltic states); and a continuing strong growth of imports (11 per cent in eastern Europe, 24 per cent in the Baltic states), although these increases were also smaller than those in 1995.

The slowdown in export growth appears to be due to a number of external and domestic factors. Slower growth in western Europe was an obvious factor with the increase of import demand falling by nearly two thirds between 1995 and 1996.[4] However, this affected the eastern countries differently and some of them (Estonia, Hungary, Latvia and Poland) were still able to achieve export *volume* growth of 6-9 per cent, more than the growth of world trade and despite their now considerable dependence on the much weaker west European market.

As suggested above, in the summary of price developments, a number of domestic factors have moved in the direction of weakening the international competitiveness of the tradeable goods sectors. Rising unit labour costs and appreciating foreign exchange rates (especially against the deutsche mark) have been squeezing exporters' profit margins and increasing the competitiveness of foreign suppliers on the eastern markets. And delays in restructuring industrial enterprises, especially the larger ones – many of which, but not all, remain under state control – have led to smaller than expected gains in productivity. Another factor is that much of the slack in export capacity created by the collapse of domestic and CMEA demand in the immediate aftermath of the shocks of 1989 has now been taken up and a continued rapid growth of exports will depend increasingly on the creation of new capacities and the upgrading of existing ones. This would also suggest

that, in present circumstances, devaluation of the exchange rate might not be very effective in boosting exports. Moreover, both east European exports as well as the restructuring process itself have a high import content and so devaluation could prove to be a two-edged sword. Devaluation may sometimes be appropriate, as decided in Hungary last year, if the exchange rate is estimated to be significantly above its equilibrium value (invariably a fragile and controversial calculation) and if there is spare productive capacity. But both from the point of view of reducing the underlying rate of inflation and improving external balance, manipulating the exchange rate cannot be a substitute for structural changes in the real economy.

A question that will increasingly preoccupy policy makers in eastern Europe and the Baltic states is whether the growing external imbalances can continue to be financed easily or whether they are likely to become a constraint on growth. As most of the east European and Baltic countries are in surplus on transfers and services (remittances from citizens working abroad and tourism being among the most important sources of net income) their current account deficits are less dramatic than those for trade. Nevertheless the deterioration on current account was still considerable, from an aggregate deficit of $1.4 billion in 1995 to $13.5 billion in 1996 for eastern Europe, and from $0.8 billion for the three Baltic states to over $1 billion (full-year data for 1996 are not yet available for the latter).

Russia's current account *surplus* is much smaller than its trade surplus, partly because of a deficit on foreign travel, and it increased less than the trade surplus, from $9.3 billion in 1995 to $10.2 billion in the first three quarters of 1996. Data for the other CIS members are limited, but Belarus, the Republic of Moldova and Ukraine were all in current account deficit in 1996, although only in the first two was this a significant proportion of GDP (8 and 11 per cent respectively).

Despite the deterioration in east European and Baltic current accounts in 1996, so far there has been no problem in financing them, mainly from private sources. The foreign capital inflow into eastern Europe last year is estimated at just over $15 billion which was sufficient to finance the deficits without adding to aggregate reserves. (In 1995 the inflow had been $24 billion and several governments considered the addition to reserves to have been excessive.) Foreign direct investment (46 per cent of the total) and medium- and long-term borrowing (22 per cent) are the most important sources of financing, while portfolio and short-term funds account for some 16 per cent. However, if "errors and omissions" are assumed to contain mainly short-term flows, the short-term component (i.e. including portfolio and capital identified as short-term) is not insignificant (31 per cent).

When current account deficits rise above 5 per cent of GDP for any length of time it is conventionally taken

---

[4] This refers to the decline in the volume growth of merchandise imports (see chap. 2.5).

as a warning signal for corrective policies to be introduced. All the Baltic states and many of the east European countries have now passed this threshold, several by a large margin, and current forecasts suggest a further deterioration in some of them in 1997 (Croatia, Czech Republic, Poland, Slovakia, for example). Does this mean that governments will be forced to take action to check the rising external deficits and as a result cut back their growth rates of GDP? Like all rules of thumb, however, the 5 per cent signal is only a first warning to pay more attention to what may be happening and the possible risks involved.

The sustainability of current account deficits (or foreign borrowing) basically rests on whether foreign investors believe a country will eventually be able to generate a stream of net export earnings sufficient to service and repay its debts. If foreign investors do believe that, then they will continue to sustain a country's deficit – and there are many examples of countries running deficits of over 5 per cent of GDP for many years (including Ireland and Israel among the ECE member countries). The key points here are whether the country has plenty of investment opportunities and whether the domestic environment is favourable to investors taking advantage of them. A recent study[5] has stressed that a current account deficit of more than 5 per cent is likely to become a problem only when a range of conditions apply: when the export sector is small, when debt service is large, domestic savings are low, and the financial sector is weak and dominated by banks subject to weak regulation.

As far as the fast reforming east European and Baltic economies are concerned, their performance according to all these criteria is mixed but by no means bad. In the first place, they do have considerable investment opportunities, and fixed investment, both domestic and foreign, is rising strongly in some of them and in several cases more rapidly than consumption, especially in central Europe. It should also be stressed that the available data do not suggest that east European current account deficits are being used to finance a boom in imports of consumer goods. In 1995, capital goods accounted on average for about 27 per cent of total imports and manufactured intermediate goods for another 40 per cent; manufactured consumer goods accounted for about 19 per cent.[6] Between 1993 and 1995 manufactured, final consumer goods accounted for only 17 per cent, on average, of the rise in total imports. While the greatest importance is often attached to imports of capital goods, it should be stressed that imports of manufactured intermediates also play a major role in the modernization of production structures, upgrading the quality of output, and generating positive spillovers to the rest of the economy.[7] The import structure of the east European transition economies thus appears to be broadly consistent with efforts to restructure their economies.

On the other criteria mentioned above, most of the east European and Baltic economies are open with relatively large export sectors: in relation to GDP, the share of exports ranged from 19 per cent (in Poland) to just under 50 per cent (in Slovakia and Estonia) in 1996.[8] As far as foreign indebtedness is concerned, eastern Europe is now classified as a comparatively low-debt area with gross debt to export ratios averaging about 100 and net debt-export ratios much lower at about 56 per cent. In the Baltic states the ratios are very much lower. The ratios of the most highly-indebted countries – Albania, Bulgaria, Hungary and Poland – have fallen considerably in the last few years, although there was a deterioration in Bulgaria last year and there has also been a sharp rise in those for The former Yugoslav Republic of Macedonia. Thus existing levels of foreign debt and debt servicing are not problematic for the majority of east European and Baltic countries.

The financial system, however, is another matter and it is now widely recognised that this is one of the most fragile foundations of the whole transition process. The crises of the last year or so in Albania and Bulgaria have highlighted the issues in a dramatic fashion, but many of the constituent elements of the problems in those countries – low quality of banks' asset portfolios, "cosy relationships" between banks, state owned enterprises and ministries, weak or non-existent regulation, and a widespread, popular mistrust of the banking system – are present in varying degrees elsewhere in eastern Europe and in the CIS.[9] If the confidence of foreign investors is to be maintained, then the governments of transition economies will have to be seen to be making progress in reforming their financial and banking systems as well as restructuring their productive capacities. If this can be done – and it might also help if governments were to set out their medium-term programmes for the eventual (and inevitable) conversion of current account deficits into surpluses – then balance of payments crises, with sudden

---

[5] G.M. Milesi-Ferretti and A. Razin, "Sustainability of persistent current account deficits", *NBER Working Paper,* No. 5467 (Cambridge, MA), 1997.

[6] Based on trade data from the United Nations COMTRADE Database for Bulgaria, Czech Republic, Hungary, Poland, Romania and Slovakia, prepared for an ongoing study by the UN/ECE secretariat. The analysis is based on the data for the exports of 19 OECD countries to their east European partners.

[7] D.T. Coe, E. Helpman and A.W. Hoffmaister, "North-south R&D spillovers", *The Economic Journal,* Vol. 107, No. 440 (London), January 1997, pp. 134-149. Essentially, imports of capital and intermediate goods provide the importing country with access to the benefits of R&D in the country of origin. The more imports are sourced from high R&D countries and the higher the level of the human capital stock in the importing country, the greater are these spillovers likely to be.

[8] The smallest export sectors were in Albania (8 per cent of GDP) and Yugoslavia (13 per cent).

[9] See chap. 3.1(ii) and (iii).

and painful cuts in domestic absorption, can probably be avoided. It also needs to be emphasized that, unlike the west European countries in the late 1940s and 1950s which benefited from Marshall aid and other official support for financing their current account deficits over a number of years and thus sustaining growth and reconstruction, the transition economies are largely dependent on private financial flows which are much more nervous and fickle than official commitments. It is therefore all the more vital that the programmes for reform and adjustment are kept on track and that they be perceived to be coherent and transparent by those financing the deficits. Any signs that reforms are being stalled – or that demand management is being relaxed, for electoral reasons for example – could quickly reduce the net inflow of funds and compromise the country's investment grade status on the international capital markets.

(e) South-eastern Europe

The two major crises that broke out in Bulgaria and Albania in 1996 – and which are still unresolved, although progress has been made in Bulgaria – underlined in a dramatic fashion the fragility of the transition process in some countries and demonstrated how misleading and dangerous it can be to focus on changes in a few macroeconomic variables as indicators of progress and success. The growth rate in Albania of 9 per cent and more between 1993 and 1995 and the reduction in the annual average inflation rate from nearly 200 per cent in 1992 to 8 per cent in 1995 were naturally seen as impressive; inflation in Bulgaria remained much higher but still fell from over 300 per cent in 1991 to 62 per cent in 1995 while GDP was rising modestly at some 2 per cent between 1993 and 1995. However, reducing inflation from three-digit rates to two digits or raising output rapidly from the depths of an exceptional slump is, as has been pointed out in previous issues of this *Survey*, the relatively easy, although still very important, part of the recovery process. But bringing about a sustained recovery in a market economy presupposes that the basic institutional framework required to support market based activity is in place and functioning effectively. The immediate events that triggered the crises, as well as the more general context in which they occurred, obviously differed substantially in each country. But both crises originated essentially in the "institutional hiatus" that arose between the speedy destruction of the former system of economic organization and the much longer time required to build the new institutions required to support a market based system of coordination. Developing this new system takes time and this helps to explain not only why the transitional recession was so deep and long in most transition economies but also why rent-seeking, corruption and criminal activities have flourished in the vacuum. The underdevelopment of the financial sector is a particularly important example of this general problem: in the absence of an effective institutional and regulatory framework for banking and other financial activities, credit risks will not be undertaken on the basis of a commercial evaluation of risks and returns, and this lack of financial discipline leads to the financial sector being an obstacle rather than a support to sustained economic growth and the creation of a modern capitalist economy. This lack of progress towards the creation of an effectual framework for sustained development also means that the effects of stabilization tend to be short-lived: some of the macroeconomic conditions for growth are put in place but growth does not ensue. For the majority of the people, the costs of adjustment are prolonged and the promised benefits are postponed, except for the minority which is enabled to enrich itself through non-market activity. This is an obvious recipe for social unrest and political instability.

Because the Bulgarian and Albanian crises were so extreme – and especially the latter, in its consequences – there is a natural temptation to set them aside as special cases. But most of the basic elements of these crises are present in one form or another in all the transition economies and especially in some of the other countries in south-eastern Europe – the successor states to the former SFR of Yugoslavia (except Slovenia) and Romania – as well as most of the CIS. In the central European countries, especially the CEFTA members, there are still problems with certain missing or ineffective or underdeveloped institutions, but their general institutional and economic development has advanced to a degree where the risks of a serious breakdown are now fairly low, and considerably less than in south-east Europe.

The timing and form of any particular crisis always comes as a surprise, but the dangers and risks created by the existence of the "institutional hiatus" have been pointed out in earlier issues of this *Survey*[10] and warnings that a "two-speed" transition process was emerging in eastern Europe, with countries such as Albania, Bulgaria and Romania in the slow lane, were being made from the end of 1991.[11] These countries all had much greater difficulties than the CEFTA countries in breaking with their communist past, in establishing the democratic legitimacy of their governments and in getting effective programmes of reform and stabilization underway.[12] It was tempting then – and is tempting now – for many

---

[10] For example, UN/ECE, *Economic Survey of Europe in 1989-1990*, chap. 1; and UN/ECE, *Economic Survey of Europe in 1992-1993*, chap. 1.

[11] UN/ECE, *Economic Bulletin for Europe*, Vol. 43 (1991), pp. 9-10, and subsequent issues.

[12] UN/ECE, *Economic Survey of Europe in 1990-1991*, chap. 4.5 ("Late starters": Bulgaria, Romania), pp. 164-168.

commentators to say that these failures simply reflect a lack of commitment to reform and policy discipline on the part of governments. That may be true in a proximate sense – but it begs the question of why governments do not persist with policies which they – and their electorates – often believe to be necessary for their future well-being.[13] At least part of the answer lies in the *ex ante* uncertainty of *individuals* as to whether they, personally, will actually benefit from reform even if they still believe that the majority will do so.[14] Such individual uncertainty may actually be greater the poorer people are, simply because they are closer to the borderlines of penury than the better off. The bias towards the status quo is therefore stronger in such countries.

The countries of south-eastern Europe are generally very much poorer than the other transition economies of eastern Europe, although one of the results of the former socialist systems is that on many development indicators (such as life expectancy and literacy) there is no significant difference between the two groups,[15] and even the poorest countries of south-eastern Europe have considerably higher values for these indicators than most other countries at similar levels of income per head. But in terms of nominal GDP per head, the five CEFTA countries were nearly 3 times better off than the south east in 1994 and although the difference is reduced to 1.6 times when adjusted for purchasing power parity there still remains a significant[16] difference in their levels of income per head.

The transitional slump in output in the poorer countries of south-east Europe was much deeper than that in central Europe: in Albania, Bulgaria, Croatia, Romania, The former Yugoslav Republic of Macedonia and Yugoslavia, the average fall in GDP between 1989 and 1992 or 1993 was nearly 40 per cent,[17] compared with an average of about 18 per cent for the five CEFTA countries. The slump in output in south-east Europe, although not as bad as in the Baltic states, was closer to that of the CIS than to the rest of eastern Europe.

The other "initial conditions" from which most of the south-eastern countries embarked on the construction of market economies were also more unfavourable than those facing the CEFTA countries. Bulgaria's dependence on trade with the Soviet Union was much greater than other east European countries and this not only worsened the slump in output but presented relatively greater difficulties in adapting production and export structures to meet demand in the western market economies. Albania's long period of isolation from the rest of the world also left it with considerable problems of adaptation, and Albania, Bulgaria and Romania all had relatively unfavourable export structures (high shares of agricultural products and/or of "sensitive products") for taking advantage of the new markets of western Europe. Bulgaria also started the transition with high levels of debt, and quite early in the process all the countries in the region were affected, to varying degrees, by economic sanctions imposed on Iraq and on Yugoslavia. In addition to the economic shocks, the legacy of communist rule in these countries was probably much more unfavourable than in central Europe: more ruthless regimes ensured that there was little if any reform before 1989 and there was little opportunity for the emergence of opposition parties which might have developed coherent alternative programmes when eventually presented with the possibility of taking over government.

Very little special allowance was made for this historical legacy when the international community first responded with assistance for the transition economies. The relations of the south-eastern countries with the international financial institutions were slow to develop[18] and they lagged behind the other transition economies in obtaining more open markets in the western economies. In the early 1990s, pledges of assistance to Bulgaria and Romania not only frequently fell short of IMF estimates of their short-term financing needs for the current account, but agreed amounts were often delivered behind schedule, thus creating liquidity shortages and leading to unnecessary cuts in essential imports.[19] The explanations for these delays usually focused on the slow pace of reform and doubts about even the commitment of the various governments to both economic and democratic reform. Although there were certainly foundations for these doubts, they ignored the dangers of countries falling into a transition trap due to the "institutional hiatus": the

---

[13] D. Rodrik, "Understanding economic policy reform", *Journal of Economic Literature*, XXXIV, March 1996, pp. 9-41.

[14] UN/ECE, *Economic Survey of Europe in 1992-1993*, p. 12.

[15] Life expectancy in five south-east European countries (Albania, Bulgaria, Croatia, Romania and The FYR of Macedonia) in 1994 averaged 72 years, and in five CEFTA economies (Czech Republic, Hungary, Poland, Slovakia and Slovenia) 72.2 years. However, the difference between all the transition economies and the average for western Europe (76.8 years) is significant (on the basis of a t-test at the 5 per cent level of significance). Basic data taken from the World Bank, *World Development Report 1996* (Washington, D.C.), 1996, table 1.

[16] That is, the difference between the means was significant at the 5 per cent confidence level for the t-test. The basic GDP data were also taken from World Bank, op. cit., table 1. On the World Bank's classification, Albania falls in the category of "low income" countries in 1994 and is situated between the Central African Republic and Ghana; Bulgaria, The FYR of Macedonia and Romania are all in the bottom half of the "lower-middle" income group and fall between Bolivia and Ecuador; Croatia is in the upper half of the "lower-middle" income group and ranks above Poland and Slovakia; Bosnia and Herzegovina and Yugoslavia are not classified.

[17] Excluding Croatia and Yugoslavia, affected by the war, only reduces the average to 35 per cent.

---

[18] Although Romania had been a member of the IMF since the 1970s.

[19] UN/ECE, *Economic Bulletin for Europe*, Vol. 43 (1991), pp. 9-10 and Vol. 44 (1992), pp. 8, 99-100; and UN/ECE, *Economic Survey of Europe in 1992-1993*, pp. 249-251.

old institutions were discredited and unable to perform, but the domestic forces for reform were still too weak to establish viable alternatives. The rent-seekers and vested interests, criminal and non-criminal, who were able to exploit the institutional incoherence, thus made it even more difficult for the transition process to move forward.

The transition process in the south-eastern economies is now in a very vulnerable position. A significant degree of liberalization of trade and prices, including interest rates and credit allocation, has been achieved in several of them, but the institutional structure required for these newly released market forces to operate in a socially optimal manner is still highly deficient in many crucial respects. Extensive legal reforms have been implemented in most countries but their administration and judicial enforcement is often weak; small-scale privatization has proceeded rapidly and extensively, but the transfer of large state owned enterprises to private ownership has been very slow and there has been little effort made to break up and restructure dominant enterprises. Competition policies are still embryonic and there are widespread restrictions on new entrants to different markets. All these problems are common to most transition economies but they are generally more severe in the south-east than in central Europe.[20]

One of the most crucial institutional weaknesses, however, lies in the banking and financial systems. The financial "pyramid" schemes in Albania were only exceptional in the scale on which the population invested in them and in the dramatic and widespread economic, social and political consequences of their collapse. But essentially they were a symptom of the "institutional hiatus" – the lack of sound financial institutions and instruments for the placement of savings and a lack of effective regulation designed not only to proscribe certain types of financial practice but also to limit the systemic effects of financial failure. The naive and the credulous, willing to gamble their savings on promises of gain which exceed all reasonable degrees of probability, are not concentrated in south-east Europe; they are to be found not only in all the transition economies but in all the western market economies as well. The real difference between these two sets of countries lies not so much in the propensities of their peoples to indulge in dubious speculation but in the nature and efficiency (and depth) of their financial and regulatory institutions.

The description of the banking problems in Bulgaria later on in this *Survey* applies more or less to most of the countries in south-east Europe. Most of the banks are burdened with bad debts (non-performing loans) and many are insolvent by western banking standards; loans continue to be made to loss-making enterprises in the state sector, partly under political pressures or on the basis of personal relationships but also to prevent balance sheet problems from emerging into the open; and there is generally inadequate regulatory and supervisory power in the hands of the central banks. In addition there is widespread distrust among the general population of financial instruments and institutions – this is of long standing in some of the states of the former SFR of Yugoslavia where foreign currency deposits were frozen in the early 1990s, but the credibility of financial institutions and the banks has clearly worsened as a result of last year's crises. Incapable of playing a constructive role in the transition process, the banking system in much of south-east Europe is actually a major obstacle to progress.

Reform of the financial system is thus at the top of most programmes for reform and the financial and technical assistance programmes of the international financial institutions. But the fragility of the banking and financial systems is closely bound up with the slow development of the private sector and the lack of other institutions required to support the market economy. Action will therefore be needed on a wide front and reforms will need to fit together in a coherent framework. And, in addition, urgent measures are required to restore macroeconomic stability. The question therefore arises as to whether the traditional responses of the international community will be sufficient to ensure success this time. Romania has already reached four stand-by agreements with the IMF since April 1991 and all of them were suspended because the Fund's conditions could not be met by the Romanian government. The difference now is that there is a new government in power elected by a popular vote in favour of reform. However, while the electorates will be keen for quick results, the development and transition problems of Romania and the other countries of south-east Europe are considerable and cannot be solved quickly – the crucial institutional changes cannot be brought about through "shock therapy".

Recent developments have highlighted not only the lack of mechanisms and institutional safeguards to deal with serious shocks and upsets, but also the danger that, in the absence of such safeguards, setbacks in the economy can rapidly degenerate into a broader social and political crisis which is extremely difficult to control. The two open crises of the last year raise important issues not only for policy makers in all the transition economies[21]

---

[20] For a useful survey of progress in these areas see EBRD, *Transition Report 1995*, table 2.1 and *Transition Report 1996*, table 2.1, pp. 11-13 (London).

[21] The warnings have not been ignored in the CIS. In the Republic of Moldova, Deputy Prime Minister I. Gutu has warned that in today's statistics Moldova resembles Bulgaria a year ago and stressed the need to undertake necessary reforms over the "next few months" (OMRI, *Daily Digest*, 26 March 1997). Prime Minister Lazarenko of Ukraine has also

but also for those countries and institutions in a position to provide assistance. These transition failures in southeast Europe also bring again to the forefront of public debate a number of questions related to the goals and methods of providing assistance to countries which still face considerable problems not only in escaping from the transition trap but also in nurturing democratic institutions and forms of government.

Overcoming these problems requires concerted action and long-term commitment both on the part of policy makers in the transition countries themselves and of the international community. It will also require a more generous, better focused, and much speedier delivery of assistance – technical and financial – which is not available domestically. Speedy assistance is likely to be more effective in helping governments with their stabilization policies while aid delayed can actually have perverse effects.[22] Effective assistance may also help to reduce the *ex ante* uncertainty of individuals which can be a brake on the reform process. Such assistance should not be regarded as an act of generosity subject to the fiscal restraints imposed in the west European and other economies, but a form of public investment guided by enlightened self-interest. Among the future returns to the donors of prompt and appropriate assistance should be increased economic and political stability in Europe and the ECE region as a whole. Without such assistance, the results of further crises may be even more chaos and destruction with severe spillover effects on neighbouring countries, in the first instance, and eventually on the rest of Europe.

(iii) Western Europe and North America in 1996

Economic growth in western Europe strengthened during the course of 1996, but, for the year as a whole, the average rate of increase of just under 2 per cent was still much lower than that in 1995. (Allowing for revisions in the 1995 national accounts data, the outcome was a little bit better than expected in this *Survey* last year.) Of the four large European economies, only the United Kingdom, which has been cyclically detached from mainland Europe since 1993, maintained a significant rate of expansion, although growth in many of the smaller economies was some way above the west European average. Although most of the data are still provisional, the recovery in continental Europe appears to have lost momentum in the last quarter of 1996 while in the United Kingdom there was an acceleration in the closing months of the year.

The main source of west European growth last year was rising exports, mainly to regions outside western Europe such as the United States, the transition countries of eastern Europe, and the developing countries. Many countries benefited from the competitive advantages resulting from the appreciation of the dollar and the stronger growth of demand in countries such as the United States. But the impulse from net exports was not strong enough to give a significant boost to total demand. In general *private consumption expenditure* in western Europe rose slightly more than 2 per cent, not much more than in 1995 and still restrained by modest growth in wages and salaries and in employment. Government consumption expenditure increased very little (1.4 per cent on average, the only large increase among the larger economies being in Germany) as a result of the continuing efforts to meet the Maastricht criteria for entry into the EU.

As part of the same efforts at fiscal retrenchment, *fixed investment* by governments actually fell quite sharply in a number of countries and for the second year in succession in France, Germany and the United Kingdom. Business fixed investment showed little buoyancy and investment plans were cut back during the year as a response to expectations of relatively weak output growth. What investment did take place appears to have been focused more on rationalization rather than expanding capacity. The total volume of fixed investment rose by just 2 per cent in 1996 and was only 2.5 per cent higher than in 1990 (only 1 per cent if Turkey is excluded). The weak performance of fixed investment is a matter of some concern as it implies some slipping behind in the rate of technological progress in the west European economy and, ultimately, a slowdown in the creation of new jobs. Since costs of capital have fallen and rates of return have risen, the main restraint on investment would appear to be the weak prospects for economic growth.

In sharp contrast to western Europe, the United States economy continued to expand for the sixth successive year (the present upswing began in early 1991). Growth actually strengthened through the year and for 1996 was about 2.5 per cent higher than in 1995. The expansion was broadly based but rising employment, higher real incomes, and rising consumer confidence meant that households' expenditure (up 2.5 per cent) was the main support. Nevertheless, business investment also rose strongly, encouraged by high capacity utilization, rising profits and confident expectations of continued growth. In contrast to western Europe, American business was also expanding productive capacities as well as spending heavily on new technology. Strong domestic demand in the United States also gave a boost to the world economy in 1996, imports of goods and services rising nearly 6.5 per cent and creating a slight net drag on the domestic growth rate.

---

been quoted as saying that "the terrible Bulgarian lesson will not be repeated in Ukraine" (*Financial Times*, 13 March 1997).

[22] A. Casella and B. Eichengreen, "Can foreign aid accelerate stabilization?", *The Economic Journal*, Vol. 106 (London), May 1996, pp. 605-619.

In many ways the performance of the Canadian economy last year was closer to that of western Europe than the United States: annual growth averaged some 1.5 per cent, rather less than in 1995; relatively weak labour markets and increased taxes held back the growth of personal disposable income, although consumption picked up to a large extent because of a steep fall in the savings ratio; and business fixed investment is also more concerned with rationalization rather than expanding capacities.

Despite the tighter labour markets and strong consumer demand in the United States and the United Kingdom, inflation rates in both countries remain subdued and there were only sporadic signs of incipient inflationary pressures, although in late March 1997 the Federal Reserve considered that these were sufficient to justify a small increase in the federal funds rate. In continental Europe, inflation rates have continued to decelerate and this development led to an easing of monetary policy in a number of countries; subsequent expectations about the likely movement of interest rates in favour of United States and United Kingdom assets strengthened the dollar and the pound sterling against the continental currencies during much of the year, improving the export competitiveness of the continental economies. European rates of consumer price inflation continued to fall through most of 1996 and the average rate for the year as a whole was just under 2.5 per cent, its lowest rate in more than 30 years. Little change is expected in average rates of inflation in the near future: some slight increases might occur in the United States and the United Kingdom, but in the rest of Europe, if anything, a further deceleration would seem more likely than a rise.

Low rates – and weak expectations – of economic growth in western Europe have meant there was virtually no increase in the average level of employment. There was some creation of new jobs in the United Kingdom, especially in the second half of the year, and in some of the smaller economies, but in general the increases were less than in 1995. Most of the new jobs that were created were in services. But whereas in Europe the numbers in employment increased by just 0.3 per cent on average, in the United States job creation was still robust, rising by 1.4 per cent despite still larger increases in the previous three years.

Weak output and employment growth in western Europe mean that little progress has been made in reducing the persistently high rates of unemployment throughout the region. The average rate in 1996 was 10.3 per cent, somewhat higher than in 1995 and about the same as in 1993. Among the four large economies it varied in 1996 from 8.2 per cent in the United Kingdom, the only one of the four where the rate fell, to 12.3 per cent in France. There were a few small reductions in some of the smaller economies where unemployment rates ranged from 3 per cent in Luxembourg to over 20 per cent in Spain.

Despite the considerable deregulation of European labour markets in recent years, the weakness of demand and output growth continue to hold back any significant reduction in unemployment. The major exception to this is the United Kingdom where output has grown relatively strongly since early 1993. However, the prospects for any significant reduction in European unemployment are still rather poor. The stance of economic policy is dominated by concerns to meet the Maastricht criteria for general government budget deficits and this implies tight fiscal policies in the coming year. Although monetary policy was loosened somewhat in 1996, it can still be expected to be kept relatively tight given the ambitious targets for inflation now being set, as shown by the Bundesbank's decision to lower its target rate to a range of 1.5-2 per cent. It may therefore still be some time before west European growth can rise again to rates that would bring about a significant reduction in unemployment.

Chapter 2

THE WESTERN MARKET ECONOMIES

2.1 The general context

Economic activity picked up again in western Europe in the course of 1996, although the average annual output growth was significantly lower than in 1995. There was a similar pattern in Canada. In the United States, economic activity rebounded significantly in the second and final quarters of 1996 and, in contrast to western Europe, the growth of real GDP was not only higher than in 1995 but also somewhat above the estimated rate of potential output growth. Cyclical growth forces still appear to be rather fragile in Japan; there are continuing balance sheet adjustment problems in the private sector, which were partly offset by the fiscal stimulus of late 1995.

Financial markets throughout the market economies continued to rise against a background of falling short-term interest rates, low inflation rates and moderate inflationary expectations. Despite the warnings of the chairman of the United States Federal Reserve against "irrational exuberance", there was a considerable surge in share prices, in many countries to record levels.

In western Europe, these developments contrasted sharply with moderate output growth and the bleak situation in the labour markets.

The spending and production decisions of economic agents are mainly determined by expectations concerning future incomes, profits and sales. These expectations are also shaped by the regularly published forecasts of economic growth by various national and international institutions.

As can be seen from chart 2.1.1, forecasts of real GDP growth in 1996 have been significantly lowered since the second half of 1995; a trough was reached at the beginning of the third quarter of 1996, when expectations stabilized at a significantly lower level compared with a year earlier. In the United States, the consensus of forecasts for economic growth in 1996 was revised sharply downwards in February 1996 against the background of a perceived weakening of cyclical growth forces. But, in striking contrast to western Europe, the dominating feature has been for an upward revision of forecasts since the spring of 1996, and the gap between the actual outcome and earlier expectations is less pronounced than in western Europe.

Economic activity in western Europe was supported by easier monetary conditions in 1996. The stance of monetary policy was increasingly relaxed and both short- and long-term interest rates fell to low levels. In addition, the dollar's appreciation improved the price competitiveness of west European firms both at home and in overseas markets. Within western Europe, the strengthening of the Italian lira, the peseta and the pound sterling benefited exporters trading with Italy, Spain and the United Kingdom, but the net impact on economic growth in the region as a whole is more difficult to gauge. Overall, however, the easing of monetary conditions provided an offset to the continuing tight stance of fiscal policy.

Although output growth rates converged between western Europe and the United States in the third quarter of 1996, the significant difference in their cyclical positions should be emphasized. The United States is on the plateau of a long cyclical upswing whereas in many west European countries, notably France, Germany and Italy, cyclical growth forces remain weak and a sustained upswing is not yet in sight.

Viewed from the demand side, there continues to be a split cycle in western Europe: economic growth is largely export led – mainly reflecting impulses from outside the region – while domestic demand growth has been weak. In the United States, the ongoing recovery continues to be broadly supported by both foreign and domestic demand. A matter of concern in western Europe is the persistent overall weakness of fixed capital formation which has dampened potential output growth in recent years. Thus, it is estimated that in Germany potential output rose by only 1.5 per cent in 1995-1996. The negative impact of this slowdown on potential job creation has been accentuated by the strong emphasis on rationalization and modernization in the investment process.

Economic activity continued to rise strongly in the developing countries as a group in 1996 reflecting, *inter alia,* continuing robust growth in Asia and a significant rebound in Latin America.

CHART 2.1.1

Consensus forecasts of annual changes in GDP in western Europe and the United States, 1996 and 1997

(Percentage change over preceding year)

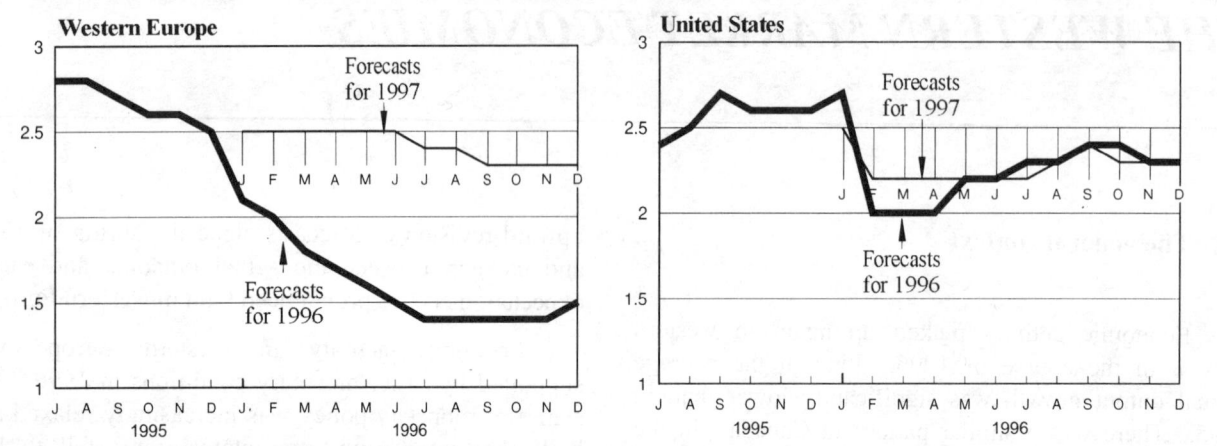

Source: Consensus Economics Inc., *Consensus Forecasts* (London), various issues.

Note: Monthly evolution of consensus forecasts of annual changes in real GDP from July 1995 to December 1996. Data for western Europe are weighted averages of consensus forecasts for nine countries: Belgium, France, Germany, Italy, the Netherlands, Spain, Sweden, Switzerland and the United Kingdom.

In the transition countries there was a further fall in aggregate output levels in 1996 but the underlying performance in the main subgroups is quite diverse. There was a slowdown in economic growth in eastern Europe, higher growth in the Baltic states and a return to growth in several of the CIS countries but not in Russia or Ukraine. In the event, world output growth accelerated slightly in 1996 compared with 1995.

Despite this slight acceleration in world output growth, there was a marked slowdown in the volume growth of world merchandise trade in 1996. For the year as a whole, it increased by some 4.5 per cent compared with 1995, when it rose by about 8.5 per cent. There was a pervasive weakening in import demand, which was quite pronounced in western Europe, Japan and also the developing countries of Asia. There was also a slowdown in the growth of import demand originating in the transition economies in 1996, but the average increase of more than 10 per cent exceeded by far the average growth of world trade. The bulk of this demand was addressed to west European countries and has accordingly helped to provide some support to higher activity levels in those economies.

Economic growth in western Europe in 1996 was too weak to alleviate the grave situation in the labour markets. On average, there was only a slight increase in employment, while the rate of unemployment was somewhat above 10 per cent and broadly unchanged from 1995. This poor performance continues to contrast with significant employment gains in Canada and the United States. In fact, in the United States the unemployment rate fell to only 5.4 per cent in 1996, which is about half the west European average.

Inflationary pressures remained subdued in 1996. Industrial raw material prices fell, and the marked rise in crude oil prices petered out in the final months of the year. In any case, the inflationary consequences of higher energy costs have been much reduced as a result of significant gains in energy efficiency compared with 1973, the year of the first oil price shock. In western Europe, the average annual inflation rate (measured by the consumer price index) fell to 2.4 per cent, its lowest rate in 30 years. In the United States, the inflation rate rose only slightly to 3 per cent in 1996, which is the more remarkable in view of the high rates of resource utilization in the United States economy.

The consensus of forecasts is for only a moderate strengthening of the recovery in western Europe in 1997 and for little if any improvement in the labour markets. Economic developments within the year will be closely scrutinized in order to assess which countries will be likely to meet the various criteria for admission to stage three of economic and monetary union (EMU). The decision as to which countries, if any, will introduce the single currency from the beginning of 1999 will be taken in the second quarter of 1998.

In the United States, the upswing has entered its seventh year and there is as yet no sign of a weakening of cyclical growth forces.

A marked acceleration in economic growth is forecast for Canada. In contrast, a significant deceleration in economic growth is generally expected for Japan.

In the event, real GDP in the industrialized countries is forecast to rise by some 2¼ per cent in 1997, the same rate as in 1996.

CHART 2.1.2

Comparison of actual outcomes and medium-term projections of GDP growth, 1988-2000
(Per cent per annum)

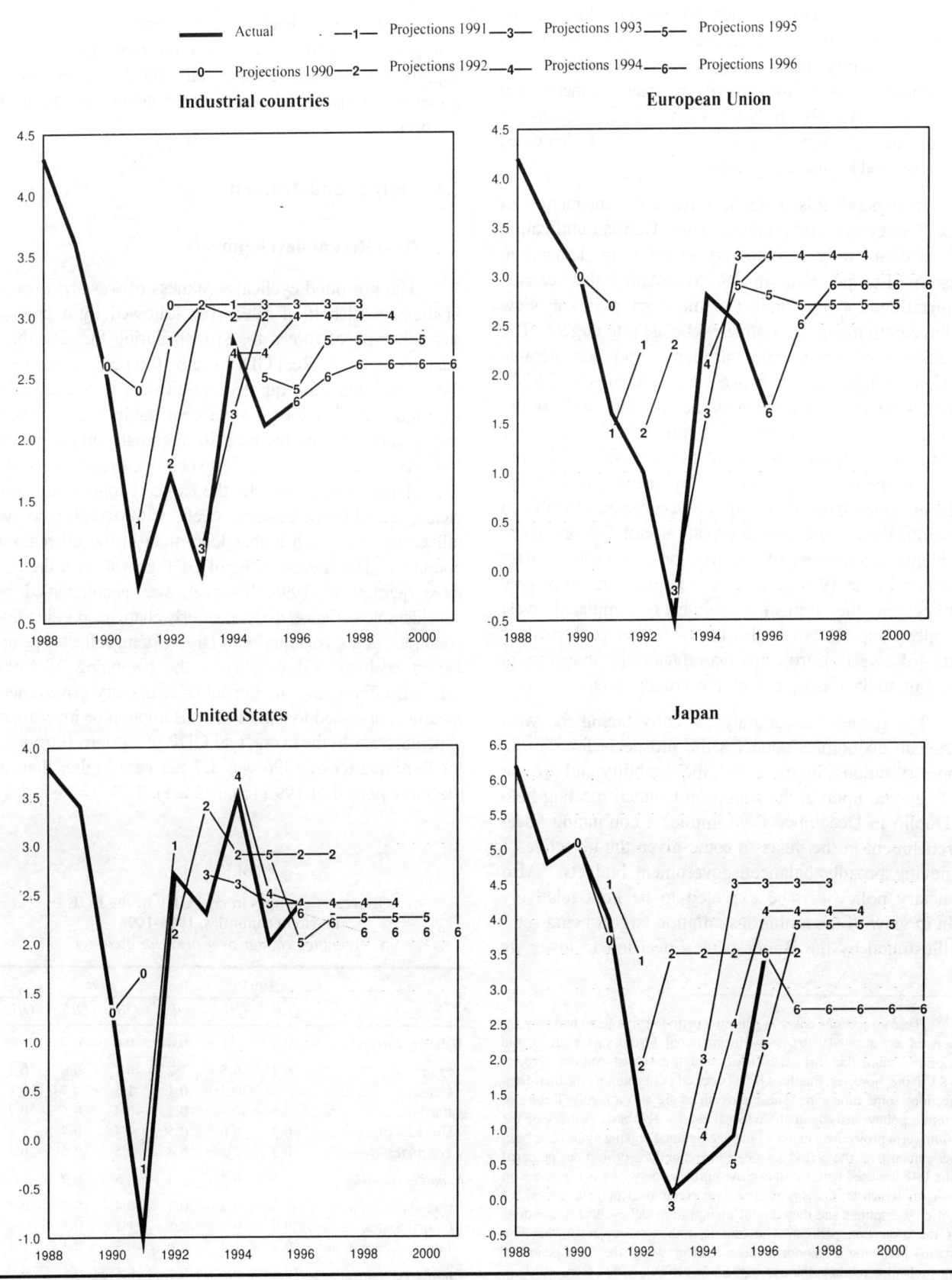

Source: IMF, *World Economic Outlook* (Washington, D.C.), October issues, 1990-1996.

The overall economic performance in western Europe has been disappointing for most of this decade, especially when it is compared to the optimism prevailing at the end of the 1980s in anticipation of the positive effects of the single European market, which entered into force at the beginning of 1993. In fact, annual output growth has been persistently lower than successive medium-term projections.[23] Over time, growth expectations themselves have been steadily reduced (chart 2.1.2) against a background of tight economic policies and depressed consumer and business confidence.

In general, this outcome reflects the interaction of shocks and economic policies. Thus, German unification provided not only a temporary stimulus to demand in western Europe but, in its aftermath, also carried implications for the conduct of monetary policy in view of the commitment to existing exchange rate targets. The result was exchange rate turbulence and considerable tensions within the exchange rate mechanism (ERM), which were also exacerbated by the volatility of the dollar exchange rate. The criteria and timetable for economic convergence, agreed upon in the Maastricht Treaty, entailed, moreover, restrictive monetary and fiscal policies which have also restrained economic growth. In addition, there were severe private sector balance sheet problems in a number of countries originating from asset deflation in the 1991 recession. The opening up of new markets in the transition countries stimulated west European exports but it also led to further problems for parts of west European manufacturing industry in adjusting to the emergence of new competitors.

The greatest uncertainty currently facing the west European economies concerns the prospects for starting monetary union. In any event, the "stability and growth pact" agreed upon at the European Council meeting held in Dublin in December 1996 implies a continuing fiscal retrenchment in the years to come given the objective of achieving broadly balanced government budgets. Also monetary policy can be expected to be kept relatively tight in view of the ambitious inflation targets being set — as illustrated by the Bundesbank's decision to lower its implicit annual inflation target to 1.5 per cent — and the need for the new European Central Bank (ECB) to establish its (anti-inflation) credibility in international financial markets.

Against this backdrop, the return of the west European economies to a medium-term growth path slightly below 3 per cent, as envisaged in most recent projections (chart 2.1.2), could, once again, turn out to be too optimistic.

2.2 Output and demand

(i) Recent developments

The profound cyclical weakness of *western Europe* in the second half of 1995 was followed by a gradual strengthening of forces for growth during the first three quarters of 1996. Real GDP rose by 0.6 per cent between the second and third quarters, equivalent to a seasonally adjusted annual rate of 2.4 per cent (table 2.2.1). There was a simultaneous increase in aggregate output in the four major economies, the first such occurrence since the second quarter of 1995. But the United Kingdom remains clearly ahead in the business cycle, with overall resource utilization at a much higher level than in the other three countries. The strengthening of GDP growth over the first three quarters of 1996, however, was accompanied by considerable variation in the quarterly changes in individual countries. This volatility, to a large extent, reflects special factors such as bad weather at the beginning of 1996, calendar effects, and the impact of temporary government measures designed to stimulate consumption or investment expenditures. In the event, real GDP in western Europe in the third quarter of 1996 was 1.7 per cent higher than in the same period of 1995 (chart 2.2.1).

TABLE 2.2.1

Quarterly changes in real GDP in the ECE market economies, 1995-1996
(Percentage change over previous quarter)

	1995		1996			
	QIII	QIV	QI	QII	QIII	QIV
Western Europe	0.3	–	0.4	0.6	0.7	..
France	0.2	-0.5	1.2	-0.1	0.8	0.2
Germany	–	-0.3	-0.1	1.5	0.7	0.1
Italy	0.6	0.3	0.2	-0.4	0.5	-0.2
United Kingdom	0.3	0.6	0.5	0.4	0.4	1.1
4 countries above	0.3	–	0.4	0.5	0.6	0.3
8 smaller countries	0.5	0.2	0.4	0.8	0.7	..
Canada	0.3	0.2	0.3	0.4	0.8	0.7
United States	0.9	0.1	0.5	1.1	0.5	0.9
Japan	0.3	1.3	2.0	-0.3	0.3	1.0
7 major economies	0.6	0.3	0.7	0.6	0.5	0.7

Source: National statistics.
Note: Data are seasonally adjusted.

---

[23] These projections cover in general a period of five years and they are elaborated on a regular basis by international institutions such as the European Union, the IMF, the OECD and the United Nations (Project LINK). Note, however, that the OECD stopped publishing its medium-term projections some time ago. These projections are also a regular feature of economic policy analysis at the national level. The basic features of the medium-term projections made by the international organizations have been rather similar over the period considered here, and the projections prepared by the IMF are used here for illustrative purposes only. As is the case with short-term forecasts, these projections are conditional upon a number of specific assumptions and they do not attempt to model cyclical fluctuations over the projection period. The idea is, rather, to project a "baseline scenario" involving a smooth medium-term growth path. In a period of weak cyclical conditions the assumption, in general, is that there will be a return to normal levels of factor utilization with a concomitant rise in output growth. The same holds, *mutatis mutandis*, when projections are made from a cyclical peak in the base period.

CHART 2.2.1

Quarterly changes in real GDP in the western market economies, 1993-1996
(Percentage change over same period of preceding year)

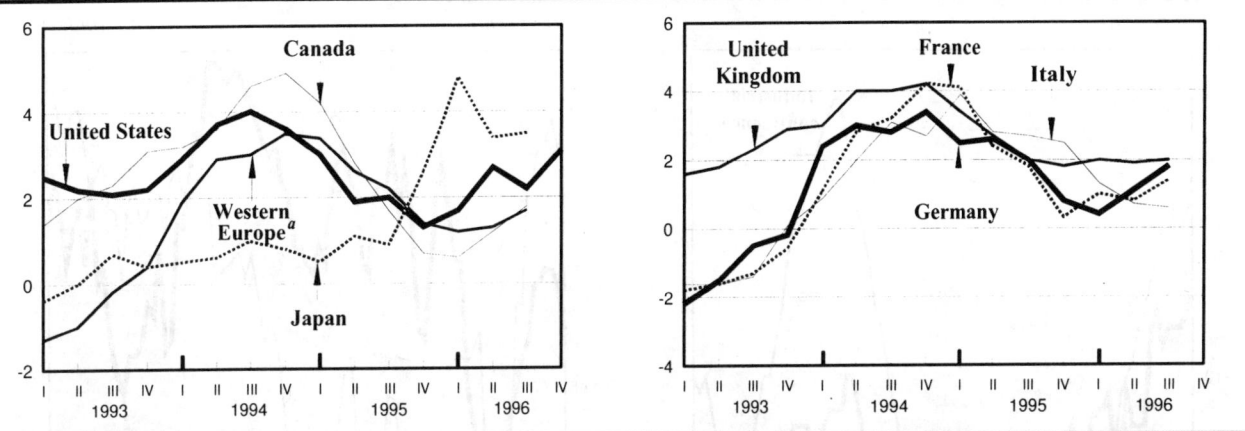

Source: National statistics.
Note: Data are seasonally adjusted.
a Twelve countries: Belgium, Denmark, Finland, France, Germany, Italy, the Netherlands, Norway, Spain, Sweden, Switzerland and the United Kingdom.

CHART 2.2.2

Industrial output in the western market economies, 1992-1996
(Indices, 1990=100)

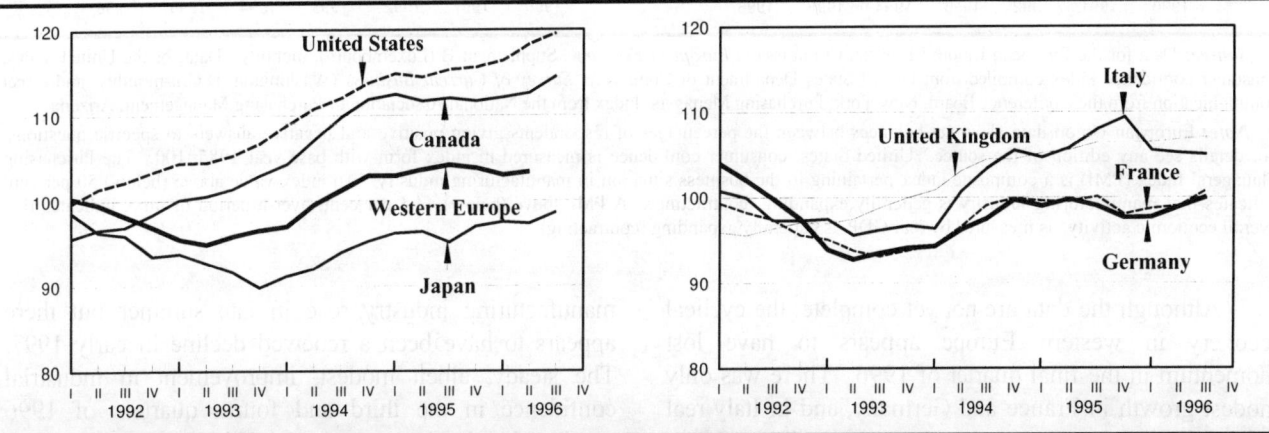

Source: OECD, *Main Economic Indicators* (Paris), various issues; national statistics.
Note: Data are seasonally adjusted.

The moderate cyclical upturn has rested mainly on rising industrial activity, which, in turn, was largely stimulated by the favourable development of foreign demand, notably for investment goods. It should be underlined that the profile of the international business cycle continues to be largely determined by fluctuations in industrial activity, even though this sector now accounts for only some 25 per cent of total output in western Europe (chart 2.2.2). There was a significant rebound in the growth of industrial output in western Europe between the second and third quarters of 1996.

The improvement in business conditions is mirrored in a rise of industrial confidence in the second half of 1996 (chart 2.2.3). Industrial confidence was, on average, quite depressed in western Europe throughout the period 1990 to 1996. In fact, the overall assessment was improving in 1994 but there was only a brief period when the net balance of assessments was significantly positive: this was in late 1994 when the dominant feeling was that the recovery from the 1993 recession was gaining strong momentum. But this rise in confidence did not last for long: it foundered under the impact of exchange rate turbulence in the early months of 1995. Consumer confidence has been even weaker: the net balance of responses pertaining to the overall assessment of the economic environment faced by private households has not been positive for a single month in the past seven years. In the main this reflects the calamitous situation in the west European labour markets and the associated uncertainty surrounding incomes and jobs.

CHART 2.2.3

Business cycle indicators for the European Union and the United States, January 1990-January 1997

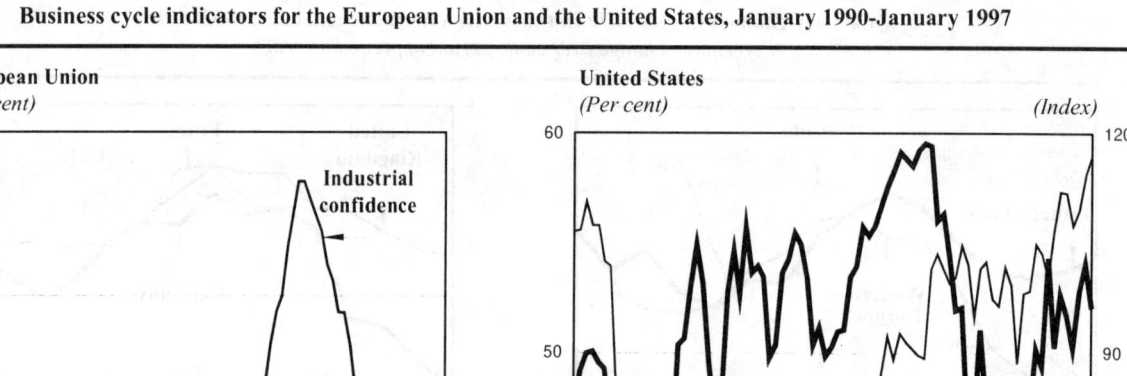

Source: Data for the European Union: European Commission, *European Economy*, Supplement B (Luxembourg), monthly. Data for the United States: consumer confidence index compiled from United States Department of Commerce, *Survey of Current Business* (Washington, D.C.), monthly, and direct communication from the Conference Board, New York; Purchasing Managers' Index from the National Association of Purchasing Management, Arizona.

Note: European Union data show net balances between the percentages of respondents giving positive and negative answers to specific questions. For details see any edition of the source. United States: consumer confidence is measured in index form with base year 1985=100. The Purchasing Managers' Index (PMI) is a composite index pertaining to the business situation in manufacturing industry. An index value above (below) 50 per cent indicates that manufacturing industry is generally expanding (contracting). A PMI above (below) 44.5 per cent, over a period of time, indicates that overall economic activity, as measured by real GDP, is generally expanding (contracting).

Although the data are not yet complete, the cyclical recovery in western Europe appears to have lost momentum in the final quarter of 1996. There was only modest growth in France and Germany, and in Italy real GDP fell slightly compared with the preceding quarter. In contrast, the rate of economic expansion in the United Kingdom accelerated markedly in the final months of the year. In the four major economies combined real GDP rose by only 0.3 per cent in the fourth quarter compared with the third quarter, when the equivalent growth rate was 0.6 per cent.

Industrial output, data for which are available in most countries up to November 1996 at the time of writing, suggest that activity in western Europe had more or less stagnated since August 1996. On average, industrial output probably declined between the third and final quarters.

The composite index of leading indicators, a time series which has a relatively close association with the industrial cycle, fell in December 1996, following a steady rise between February and November of 1996. In the European Union, capacity utilization rates in manufacturing industry rose in late summer but there appears to have been a renewed decline in early 1997. The steady, albeit modest, improvement in industrial confidence in the third and fourth quarters of 1996 continued in the first two months of 1997.

For the year as a whole, the level of west European industrial output was broadly unchanged compared with 1995 (table 2.2.2). Activity in the construction sector was generally depressed, reflecting weak residential investment, a reluctance of firms to engage in capacity augmenting expenditures (which are normally associated with construction projects), and tight government budgets which restrained infrastructure investment.

Real GDP in western Europe in 1996 is currently estimated to have increased by 1.9 per cent compared with 2.7 per cent in 1995 (table 2.2.3). The outcome is somewhat lower (an increase by 1.6 per cent), when Turkey, which has a relatively high weight in the regional aggregate and quite volatile output growth, is excluded from the regional aggregate. But the extent of the slowdown in growth between 1995 and 1996 is about the same for both aggregates.

TABLE 2.2.2

Annual changes in industrial production in the ECE market economies, 1993-1996

(Percentage change over previous year)

	1993	1994	1995	1996*
Western Europe	-2.6	4.4	3.9	0.6
4 major countries	-3.5	4.6	3.0	–
France	-3.8	3.8	1.8	0.4
Germany	-7.4	3.8	2.1	0.2
Italy	-2.1	6.3	6.1	-1.8
United Kingdom	2.1	5.0	2.5	1.2
13 smaller countries	-0.6	4.0	6.0	2.2
North America	3.5	5.2	3.3	2.7
Canada	4.4	7.0	3.4	1.6
United States	3.4	5.0	3.3	2.8
Total above	0.1	4.8	3.6	1.5
Memorandum item:				
Japan	-4.2	1.2	3.3	2.6
Total above, *including Japan*	-0.7	4.1	3.6	1.7

Source: OECD, *Main Economic Indicators* (Paris), March 1997.

TABLE 2.2.3

Annual changes in real GDP in the ECE market economies, 1994-1997

(Percentage change over previous year)

	1994	1995	1996	1997[a]
Western Europe	2.5	2.6	1.9	2.4
4 major countries	2.9	2.3	1.4	2.1
France	2.8	2.2	1.3	2.2
Germany	2.9	1.9	1.4	2.2
Italy	2.1	3.0	0.7	1.2
United Kingdom	3.8	2.5	2.1	3.0
17 smaller countries[b]	1.7	3.2	2.8	2.9
Austria	3.0	1.8	0.8	1.2
Belgium	2.3	1.9	1.4	2.1
Cyprus	6.1	5.3	2.4	3.5
Denmark	4.4	2.8	2.2	2.9
Finland	4.5	4.5	3.2	3.5
Greece	1.5	2.0	2.2	2.5
Iceland	3.5	2.1	5.4	3.3
Ireland	6.5	10.3	6.3	5.5
Israel	6.8	7.1	3.8	4.0
Luxembourg	4.2	3.7	2.4	3.1
Malta	4.0	9.0	4.0	4.0
Netherlands	3.4	2.1	2.7	2.8
Norway	5.0	3.3	4.8	3.1
Portugal	0.8	2.4	2.6	2.9
Spain	2.1	2.8	2.3	3.0
Sweden	3.3	3.6	1.1	2.0
Switzerland	1.0	0.1	-0.7	0.5
Turkey	-5.5	7.3	7.5	5.0
North America	3.5	2.1	2.4	2.6
Canada	4.1	2.3	1.5	3.3
United States	3.5	2.0	2.4	2.5
Total above	3.0	2.3	2.1	2.5
Memorandum item:				
Japan	0.5	0.9	3.7	1.6
Total above, *including Japan*	2.6	2.1	2.3	2.3

Source: National statistics; UN/ECE secretariat estimates.
[a] Forecasts.
[b] Excluding Israel.

The average growth rate in 1996 masks a wide variation in the economic performance of individual countries. Among the four major economies, there was modest growth (1.5 per cent and less) in 1996 in France, Germany and Italy, while real GDP rose by about 2 per cent in the United Kingdom. Among the 18 smaller economies, growth was 1.5 per cent or less only in Austria, Belgium, Sweden and Switzerland. In fact, in the latter country real GDP actually fell in 1996. All the other countries grew at rates of at least 2 per cent and, in some cases, significantly more (table 2.2.3).

Among the four larger economies, the United Kingdom continues to enjoy a sustained and, so far, balanced economic upswing. Among the smaller economies, rapid economic growth has been a feature for most of this decade in Cyprus, Ireland, Israel, Luxembourg, the Netherlands, Norway and Turkey. Ireland continues to grow significantly above average, the major stimulus being foreign direct investment in high-tech industries which at the same time has led to buoyant export growth, although over the past year or so there has been increasing support from domestic demand. In Norway, the strong performance in 1996 reflects in the main a surge of activity in the oil and gas sector.

In Finland, three years of consecutive and robust output growth (1994-1996) have nearly offset the large cumulative fall in GDP of more than 11 per cent (on an annual basis) between 1991 and 1993. This deep recession was, *inter alia,* due to the collapse of established markets in the former Soviet Union and central and eastern Europe and the build-up of severe imbalances in private sector balance sheets.[24] There were also three consecutive years of falling output in Sweden from 1991 to 1993, but the recession was much less severe than in Finland; and two years of robust GDP growth in 1994 and 1995 were sufficient to offset the previous decline. In Switzerland, the last year of significant output growth was 1990. The economy was in recession in the following three years and after a modest increase in 1994-1995, output fell again in 1996. Overall activity levels continue to be restrained by persistent imbalances in the real estate sector and the strength of the Swiss franc from 1993 to mid-1996.

In *France,* a surprising feature of economic developments in 1996 was the strength of private

---

[24] S. Honkapohja, E. Koskela and J. Paunio, "The depression of the 1990s in Finland: an analytical view", *Finnish Economic Papers*, Vol. 9, No. 1 (Helsinki), Spring 1996, pp. 37-54.

spending. As there was hardly any growth in real disposable incomes this rise in expenditures was largely based on reduced savings, which in turn may be partly due to government incentives designed to encourage the purchase of new cars and to reduce contractual savings. Business fixed investment picked up in the third quarter of 1996, but broadly stagnated for the year as a whole. Construction activity, both in the private and public sectors, remained weak. Changes in real net exports continued to support domestic activity levels in 1996.

In *Germany*, the most striking feature of 1996 was the significant slowdown in economic growth in the new eastern *Länder*, where real GDP rose by only 2 per cent compared with 1.3 per cent in west Germany. The growth of the east German economy had been quite impressive between 1991 and 1995, real GDP rising at an average annual rate of 6.1 per cent. This marked deceleration does not augur well for the underlying strength of the east German economy, which depends on the manufacturing sector. Economic growth in east Germany had been boosted since 1991 with the help of massive transfer payments and generous investment incentives, which triggered a wave of fixed investment, mainly related to privatization of enterprises and the creation of new firms. In addition, the German authorities considerably boosted expenditure on public infrastructure and the housing stock, which had been neglected for decades. The impetus of these spending sprees has now faded without, however, generating a self-sustained recovery (see section 2.2(iii) below). In the pan-German economy, real GDP rose by 1.4 per cent in 1996. There was modest growth in household expenditures which rose in line with real disposable incomes, and government spending was pushed up by rising expenditures on social security. Construction investment fell in all major areas, i.e. residential investment, industrial buildings and government fixed investment; fixed investment in machinery and equipment rose only modestly in view of reduced sales prospects. The weak investment performance is a noteworthy feature of this recovery. Exports of goods and services were a major support to overall economic activity in 1996, but their annual growth rate was significantly lower than in 1995. Exports were stimulated mainly by stronger demand outside the European Union notably associated with improved price competitiveness on account of the real effective depreciation of the deutsche mark. Changes in stockbuilding ceased to be a major restraint on output growth in the course of 1996.

In *Italy*, real GDP rose by a meagre 0.7 per cent in 1996, the combined effect of a marked slowdown in domestic demand and stagnating exports of goods and services. Domestic demand was restrained by tight fiscal and monetary policies, while stockbuilding subtracted from output growth. The export boom of 1993-1995 faded against the background of an appreciating lira and weak cyclical demand in major west European markets. Import demand fell in 1996 compared with the preceding year. Industrial employment fell and employment in the total economy rose by only 0.1 per cent in 1996. Fixed investment in machinery and equipment was stimulated by tax incentives (Tremonti Law) until April 1996, but falling capacity utilization and high real interest rates have generally dampened fixed capital formation.

In the *United Kingdom*, the expansion has now lasted for about four years. Real GDP continued to rise more or less steadily and at a relatively moderate rate in the first three quarters of 1996 but there was a marked strengthening of growth forces in the final quarter (table 2.2.1). A striking feature has been the overall sluggishness of manufacturing activity in 1996. This partly reflects weak demand in the major west European markets but it also reflects a shift in domestic demand towards foreign goods and domestic services relative to tradeables. For the year as a whole real GDP is estimated to have increased by somewhat more than 2 per cent. The expansion in 1996 was supported in the main by household expenditure, fixed investment and exports, although the contribution of exports was more than offset by the surge in imports associated with the strengthening of domestic demand. This is in marked contrast to the two preceding years, when real net exports were the mainstay of the recovery. Households' consumption expenditure was supported by gains in incomes and employment which boosted consumer confidence. The savings ratio was broadly unchanged in the course of 1996 but was higher than in 1995. The strong growth in business fixed investment masks a sharp fall in manufacturing investment, where output was very sluggish in 1996. Housing investment edged upwards against the background of rising real house prices. Public investment plummeted, but in aggregate there was relatively robust investment growth in 1996. Strong annual export growth masks a weakening of foreign demand in the second half of the year as companies started to feel the restraining effects of the appreciation of the pound sterling. Stockbuilding started to support economic activity in the second half of 1996 and household expenditure strengthened against a background of robust consumer confidence and favourable developments in the labour market.

Looking at the *major expenditure items* in western Europe, exports were the single most important driving force behind the gradual recovery of economic activity in the course of 1996. Foreign demand was strong mainly outside the region, viz. the United States, the developing countries and the transition economies. The impulses from stronger export growth have started to feed through to domestic demand and this has helped to support intraregional trade and to raise domestic activity levels. The concomitant rise in import demand tended to dampen growth in domestic output, but, on average, changes in real net exports have supported economic activity. Many countries benefited from improved price competitiveness not only on account of the appreciation of the dollar but also the stronger Italian lira and pound sterling, which "redistributed" output within western Europe. In general, however, the impulses emanating from foreign demand were too weak to trigger a significant strengthening of domestic demand in western Europe in 1996. This was especially true for fixed investment which remained rather lacklustre and disappointed expectations that it would increasingly add momentum to the nascent upswing. The process of stock adjustment, which had been restraining output growth since the second half of 1995, abated in the course of 1996, but private household expenditure, the single most important component of demand, remained sluggish.

For the year as a whole, real private consumption expenditures in western Europe rose by somewhat more than 2 per cent compared with 1995 (table 2.2.4). Moderate wage increases and weak employment growth restrained the growth in aggregate personal incomes, as did widespread fiscal retrenchment. In many countries falling savings ratios helped to support expenditure levels. In addition, the rise in financial wealth associated with the strong increase in share prices may have strengthened expenditures. In some countries (Finland, Norway, United Kingdom) households' net wealth position also improved on account of increases in house prices.[25] Consumer confidence improved somewhat in the final months of 1996 but overall the level of confidence in early 1997 is still quite depressed. But this average masks some marked differences among countries. Where there has been a significant improvement in economic performance households tended to be more optimistic. This was notably the case in the United Kingdom, Finland, Ireland, the Netherlands and Norway. But, in general, the bleak labour market prospects have depressed spending propensity particularly for "big ticket" items. The volume of retail sales generally stagnated in 1996 as compared with 1995.

Government consumption expenditures have continued to rise moderately, but the major focus of fiscal policy has been on trying to meet the Maastricht budget deficit criteria. This has also restrained government fixed investment.

Business fixed investment did not develop any significant momentum in 1996. Investment plans in industry were cut back in the course of the year against a background of moderate growth expectations. A major aim of fixed investment remains the replacement of old equipment, i.e. rationalization and modernization. Expenditure on new industrial buildings, which in general is associated with increases in productive capacity, rose only slightly. To some degree this also reflects persistent excess supplies in this market in several countries. Residential investment was restrained by unfavourable income and employment prospects and, in some countries, by high vacancy rates of dwellings, which depress actual and prospective rates of return. Furthermore, the continuing effort to consolidate budget deficits has left little scope, if any, for increased government investment (table 2.2.5). Construction investment fell in Austria, France, Germany and Switzerland in 1996; in contrast there was rather strong growth in Denmark, Ireland and Portugal.

In sum, total real fixed capital formation in western Europe rose by only about 2 per cent in 1996, although this average masks quite buoyant growth in a number of smaller economies (table 2.2.4).

The weakness of investment in western Europe remains a matter of concern. Although supply-side conditions have tended to improve significantly (low interest rates, increased profitability and rates of return), the overall propensity to invest has continued to be restrained by weak sales prospects and the presence of relatively large margins of spare capacity. Fixed investment is the vehicle of technical change and a major determinant of potential output and, therefore, of job creation in the economy. Fixed investment fell by 6 per cent between 1990 and 1993 and has since recovered only gradually. The level of fixed investment in 1996 was only some 2.5 per cent higher in real terms than in 1990.[26]

Although exports of goods and services picked up in the course of 1996, for the year as whole there was a significant deceleration in their growth as compared with 1995. Given the weakness of domestic demand and the close trade links among the west European countries, there was a concomitant slowdown in the growth of import demand (table 2.2.6).

---

[25] Changes in asset prices affect the overall economic performance via the impact on household wealth, which is an important determinant of consumption expenditures. The stimulus of rising asset prices on economic activity may be offset, however, by rising short-term interest rates as a result of a tightening of monetary policy and a concomitant rise in long-term interest rates. This would dampen the growth of private sector investment. Model simulations suggest that the net effect of rising share prices may have been to increase real GDP in the United States by nearly 1 percentage point in 1996 compared with 1994. See Statistics Norway, *Economic Survey*, Vol. 6 (Oslo), April 1996, p. 6. Note also that the coefficient on asset values (especially housing) in the consumption function may have fallen as a result of the fall in house prices, which in the United Kingdom between 1989-1993 was unprecedented in the post-Second World War period. The wealth effect seems to have weakened in the United States in 1996, which could perhaps be associated with an increase in middle-class, professional insecurity.

[26] If Turkey is excluded from the aggregate the increase is only 1 per cent.

TABLE 2.2.4

Annual changes in major domestic expenditure items in the ECE market economies, 1995-1996

(Percentage change over previous year)

	Private consumption		Government consumption		Gross fixed investment		Final domestic demand		Changes in stockbuilding[a]		Total domestic demand	
	1995	1996	1995	1996	1995	1996	1995	1996	1995	1996	1995	1996
Western Europe	2.0	2.2	1.3	1.4	3.9	1.8	2.2	2.0	0.5	-0.3	2.7	1.7
4 major countries	1.8	1.8	1.1	1.2	2.4	0.1	1.8	1.3	0.2	-0.3	2.0	1.0
France	1.8	2.3	0.9	1.5	2.6	-0.6	1.8	1.6	0.3	-0.6	2.0	1.0
Germany	1.8	1.3	2.0	2.4	1.5	-0.8	1.8	1.1	0.3	-0.3	2.1	0.8
Italy	1.7	0.7	-0.5	-0.4	5.9	1.2	2.0	0.6	0.3	-0.4	2.3	0.2
United Kingdom	1.9	3.0	1.5	0.8	-0.2	1.0	1.5	2.2	0.1	-0.2	1.5	2.0
17 smaller countries[b]	2.5	3.0	1.7	1.9	6.7	5.3	3.2	3.4	1.0	-0.2	4.2	3.0
Austria	1.9	1.2	2.1	0.5	2.3	1.4	2.0	1.1	0.8	-0.3	2.8	0.8
Belgium	1.2	1.4	1.0	1.8	3.0	2.2	1.5	1.6	0.1	–	1.6	1.6
Cyprus	11.5	2.5	2.9	9.0	1.8	2.4	8.0	3.5	1.5	0.3	9.1	3.5
Denmark	-2.3	2.4	0.8	1.0	10.2	5.6	3.2	2.6	1.3	-1.3	4.6	1.1
Finland	3.7	3.3	1.7	2.9	8.5	5.5	4.1	3.6	-0.1	-0.7	3.8	2.8
Greece	1.6	1.9	0.6	3.6	5.8	9.2	2.3	3.6	1.3	0.2	3.4	3.7
Iceland	4.6	6.8	2.0	2.8	2.2	21.4	3.6	8.2	0.6	-0.4	4.3	7.8
Ireland	3.9	6.3	3.0	3.0	10.1	10.3	4.8	6.4	1.0	0.2	6.0	6.7
Israel	7.3	6.0	0.6	4.2	8.2	8.1	6.0	6.1	0.7	-0.4	6.5	5.7
Luxembourg	2.4	2.1	2.2	2.0	3.5	5.9	2.6	2.9	0.2	1.3	2.7	4.0
Malta	11.2	9.4	7.5	5.9	21.2	-2.4	12.7	5.8	0.7	1.3	13.4	7.1
Netherlands	1.8	3.0	0.6	0.6	6.7	5.3	2.6	3.1	-0.4	-0.7	2.2	2.4
Norway	2.6	4.7	0.2	1.6	4.5	3.1	2.5	3.6	1.2	-0.9	3.7	2.5
Portugal	1.7	2.0	1.8	1.8	4.6	5.8	2.5	3.0	0.2	-0.1	2.6	2.8
Spain	1.5	1.8	1.3	0.9	8.2	1.0	2.9	1.5	0.2	0.2	3.1	1.7
Sweden	0.8	1.5	-1.0	-1.8	10.9	4.7	1.8	1.1	0.5	-1.0	2.3	–
Switzerland	0.7	0.2	-0.1	-0.5	2.3	-0.1	1.0	–	0.9	-0.6	1.9	-0.5
Turkey	7.6	9.6	6.7	7.6	8.3	18.2	7.7	11.7	5.3	0.1	13.3	11.6
North America	2.3	2.5	-0.4	0.3	4.8	6.1	2.3	2.8	-0.3	-0.3	1.9	2.4
Canada	1.4	2.4	-0.7	-1.8	-0.1	6.3	0.6	2.5	0.3	-0.9	1.0	1.6
United States	2.3	2.5	-0.3	0.5	5.2	6.1	2.4	2.8	-0.4	-0.3	2.0	2.5
Total above	2.1	2.3	0.5	0.8	4.3	4.0	2.2	2.4	0.1	-0.3	2.3	2.1
Memorandum item:												
Japan	1.7	2.9	2.0	2.3	0.9	9.0	1.5	4.7	0.2	–	1.7	4.6
Total above, including Japan	2.1	2.4	0.7	1.1	3.8	4.8	2.1	2.7	0.1	-0.3	2.2	2.5

Source: National statistics; UN/ECE secretariat estimates.

[a] Contribution to GDP growth (percentage points).

[b] Excluding Israel.

In terms of the contributions of the individual expenditure items to the growth in real GDP, it can be seen (table 2.2.7) that exports were the single most important driving force of economic activity in western Europe in 1996. In fact, exports contributed 1.5 percentage points to the annual increase in real GDP, slightly more than all the components of domestic expenditure combined. The same dominance of exports was also true of 1995. The very modest and declining contribution to growth from fixed capital formation in 1996 as compared with 1995 is noteworthy. Domestic expenditures, however, also include spending on imported goods and services, which means that the impact of domestic spending on domestic activity will depend on the share of imports in each of the various expenditure items. Direct measurement of this leakage into imports is not possible at the level of each expenditure component, given the lack of statistics, but in aggregate, real imports of goods and services subtracted 1.3 percentage points from GDP growth in 1996. This largely offset the stimulus provided by exports to increased domestic economic activity in western Europe, although it should be noted that the relative importance of changes in real net exports varied considerably among the individual countries (table 2.2.6).

In the *United States*, the cyclical upswing, which started in the second quarter of 1991, continued in 1996. Against a background of relatively relaxed monetary policy, the pace of economic expansion, however, was quite uneven in the course of the year. Moderate growth in the first and third quarters (compared with the preceding quarters) was followed by a significant rebound

TABLE 2.2.5

Annual changes in government fixed investment in the ECE market economies, 1992-1996

(Percentage change over previous year)

	1992	1993	1994	1995	1996
France	3.9	-0.4	-0.4	-1.1	-4.0*
Germany	9.8	-3.4	1.4	-4.4	-5.5
Italy	..	..	..	..	..
United Kingdom[a]	13.2	3.0	1.7	-6.8	-12.0*
Denmark	11.8	0.8	0.7	-1.6	0.2
Finland	-2.0	-17.8	1.6	-2.8	2.8
Greece	11.0	-0.7	0.5	..	..
Netherlands	7.4	-1.1	-2.8	2.2	1.5
Norway	4.3	-9.6	1.6	-0.5	4.9
Sweden	0.6	6.8	11.1	-0.5	-3.0
Canada	-0.1	0.7	6.4	2.7	-2.0*
United States	3.4	-1.1	-1.3	1.9	2.7
Japan[a]	14.5	15.7	2.8	0.7	13.6

Source: National statistics.

[a] Including public enterprises.

TABLE 2.2.6

Annual changes in the real foreign balance in the ECE market economies, 1995-1996

(Percentage change over previous year)

	Goods and services				Net exports[a]	
	Exports		Imports			
	1995	1996	1995	1996	1995	1996
Western Europe	7.5	4.9	7.8	4.5	-0.1	0.1
4 major countries	7.7	3.7	6.4	2.6	0.2	0.2
France	6.0	3.6	5.3	2.2	0.2	0.4
Germany	5.9	4.9	6.4	2.6	-0.1	0.6
Italy	11.6	-0.3	9.6	-1.7	0.1	0.1
United Kingdom	8.0	6.3	4.4	7.8	0.9	-0.3
17 smaller countries[b]	7.0	7.3	10.7	8.3	-0.9	-0.3
Austria	5.4	3.7	7.3	3.4	-1.1	—
Belgium	5.0	4.4	4.8	4.6	0.4	—
Cyprus	4.5	4.6	11.2	6.3	-3.5	-1.2
Denmark	3.7	3.3	8.2	1.0	-1.5	1.1
Finland	8.2	3.8	6.9	4.5	0.9	0.1
Greece	1.8	1.7	6.2	6.5	-1.8	-2.1
Iceland	-2.3	5.2	3.8	12.8	-2.0	-2.1
Ireland	17.0	10.0	12.3	11.5	5.1	0.6
Israel	10.9	3.7	8.4	7.6	0.8	-3.0
Luxembourg	4.4	2.8	3.3	4.5	0.9	-1.7
Malta	4.7	-5.7	10.3	-1.7	-3.5	-2.9
Netherlands	6.9	3.9	7.7	3.6	0.1	0.5
Norway	3.8	8.2	5.1	2.5	-0.1	2.4
Portugal	11.4	9.6	8.6	7.6	-0.8	-0.8
Spain	8.2	10.3	8.8	7.5	-0.4	0.6
Sweden	12.6	5.6	10.3	3.5	1.4	1.1
Switzerland	3.0	1.9	6.6	1.9	-1.9	-0.1
Turkey	6.7	15.0	30.0	29.6	-5.1	-4.6
North America	9.2	6.3	8.1	6.3	0.1	-0.1
Canada	12.0	4.5	8.7	5.1	1.2	-0.2
United States	8.9	6.5	8.0	6.4	—	-0.1
Total above	8.3	5.6	7.9	5.4	—	—
Memorandum item:						
Japan	5.0	2.1	13.5	11.4	-0.7	-0.9
Total above, including Japan	7.8	5.1	8.8	6.3	-0.1	-0.2

Source: National statistics.

[a] Contribution to GDP growth (percentage points).

[b] Excluding Israel.

in output and demand in the second and fourth quarters of 1996, when real GDP rose at unsustainably high rates. For 1996 as a whole, real GDP was 2.4 per cent higher than in 1995, when it increased by 2 per cent. The degree of resource utilization remained very high, the gap between potential and actual output being virtually closed. But inflationary pressures have remained surprisingly moderate given the advanced stage of the United States business cycle.[27]

Industrial output growth accelerated again in the final months of 1996, as domestic and foreign demand rebounded. The relatively high level attained by the Purchasing Managers' Index signalled continuing expansion of output in the manufacturing sector and in the whole economy throughout the year (chart 2.2.2). Aggregate industrial output rose by 2.7 per cent in 1996 compared with the previous year (table 2.2.2), supported in the main by strong demand for investment goods. Capacity utilization rates in manufacturing industry remained high: on average they were 1 percentage point above their long-run average, but 3.5 percentage points lower than at their previous cyclical high in 1988-1989.

The upswing in the United States economy continued to be broadly based, supported by private consumption, fixed investment (table 2.2.4) and exports (table 2.2.6). Private households' expenditure on goods and services remained the mainstay of domestic demand. Consumer confidence rose to very high levels in the second half of 1996 bolstered by low unemployment, rising demand for labour and higher real incomes (chart 2.2.3). Households' expenditure was also stimulated by the favourable wealth effects associated with the considerable rise in share prices, although some restraint effects on spending may have originated in the rise in debt interest payments to a relatively high proportion of personal income. Increasing financial imbalances also became apparent in the number of credit card defaults which rose to record levels.[28] There was a slight rise in the average annual savings ratio in 1996 compared with 1995.

---

[27] It should be recalled in this connection that services account for a rather large share of total GDP in the industrialized countries and that capacity measures for this sector are probably more ambiguous than in manufacturing industry.

[28] Although credit card debt rose strongly between 1992 and 1995, its share in total household debt remained small. See "Family finances in the U.S.: recent evidence from the survey of consumer finances", *Federal Reserve Bulletin*, Vol. 83, No. 1 (Washington, D.C.), January 1997, pp. 1-24.

TABLE 2.2.7

Contribution of major expenditure items to annual changes of GDP in the ECE market economies, 1995-1996
(Percentage points)

	Western Europe[a]		United States		Japan	
	1995	1996	1995	1996	1995	1996
Household consumption	1.1	1.1	1.6	1.7	1.0	1.7
Government consumption	0.2	0.2	-0.1	0.1	0.2	0.2
Gross fixed investment	0.7	0.3	0.9	1.1	0.3	2.7
Changes in stockbuilding	0.3	-0.3	-0.4	-0.3	0.2	–
Total domestic demand	2.3	1.3	2.0	2.5	1.6	4.6
Exports	2.3	1.5	1.0	0.7	0.6	0.3
Imports	2.1	1.3	1.0	0.8	-1.3	-1.1
Net foreign balance	0.2	0.3	–	-0.1	-0.7	-0.9
GDP	2.5	1.6	2.0	2.4	0.9	3.7

Source: UN/ECE secretariat, based on national statistics.

[a] Excluding Turkey.

Business fixed investment, stimulated by increased profits growth and favourable sales expectations, remained buoyant, with expenditure on information processing equipment continuing to be the major driving force. Against a background of high capacity utilization rates, there was a strong growth of expenditure on structures, an indication that increasing capacity was an increasingly important reason for investment. Residential investment weakened in the second half of 1996, but for the year as whole there was significant growth, notably for single-family housing. Changes in business inventories dampened domestic output growth in 1996, but the downward adjustment, underway since the beginning of 1995, appears to have run its course. Strong domestic demand was reflected in buoyant import demand, and changes in real net exports were a slight drag on overall output growth in 1996.

Economic recovery took hold in *Canada* in the second half of 1996. Growth accelerated markedly in the third quarter and real GDP rose by 0.8 per cent compared with the preceding period (table 2.2.1). Some of this momentum was lost in the final quarter when exports, which were the mainstay of economic growth in the second and third quarters, declined, partly reflecting the impact of strikes on automobile output. The average annual increase in real GDP was 1.5 per cent in 1996. Households responded to low financing costs and there was strong growth in purchases of consumer durables and houses. In contrast, weak labour markets and fiscal pressures constrained the growth of disposable incomes, which is reflected in sluggish demand for non-durable goods and for services. Household expenditures were supported by a significant decline in savings, the savings ratio falling to only 5.1 per cent in the third quarter of 1996. This compares with a previous peak of 11 per cent in the second quarter of 1992.[29] Business investment in machinery and equipment rose strongly, stimulated by low interest rates and the drive for modernization and rationalization. In contrast, investment in business structures remained rather sluggish, which is not surprising in view of the early stage of the recovery and significant margins of spare capacity. The more optimistic business outlook led firms also to rebuild stocks in the second half of the year. Fiscal retrenchment was reflected in falling government consumption and fixed investment.

In *Japan*, the stimulus to fixed investment from the September 1995 fiscal package and the reconstruction programmes related to the Kobe earthquake waned after the first quarter of 1996, when there was an exceptionally strong rebound of economic activity. The priority of fiscal policy shifted towards budget consolidation against the backdrop of a significant deterioration in public finances. Economic activity broadly stagnated in the second and third quarters of 1996 mainly reflecting the weakness of domestic demand. But there was a significant acceleration in economic activity in the final quarter of 1996. Exports, which had already picked up in the third quarter, strengthened considerably in the final quarter, benefiting from the recovery in world trade and the depreciation of the yen, and this supported industrial activity. In addition, the announced increase in the sales tax as from April 1997 appears to have stimulated households to bring forward purchases of cars and houses. This led to a significant rise in private consumption in the final quarter of 1996 following declines in the two preceding quarters. In the event, real GDP rose by 3.7 per cent in 1996, the highest annual increase among the industrialized countries, but this greatly exaggerates the underlying momentum of the economic recovery.

In aggregate, real GDP in the *industrialized countries* rose by 2.3 per cent in 1996 compared with 2.1 per cent in 1995. A slowdown in western Europe was more than offset by faster growth in North America (the United States) and Japan.

Table 2.2.7 indicates the relative importance of the various components of expenditure in overall economic growth in western Europe, the United States and Japan in

---

[29] It is noteworthy that this was the first time in 25 years that the savings rate in Canada was below the United States personal savings rate, which was 5.3 per cent in the same period. There is a more general point to be made here, which applies also to other countries, where consumption was supported by falling savings. The contractionary fiscal policy is regarded as something positive as it entails, *ceteris paribus*, a rise in national savings. But the fall in private household savings, which is, to some extent, linked to the contractionary impact of fiscal policy, is viewed rarely as something negative. It is viewed predominantly from the short-term demand perspective. But if it is correct that the rise in budget deficits has put upward pressure on long-term interest rates then the same should hold for the decline in private sector savings. See also Deutsches Institut für Wirtschaftsforschung (DIW), "Grundlinien der Wirtschaftsentwicklung", *Wochenbericht 1-2/97* (Berlin), 9 January 1997, pp. 24-26.

TABLE 2.2.8
Average annual growth rates of real GDP in the ECE market economies, 1960-1996
(Percentage change over previous year)

	Per cent				Rank			
	1960-1973	1973-1989	1989-1996	1960-1996	1960-1973	1973-1989	1989-1996	1960-1996
Western Europe	4.8	2.4	1.7	3.2	..	..	..	..
France	5.3	2.4	1.3	3.2	8	13	15	11
Germany[a]	4.2	2.0	2.2	2.8	16	19	6	18
Italy	5.2	2.8	1.2	3.4	9	9	16	10
United Kingdom	3.1	2.0	1.2	2.3	22	18	18	21
Austria	4.8	2.3	2.1	3.2	12	15	8	12
Belgium	4.8	2.1	1.6	3.0	11	16	13	17
Denmark	4.2	1.8	2.0	2.7	17	21	10	19
Finland	4.8	3.0	-0.1	3.1	10	7	22	15
Greece	7.4	2.5	1.1	4.0	2	12	19	6
Iceland	5.4	3.9	1.5	4.0	5	2	14	7
Ireland	4.3	3.8	5.6	4.3	15	3	1	3
Luxembourg	4.0	2.4	2.6	3.0	21	14	4	16
Netherlands	4.7	2.0	2.4	3.1	13	17	5	14
Norway	4.2	3.4	3.4	3.7	18	5	3	8
Portugal	6.6	3.0	2.0	4.1	4	8	9	4
Spain	7.0	2.5	1.8	4.0	3	11	12	5
Sweden	4.0	1.9	0.7	2.4	20	20	20	20
Switzerland	4.3	1.3	0.2	2.2	14	22	21	22
Turkey	5.4	4.2	4.6	4.7	6	1	2	2
Canada	5.3	3.4	1.2	3.7	7	6	17	9
United States	4.2	2.8	1.9	3.1	19	10	11	13
Memorandum item:								
Japan	9.2	3.6	2.1	5.3	1	4	7	1
Total above, *including Japan*	5.3	2.8	1.8	3.5	..	..	..	..

Source: UN/ECE secretariat, based on national statistics.

[a] West Germany.

1995 and 1996. It shows that the cyclical upswing in the United States continued to be broadly based on both domestic and foreign demand. In contrast, growth in western Europe continued to be held back in 1996 by the weakness of domestic demand, notably fixed investment and private consumption. The surge in real GDP in Japan is fully accounted for by the rebound in domestic demand, which was partly offset by adverse changes in the real foreign balance.

(ii) Growth and volatility

The period since 1989, the last general cyclical peak, stands out as a period of relatively low growth when seen in a longer term perspective of nearly forty years (table 2.2.8). In western Europe, the average annual growth rate of real GDP between 1989 and 1996 was only 1.7 per cent compared with 2.4 per cent for the period 1973-1989 and 4.8 per cent between 1960 and 1973. A similar slowdown can be observed in North America, while the deceleration is even more pronounced in Japan.

Looking at the most recent period, 1989-1996, it is noteworthy that there is no significant difference in the average growth performance of western Europe relative to the United States despite the significant cyclical desynchronization over this period. Among the four major economies, France, Italy and the United Kingdom experienced below average growth during this recent period. The better west German performance reflects in the main the temporary boom associated with reunification. Among the smaller economies the rapid growth of Ireland, Turkey and Norway stands out. But economic growth was also significantly above average in Luxembourg and the Netherlands. These relatively favourable developments contrast with broad stagnation in Finland and Switzerland and only little growth in Sweden. In Finland and Sweden this reflects the severe recession of 1991-1993, while Switzerland has remained more or less in recession since the beginning of the current decade. Outside western Europe, economic growth was also relatively low in Canada in the 1990s. This can be partly attributed to the slump in economic activity in 1990-1991, when real GDP fell by a cumulative 2 per cent, and to monetary policy which was tightened to keep inflation at very low levels.[30]

---

[30] P. Fortin, "Presidential address: the great Canadian slump", *Canadian Journal of Economics*, Vol. XXXIX, No. 4, November 1996, pp. 761-787.

TABLE 2.2.9

Volatility^a of year-to-year changes in real GDP in the ECE market economies, 1961-1996

(Percentage change over previous year)

	Per cent				Rank			
	1961-1973	1973-1989	1989-1996	1961-1996	1961-1973	1973-1989	1989-1996	1961-1996
France	0.84	1.45	1.55	2.00	22	21	14	20
Germany^b	1.89	1.86	2.25	2.24	14	18	5	15
Italy	1.91	2.31	1.29	2.46	12	10	20	10
United Kingdom	1.67	2.51	1.77	2.11	15	7	10	18
Austria	1.37	1.78	1.30	1.97	19	20	19	21
Belgium	1.20	2.09	1.47	2.17	21	14	17	16
Denmark	2.17	2.12	1.24	2.32	9	13	21	13
Finland	2.51	2.04	4.15	3.18	4	16	2	6
Greece	2.39	2.85	1.62	3.62	6	5	13	3
Iceland	5.07	3.01	2.37	4.08	1	4	4	1
Ireland	1.92	2.34	2.63	2.43	11	9	3	11
Luxembourg	3.00	3.29	1.80	2.89	2	3	9	8
Netherlands	1.98	1.81	1.18	2.15	10	19	22	17
Norway	1.31	2.08	1.54	1.76	20	15	15	22
Portugal	2.21	3.55	1.98	3.29	8	1	7	5
Spain	2.27	2.17	1.70	3.09	7	12	11	7
Sweden	1.66	1.41	2.09	2.11	16	22	6	19
Switzerland	1.55	2.66	1.53	2.69	18	6	16	9
Turkey	2.64	3.41	4.79	3.59	3	2	1	4
Canada	1.58	2.29	1.70	2.41	17	11	12	12
United States	1.89	2.43	1.33	2.24	13	8	18	14
Memorandum item:								
Japan	2.44	1.89	1.93	3.74	5	17	8	2

Source: UN/ECE secretariat, based on national statistics.

a Standard deviations.

b West Germany.

Over the "long run", variations in average growth rates in western Europe and the United States tend to offset each other: their average annual growth rates over the period 1960-1996 are virtually identical. Among the west European countries, a notable feature is the favourable long-run growth performance of several of the smaller economies with mean growth rates of around 4 per cent or above (Greece, Iceland, Ireland, Norway, Portugal, Spain, Turkey). Differential growth performance reflects always a host of factors, but the well-known "catch-up" effect observed in countries with below average GDP per head is clearly present here.[31] The exception is Norway, where the growth performance has been influenced by the favourable development of the oil and gas sector. In fact, there is a significant negative correlation between average annual economic growth in 1960-1996 and GDP per head in 1960 for the given sample of countries.[32] Among the four major west European economies, long-run growth of output was equal to the regional average in France and slightly higher in Italy, but below average in west Germany and, notably, the United Kingdom.

The rankings of individual countries in the "growth league" vary, on occasion considerably, from period to period (table 2.2.8). There is, however, a rather close association between the rankings in 1960-1973 with those in 1973-1989 but not with those in the most recent period 1989-1996.[33]

Countries differ not only with regard to their longer-term growth performance but also, in the short term, with regard to their position in the international business cycle and the volatility of their own business cycle. A traditional measure of volatility is the standard deviation of annual growth rates of real GDP over a longer period (table 2.2.9), and this shows that there exist significant differences in volatility among countries and also over time.

Thus, over the whole period considered here (1960-1996), France stands out among the major economies as having a rather low volatility, while in Japan it is very high. On the other hand, the average annual growth of

---

[31] As regards the member countries of the European Union, this "catching up" has been supported by regional assistance schemes.

[32] The correlation coefficient is -0.74.

[33] The Spearman rank correlation coefficient between the rankings in 1960-1973 and 1973-1989 is 0.53, which is significant at the 5 per cent level. In contrast, the coefficient between the rankings in 1960-1973 and 1989-1996 is slightly negative and not statistically significant. There is also no statistically significant relationship between the rankings in 1973-1989 and 1989-1996.

CHART 2.2.4
Growth and volatility in the industrialized countries, 1960-1996
(Per cent)

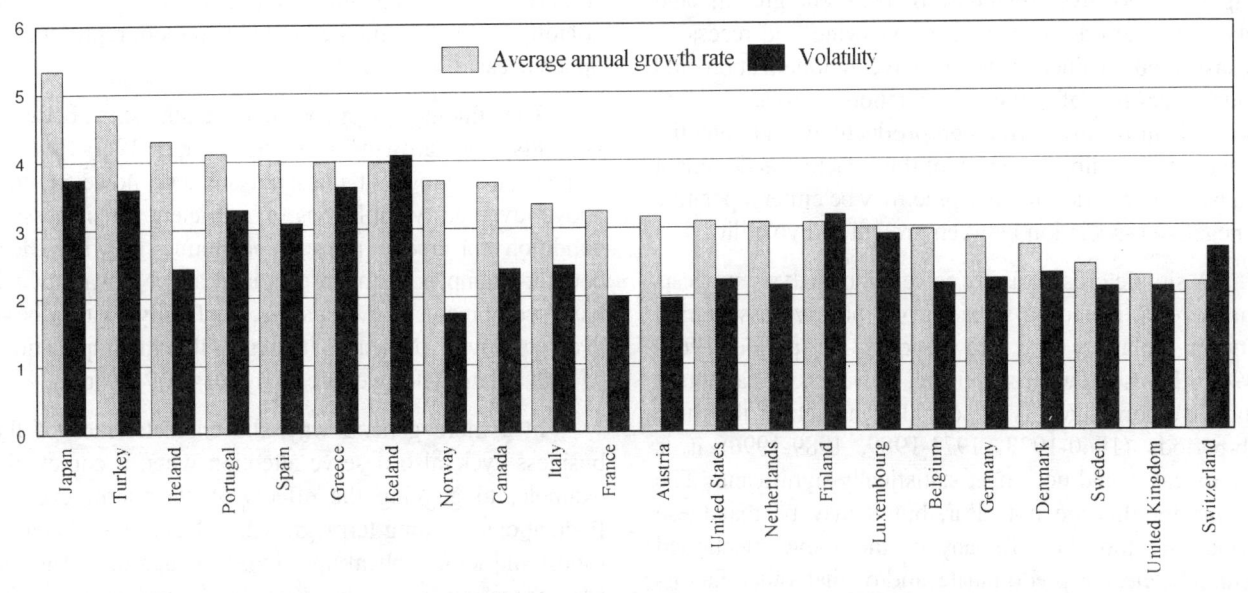

Source: UN/ECE secretariat.

Note: Average annual growth rates of real GDP in 1960-1996 and actual standard deviations of year-to-year changes in real GDP, 1961-1996. Germany refers to west Germany only.

output in Japan was much higher than in France. An interesting question is whether there is a significant relationship between average, longer-term rates of economic growth and their volatility. Chart 2.2.4 depicts the association between growth and volatility in individual countries.

In a more general way, this question leads back to the important issue of the relationship between the business cycle and the trend rate of growth. Traditionally, business cycle analysis has been mainly concerned with the causes of fluctuations in demand, while the analysis of long-term growth has focused more on the supply side, notably on the role of technical progress, population growth and human capital formation, than on demand factors.

Long-term growth is nothing but the cumulation of the short-term cyclical variations of economic activity,[34] but there is no simple unidirectional relationship between short-term cyclical variations and long-term growth, rather, the relationship is a complex one, characterized by mutual interdependence. The specific nature of short-term cyclical fluctuation impinges on the basic transformation of an economy (e.g. changes in production methods, development of new products, international division of labour) and, consequently, long-term growth. At the same time, long-term growth factors (technical progress, the supply of capital and labour) will also influence the characteristics of cyclical short-term fluctuations.[35] The analysis of the complicated interaction between short-run cyclical fluctuations and long-run growth goes far beyond the intended purpose of this short note. Rather, the examination is limited to the question of whether there is a significant statistical relationship between long-run growth and short-run cyclical volatility and whether this association is positive or negative.

From a theoretical point of view, there are good reasons to believe that there could be either a positive or negative relationship between long-run growth and short-run volatility.[36] One possible channel for a negative relationship is that higher volatility entails increased uncertainty for business investment decisions, which in turn may hold back investment expenditures. The ensuing lower rate of investment will therefore dampen output growth. An alternative negative transmission channel could stem from the role which human capital formation ("learning by doing") has via its impact on productivity and long-term growth. This rests on the assumption that periods of cyclical downswing lead to

---

[34] I. Svennilson, "Long-term changes manifest themselves in the disguise of short-term cyclical variations", *Growth and Stagnation in the European Economy* (United Nations publication, Sales No. 1954.II.E.3), p. 4.

[35] Op. cit., p. 12.

[36] G. Ramey and V. Ramey, "Cross-country evidence on the link between volatility and growth", *American Economic Review*, Vol. 85, No. 5 (Nashville), December 1995, pp. 1138-1151; P. Martin, "Faut-il aller contre le cycle?", *La Lettre du CEPII*, No. 149 (Paris), September 1996.

higher levels of unemployment and therefore to a loss of human capital formation, which is only partly offset in periods of economic upswing. On the other hand, the Schumpeterian process of "creative destruction" might suggest a positive relationship between growth and volatility. Periods of cyclical downswing and recession are assumed to encourage intensive rationalization and modernization of the production process with concomitant positive effects on productivity and potential growth. Depending on which of these factors is dominant in a particular period the outcome may be either a positive or negative association between growth and volatility.[37]

A simple cross-country correlation analysis suggests that there is, in fact, a rather strong positive association between volatility and growth over the period 1960-1996.[38] It is somewhat surprising, however, that although a positive correlation is present also in each of the three sub-periods (1960-1973, 1973-1989, 1989-1996) it is much weaker and no longer statistically significant. The reasons for this are not clear, but it may be that these periods are too short for any of the above mentioned factors to clearly predominate and/or that other factors (e.g. shocks such as the oil price crises in 1973 and 1979) obfuscate the relationship between long-term growth and economic fluctuations.

The finding of a positive relationship for the period 1960-1996 contrasts also with the results of a study[39] undertaken recently, which obtains a negative relationship between growth and volatility in OECD countries but over a different period, namely 1952-1988. This study uses regression analysis in which long-term growth is "explained" by a number of factors such as volatility, the average investment-output ratio, initial GDP per capita and the initial human capital stock. An interesting finding is that the (negative) impact of volatility on growth is not significantly altered after controlling for the influence of fixed investment.

A broadly similar approach is adopted here, but with a somewhat simplified regression analysis, in which volatility, the average investment share in GDP and initial GDP per capita were used as "control variables". The result is that there still appears to be a relatively strong positive link between volatility and growth over the period 1960-1996.[40] Chart 2.2.5 displays the positive relationship between average economic growth and the estimated standard deviations, after removing the effects of the investment-output ratio and GDP per capita. It can be seen that using this approach Japan and Turkey are the countries with both the highest output growth and volatility. Sweden and the United Kingdom represent the opposite cases.

The finding of a positive relationship between volatility and growth over the period 1960-1996 is somewhat striking. It does not suggest, however, that above average volatility is a sufficient or necessary condition for above average economic growth. Clear counter-examples to this are Ireland and Norway, which have combined below average volatility with above average growth. Nor does it suggest that the implications of high volatility for welfare are necessarily positive.

In a more general way, the characteristics of the business cycle also deserve attention when it comes, for example, to gauging the effects of short-term cyclical fluctuations on long-term growth. Thus, a sufficiently robust and long cyclical upswing, in which output in the next recession is expected to be higher than in the preceding boom, can be deemed to be more supportive of long-term growth factors compared with a situation where the output level in the next recession is the same or lower than in the preceding boom.[41] It would, however, still be difficult to determine the expected sign of the relationship. In a similar vein, if the periodicity and amplitude of the cycle are fairly regular – and are perceived to be so – the effect on uncertainty (and investment) will be less than if these two characteristics of the fluctuations are themselves volatile. One reason for regular amplitude and periodicity might be the effect of government policy – investors might be confident that when a negative deviation from trend reaches a certain point (say, unemployment at a certain rate) government would act to restore output to its trend. This was more or less the impact of Keynesian demand management on expectations and, to the extent that such policies were successful, would be suggestive of a negative association between long-term growth and volatility.

The purpose of this short digression is to illustrate that the traditional dichotomy between business cycles and long-term growth is questionable. There is, in fact, clear empirical evidence suggesting a more or less close

---

[37] P. Martin, loc. cit.

[38] The Pearson correlation coefficient between volatility and average output growth over the period 1960-1996 is 0.62, which is statistically significant at the 1 per cent level. A non-parametric test, a Spearman rank correlation, leads to a similar finding. This correlation coefficient is 0.51 (significant at the 5 per cent level).

[39] G. Ramey and V. Ramey, loc. cit.

[40] The maximum-likelihood method was employed for joint estimation of regression coefficients and variances (i.e. the standard deviations which gauge volatility) in a panel of 792 observations. The estimated regression equation is as follows: $y = 1.78 + 0.52 v + 0.06 I/Y$ $- 0.0007$ gdpc, where y = average output growth, v = standard deviation of output growth (volatility), I/Y = average investment-output ratio, and gdpc = GDP per capita in 1960. The estimated coefficients on volatility and GDP per capita were significant at the 5 per cent and 1 per cent levels, respectively.

A different method, using actually "observed" standard deviations led to broadly the same results: $y = 2.43 + 0.37 v + 0.05 I/Y - 0.0007$ gdpc. Note that the estimated coefficient on volatility is only significant at the 10 per cent level in this case.

[41] I. Svennilson, op. cit., p. 12.

CHART 2.2.5
Volatility and growth in the industrialized countries, 1960-1996

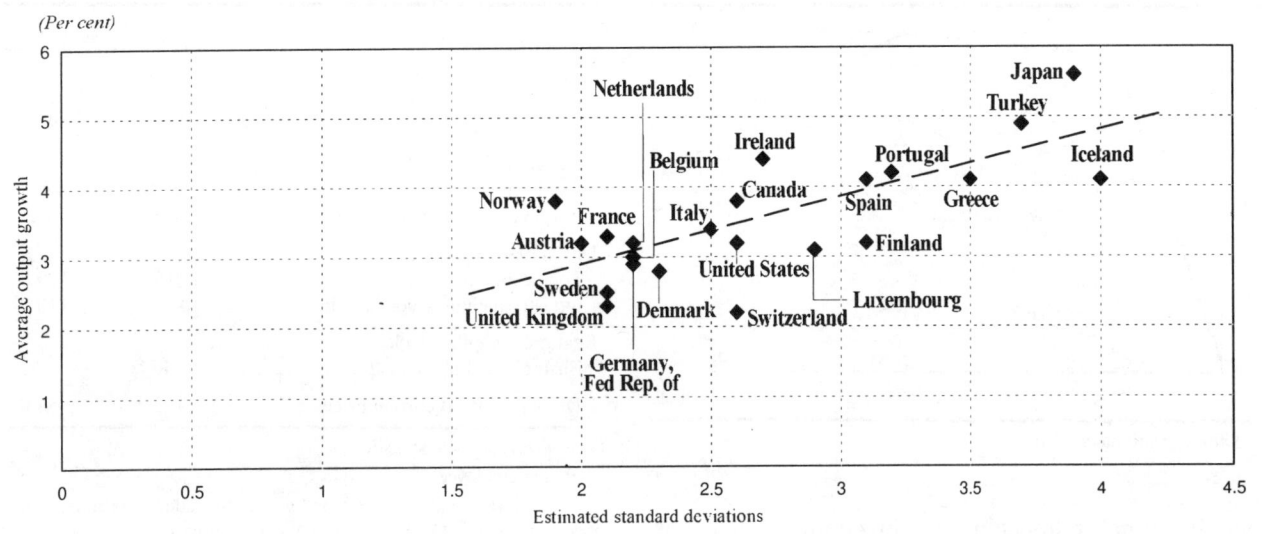

Source: UN/ECE secretariat.
Note: Based on regression analysis (see text).

relationship between the two, even if the sign of the relationship appears to be controversial. To understand and isolate the influence of short-term economic fluctuations on long-term growth is an important but difficult question which requires more research before any confident answers can be made.

(iii) Germany's transition economy: development of the eastern *Länder*, 1990-1996

The economic reconstruction of the new eastern *Länder* of united Germany has turned out to be a much more difficult task than generally expected when German economic and monetary union started in July 1990. Since then the east German economy has undergone a radical transformation. There has been a significant improvement in real incomes of the population supported by fiscal transfer payments but the overall productivity gap between east and west Germany is still very large. There are major structural problems which continue to restrain economic growth.

Overall economic output in east Germany fell sharply in 1990-1991. But economic activity picked up significantly in the following years and real GDP rose by nearly 40 per cent between 1991 and 1996. Economic growth was buoyant in 1992-1994 with annual growth rates of 8-10 per cent, reminiscent of the dynamic economies of Asia. This favourable performance fostered hopes of a rapid convergence towards west German incomes and productivity levels. But economic growth slowed down in 1995 and there was a further deceleration to only 2 per cent in 1996 (chart 2.2.6). On current forecasts, real GDP is set to rise by only 1 per cent in 1997 compared with 2 per cent in west Germany. If these forecasts prove correct this will be the first year in which the process of economic convergence, underway since 1991, would be partly reversed. This is certainly a matter of concern for German policy makers given that the process of convergence is far from being completed: in 1996 GDP per head in the east was still only 54 per cent of the west German level (table 2.2.10).

The significant loss of growth momentum over the last two years can only be partly explained by cyclical factors.[42] Closer integration with the west German economy and a stronger orientation towards west European markets obviously entailed greater sensitivity to the west European business cycle. Output growth in east Germany was therefore bound to be affected by the recent weakening of demand in both west Germany and abroad.

A more important role in the recent slowdown, however, was played by adverse changes in the special exogenous factors which have been the major engine of growth since 1991. On the demand side, the growth of fixed investment associated with the modernization of public infrastructure, the privatization of firms and massive government incentives designed to stimulate private sector investment was petering out, albeit at a high level, in 1996. Major infrastructure projects have been completed and new government investment is now held

---

[42] Sachverständigenrat zur Begutachtung der gesamtwirtschaftlichen Entwicklung, Reformen voranbringen, *Jahresgutachten 1996/97* (Stuttgart, Metzler-Poeschel, 1996), p. 71, para. 79; Deutsches Institut für Wirtschaftsforschung (DIW), "Gesamtwirtschaftliche und unternehmerische Anpassungsfortschritte in Ostdeutschland", Vierzehnter Bericht, *Wochenbericht 27/96* (Berlin), 4 July 1996.

CHART 2.2.6

Real GDP in east and west Germany, 1991-1996
(Percentage change over preceding year)

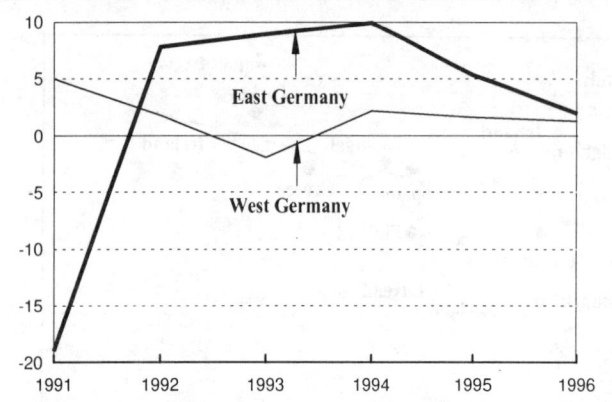

Source: National statistics.

TABLE 2.2.10

Convergence indicators for east Germany, 1991 and 1996
(West Germany=100)

	1991	1996
Average monthly wages[a]		
Gross	48.3	76.7
Net	54.8	82.4
Productivity[b]		
Total economy	31.0	56.7
Industry including construction	28.8	61.2
Unit labour costs[c]		
Total economy	150.7	129.8
Industry including construction	147.3	110.3
Real gross capital stock		
Business sector,[d] per capita	46	61
GDP per capita at current prices	31.3	54.0

Source: National statistics.
[a] Per employee.
[b] GDP (for total economy), and gross value added (for industry), per person employed. Output is valued at current prices.
[c] Total labour costs per employee relative to productivity.
[d] Excluding dwellings.

back by fiscal retrenchment. In addition, generous incentive schemes designed to support housing investment are coming to an end.

Hopes that the initial surge in investment would generate a virtuous growth circle, leading into a self-sustained upswing, have been disappointed. This has led to a more sober assessment of the prospects for a rapid closing of the productivity gap between the east and the western parts of the country and a concomitantly greater self-reliance of the east German economy.

(a) The basic strategy

The federal government's strategy[43] for the reconstruction of the east German economy was built on the following main pillars:

- rapid privatization of state enterprises;

- rapid modernization and extension of the public infrastructure;

- a multitude of subsidies and other financial incentives designed to stimulate private sector investment;

- a generous social safety net designed to cushion the adverse labour market consequences of "shock therapy", including extensive measures for retraining and improving the qualifications of the labour force.

Privatization proceeded very swiftly. The Treuhand, the organization in charge of privatization, completed its assignment after a short period of about four years by the end of 1994 and was subsequently closed down. There has also been a considerable improvement in the public infrastructure, notably as regards telecommunications and the railway system. And private investors responded strongly to generous investment stimuli offered by the German authorities. Another important factor in the strong investment growth was the contractual investment pledges which the Treuhand arranged with the investors, predominantly from west Germany, who bought east German firms. In addition, there was a considerable surge in the creation of new – mainly small – firms.

Fixed investment was, indeed, the major engine of growth over the period 1991-1996, although it weakened in 1995-1996. Aggregate investment expenditures per head of the eastern population rose significantly above the corresponding west German levels. The share of fixed investment in GDP (investment-output ratio) amounted to 52.5 per cent in east Germany in 1994 compared with 18.3 per cent in west Germany. In the business sector, the real gross capital stock was radically modernized. It is estimated that only one quarter of the manufacturing industry's capital stock in 1994 existed at the time of the former German Democratic Republic (GDR), that is, in 1989.[44]

---

[43] Bundesministerium für Wirtschaft, Jahreswirtschaftsbericht 1991 der Bundesregierung (Bonn), 8 March 1991; Jahreswirtschaftsbericht 1992 der Bundesregierung, *Deutscher Bundestag, Drucksache 12/2018* (Bonn), 30 January 1992, pp. 15-23. The government strategy has been broadly in line with the so-called "growth-oriented approach" advocated by the German Council of Economic Experts. See Jahresgutachten 1991/92 des Sachverständigenrates zur Begutachtung der gesamtwirtschaftlichen Entwicklung, *Deutscher Bundestag, Drucksache 12/1618* (Bonn), 18 November 1991, p. 167, para. 262; Jahresgutachten 1992/93 des Sachverständigenrates zur Begutachtung der gesamtwirtschaftlichen Entwicklung, *Deutscher Bundestag, Drucksache 12/3774* (Bonn), 18 November 1992, pp. 190-194, paras. 290-312.

[44] The corresponding share for the total business sector (excluding housing) amounted to some 40 per cent. See DIW, "Gesamtwirtschaftliche und unternehmerische Anpassungsfortschritte in Ostdeutschland", Zwölfter Bericht, *Wochenbericht 3/95* (Berlin), 19 January 1995, p. 76.

But it takes a long time to rebuild the largely obsolete capital stock of an entire economy. The east German capital stock relative to the total population is therefore still rather small. It corresponded to 61 per cent of the west German capital stock per capita in 1996, up from only 46 per cent in 1991.[45]

The wave of fixed investment in the business sector led to significant capital-labour substitution, partly as a response to the available investment incentives but also because of the rapid wage growth, far in excess of productivity growth, in the immediate aftermath of unification. Consequently, there was massive labour shedding, which also partly reflected the high degree of hidden unemployment in the former GDR. The official unemployment rate was some 15 per cent in 1996, but taking the persons benefiting from special labour market measures[46] into account the actual rate was closer to 25 per cent.

Public sector financial support for the east German economy reached unprecedented levels, especially when compared with international economic assistance to disadvantaged regions and the other transition economies. Transfer payments were designed not only to support investment but also incomes, consumption expenditures, and the social security system at large.[47] This assistance amounted on average to 4.5 per cent of west German GDP (or 44.5 per cent of east German GDP) per annum over the period 1991-1995.

This dependence on considerable net fiscal transfers has hardly diminished. It is mirrored in the huge gap between output and domestic demand, which is tantamount to the net transfer of resources required to sustain expenditure levels or, in other words, to finance the implicit current account deficit of east Germany with the west. Domestic demand exceeded total GDP (measured at current prices) by some 75 per cent in 1994, the last year for which official data on GDP by type of expenditure in the east were compiled by the federal statistical office. Imports of goods and services alone were equal to total GDP in 1994. These proportions have probably not changed significantly in the last two years. (For comparison, total domestic demand in west Germany corresponded to some 93 per cent of GDP in 1994; imports were equivalent to 29 per cent of GDP.)

Apart from the imbalances on the demand side, the supply side of the east German economy remains weak.[48] This pertains, *inter alia,* to the relative level of costs, the small export base, and the ownership structure of firms which is reminiscent of a "subsidiary economy", that is, one in which production or assembly takes place in the plants of large, often multinational, companies, but where strategic management, R&D, etc., are conducted by the parent company elsewhere.

(b) High unit labour costs

A key problem has been the rapid growth in wages designed to bring about swift convergence with west German wage levels. Average monthly gross wages per employee corresponded to about 77 per cent of west German wages in 1996, compared with somewhat less than 50 per cent in 1991. The rate of convergence was broadly the same for average monthly net wages, but the gap in comparison with west German levels has been smaller on account of lower income tax deductions. Wages rose much faster than productivity in the immediate aftermath of unification with the result that, on average, unit labour costs – in general the most important cost component – have been significantly above west German levels for some years now (table 2.2.10). Not surprisingly this has eroded profitability in large segments of the business sector, notably manufacturing industry. On average, revenues in large parts of the business sector do not cover costs, although the average outcome masks considerable variation among sectors and companies.[49] The depressed profitability facing many firms has had a negative impact on investment activity and job creation, which the government, in turn, has tried to offset by generous investment subsidies. It is noteworthy that, in order to survive, an increasingly large number of east German firms have de facto withdrawn from collective bargaining agreements and pay wages which are below those agreed between the trade unions and employers associations.

(c) Weak industrial export base

The other major matter of concern has been the erosion of the industrial base in east Germany, which may be difficult to reverse. Manufacturing output collapsed when, overnight, the east German economy was exposed to fierce competition from western, notably west German, firms in both the domestic market and in the GDR's

---

[45] Sachverständigenrat zur Begutachtung der gesamtwirtschaftlichen Entwicklung, Reformen voranbringen, *Jahresgutachten 1996/97* (Stuttgart, Metzler-Poeschel, 1996), p. 57, table 12.

[46] Op. cit., p. 111, table 32.

[47] For an overview of public assistance schemes see Sachverständigenrat zur Begutachtung der gesamtwirtschaftlichen Entwicklung, *Im Standortwettbewerb, Jahresgutachten 1995/96* (Stuttgart, Metzler-Poeschel, 1995), pp. 150-153, paras. 204-207.

[48] DIW, "Gesamtwirtschaftliche und unternehmerische Anpassungsfortschritte in Ostdeutschland", Dreizehnter Bericht, *Wochenbericht 27-28/95*, 13 July 1995 and Vierzehnter Bericht, *Wochenbericht 27/96*, 4 July 1996 (Berlin).

[49] For details on profitability and financing in the enterprise sector see "Ertragslage und Finanzierungsverhältnisse ostdeutscher Unternehmen im Jahre 1994", Deutsche Bundesbank, *Monatsbericht* (Frankfurt am Main), July 1996, pp. 49-62.

traditional export markets in the former CMEA. Although manufacturing output has risen strongly since 1991, the share of manufacturing activity accounted for only 14 per cent of total economy gross value added (at current prices) in 1995. This compares with a corresponding manufacturing output share of 27 per cent in west Germany. The respective employment shares are 16 per cent for east Germany and 28 per cent for west Germany.

Moreover, the weak export performance of the manufacturing sector is striking. There is an overwhelming orientation towards the regional, i.e. local, markets. Manufacturing exports accounted for only 12 per cent of total sales in 1995 which compares with a share of about 30 per cent in west German manufacturing industry.[50] There has been progress in building up a competitive export sector – but overall it is still quite small. Difficulties were encountered in developing new products and finding markets for them (also in west Germany) and in resisting the intense competitive pressures from the west. Export performance has been notably disappointing in the markets of central and eastern Europe, where east German firms were always perceived to have a traditional comparative advantage. This contrasts with the large rise in exports from western Germany to eastern Europe. In fact, east German merchandise exports accounted for only 2 per cent of total German exports in 1995.

It is well established in regional economic analysis that the growth potential of a region is strongly dependent on its openness and integration with other regions. Given the limited scope of the regional market relative to more distant markets (in the domestic economy and abroad) this points to the key role of the tradeables sector, and hence of exports, in the regional development process.

Evidently, export orientation is a feature which can be found to varying degrees across a wide range of economic activity, e.g. agriculture, certain business and financial services and tourism. But, in general, manufacturing constitutes the core of a region's export base and it is the manufacturing industry which is the main engine of growth of an economy. It is this sector which generates the incomes on which other, inward-oriented sectors such as construction, transport and communications, and a wide range of local services are largely dependent.

The implication of this argument is that the process of convergence between east and west Germany will have to be driven mainly by the east German manufacturing sector, which is the largest potential source of exports. Unless there is sustained and strong export-led growth, the east German economy will not catch up with the west. This points to the need for a sufficiently large and competitive export base in the regional economy. But the question, however, is whether the manufacturing sector is now too small to fulfil that role.[51]

(d) A subsidiary economy

A conspicuous feature of swift privatization was that the large majority of east German firms (some three quarters) was taken over by existing west German companies. Thus, the east German firms became "subsidiaries" of firms having their headquarters in the western part of the country. It is also worth emphasizing that the size distribution of companies is more dominated by small- and medium-sized enterprises than in west Germany. This has largely been a consequence of the privatization process, which involved the breakup of the large state monopolies of the former GDR into much smaller units deemed to be viable by the Treuhand and potential investors and the closure of the non-viable parts. In addition, newly created companies fall into the category of small- and medium-sized enterprises. While it is true that these enterprises play an important role in regional economic development, it is also the case that it is their involvement in the production networks organized by larger firms that facilitates their integration into the broader division of labour within the domestic economy and the international economy at large.[52] There are, moreover, two other factors which have been restraining the potential impact of small- and medium-sized firms on economic growth in east Germany. First, many of the newly created firms are facing the same normal start-up difficulties as new firms in western market economies. These include a weak equity base, the reluctance of banks to lend funds and the time needed to find sufficiently large market niches and develop stable relationships with clients. As a result, therefore, a large proportion of these new firms are still dependent on direct and indirect government support. Second, the fact that many of the privatized firms are subsidiaries of west German companies could have negative repercussions on the long-run growth potential of the east German region.[53]

Not only are subsidiaries more exposed to the risk of closure in times of economic crisis, but they often tend to buy fewer inputs in the regional market than indigenous firms. The dominance of such plants also has repercussions on the skill structure of jobs. Upper managerial, technical and R&D staff will, in general, tend

---

[50] Sachverständigenrat zur Begutachtung der gesamtwirtschaftlichen Entwicklung, Reformen voranbringen, *Jahresgutachten 1996/97* (Stuttgart, Metzler-Poeschel, 1996), p. 75, para. 83.

[51] W. Mieth, "Die Wachstumsschwäche des verarbeitenden Gewerbes in Ostdeutschland", *Jahrbücher für Nationalökonomie und Statistik*, Vol. 214, No. 1 (Stuttgart), 1995, pp. 11-20.

[52] DIW, "Gesamtwirtschaftliche und unternehmerische Anpassungsfortschritte in Ostdeutschland", Dreizehnter Bericht, *Wochenbericht 27-28/95* (Berlin), 13 July 1995, pp. 470-472.

[53] Op. cit., pp. 474-476; see also W. Mieth, loc. cit.

to remain in headquarters and the overall regional demand for labour will accordingly be biased towards less skilled workers. The absence of the headquarters of large companies may also have a negative impact on the attractiveness of the region to service industries which cater for wider national or international markets, such as business and financial services, and which require close contact with each other and with the headquarters of large industrial firms.

In general, the existing high concentration of economic activity in west Germany could in itself be a major constraint on the development potential of the east German region. Increasing returns to scale of production, close integration of firms in "industrial networks", agglomeration advantages, and a favourable spatial allocation of economic activity could all support a cumulative and virtuous causation process in the more advanced west German region, which in turn could restrain the potential for eastern convergence on the west.

(e) Where now?

There has been a notable degree of regional economic convergence between east and west Germany since 1991. But the growth process in the east has been driven in the main by the expansion of the non-tradeable sector, i.e. activities such as construction and local services, which are not exposed to international competition. In fact, the construction sector has attained a disproportionately large size in the east German economy, accounting for some 18 per cent of gross valued added generated in the region in 1995, compared with a share of about 5 per cent in west Germany. The construction boom, however, peaked in 1996. A major downsizing of the sector now lies ahead, with negative implications for the labour market. The crucial issue now is whether the tradeable sector, and especially manufacturing industry, can take over as the main engine of growth. It is only via a stronger export base that the east German region will be able to prosper and that the need for large net transfer payments can be reduced and, ultimately, eliminated.[54]

The exceptional challenge of integrating a backward economic region into one of the most advanced economies in the world has been met by the German authorities with exceptionally large amounts of financial and technical assistance. There can be no doubt that such an effort was justified given the severe locational disadvantages of the east German region vis-à-vis west Germany at the start of economic and monetary union in July 1990. The dilemma which the government faces now, however, is that endogenous growth forces are still weak and that the competitive export base appears too small to generate sufficient momentum for self-sustainable growth. Thus, there appears to be no alternative but to continue with substantial financial assistance to east Germany. The risk is that the various support schemes, which have already been available for a long time, will not only continue to encourage rent-seeking, but, more importantly, to reinforce a "subsidy mentality", i.e. an economic environment where incentives are blunted and dependence feeds on itself.

With hindsight, hopes that the reconstruction process in east Germany needed only a strong initial push from the government after which the rest could be left to the market can now be seen to have contained a large measure of wishful thinking.

Thus, the special support schemes designed in 1991 for east Germany (*Sonderförderung Ost*) such as investment allowances and special depreciation schemes, were initially expected to be phased out in 1996, the assumption being that by then viable economic structures would be in place. When it became clear that this was too optimistic, the authorities extended the support measures to the end of 1998. Yet again, against the recent backdrop of disappointing economic developments in the eastern *Länder*, the government has announced that financial assistance and incentives will be prolonged, at a high level, for another five years or so.

The government is therefore faced with the need to rethink its regional and, notably, industrial policy, which has been a prominent feature throughout the reconstruction process so far.[55] The basic issue, however, is not a choice between industrial policy or market forces, but, rather, getting the right balance and emphasis between the two.[56]

The general direction will probably be to put major emphasis on further improving the public infrastructure, notably local area transport networks, and to streamline the range of instruments designed to stimulate business investment, the creation of new firms, including measures to fortify their equity endowment, and human capital formation. In view of the weak export base of the east German economy, preference is likely to be given to strengthening manufacturing and other sectors exposed to international competition, including making the conditions for foreign direct investment more attractive. In addition, prolonged wage restraint will be required to improve business profitability.

---

[54] A. Boltho, W. Carlin and P. Scaramozzino, "Will east Germany become a new Mezzogiorno?" Centre for Economic Policy Research, *Discusssion Paper*, No. 1256 (London), February 1996.

[55] For an evaluation of the various policy measures designed to promote the reconstruction and competitiveness of the east German economy since unification see DIW, "Gesamtwirtschaftliche und unternehmerische Anpassungsfortschritte in Ostdeutschland", Fünfzehnter Bericht, *Wochenbericht 3/97*, 16 January 1997.

[56] See also A. Boltho, "The Italian Mezzogiorno: markets or policies?", *Banca Nazionale del Lavoro Review*, No. 175 (Rome), December 1990, pp. 431-439.

Germany is now facing a major regional development problem in the new eastern *Länder*. The greatest challenge is to foster adequate endogenous regional growth forces which would ensure a sustained rise in regional incomes and output and which, in addition, must be strong enough to steadily close the real income gap with the western *Länder*. A crucial requirement is to improve the net export performance of the region. It will also be important to foster human capital formation and to ensure that conditions are attractive enough to encourage the young and skilled to stay and not migrate to the west.

2.3 Labour markets

(i) Employment

Preliminary data for *western Europe* suggest that the modest boost to employment growth in 1995 has not been sustained and net job creation appears to have virtually stagnated in 1996 (table 2.3.1). Among the four major economies, average employment actually fell slightly, largely reflecting a sharp deterioration in Germany. Although the demand for labour has generally been more robust in the smaller economies in 1996, reflecting more favourable output growth (table 2.2.3), there was a slowdown in employment growth in many of these countries compared with the previous year. Thus, for the west European countries combined, employment grew by only 0.3 per cent in 1996 compared with nearly 1 per cent in 1995.

In *France*, employment stagnated in the first half of 1996 compared with a year earlier and fell in the second half, a lagged response to the slowdown in output growth. A fall in employment of 0.5 per cent is expected for the year as a whole, although the job losses would probably have been greater but for policy measures designed to reduce employers' labour costs. Employment will probably stabilize by mid-1997, but there is unlikely to be any growth for the year as a whole.

Annual average employment fell by a little over 1 per cent in *Germany* in 1996. With the year-on-year decline steady at some 1 per cent in each quarter, it is clear that the economic upturn had yet to have a positive impact in the labour markets by the turn of the year. Employment fell again in the new *Länder* in 1996, after expanding in 1994 and 1995. Furthermore, the rate of job losses in Germany in January 1997 appeared to have accelerated more than might be expected on the basis of seasonal factors alone. For the year as a whole employment is forecast to fall by some 0.5 per cent.

There was a short-lived boost to employment in *Italy* in the early months of 1996, but it tailed off quickly and employment growth was virtually stagnant by year-end. The situation in the labour market is expected to improve only slightly if at all in 1997.

In the *United Kingdom*, there was no net employment growth during the first half of 1996, but a notable surge occurred in the third quarter, when seasonally adjusted employment grew by some 200,000 persons, the largest quarterly increase since the beginning of the current recovery, and a rise of more than 1 per cent over the third quarter of 1995. Labour force survey data provide further evidence of strengthening employment growth from mid-year, and indicate that much of the growth was in full-time jobs. In view of expected robust output growth, the outlook for the labour market is encouraging, with employment forecast to rise by more than 1 per cent in 1997.

Among the smaller economies, preliminary data suggest that annual average employment growth was relatively strong in 1996 (at least 2.5 per cent) in Ireland, Israel, Malta, Norway and Spain (table 2.3.1). Of this latter group, however, employment growth accelerated between 1995 and 1996 only in Norway, in line with continuing robust output growth, and certain skills shortages are expected in 1997. Similarly, job creation remains very strong in Ireland, sustained by the continuing high rate of GDP growth.[57]

Short-run cyclical fluctuations in the demand for labour across broad sectors of the economy have generally occurred around the longer-run trends which have changed the employment balance between sectors. Thus, agricultural employment has continued to shrink, and the rate of decrease in this sector accelerated in many countries in 1996. In Finland, Germany and Italy, for example, the rate of job loss in agriculture was at least 6 per cent compared with the previous year. In terms of numbers of jobs lost, however, declining employment was again concentrated in the industrial sector, notably so in Germany where industrial employment (including construction) fell by 425,000 persons between 1995 and 1996, or 3.4 per cent compared with a fall of 1.4 per cent in the previous year, and manufacturing employment was 4.5 per cent lower in September 1996 compared with the same month of 1995.[58] Industrial employment (including construction) also fell in France, by 1.7 per cent year-on-year in the first three quarters of 1996, with a steady deterioration in each quarter.

In contrast, there was only a slight decline in annual average industrial employment (including construction) in Italy in 1996, reflecting net losses in the second half of

---

[57] The employment growth rate of 3.6 per cent in 1996 probably understates the actual rate as the survey week in April coincided with the BSE-related temporary shutdown of plants in the meat processing industries. Economic and Social Research Institute, *Quarterly Economic Commentary* (Dublin), December 1996, p. 24.

[58] Deutsche Bundesbank, *Monthly Report* (Bonn), December 1996, p. 54.

TABLE 2.3.1

Employment in the ECE market economies, 1993-1996
(Percentage change over previous year)

	1993	1994	1995	1996[a]
Western Europe	-1.6	–	0.9	0.3
4 major countries	-1.8	-0.4	0.3	-0.3
France	-1.2	-0.1	1.2	-0.5
Germany	-1.7	-0.7	-0.3	-1.1
Italy[b]	-2.9	-1.5	-0.4	0.1
United Kingdom[c]	-1.5	0.6	1.1	0.5
17 smaller countries[d]	-1.2	0.6	1.7	1.2
Austria	–	0.2	-0.4	-0.7
Belgium[c]	-1.1	-0.7	0.4	-0.1
Cyprus	0.1	1.8	2.2	..
Denmark	-1.0	-0.6	1.6	1.2
Finland	-6.5	-1.1	1.5	1.1
Greece	0.9	1.9	0.9	0.8
Iceland[b]	-0.8	1.2	0.9	1.7
Ireland[e]	0.6	3.1	4.8	3.6
Israel	6.1	6.9	5.2	2.5
Luxembourg	1.8	2.3	2.5	2.3
Malta[f]	0.5	1.2	4.2	2.6
Netherlands[b]	-0.1	0.1	1.9	2.0
Norway	0.2	1.2	2.1	2.7
Portugal	-2.0	-0.1	-0.6	0.6
Spain	-4.3	-0.9	2.7	2.8
Sweden	-5.2	-1.0	1.5	-0.5
Switzerland	-0.6	-0.2	0.2	–
Turkey[g]	0.9	2.5	2.5	2.4
North America	1.8	2.3	1.6	1.4
Canada	1.3	2.1	1.6	1.3
United States[b]	1.8	2.3	1.6	1.4
Total above	-0.2	1.0	1.2	0.8
Memorandum item:				
Japan	0.4	0.1	0.1	0.6
Total above, including Japan	-0.1	0.8	1.0	0.7

Source: National statistics; OECD, *National Accounts Detailed Tables, Vol. II, 1982-1994,* 1996 and *OECD Economic Outlook,* December 1996 (Paris); UN/ECE secretariat estimates.

Note: National accounts statistics, where available; otherwise annual labour force surveys.

[a] Provisional.
[b] Full-time equivalent data.
[c] June of each year.
[d] Excluding Israel.
[e] April of each year.
[f] End of year.
[g] Civilian employment.

CHART 2.3.1

Employment expectations in the European Union, 1991-1996
(Balances,[a] seasonally adjusted)

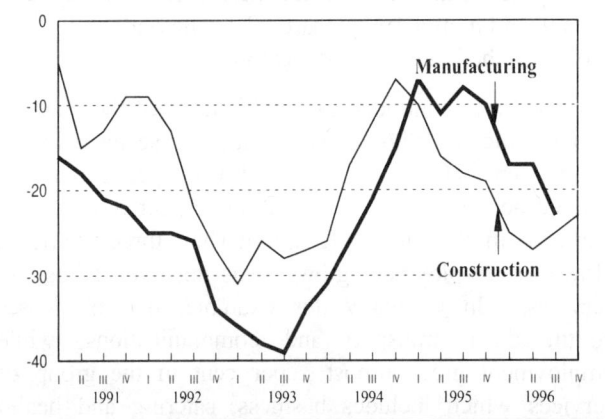

Source: European Commission, *European Economy*, Supplement B (Luxembourg), various issues.

Note: Business survey question: employment expectations for the months ahead: up, unchanged, down?

[a] Balances are the differences between the percentages of respondents giving positive and negative replies.

Available data at the time of writing show that manufacturing employment fell between the first three quarters (or halves) of 1995 and 1996 in Austria, Denmark, the Netherlands, Portugal, Spain and Switzerland. In Sweden there was growth in the first half, followed by a decline from mid-year. In sharp contrast, manufacturing employment rose by 5 per cent between the first three quarters of 1995 and 1996 in Ireland, while the full-year rise in Norway was 2.1 per cent.

Business surveys indicate that, in general, employment expectations of manufacturing firms weakened further in the first three quarters of 1996 (chart 2.3.1). Given that these employment expectations look forward several months, it seems likely that the demand for labour in the manufacturing industries will continue to fall in the first half of 1997.

Construction employment data are not readily available for all countries, but in some at least there was a fall in 1996 compared with 1995, in line with the slowdown in activity in the sector (Austria, France, Germany and Switzerland). In many countries, new house building has been held back by stagnant prices and a stock overhang, corporate investment in construction has been very sluggish, and fiscal restraint has led to a decline in government investment in infrastructure. Employment in the sector also fell between the first three quarters of 1995 and 1996 in Denmark and Finland, but there was a pick-up during the year which may be partly attributable to the resurgence of growth in government fixed investment (table 2.2.5). In contrast, there was year-on-year jobs growth in construction in Norway and Spain, though at a slower pace than a year earlier, and in

the year. The deterioration from mid-year was especially evident in larger firms (more than 500 employees). In the United Kingdom, the decline in the number of industrial employees (excluding construction) was also small – less than 0.5 per cent in the first three quarters of 1996 compared with a year earlier. Most of the fall was accounted for by mining and utilities industries, while employment in manufacturing was broadly unchanged.

Demand for labour in the industrial sector was also weak in many of the smaller economies in 1996.

Sweden, largely the result of central government subsidies to the industry; it is likely that there was also an increase in Ireland. While the sharp deterioration in employment expectations since the end of 1994 may have been halted in mid-1996 (chart 2.3.1), the outlook for jobs growth in the industry remains gloomy.

Services continued to provide most of the new job opportunities in 1996, employment in the sector rising by 0.4 per cent in Germany compared with a year earlier, 1.5 and 3.5 per cent in Italy and Norway respectively, and 1.6 per cent in the United Kingdom (first three quarters). However, employment growth was not even across all services. In Germany, for example, net job losses continued in transport and communications, while employment grew almost 3 per cent in the group of services which includes business, catering and health care. A lack of uniformity in presentation of the data makes cross-country comparisons rather difficult, but available data suggest that most of the employment growth was in services provided in the private sector, while state sector employment declined, especially in government administration.

The modest output growth in western Europe in 1996 was, in general, met from existing productivity reserves, provisional data showing that average annual labour productivity grew by almost 1.5 per cent (table 2.3.2). This was a slight slowdown compared with the previous year, but there were wide differences between countries. In Italy, for example, there was a sharp deceleration which may reflect the optimism of firms about recovery prospects in 1997, and hence their decisions not to shed labour at a rate commensurate with the current contraction in output and demand.[59] In contrast, productivity growth accelerated in France and, slightly, in Germany and the United Kingdom. The comparatively slow growth in productivity in the latter in 1995 and 1996, particularly in manufacturing, partly reflects the relatively strong growth of part-time employment. It is likely that there has also been some labour hoarding in anticipation of a pick-up in demand,[60] and in response to reports of certain skill shortages.

Cross-country comparisons of labour productivity must be interpreted with caution, however, as there are likely to be considerable differences in the national compilation of employment data. A problem arises, for example, when part-time employment is reported as full-time equivalent employment in some countries but not in others (see table 2.3.1). The relative importance of part-time employment across countries and by gender is shown in table 2.3.3, although it should be noted that

---

[59] Banca d'Italia, *Economic Bulletin*, No. 23 (Rome), October 1996, p. 24.

[60] Office for National Statistics, *Economic Trends*, No. 519 (London), January/February 1997, p. 60.

TABLE 2.3.2

Labour productivity[a] in the ECE market economies, 1993-1996
(Percentage change over previous year)

	1993	1994	1995	1996[b]
Western Europe	1.5	3.2	1.7	1.4
4 major countries	1.4	3.3	2.0	1.7
France	-0.2	2.9	1.0	1.8
Germany	0.6	3.6	2.3	2.5
Italy	2.2	3.7	3.4	0.6
United Kingdom	3.6	3.2	1.4	1.7
16 smaller countries	1.7	3.0	1.1	0.7
Austria	0.4	2.8	2.2	1.5
Belgium	-0.3	3.1	1.5	1.3
Cyprus	1.7	4.2	3.0	..
Denmark	2.5	5.0	1.1	1.0
Finland	5.7	5.8	2.9	2.1
Greece	-1.8	-0.4	1.1	1.4
Iceland	1.7	2.3	1.2	..
Ireland	2.4	3.3	5.2	2.5
Israel	-2.5	-0.1	2.0	..
Luxembourg	-1.7	1.8	1.2	0.1
Malta	4.0	2.7	4.7	1.4
Netherlands	0.9	3.3	0.2	0.7
Norway	2.6	3.8	1.2	2.0
Portugal	0.9	0.9	3.0	2.0
Spain	3.2	3.1	0.1	-0.5
Sweden	3.2	4.4	2.0	2.0
Switzerland	-0.2	1.2	-0.1	-0.7
Turkey	7.1	-7.8	4.7	..
North America	0.5	1.2	0.4	1.0
Canada	0.9	1.9	0.7	0.2
United States	0.5	1.2	0.4	1.0
Total above	1.0	2.2	1.1	1.2
Memorandum item:				
Japan	-0.1	0.5	1.3	3.0
Total above, *including Japan*	0.8	1.9	1.1	1.5

Source: National statistics; UN/ECE secretariat estimates.
Note: Regional aggregates exclude Israel and Turkey.
[a] Real GDP per person employed in the total economy.
[b] Provisional.

there are many limitations on the comparability of the data in this table. In many countries, however, it appears that part-time employment has risen for both genders, both in recent years and in comparison with the early 1980s. This expansion must also be borne in mind when assessing productivity performance in a given country over time.

In *North America*, average annual growth in the demand for labour slowed a little in 1996 (table 2.3.1). Year-on-year employment growth in *Canada* was steady at around 1.3 per cent in each quarter of the year, but there were noticeable differences between 1995 and 1996 on a sectoral basis. The rapid rate of growth of employment in manufacturing in 1995 slowed to an increase of some 1 per cent in 1996, but there were fewer job losses in construction. In the services sector, employment fell in financial services, but there was a

TABLE 2.3.3

Part-time employment in the ECE market economies, 1983, 1993 and 1995

(Per cent of employment by gender)

	Male			Female		
	1983	1993	1995	1983	1993	1995
France	2.5	4.1	5.0	20.1	26.3	28.9
Germany	1.7	2.9	3.6	30.0	32.0	33.8
Italy	2.4	2.5	2.9	9.4	11.0	12.7
United Kingdom	3.3	6.6	7.7	42.4	43.9	44.3
Austria	1.5	1.7	4.0	20.0	22.8	26.9
Belgium	2.0	2.3	2.8	19.7	28.5	29.8
Denmark	6.6	11.0	10.4	44.7	37.4	35.5
Finland	4.5	6.2	5.7	12.5	11.1	11.3
Greece	3.7	2.6	2.8	12.1	7.6	8.4
Iceland	..	9.9	12.4	..	47.5	51.4
Ireland	2.7	4.8	5.1^a	15.5	21.3	21.7^a
Luxembourg	1.2	1.0	1.1	18.0	18.3	20.3
Netherlands	6.9	15.3	16.8	50.3	64.5	67.2
Norway	11.5	9.7	9.4	54.9	47.6	46.6
Portugal	2.5^b	4.5	4.2	16.5^b	11.1	11.6
Spain	..	2.4	2.8	..	14.8	16.4
Sweden	6.3	9.1	9.4	45.9	41.4	40.3
Switzerland	..	8.6	8.6	..	54.1	54.7
Canada	8.7	11.0	10.6	28.1	28.8	28.2
United States	10.8	11.0	11.0	28.1	25.5	27.4
Memorandum item:						
Japan	7.3	11.4	10.1	29.8	35.2	34.9

Source: OECD, *Employment Outlook* (Paris), July 1996, p. 192, table E.

Note: Data for 1983 are not directly comparable with later years for most countries because of changes in methodology; comparability between countries is limited by the variation in definitions of part-time employment. For more details, see the note to the above source table.

^{a} 1994.
^{b} 1979.

notable pick-up in the distributive trades, and net recruitment in public administration following a 7.5 per cent reduction in the previous year.

In the *United States*, preliminary estimates of full-time equivalent civilian employment indicate an average annual growth rate of almost 1.5 per cent. The number of persons in non-farm payroll employment rose a little faster at some 2 per cent, which was less than 1995, but year-on-year growth strengthened in each quarter of 1996 suggesting continued tightening in the labour markets. Job creation in private services remained buoyant at some 3 per cent, with business services growing at almost twice that rate[61] and continuing strength in health services employment. In construction, average annual employment growth increased in 1996 – 4.8 per cent compared with 3.4 per cent in 1995 – while manufacturing employment actually fell by 1 per cent. The shake-out in federal government employment accelerated, in line with policy objectives related to the federal budget deficit.

Employment prospects for western Europe are a little brighter in 1997 than 1996, with growth of 0.6 per cent forecast for the European Union.[62] Employment expectations in North America remain more optimistic than in western Europe. In the United States this is associated with the continuing robust output growth, which is even expected to strengthen somewhat, on average, in 1997 compared with the previous year. There are also good prospects of relatively buoyant GDP growth in Canada and employment is expected to strengthen accordingly.

(ii) Unemployment

The weakness of employment growth in *western Europe* in 1996 was reflected in the average annual unemployment rate for the region which remained stubbornly above 10 per cent (table 2.3.4). Among the four major economies, the standardized unemployment rate rose in France and Germany, to 12.3 and 9 per cent respectively, reversing the improvements of the previous year, while in Italy the rate remained at around 12 per cent. In contrast, it fell in the United Kingdom to 8.2 per cent, the fourth successive annual decrease. For a brief review of the methodology of standardized unemployment rates see box 2.3.1.

Unemployment rose particularly sharply in *Germany* where there were nearly 4 million people unemployed on average in 1996, an increase of nearly 10 per cent over 1995. This was a major reversal as the rise in unemployment in Germany had slowed in 1994 compared with the preceding year and there was actually a slight fall in 1995, but the rapid deterioration in 1996 was to a level of unemployment unprecedented since reunification. Just over half of those entering the unemployment registers were previously employed, with a notable increase in construction workers and many from industry. Measured unemployment would be much higher but for the programme of job creation and training measures. In total, there were 895,000 people engaged in these schemes in Germany at the end of November 1996, and they continued to be of particular importance in the eastern *Länder*. In addition, the federal government was assisting over 200,000 short-time employees.[63]

In *France*, falling employment was also partly offset by active labour market measures in 1996, but the unemployment rate still rose. Employment policies are expected to have a smaller impact in 1997 as there will be fewer training places and a slowdown in the rate of early retirements; in addition the numbers in national service will be reduced, and it is likely that the unemployment rate will rise to 13 per cent by mid-year. There was little change in the high unemployment rate in *Italy* in 1996, and no improvement is forecast before 1998.

---

[61] Within business services, computer and data processing services employment rose 11 per cent in 1996 compared with 1995, while jobs in help supply services, which provide temporary workers, rose 6.5 per cent.

[62] European Commission, *European Economy*, Supplement A (Luxembourg), December 1996, p. 20, table 16.

[63] Deutsche Bundesbank, op. cit., pp. 54-55.

TABLE 2.3.4

Standardized unemployment rates[a] in the ECE market economies, 1993-1996

(Per cent of labour force)

	1993	1994	1995	1996*
Western Europe	10.3	10.6	10.2	10.3
4 major countries	9.8	10.2	9.8	10.1
France	11.7	12.3	11.6	12.3
Germany	7.9	8.4	8.2	9.0
Italy	10.3	11.4	11.9	12.0
United Kingdom	10.5	9.6	8.8	8.2
17 smaller countries[b]	10.9	11.4	10.8	10.6
Austria	4.2	3.6	3.8	4.1
Belgium	8.9	10.0	9.9	9.8
Cyprus	2.7	2.7	2.6	3.0
Denmark	10.1	8.2	7.1	6.0
Finland	17.6	17.9	16.6	15.7
Greece	9.7	9.6	10.0	10.1
Iceland	4.4	4.8	5.0	4.5
Ireland	15.6	14.3	12.4	12.3
Israel	10.0	7.8	6.9	6.6
Luxembourg	2.7	3.2	2.9	3.1
Malta[c]	4.5	4.0	3.6	3.7
Netherlands	6.6	7.1	7.0	6.6
Norway	6.1	5.5	5.0	4.8
Portugal	5.7	7.0	7.3	7.3
Spain	22.8	24.1	22.9	22.3
Sweden	9.5	9.8	9.2	10.0
Switzerland	4.5	4.7	4.2	4.7
Turkey	7.7	8.1	7.5	7.2
North America	7.3	6.5	6.0	5.8
Canada	11.2	10.4	9.5	9.7
United States	6.9	6.1	5.6	5.4
Total above	9.0	8.9	8.4	8.4
Memorandum item:				
Japan	2.5	2.9	3.1	3.4
Total above, including Japan	8.0	7.9	7.6	7.6

Source: National statistics; OECD, *Quarterly Labour Force Statistics*, No. 4, 1996, *Main Economic Indicators*, February 1997 and *OECD Economic Outlook*, December 1996 (Paris); UN/ECE secretariat estimates.

Note: Comparisons with previous years are limited by changes in methodology in Austria (1995), Norway (1996) and the United States (1994).

[a] Adjusted for comparability between countries except Austria (1993-1994), Cyprus, Greece, Iceland, Israel, Malta, Switzerland and Turkey.

[b] Excluding Israel.

[c] End of year.

In the *United Kingdom*, the unemployment rate fell again in 1996, the third successive annual decrease. This was in spite of the continuing reduction in the number of persons in work-related, government-supported training. There is a sharp contrast in labour market performance since 1993 between the United Kingdom and the other major west European economies which reflects to some degree the relative strength of GDP growth in the former. The structural reforms of the 1980s may also have improved the functioning of the labour markets in the United Kingdom.[64]

Unemployment rates fell in some of the smaller west European economies in 1996, but generally only slightly. The most notable improvement was in Denmark, where the rate decreased steadily during the year. It also fell in Finland, although largely because the growth of the labour force was dampened by an increase in the number of vocational training opportunities for young people and by a 30 per cent rise in participation in other job training schemes; without these programmes, the fall in the unemployment rate would have come to a halt in 1996.

In some countries (Sweden, Switzerland) improvements in the unemployment rate in 1995 were reversed in 1996. In Sweden this deterioration occurred in spite of a 12 per cent increase in the number of persons benefiting from labour market policy measures. Even in those few countries where there was buoyant job creation in 1996 there were not necessarily matching reductions in unemployment. In Ireland and Norway, for example, employment growth was offset to a large extent by rising labour force participation, as in the previous three years.

In North America, the sharp contrast remains between the unemployment rate in *Canada*, which has not been below 9.3 per cent since the final quarter of 1990, and that in the *United States*, where it has continued to fall, reaching 5.3 per cent in the final quarter of 1996.

The very high level of unemployment in western Europe and Canada represents a massive waste of economic resources. As discussed in box 2.3.1, standardized unemployment rates attempt to measure this in accordance with the ILO guidelines. To the extent that people are working in the shadow economy, however, unemployment rates will overstate the problem, but there are also factors working in the opposite direction. As illustrated above, extensive government sponsored training and work creation programmes are keeping many people off the unemployment registers. In addition, a significant proportion of part-time working is involuntary; many part-time workers would take full-time employment if it were available.[65] And there are unknown numbers of discouraged workers who have abandoned the search for

---

[64] A recent study of the last two economic recoveries in the United Kingdom found some evidence of greater labour market flexibility which may be linked to the reforms of the 1980s. It also concluded, however, that there were other factors which were contributing to the favourable labour market developments in the 1990s, including lower participation and a reduced regional bias in the recession and recovery. J. Morgan, "What do comparisons of the last two recoveries tell us about the UK labour market?", NIESR, *National Institute Economic Review*, No. 156 (London), May 1996, pp. 80-92.

[65] The EU Community labour force survey of 1995 found that 38 per cent of men and 18 per cent of women across the Union working part-time would have preferred a full-time job if they could have found one. For men, the figure was over 50 per cent in Finland, France, Greece, Ireland and Italy. European Commission, *Employment in Europe, 1996* (Luxembourg, Office for Official Publications of the European Communities, 1996), p. 54.

> **BOX 2.3.1**
>
> **Standardized unemployment rates**
>
> Unemployment rates published by national statistical offices may not be comparable between countries because of differences in data and methodology. International agencies attempt to resolve this problem by calculating unemployment rates for different countries on the basis of a common or "standardized" methodology. For some countries the standardized unemployment rates differ quite markedly from the rates published by national statistical offices. For example, in Germany the official unemployment rate for 1996 is 10.3 per cent but standardization lowers this to 9 per cent. The difference arises in this case largely because the official rate is computed as a percentage of the dependent labour force – the self-employed are excluded from the labour force count. In the United Kingdom, standardization raises the rate for 1996 from the official rate of 7.5 per cent to 8.2 per cent. This adjusts for the fact that the national computation is based on the claimant count of the unemployed, i.e. those registering as unemployed at state employment offices, which does not take account of non-registrants who may also be actively seeking work.
>
> In previous editions of this *Survey*, the Economic Commission for Europe has used standardized unemployment rates compiled by the OECD, where available, for comparative analysis of labour markets in the ECE market economies. Recently the OECD has begun to collaborate with the Statistical Office of the European Communities (Eurostat) on the compilation of these rates, and Eurostat now calculates them for its member states (with the exception of Greece), while the OECD remains responsible for its non-Eurostat member countries. The adoption of a common methodology has entailed some slight revisions to the series previously compiled by the OECD, which are reflected in the data published here (table 2.3.4 and appendix table A.10).[1]
>
> Standardized unemployment rates express the number of unemployed persons as a percentage of the total labour force, where the measurement of each of these variables should ideally conform to the definitions established by the 13th Conference of Labour Statisticians (commonly referred to as the ILO guidelines). Thus the total labour force consists of civilian employees, the self-employed, unpaid family workers, professional and conscripted members of the armed forces, and the unemployed. However, Eurostat estimates of the total labour force are derived from labour force surveys which cover only *private households*, so that persons living in institutions, such as the armed forces, are excluded. This slight departure from the ILO guidelines has been adopted by the OECD, entailing some revisions to the time series.
>
> The ILO definition of unemployment covers those of working age who, in a specified time period, are without work, are available for work, and are seeking work. As noted above, however, national unemployment rates are often derived from monthly data of persons *registered* as unemployed at employment offices, which may differ from the ILO guidelines in a number of respects. For the computation of standardized rates, member countries of the EU provide Eurostat with annual estimates of unemployment based on a precise definition and derived from a common set of questions incorporated in their national labour force surveys. Eurostat uses these annual estimates to adjust national monthly unemployment data to achieve comparability between countries. In countries which carry out monthly labour force surveys (Finland and Sweden) this is a one stage process in which the monthly survey results are adjusted directly to the EU common format. In the other EU countries, which conduct labour force surveys on a quarterly or annual basis, there are two stages. First, monthly data on *registered* unemployment are adjusted on the basis of the national survey estimates, and then there is a further adjustment based on the EU common format. The Eurostat standardized unemployment series thus differs slightly from the series previously compiled by the OECD, and the standardized unemployment rates published by the OECD have been revised accordingly.
>
> Unemployment data for Canada, Japan and the United States are derived from monthly labour force surveys which already conform to ILO guidelines. However, to bring the standardized unemployment rates for these countries into line with the new methodology the OECD no longer adjusts labour force data to include the armed forces. Standardized unemployment rates are also available for Norway on a quarterly basis derived from a continuous labour force survey.
>
> ---
>
> [1] For a more detailed discussion of the methodological differences see OECD, *Quarterly Labour Force Statistics*, No. 4 (Paris), 1996, pp. 84-85.

work in the belief that further attempts are futile.[66] These stubbornly high levels of unemployment have a major social cost as well, particularly in the way they impact on the lives of young people (table 2.3.5).

---

[66] These people may be drawn into the labour force if improving economic conditions make them more optimistic about their job prospects. This partly accounts for the situation where rising employment in a recovery is not matched by a commensurate fall in the unemployment rate.

(iii) Labour market policies

The high and persistent unemployment in western Europe has given rise to widespread debate over the appropriateness and efficacy of measures to combat it. At a general level, there is a distinction to be drawn between arguments which regard demand deficiency as the main underlying cause of this exceptional labour market slack and those which see the problem more in terms of various structural rigidities in labour market relationships which raise the cost of the labour input.

TABLE 2.3.5

Youth[a] unemployment rates in the ECE market economies, 1994-1996

(Per cent)

	1994	1995	1996[b]
France	29.0	27.3	28.4
Germany	8.7	8.8	9.6
Italy	32.3	33.2	33.5
United Kingdom	17.0	15.9	15.4
Austria	..	5.6	6.0
Belgium	24.2	24.4	24.9
Denmark	11.0	10.1	9.6
Finland	42.2	38.2	28.4
Greece	27.7	27.9	..
Ireland	22.8	19.5	19.0
Luxembourg	7.3	7.1	7.7
Netherlands	11.4	11.6	10.1
Norway	8.9	9.4	..
Portugal	15.1	16.6	17.0
Spain	45.0	42.5	41.2
Sweden	22.6	19.4	21.3
Canada	16.5	15.6	15.8
United States	12.5	12.1	12.2
Memorandum item:			
Japan	5.5	6.1	6.7

Source: National statistics; Eurostat, *Eurostatistics*, No. 1 (Luxembourg), January 1997; OECD, *Employment Outlook* (Paris), July 1996; UN/ECE secretariat estimates.

[a] Less than 25 years old.

[b] Average of January-September monthly rates, except Italy (January-July) and Netherlands and United States (January-August).

Against a background of tight fiscal retrenchment, recent labour market policy initiatives have tended to be oriented towards the reform of perceived structural rigidities. More flexible working time arrangements — whether in terms of the pattern of hours worked in a firm or the length of a person's working time — have been advocated as a way of lowering labour costs and creating more employment opportunities. In France, for example, some agreements have been reached in the metalworking industry for a shorter working week and a more flexible shift system. In addition, recent legislation offers employers one-off reductions in social security contributions in return for a cut in working hours and a matching rise in new recruitment. In Germany, more flexible working time has also been introduced recently in the metalworking and electrical engineering industries to secure employment in a downturn (weekly working hours can be reduced by up to five hours, with correspondingly lower pay), while in the construction industry, weekly working hours have been reduced in the winter and extended in the summer to take account of the importance of seasonal and weather conditions.[67] A recent study of enterprises at the establishment level suggests that such measures are part of a more general trend in the 1990s towards greater working time flexibility in Germany,[68] where, apart from changes in the pattern of working hours, there has also been a rise in the incidence of part-time employment.

Wage inflexibility is frequently cited as a contributory cause of the high levels of unemployment in western Europe, and particularly the erosion of pay differentials which may have led to reduced employment opportunities for workers at the lower end of the pay scale — especially young workers. The impact of statutory minimum wages is an issue of current debate, partly because the perceived negative effects have been challenged by new research. A recent study which focused on the minimum wage systems in France, the Netherlands, Spain and the United Kingdom found "no general evidence that minimum wages reduced employment, except perhaps for young workers".[69] And a study of the United States has found that not only have recent increases in minimum wages not had negative employment effects, but that there may actually be positive employment effects from minimum wages.[70]

Germany is sometimes given as an example of a country where the deterioration in the labour markets can be linked, at least in part, to pay rigidities which are detrimental to competitiveness. There is evidence, however, that wage flexibility has increased over time in western Germany, with a widening of wage differentials, a decline in the proportion of establishments bound by collective agreements, and a decline in the wage gap between actual pay and collectively agreed rates.[71] A decline in the wage gap has also been reported in the new *Länder* where problems of competitiveness are particularly acute. And further evidence of flexibility can be found in the recent agreement in the chemical industry to ease pay rates at the lower end of the scale to facilitate the recruitment of the long-term unemployed and trainees.

While working time and wage flexibility are issues which are largely internal to individual enterprises, other structural rigidities may arise from the general legal and

---

[67] Deutsche Bundesbank, op. cit., p. 56.

[68] L. Bellmann, H. Düll, J. Kühl, M. Lahner and U. Lehmann, "Patterns of enterprise flexibility in Germany. Results from the IAB establishment panel, western Germany 1993-95", Country Report for the OECD project, "Technological and organizational change and labour demand, flexible enterprise: human resource implications", on behalf of the Federal Ministry of Education, Science, Research and Technology, Institute for Employment Research (Nuremberg), 1996.

[69] J. Dolado, F. Kramarz, S. Machin, A. Manning, D. Margolis and C. Teulings, "The economic impact of minimum wages in Europe", *Economic Policy*, No. 23 (London), October 1996, pp. 319-372.

[70] D. Card and A. Krueger, *Myth and Measurement: The New Economics of the Minimum Wage* (Princeton, NJ, Princeton University Press, 1995).

[71] L. Bellman et al., op. cit. One conclusion of this study is that "if flexibility of working hours and wages at the firm level are taken together there is a considerable and increasing flexibility in collective bargaining and in the employment system".

institutional framework of an individual economy. The social benefit system, for example, may reduce the demand for labour by raising employers' labour costs, but may also have supply-side effects to the extent that unemployment benefits and other types of income support relative to wage income may act as a disincentive to work. Modifications to the social benefit system have been undertaken in an attempt to alleviate these negative effects. In France, for example, employers have been offered temporary reductions in social contributions which are linked to job creation obligations. And in the United Kingdom, disincentive effects have been partly addressed by the replacement of unemployment benefits with a job seekers allowance in October 1996. However, the empirical evidence linking unemployment to labour market regulation may not be very strong.[72]

In the public debate on the current problems besetting west European labour markets close attention has been paid to the detailed aspects of labour market flexibility, but the negative effects in the labour markets of disinflationary policies have also been widely discussed. Explanations for the high and persistent unemployment based on insufficient demand have been emphasized in previous editions of this *Survey*;[73] unfortunately, little improvement is foreseen for the west European labour markets given the current forecasts for economic growth.

2.4 Costs and prices

(i) World commodity prices

The average price of raw materials in dollar terms on world markets moved sharply upwards during 1996, heavily influenced by increases in the price of crude oil, and masking the continuing downward trend in the prices of non-energy raw materials underway since mid-1995 (chart 2.4.1). The HWWA index[74] of industrial raw materials prices fell by some 11 per cent in 1996 compared with 1995 (table 2.4.1), a reflection, in part, of sluggish industrial output growth, particularly in western Europe. There was a steep decrease in the first half of the year which levelled off in the remaining months.

While the general volatility of energy prices can be seen in chart 2.4.1, there was a notable surge in crude oil prices in the second half of 1996. The price of Brent Blend, the crude oil which serves as a global price benchmark, rose to a post-Gulf war high of just over $25

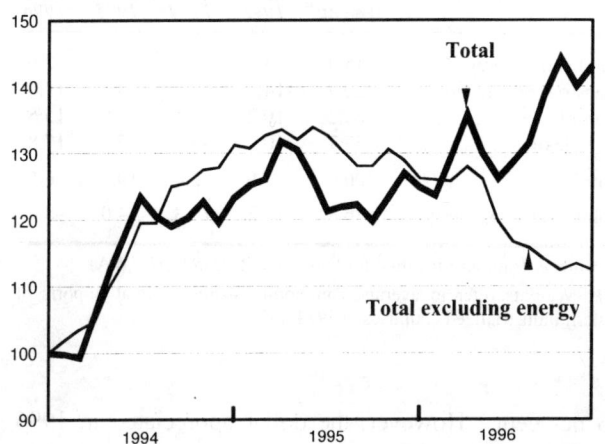

CHART 2.4.1

World commodity prices, 1994-1996
(Indices, January 1994=100)

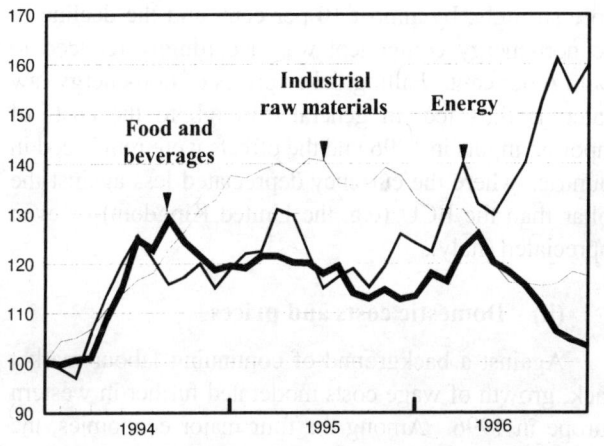

Source: Hamburg Institute for Economic Research (HWWA).
Note: Indices calculated on the basis of dollar prices.

a barrel in mid-October, and the HWWA crude oils index was 46 per cent higher in that month than a year earlier. These increases reflected concerns about low stock levels of crude oil and some refined products such as heating oil. Supply uncertainties were further compounded when, in September, the United Nations suspended the planned food-for-oil programme for Iraq.[75] In the event, the difficulties were resolved more quickly than had been expected and Iraqi oil returned to the market in January 1997 on a limited basis. Forecasts for crude oil prices in the rest of the year are sharply divergent.

In 1996, the total HWWA raw materials price index rose by 6.5 per cent in dollar terms compared with the previous year but this increase was strongly influenced by an almost 18 per cent rise in the price of crude oil. Excluding energy prices, the total index actually fell by

---

[72] S. Holden, "The unemployment problem – a Norwegian perspective", *OECD Working Paper*, No. 172 (Paris), 1997, p. 34.

[73] See, for example, UN/ECE, *Economic Survey of Europe in 1995-1996*, p. 4.

[74] Produced by the Institute for Economic Research, Hamburg (HWWA), this index weights world market prices (in dollars) by the relevant commodity shares in total imports of the western industrialized countries in 1974-1976.

[75] Buyers who had delayed purchases in the expectation that renewed exports from Iraq would drive prices lower were forced into the market in October to make up shortfalls in stocks.

TABLE 2.4.1

World commodity prices in dollars, 1993-1996
(Percentage change over previous year)

	Weights[a]	1993	1994	1995	1996
Food and beverages	15.9	3.2	24.2	3.4	-2.6
Industrial raw materials	20.9	-14.3	17.1	19.5	-11.4
Energy	63.2	-10.2	-5.1	7.9	15.5
Crude oil	57.7	-10.9	-6.2	7.7	17.8
Total	100.0	-9.9	2.4	10.1	6.5
Total, *excluding energy*	36.8	-9.2	19.3	14.0	-8.6

Source: Hamburg Institute for Economic Research (HWWA).

[a] Weights refer to average commodity shares in total imports of western industrialized countries in 1974-1976.

8.6 per cent. However, the dollar appreciated in 1996 against a basket of European currencies, so measured in terms of ECUs the total HWWA index rose somewhat more strongly, by almost 10 per cent, and the decline of the non-energy component was accordingly reduced to about 6 per cent. Falling dollar prices of non-energy raw materials thus fed, in general, through to the costs of imported inputs in 1996 and the effects were reinforced in countries where the currency depreciated less against the dollar than the ECU (e.g. the United Kingdom) or even appreciated (Italy).

(ii) Domestic costs and prices

Against a background of continuing labour market slack, growth of wage costs moderated further in western Europe in 1996. Among the four major economies, the growth of average hourly earnings in manufacturing slowed to around 2 per cent in France and Italy in 1996 compared with 1995. This was most marked in the latter where there was a fifth consecutive annual deceleration. There was a slight slowdown also in the United Kingdom but from a higher rate – these earnings rose by some 4.4 per cent in the first three quarters of 1996 compared with the same period of 1995. Comparable data for Germany for 1996 were not available at the time of writing, but there was a significant slowdown in the growth of negotiated wages and salaries, which on an hourly basis rose by 3.2 per cent in 1996 down from 5.2 per cent in the preceding year. In contrast to the dominating pattern, the growth of average earnings in manufacturing rose in Spain and Sweden between the first three quarters of 1995 and 1996, to 5.5 and 6.9 per cent respectively, and there was also an acceleration between the first halves in Denmark and the Netherlands.

In North America, some wage pressures, albeit moderate, have emerged in the United States, with the annual growth in average hourly earnings in manufacturing rising from 2.5 per cent in 1995 to 3.3 per cent in 1996. It is noteworthy that this was about the same annual increase as in 1991, the year in which the present cyclical upswing started. In Canada, earnings growth accelerated to almost 3 per cent between the first three quarters of 1995 and 1996, double the rate in the same period of the two preceding years.

Unit labour costs in the total economy rose by some 2 per cent in western Europe in 1996, only slightly higher than in 1995, although the acceleration was a little faster for the four major economies combined because of a sharp rise in Italy (table 2.4.2). In France, the growth in unit labour costs was unchanged between 1995 and 1996 partly because of an easing of employers' social contributions, but there were also productivity gains (table 2.3.2). Unit labour costs actually fell very slightly in Germany in 1996 compared with 1995, reflecting relatively strong annual average productivity growth which offset the rise in labour costs. In Italy, there was only a marginal offset from productivity growth to the relatively strong growth of labour costs in 1996. The latter reflected faster growth of earnings and higher social security contributions. In the United Kingdom, there was a small rise in the rate of growth of unit labour costs in 1996.

The growth in unit labour costs in Canada was a little above 1 per cent in 1996, slightly higher than the previous year, while in the United States there was a deceleration to 2.7 per cent.

With demand only sluggish in western Europe, falling prices for non-energy raw materials and subdued growth in unit labour costs, producer prices of manufactures virtually stagnated on average in the four major economies in 1996 compared with 1995, growth decelerating sharply from some 4.5 per cent in the previous year. This masks an absolute decline in producer prices in France by 3.5 per cent.[76] In the United Kingdom, producer price inflation fell steadily in the course of 1996, in part a reflection of the appreciation of the pound sterling, and the decline continued in the first two months of 1997.[77]

Among the smaller west European economies, producer prices actually fell, year-on-year, in some months of 1996 in Austria, Finland, Ireland and Sweden, and in each month in Switzerland. There was also a notable slowdown in the growth of these prices in Belgium, Denmark, Norway and Spain.

Producer price inflation also fell sharply in Canada – from an average annual increase of nearly 8 per cent in 1995 to 0.5 per cent in 1996. The main exception to this

---

[76] This figure refers to the sales of intermediate goods only but there have also been decreases in prices of chemicals, metal products and agricultural goods. There is no aggregate comprehensive producer price index for France.

[77] The year-on-year rate of producer price inflation was 1.3 per cent in February 1997, the lowest rate of increase since October 1986.

TABLE 2.4.2

Unit labour costs[a] in the ECE market economies, 1993-1996

(Percentage change over previous year)

	1993	1994	1995	1996[b]
Western Europe	2.7	–	1.8	2.0
4 major countries	2.5	-0.3	1.4	2.0
France	3.3	-0.4	1.7	1.7
Germany	3.3	-0.4	1.1	-0.1
Italy	2.2	-0.4	1.4	5.3
United Kingdom	0.8	-0.1	1.7	2.2
15 smaller countries	3.2	0.9	2.8	2.0
Austria	3.8	0.7	1.9	1.2
Belgium	3.4	1.1	0.9	-0.2
Denmark	-0.9	-0.4	2.7	2.6
Finland	-4.6	-2.2	2.2	1.3
Greece	11.6	12.6	12.6	8.7
Iceland	-1.0	-0.3	4.0	..
Ireland	3.7	-1.8	-3.1	0.7
Israel	13.2	13.1	10.6	..
Luxembourg	7.1	2.7	1.3	1.0
Malta	5.5	4.7	2.9	3.6
Netherlands	2.1	-1.2	1.9	0.4
Norway	-0.8	0.2	2.2	2.8
Portugal	7.4	5.1	2.8	3.1
Spain	4.5	0.2	3.5	1.6
Sweden	0.4	0.1	0.8	5.3
Switzerland	1.9	0.1	2.3	0.7
Turkey	62.6	88.3	62.5	..
North America	2.0	1.5	3.0	2.6
Canada	-0.5	-1.3	0.8	1.3
United States	2.2	1.7	3.2	2.7
Total above	2.3	0.8	2.4	2.3
Memorandum item:				
Japan	2.2	1.4	0.7	-1.6
Total above, *including Japan*	2.3	0.9	2.2	1.7

Source: National statistics; UN/ECE secretariat estimates.
Note: Regional aggregates exclude Israel and Turkey.
[a] Total compensation of employees per unit of real GDP.
[b] Provisional.

general trend of weakening producer prices was the United States where there was an acceleration from just below 2 per cent in 1995 to 2.6 per cent in 1996.

Changes in producer prices might be expected to give some indication of imminent pressures on final consumer prices, but there is no simple, clear-cut causal link between changes in producer price indices and the consumer price index. One reason is that firms may choose to adjust profit margins to absorb price changes, and, in particular, increases in producer prices may not be fully passed on to final consumers if competitive pressures are very strong. Conversely, weak producer price inflation may offer an opportunity to firms to raise profit margins provided demand conditions are favourable.[78] Country studies of the link between producer price indices and the CPI have yielded different results; the relationship is obviously complex and not very stable over time. The upshot is that changes in producer price indices may have only limited use as indicators of future changes in final consumer prices.[79]

After remaining unchanged at some 3 per cent in 1995, average annual consumer price inflation fell in western Europe in 1996 to 2.4 per cent (table 2.4.3), against a background of hesitant demand and moderate output growth. This is the lowest annual inflation rate for more than 30 years. The general tendency was for consumer price inflation to moderate in the course of the year, although there was a slight upward pressure in the final months which reflected the rise in the price of petrol and heating oil.

Among the four major economies, consumer price inflation remained very subdued in France and Germany within a range of 1.5 to 2 per cent. The sharpest reduction, however, was in Italy, the year-on-year rise in the cost of living index halving from 5.8 per cent in December 1995 to 2.6 per cent in the same month of 1996. This was largely a reflection of stagnant economic activity and restrictive monetary policy, but it is notable that, with the exception of rents which continued to rise, the slowdown was considerably more marked for goods subject to price controls, particularly public utilities.

In the United Kingdom, consumer prices rose by 2.4 per cent in 1996, 1 percentage point less than in 1995. However, the average annual increase in the government's target rate of inflation (the retail prices index excluding mortgage interest payments) was 2.9 per cent, the target of 2.5 per cent being exceeded in each month of the year, and monetary policy was tightened accordingly in the autumn.

Consumer price inflation moderated in the large majority of the smaller economies in western Europe in 1996, with a decrease of at least 1 percentage point in the average annual rate in Malta, Norway, Portugal, Spain and Switzerland, and a 2 percentage point fall in Sweden.

---

[78] Another explanation lies in the structure of the indices themselves: producer price indices are in general aggregated from intermediate input prices, investment goods prices and consumer goods prices, while the CPI is a measure of changes in final consumer goods and services prices and will include indirect taxes net of subsidies. At a disaggregated level, there is likely to be a stronger direct correlation between changes in consumer goods prices measured at the factory gate and the CPI, than between changes in intermediate or investment goods prices and the CPI, partly because the latter carry less weight in the CPI than the former. Furthermore, there may be greater opportunities to offset increases in prices of intermediate or investment goods elsewhere, such as in lower unit labour costs resulting from productivity gains. H. Dellmo, "Producer and import prices and the CPI – weak aggregated relationship", Sveriges Riksbank, *Quarterly Review*, No. 2 (Stockholm), 1996, pp. 18-29.

[79] A study of the United States found that the correlation between producer price indices and the CPI appeared to be relatively weak. T. Clark, "Do producer prices lead consumer prices?", Federal Reserve Bank of Kansas City, *Economic Review*, Vol. 80, No. 3 (Kansas City), Third Quarter 1995, pp. 25-39. For a review of other country studies see H. Dellmo, loc. cit.

TABLE 2.4.3

Measures of inflation in the ECE market economies, 1993-1996

(Percentage change over previous year)

	Consumer prices				Implicit GDP deflator			
	1993	1994	1995	1996	1993	1994	1995	1996
Western Europe	3.5	2.9	3.0	2.4	3.6	2.6	2.9	2.4
4 major countries	3.2	2.7	2.9	2.4	3.5	2.3	2.7	2.4
France	2.1	1.7	1.7	2.0	2.5	1.5	1.6	1.5
Germany	4.5	2.7	1.8	1.5	3.9	2.2	2.2	1.0
Italy	4.2	3.9	5.4	3.9	4.4	3.5	5.0	4.4
United Kingdom	1.6	2.5	3.4	2.4	3.2	1.9	2.3	3.1
16 smaller countries	4.0	3.3	3.0	2.3	3.9	3.3	3.4	2.7
Austria	3.7	3.0	2.2	1.9	3.4	3.4	2.1	1.7
Belgium	2.8	2.4	1.5	2.0	3.8	2.5	1.4	1.6
Cyprus	4.9	4.7	2.6	2.8	3.5	4.3	3.0	3.0
Denmark	1.3	2.0	2.1	2.0	0.7	1.7	1.7	1.7
Finland	2.1	1.1	1.0	0.5	2.4	1.3	2.2	1.0
Greece	14.4	10.9	9.3	8.5	13.6	10.8	9.2	9.1
Iceland	4.1	1.5	1.7	2.3	2.1	2.8	2.1	3.0
Ireland	1.5	2.4	2.5	1.6	4.2	1.2	0.8	1.7
Israel	11.0	12.3	10.1	11.3	11.5	12.3	9.2	11.7
Luxembourg	3.6	2.2	2.0	1.4	4.0	1.9	2.2	1.9
Malta[a]	4.1	4.1	4.0	2.5	2.8	4.3	2.9	2.8
Netherlands	2.6	2.7	2.0	2.1	1.9	2.0	1.4	1.4
Norway	2.3	1.4	2.4	1.3	2.1	0.6	3.1	4.2
Portugal	6.5	5.2	4.1	3.0	7.4	5.2	5.0	4.3
Spain	4.6	4.8	4.6	3.6	4.3	4.0	4.9	4.0
Sweden	4.7	2.2	2.5	0.5	2.6	2.5	3.7	1.0
Switzerland	3.3	0.9	1.8	0.8	2.0	1.9	2.5	0.4
Turkey	66.1	106.3	88.0	79.8	67.8	106.5	81.6	78.0
North America	2.9	2.4	2.8	2.8	2.5	2.2	2.4	1.9
Canada[a]	1.8	0.2	2.1	1.6	1.1	0.7	1.5	1.3
United States	3.0	2.6	2.8	3.0	2.6	2.3	2.5	2.0
Total above	3.2	2.6	2.9	2.6	3.0	2.4	2.7	2.2
Memorandum item:								
Japan	1.3	0.7	-0.1	0.1	0.6	0.3	-0.6	-0.3
Total above, *including Japan*	2.9	2.3	2.4	2.2	2.7	2.0	2.1	1.8

Source: National statistics.

Note: Regional aggregates exclude Israel and Turkey.

[a] January-September 1996 over same period of 1995 for GDP deflator.

In the latter, consumer prices actually fell in the final quarter of 1996 compared with the same quarter of 1995. There was a moderate strengthening of consumer price inflation in Belgium, Iceland and Israel.

Year-on-year consumer price inflation was low and stable in Canada in the first three quarters of 1996 but strengthened in the final months of the year as food and energy prices rose, the Canadian dollar depreciated temporarily and tobacco taxes were increased. In the United States, the average annual inflation rate derived from the consumer price index edged up very slightly to 3 per cent in 1996, but it is striking that the implicit GDP deflator shows a fall in inflation from 2.5 per cent in 1995 to 2 per cent in 1996.

There has been recent discussion concerning the upward biases in the consumer price index measure of inflation. This is not a new issue – the existence of such biases has long been acknowledged – but it has been recently highlighted in the United States by a study which attempted to quantify the bias, concluding that it amounted to 1.1 percentage points, a substantial proportion of the current rate of consumer price inflation.[80] Of this, 0.4 percentage point was attributed to a substitution bias – the failure of the index to take account of substitution between goods whose relative prices have changed. The rest (0.7 percentage point) was attributed to a failure to adjust adequately for changes in the quality of goods and services, for the emergence of new goods and services, and for changes in how and where goods and services are sold. While the magnitude of these estimates has been strongly challenged,[81] the

---

[80] Final Report to the Senate Finance Committee of the Advisory Commission to Study the Consumer Price Index (Washington, D.C.), 1996.

[81] See, for example, "Briefing on the BLS reaction to the Final Report of the Advisory Commission to study the consumer price index", 19 December 1996 (http://stats.bls.gov/news.release/cpi.br121996.brief.htm); "Testimony of

CHART 2.4.2

Selling-price expectations in manufacturing industry in the major west European countries, 1993-1996

(Balances,^a seasonally adjusted)

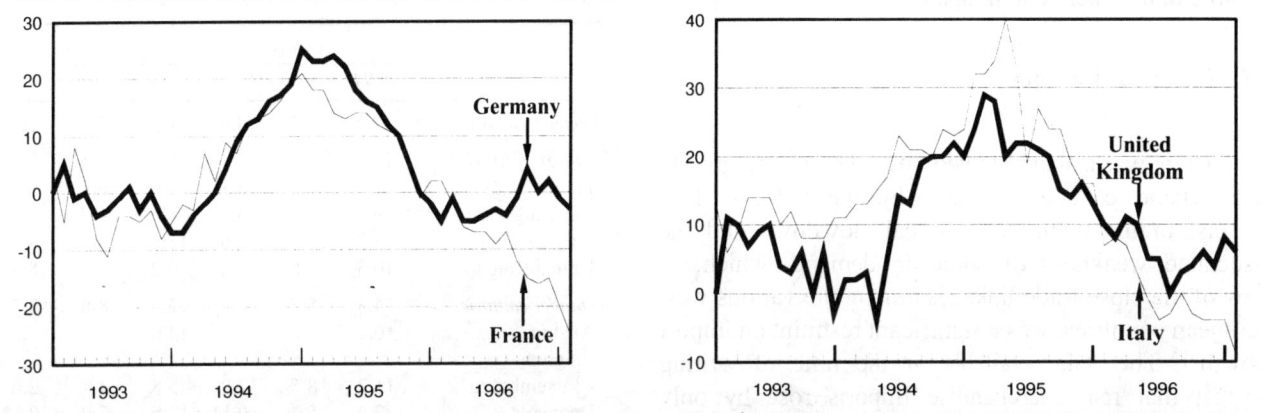

Source: European Commission, *European Economy*, Supplement B (Luxembourg), various issues.

Note: Business survey question: selling-price expectations for the months ahead: up, unchanged, down?

a Balances are the differences between the percentages of respondents giving positive and negative replies.

existence of an upward bias is generally accepted and should be borne in mind when assessing the implications for macroeconomic policy.

Irrespective of the measure used, inflationary pressures remain very low in the United States given the strength of output and demand and the continuing fall in the unemployment rate in 1996 – to 5.3 per cent in the final quarter. The unemployment rate has been below 6 per cent since the final quarter of 1994, and thus below the rate which some economists have considered to be the minimum rate consistent with non-accelerating inflation (the so-called NAIRU). The question then arises as to why inflationary pressures have remained so subdued, and the possibility that the non-accelerating inflation rate of unemployment might have fallen.

One explanation is that there have been certain favourable supply shocks that have temporarily lowered inflation for a given rate of unemployment.[82] For example, the rapid decrease in computer prices, and the move to managed health care plans which have lowered labour costs for many firms. In addition, import prices fell in 1996 because of the appreciation of the dollar and lower inflation abroad. To the extent that these supply shocks are only temporary, the fall in the NAIRU may only be transitory. But there are other possible explanations which would appear to have longer lasting effects, but which are more difficult to demonstrate quantitatively. In particular, increasing job insecurity may be dampening the bargaining behaviour of workers,

and rising competitive pressures may be modifying the pricing behaviour of firms.

More generally, the adverse supply shocks of the oil price rises in the early and late 1970s have taken a long time to work themselves out, but as the inflation of the 1970s and 1980s has been overcome, there may have been changes in expectations and in the behaviour of economic agents in the industrialized countries which have also contributed to a lowering of the non-accelerating inflation rate of unemployment.

For the western market economies combined, consumer price inflation edged down from 2.9 per cent in 1995 to 2.6 per cent in 1996 – or from 2.4 to 2.2 per cent if Japan is included in the total (table 2.4.3). Measured by the implicit GDP deflator, however, the rate of inflation for the western market economies combined was even lower in 1996, at 2.1 per cent (or 1.8 per cent, including Japan). This is largely accounted for by the differences in Canada and the United States between the CPI measure of inflation and the implicit GDP deflator.

It is likely that inflation will remain very subdued in 1997. Some indication of this is suggested by the selling price expectations revealed by business surveys of manufacturing companies in the European Union (chart 2.4.2). Forecasts suggest that there will be little change in consumer price inflation in 1997 in France and Germany: an average annual rate of some 1.6 per cent is expected in the former in the first half of the year and about 1.5 per cent for the full year in the latter. In Italy, forecasts are for a further reduction to an average annual rate of less than 3 per cent, but based on some fairly rigorous assumptions about the implementation of proposed budget measures and the containment of labour costs. In the United Kingdom, retail price inflation is expected to rise slightly to 2.8 per cent, although there is a risk that

---

K.G. Abraham, Commissioner of Labour Statistics, before the Senate Finance Committee", 11 February 1997 (http://stats.bls.gov/news.release/cpi.br21197.brief.htm).

[82] For a more detailed discussion of what follows, see "Mr. Meyer reviews the economic outlook and challenges for monetary policy in the United States", *BIS Review*, No. 4 (Basle), 21 January 1997, p. 4.

the rate will be higher if retailers try to raise profit margins. Consumer price inflation is expected to remain unchanged in Canada and the United States in 1997 but rise to around 1 per cent in Japan.

2.5 External balances

There was a sharp slowdown in the volume growth of merchandise trade in western Europe in 1996. This reflects, in the main, the cyclical slowdown and the associated weakness of domestic demand, which, in view of the close trade linkages among the various west European countries, was a significant restraint on import growth. The data available at the time of writing suggest that real merchandise imports rose by only about 2.5 per cent in 1996 compared with 1995, when there was an increase of some 7 per cent. There was a large decline in import demand in Italy (table 2.5.1). There was also a pronounced slowdown in the volume growth of exports, to only 4 per cent in western Europe in 1996. This was less than half the increase in 1995, but still higher than import growth, an indication of the importance of demand from outside the region. In fact, exports were supported by the continuing surge of import demand in the United States, central and eastern Europe, the CIS and the developing countries, although growth rates tended to weaken.

There was also a slowdown in the growth of external trade in the United States, but it was much less marked than in western Europe and Canada. Import demand decelerated sharply in Japan in 1996 to an annual growth rate of only 2.3 per cent, down from 12.5 per cent in 1995. Exports picked up in the second half of 1996 supported by the significant depreciation of the yen, but for the year as a whole the fall was nearly 1 per cent.

In total, the volume of imports into the industrialized countries rose by almost 4.5 per cent in 1996, while exports increased by nearly 5 per cent. In the preceding year, these growth rates had been close to 9 per cent (table 2.5.1).

There was also a slowdown in import demand in central and eastern Europe, the Baltic states and the CIS in 1996 (see chapter 3.5 below). But overall demand for foreign products, notably investment goods, remained quite buoyant, and for the year as a whole there was an aggregate increase in import volumes of some 12 per cent. Partial data suggest that the growth of import demand has also weakened in the developing countries, especially in Asia where several countries have adopted tighter policies designed to contain inflationary pressures and correct external balances.

Thus, the volume growth of world merchandise trade slowed down to a rate of about 5 per cent in 1996. This is a marked deceleration from the exceptionally rapid growth in the two preceding years, when international trade expanded at rates of 8.5 to 10 per cent. In general, international trade was mainly supported by rising demand for intermediate and investment goods, whereas demand for tradeable consumer goods was more sluggish.

TABLE 2.5.1

Changes in the volume of merchandise exports and imports in the ECE market economies, 1994-1996

(Percentage change over previous year)

	Exports			Imports		
	1994	1995	1996	1994	1995	1996
Western Europe	11.2	8.3	4.0	9.3	7.1	2.5
4 major countries	10.1	8.3	3.1	8.2	6.5	2.0
France	7.8	7.9	3.1*	8.6	6.0	1.6*
Germany[a]	9.3	6.3	4.2	8.3	6.7	2.5
Italy	13.3	11.9	-2.0*	11.4	8.7	-4.0*
United Kingdom	10.3	7.8	7.0	4.2	4.7	8.3
9 smaller countries[b]	14.3	8.5	6.5	12.5	8.8	3.7
Austria	10.2	..	..	13.0	..	..
Belgium-Luxembourg	11.2	8.5	2.0	5.8	4.8	2.6
Denmark	7.4	5.2	0.1*	12.9	7.0	-0.3*
Finland	13.3	6.7	4.6	17.2	6.7	8.3
Greece	..	..	..	..	..	..
Ireland	14.8	20.3	13.5*	13.2	13.9	12.5*
Netherlands	10.3	8.5	6.5*	10.0	11.6	2.2*
Norway	13.2	5.1	12.5	15.9	7.3	7.5
Portugal	..	..	..	..	..	..
Spain	20.5	9.4	9.1*	14.8	9.7	5.1*
Sweden	17.0	10.6	7.7	15.2	9.2	0.6
Switzerland	4.9	4.2	2.1	9.4	6.5	1.5
Turkey	..	..	..	..	..	..
North America	10.5	10.7	7.5	13.6	8.9	6.8
Canada	14.8	12.4	4.2	13.9	9.8	5.5
United States	10.1	10.6	7.8	13.6	8.8	6.9
Total above	10.8	9.6	5.8	11.6	8.0	4.8
Memorandum item:						
Japan	1.7	3.8	-0.7	13.5	12.5	2.3
Total above, including Japan	9.3	8.6	4.8	11.9	8.8	4.4

Source: National statistics; OECD, *Monthly Statistics of Foreign Trade, Series A* (Paris), various issues.

[a] Data up to 1995 refer to west Germany only.
[b] Regional aggregates exclude Austria, Greece, Portugal and Turkey.

Current account balances

The combination of weak domestic demand in western Europe and relatively favourable demand from outside the region was the major factor behind significantly improved trade balances in 1996. Among the four major economies, there was a conspicuous rise in the trade surplus in France, Germany and Italy, a major role being played by the weakness of domestic demand. In contrast, robust domestic demand and a deterioration in price competitiveness from the appreciation of its currency were the main reasons for a small increase in the United Kingdom's trade deficit in 1996 (table 2.5.2).

TABLE 2.5.2

Current account balances in the ECE market economies, 1994-1996
(Billion dollars and per cent of GDP)

	Merchandise trade (billion dollars)			Total (billion dollars)			Total (per cent of GDP)		
	1994	1995	1996	1994	1995	1996	1994	1995	1996
Western Europe	86.9	117.0	154.8	48.3	80.1	106.4	0.6	0.9	1.2
4 major countries	69.4	96.5	111.6	-1.8	16.9	38.5	–	0.3	0.6
France	6.8	11.2	14.3	6.5	16.3	19.8	0.5	1.1	1.3
Germany	43.7	59.5	64.9	-19.6	-20.9	-17.7	-1.0	-0.9	-0.8
Italy	35.5	44.0	51.5^a	15.1	27.4	36.5^a	1.5	2.5	3.0
United Kingdom	-16.6	-18.3	-19.1	-3.7	-5.9	–	-0.4	-0.5	–
14 smaller countries	17.5	20.5	43.2	50.1	63.2	67.8	2.1	2.2	2.4
Austria	-6.9	-7.2	-7.2	-1.8	-4.7	-4.0	-0.9	-2.0	-1.8
Belgium-Luxembourg	6.8	10.2	10.0	12.5	15.0	15.4	5.1	5.3	5.5
Denmark	7.4	6.8	7.3	3.0	1.7	1.9	2.0	1.0	1.1
Finland	6.4	10.8	9.5	1.3	5.3	4.2	1.3	4.3	3.4
Greece	-13.5	-17.1	-15.1^a	-0.1	-2.9	-4.3^a	-0.2	-3.1	-4.4
Iceland	0.3	0.2	..	0.1	0.1	-0.1	1.9	0.8	-1.9
Ireland	8.1	11.7	12.2	1.4	1.5	0.3	2.7	2.4	0.4
Netherlands	18.8	20.9	16.3^b	17.9	17.9	16.3^b	5.3	4.5	4.1
Norway	6.8	8.0	13.8	3.0	4.5	11.2	2.4	3.1	7.2
Portugal	-8.1	-8.5	..	-1.5	-0.2	..	-1.7	-0.2	..
Spain	-14.7	-17.6	-13.8	-6.8	1.3	2.8	-1.4	0.2	0.5
Sweden	8.7	14.9	17.9	0.8	4.9	6.3	0.4	2.1	2.5
Switzerland	1.6	0.9	0.8	17.9	21.1	20.3	6.9	6.9	6.9
Turkey	-4.2	-13.2	-8.6^c	2.6	-2.3	-2.3^c	2.0	-1.4	-1.3
North America	-155.2	-152.7	-161.9	-164.6	-156.3	-166.3	-2.2	-2.0	-2.0
Canada	10.9	20.7	25.3	-16.2	-8.2	-1.2	-3.0	-1.4	-0.2
United States	-166.1	-173.4	-187.2	-148.4	-148.2	-165.1	-2.1	-2.0	-2.2
Total above	-68.3	-35.7	-7.1	-116.4	-76.2	-59.9	-0.8	-0.5	-0.3
Memorandum item:									
Japan	144.1	131.2	83.8	130.5	110.4	66.0	2.8	2.2	1.4
Total above, *including Japan*	75.8	95.5	76.7	14.2	34.2	6.0	0.1	0.2	–

Source: National statistics; OECD, *Main Economic Indicators* (Paris), various issues.

^{a} January-October.
^{b} January-September.
^{c} January-June.

Partial data suggest that the aggregate trade surplus of western Europe rose to about $155 billion in 1996 compared with $117 billion in 1995.

In the United States, the customary merchandise trade deficit[83] rose to a record level of $187.2 billion in 1996 against the backdrop of robust domestic demand and a significant appreciation of the dollar. The trade balance continued to improve markedly in Canada, while in Japan there was a large fall in the trade surplus.

In general, changes in the merchandise trade balance were the major factor behind changes in current account balances. Current account positions – relative to GDP – have tended to improve in western Europe (table 2.5.2). There was also a further significant decline in the current account deficit to near balance in Canada in 1996, which contrasts with a slight rise in the United States current account deficit. The latter rose, in dollar terms, to its highest level since 1987 or to 2.2 per cent of GDP in 1996. The aggregate current account of the industrialized countries remained broadly in balance, with surpluses in western Europe and Japan offset by a deficit in North America.

2.6 The stances of economic policy

(i) Monetary policy

Monetary policy in *western Europe* has not only been concerned with short-term cyclical developments but also with preparations for stage three of European economic and monetary union, that is, the introduction of a single currency as from the beginning of 1999. Finland joined the existing exchange rate mechanism (ERM) of the European monetary system (EMS) on 14 October 1996. Italy, which had suspended its membership in

---

[83] There has been a deficit on the merchandise trade account in every year since 1976.

September 1992, returned to the ERM on 25 November 1996. A major reason for its return was to fulfil the exchange rate criterion of the Maastricht Treaty which stipulates, *inter alia,* that countries have to observe the normal fluctuation margins of the ERM for at least the two years prior to the examination which will determine which countries will join EMU. An important area of concern in the context of the single currency has been the future monetary policy strategy of the new European Central Bank, that is, whether to adopt monetary targeting – as pursued notably by the Bundesbank – or direct inflation targeting. As yet no decision has been made in this respect.[84] But it is noteworthy that in the run-up to stage three of EMU the Bundesbank decided to fix a monetary target for two years (1997-1998) instead of the traditional one-year period. In addition, the German monetary authorities lowered the implicit inflation target to a range of 1.5-2 per cent.[85] In a more general way, the extension of the time horizon and lowering of the inflation target is designed to "allow the ECB to carry on smoothly from the Bundesbank's policy of monetary targeting".[86]

While the policy strategy of the future European Central Bank has still to be agreed upon, the basic modalities of monetary and exchange rate policy cooperation between the euro area and other EU countries were agreed upon at the recent European Council held in Dublin on 13-14 December 1996.[87]

Weak cyclical conditions and moderate inflation rates generally led some shift in the stance of monetary policy in *western Europe* towards supporting economic growth in 1996.

In Germany, the Bundesbank continued to lower official interest rates gradually, a process underway since September 1992. The discount rate was reduced to the relatively low level of 2.5 per cent in April 1996 and the tender rate on repurchase agreements was fixed at 3 per cent in the summer. Since then, monetary policy has been unchanged. Other west European countries, where short-term interest rates were above German rates, exploited the scope for monetary easing provided by the decisions of the Bundesbank, and, in a number of them, the process of monetary easing continued in the second half of 1996 and in early 1997. This was notably the case in France, although after the small reduction in its key official interest rate (the *"taux d'appels d'offres"*) in January 1997 the Banque de France indicated that rates were now at a level considered commensurate with overall economic conditions.[88] In Italy, the central bank lowered the discount rate in July 1996 from 9 per cent to 8.25 per cent against a background of sluggish economic activity, falling inflation rates and an appreciating currency. The discount rate had been unchanged since May 1995. Further reductions followed in October 1996 and in January 1997, when the discount rate was fixed at 6.75 per cent.

Similarly in Spain, Sweden and Switzerland the stance of monetary policy continued to be relaxed in the second half of 1996 and in early 1997.

In contrast, in the United Kingdom, which is much more advanced in the business cycle than the majority of west European countries, the monetary authorities raised the official base rate in late October 1996 against a background of robust economic growth and fears of rekindling inflationary pressures. These concerns were increased by the fact that actual inflation remained significantly above the target rate fixed by the government.[89]

The different degree of monetary relaxation in the various countries is reflected in a concomitant decline in nominal short-term interest rates. The actual impact of monetary policy on economic activity, however, is better gauged by the real short-term interest rate, which is traditionally measured by the difference between the three-month nominal interest rate and the inflation rate (chart 2.6.1). Thus in France, real short-term rates have fallen from more than 6 per cent in early 1995 to less than 2 per cent in January 1997. But real short-term rates did not change significantly in the four major economies in the course of 1996. This is most striking for Italy, where a fall in nominal short-term rates by 2.8 percentage points in the 12 months to January 1997 was fully offset by the decline in inflation over this period.

---

[84] European Monetary Institute, *The Single Monetary Policy in Stage Three – Specification of the Operational Framework* (Frankfurt am Main), 1997.

[85] "Strategy of monetary targeting in 1997-1998", Deutsche Bundesbank, *Monthly Report* (Bonn), January 1997, pp. 17-25. The money supply target is derived from estimated changes in potential output, the underlying trend in the velocity of circulation of money and a desirable inflation rate. Traditionally, the Bundesbank has regarded an annual inflation rate of 2 per cent as satisfactory. For comparison, the United States Federal Reserve appears to have oriented its policy towards preventing inflation rising above 3 per cent. See W. T. Gavin, "The FOMC in 1995: a step closer to inflation targeting?", *Federal Reserve Bank of St. Louis Review*, Vol. 78, No. 5, September/October 1996, p. 34.

[86] Deutsche Bundesbank, loc. cit., p. 20.

[87] A new exchange rate mechanism (ERM2) will be set up with central rates defined vis-à-vis the euro. Standard fluctuation rates are intended to remain rather wide, in line with the current *modus operandi* of the ERM. For details see European Monetary Institute, op. cit., pp. 67-71.

[88] Banque de France, *Bulletin de la Banque de France*, No. 38 (Paris), February 1997, p. 6. A similar point of view with regard to interest rates in Germany appears to be held by the Bundesbank, which noted that the current economic problems facing the German economy have mainly structural causes which "can only be solved by economic policy reforms and not by a policy of cheap money". Deutsche Bundesbank, *Monthly Report* (Bonn), February 1996, p. 34.

[89] The government's target is to achieve a 12-month inflation rate of 2.5 per cent or less. The target is measured by the retail price index excluding mortgage interest payments.

CHART 2.6.1

Real short-term interest rates in the seven major economies, January 1995-January 1997

(Per cent)

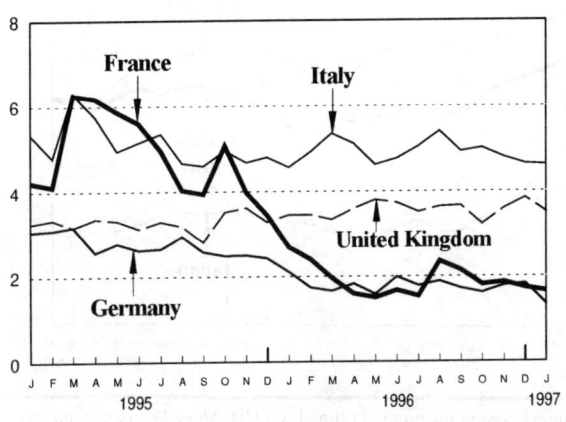
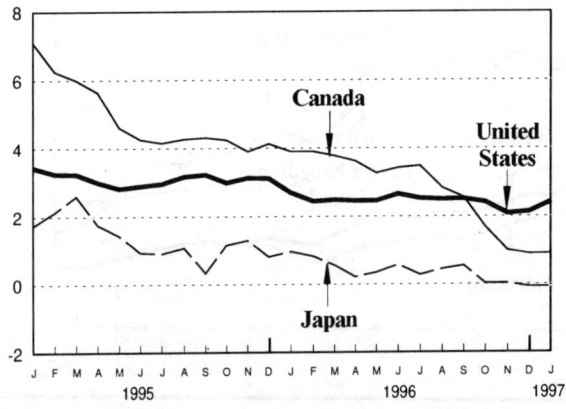

Source: OECD, *Main Economic Indicators* (Paris), various issues; national statistics.

Note: Average monthly rates. Real rates are calculated as nominal rates less the percentage change in the consumer price index in the corresponding month compared with the same month of the preceding year.

In the *United States*, the Federal Reserve lowered the federal funds rate in December 1995 and January 1996 against a background of signs of more sluggish economic activity in the final months of 1995. But the economy gained momentum in the course of 1996, shifting concern towards keeping inflationary pressures in check. The stance of monetary policy has not been altered since January 1996, although the rate of economic expansion temporarily exceeded potential output growth in the course of 1996 and the degree of factor utilization has been very high, notably in the labour market. But the factors making for low inflation have proved very resilient so far (see section 2.4).

In marked contrast to the United States, there was a substantial loosening of monetary policy in *Canada* in 1996. This policy, which succeeded in stimulating investment demand, was supported by lower inflation expectations and swift progress in fiscal consolidation. In fact, nominal short-term interest rates fell to very low levels and the gap between Canadian and United States interest rates widened substantially. Towards the end of 1996 nominal short-term interest rates in Canada were nearly 2.5 percentage points lower than in the United States, compared with a difference of only 0.1 percentage point in January 1996. In *Japan*, the discount rate was reduced to a record low of 0.5 per cent in September 1995. There was no change in this very expansionary policy stance in 1996. Real short-term interest rates changed only slightly in the United States in 1996, but they fell markedly in Canada. In Japan, real short-term rates were slightly negative in early 1997 (chart 2.6.1).

(a) Long-term interest rates

The decline in long-term interest rates in western Europe since early 1995 was temporarily reversed in the first half of 1996 (chart 2.6.2). This was generally interpreted as a spillover effect from the United States bond market. Economic indicators pointing to stronger activity in the United States economy in early 1996 raised inflationary expectations and, in anticipation of a perceived tightening of monetary policy, bond prices fell. A consequence was the upward drift of interest rates at the intermediate and longer end of the maturity spectrum. Yields on 10-year United States government bonds rose on average by 1.3 percentage points between January and June of 1996, but started to fall again when the economy slowed down in the third quarter of 1996. Nevertheless, long-term rates in December 1996 were about 0.6 percentage point higher than in January 1996, the increase rising to 0.9 percentage point in early 1997. The potentially restraining impact of higher long-term interest rates on private sector investment, however, was partly offset by the willingness of banks to ease their lending terms and by the buoyancy of the equity markets, which was tantamount to a reduction in the cost of capital to firms.[90] This development illustrates the importance of alternative channels of monetary transmission, namely, the equity price and credit channels.[91]

---

[90] Federal Reserve Board, *Humphrey-Hawkins Report*, 26 February 1997 (http://www.bog.frb.fed.us/boarddocs/hh/9702sections1.htm).

[91] Recent research on the transmission mechanism of monetary policy emphasizes the important role which changes in asset prices other than interest rates (the interest rate channel) have on economic activity. These alternative asset price channels involve changes in exchange rates and equity prices. In addition, emphasis is put on financial market

CHART 2.6.2

Nominal long-term interest rates in selected industrial countries, January 1995-January 1997

(Per cent per annum)

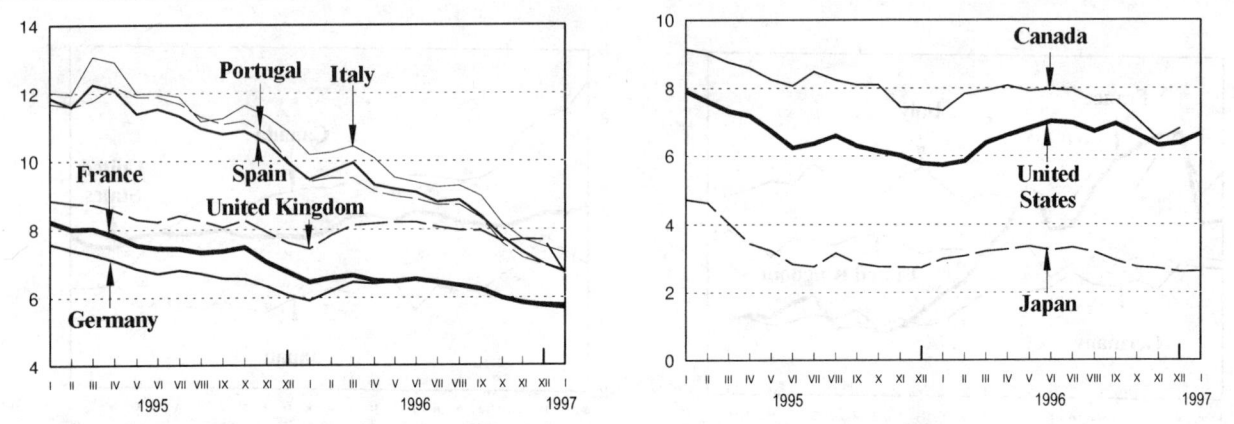

Source: Banque de France, *Bulletin de la Banque de France*, various issues and direct communications; Portugal: OECD, *Main Economic Indicators* (Paris), various issues.

Note: Average monthly yields on 10-year government bonds.

In western Europe, the decline in long-term interest rates resumed in the second half of 1996. United States nominal rates rose above German rates in the spring of 1996 and, for a time, there was some decoupling of German and United States long-term rates, which can be largely ascribed to their cyclical desynchronization. Expectations in the financial markets are for stable or falling short-term rates in western Europe and rising short-term rates in the United States. This constellation encourages shifts in portfolio preferences towards long-term bonds in western Europe.

The United States-German interest rate differential on 10-year government bond yields amounted to 1.1 percentage points in February 1997. In the same month of 1996 long-term interest rates in Germany were still above United States rates by about 0.35 percentage point.

Long-term interest rates fell, in nominal terms, to relatively low levels in many west European countries in the course of 1996, a tendency which continued in early 1997. Thus in Germany, the average yield on public bonds reached an historic low of 4.75 per cent in February 1997, while in France, yields on 10-year government bonds in early 1997 were at their lowest level in 35 years.

Changes in long-term rates since the second half of 1996 were also influenced by expectations in the financial markets about the prospects of individual countries participating in stage three of EMU from its inception. The background to such expectations were the significant efforts at fiscal consolidation being made to meet the Maastricht budget deficit criteria in countries such as Italy, Portugal and Spain. This led to a surge in demand for government bonds of these countries and a concomitant sharp decline in their long-term interest rates. In this process of "convergence trading" there was a significant narrowing of the interest rates in these countries and those in France and Germany (chart 2.6.2).

(b) Exchange rates

There was a sizeable appreciation of the dollar against the deutsche mark and other west European currencies as well as the yen in the course of 1996, a development which continued in early 1997 (chart 2.6.3). Among the factors supporting the United States currency were the expected widening of interest rate differentials in favour of dollar-denominated assets, the considerable surge in United States equity prices, and a perception that a stronger dollar was not unwelcome by policy makers in both western Europe and Japan, as this would tend to strengthen their net export growth. Nevertheless, at the meeting of the Group of Seven (G7) Finance Ministers in Berlin at the beginning of February 1997 governments signalled that they would consider a further strengthening of the dollar and a continuation of the weakening of the

---

imperfections, notably the problems of asymmetric information facing financial intermediaries. The so-called equity price channel involves the impact of changes in the relative price of capital on business investment and of wealth effects on consumption. Simply speaking, in a context of ample liquidity the public will increase its demand for equity. The ensuing rise in equity prices implies that new machinery and equipment is cheap relative to the price firms obtain for issuing new equity. Accordingly, this will stimulate investment spending. This is Tobin's q theory of investment. Within the context of the life-cycle consumption theory, a rise in equity prices bolsters net wealth, which, in turn, will encourage a rise in consumption expenditures. The credit channel emphasizes the presence of asymmetric information in financial markets, notably credit markets. For a concise overview of these and other monetary transmission channels see F.S. Mishkin, "The channels of monetary transmission: lessons for monetary policy", *Banque de France Bulletin Digest*, No. 27 (Paris), March 1996, pp. 33-43.

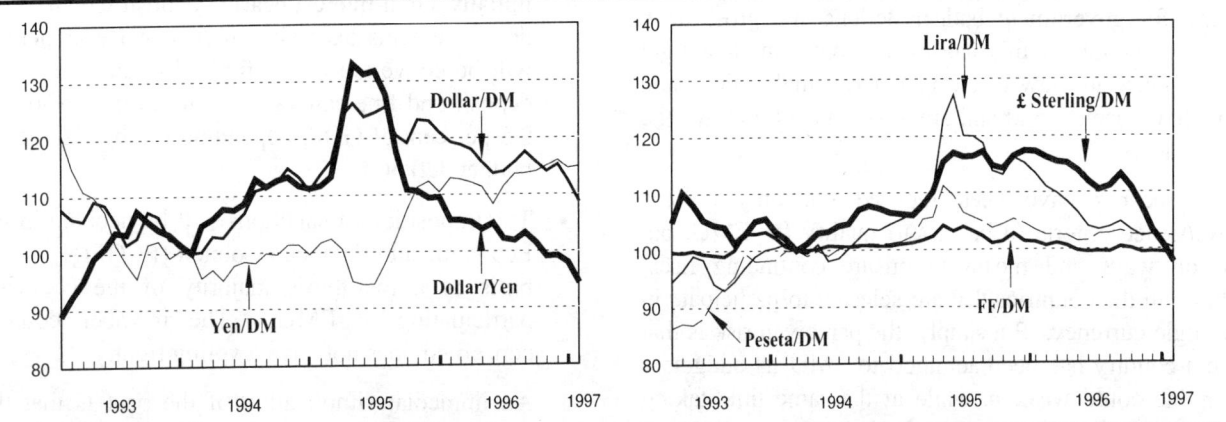

CHART 2.6.3

Average monthly bilateral exchange rates, January 1993-January 1997
(Indices, January 1994=100)

Source: Deutsche Bundesbank, *Monatsbericht* (Frankfurt am Main), various issues.

deutsche mark and the yen undesirable.[92] But the upward pressure on the dollar persisted in February.

Among the west European currencies there was a striking appreciation of the Italian lira in 1996, driven mainly by progress in curbing inflation and the above-mentioned convergence trading (chart 2.6.3). There was also a strong appreciation of the pound sterling against the deutsche mark and the French franc, an appreciation which accelerated in the final months of 1996.

To some degree the significant strengthening in the demand for pounds sterling reflects differential expectations about changes in monetary policy, that is, a temporary tightening in the United Kingdom and/or a temporary easing in France and Germany.[93] The appreciation may also reflect portfolio shifts of some investors towards assets which are denominated in a currency that will not participate, at least at the outset, in EMU.

There were pronounced changes in the pattern of *real effective exchange rates*, a synthetic measure of international price competitiveness, in the various countries in 1996.[94] Among the seven major economies there was a marked real appreciation – hence a deterioration in price competitiveness – in Italy and the United Kingdom, which amounted to some 10-11 per cent in the final quarter of 1996 compared with the same period of 1995. Over the same period, the dollar appreciated by 4 per cent in real terms. In contrast, there was depreciation – and hence a gain in price competitiveness – in France (2 per cent), Germany (4.5 per cent) and, most pronounced, in Japan (11 per cent). The real exchange rate was broadly unchanged in Canada in 1996.

Among the smaller west European countries, there was a relatively strong real appreciation in Greece and Ireland in 1996. In contrast, there was a sharp real depreciation of the Swiss franc by some 8.5 per cent in the final quarter of 1996 compared with the same period of 1995.

In general, the significant real depreciation in countries such as Germany, Japan and Switzerland in 1996 will tend to boost their net export growth and, ultimately, their domestic demand. When seen in relation to the broadly unchanged (and low) real short-term interest rates in these countries, this is tantamount to a significant easing of monetary conditions.[95] In contrast, monetary conditions tightened in the United States because of the real appreciation of the dollar.

---

[92] "Dollar has risen enough", *Financial Times*, 10 February 1997.

[93] Another factor could be actual and expected changes in fiscal policy. Lower deficits in continental Europe (to meet the Maastricht criteria) would tend to lower interest rates abroad – via the impact on national savings (up) and aggregate demand (down) in the short run. Accordingly, the exchange rate of countries progressing in fiscal consolidation would depreciate because capital would seek higher returns in other countries (unless there were a concomitant decline in the risk premium). Bank of England, *Inflation Report* (London), February 1997, pp. 46-48.

[94] Broadly speaking the real exchange rate is a bilateral nominal exchange rate adjusted for relative changes in foreign and domestic prices. The real effective exchange rate uses foreign trade weights to derive a geometric average of changes in real bilateral exchange rates. Traditional indicators for gauging relative changes in relative prices and costs in the domestic economy and abroad are unit labour costs, foreign trade prices, and wholesale and consumer prices. The measure reported here is calculated by the OECD and based on changes in consumer prices.

[95] Monetary conditions reflect the impact which changes in official interest rates have on aggregate demand and inflation via changes in other interest rates and the exchange rate. Indexes of monetary conditions can be constructed as weighted averages of changes in real short-term interest rates and the real exchange rate. They provide an approximate gauge for the overall stance of monetary policy. See B. Hansson and H. Lindenberg, "Monetary conditions index – a monetary policy indicator", Sveriges Rijksbank, *Quarterly Review* (Stockholm), March 1994, pp. 12-17.

(ii) Fiscal policy

Fiscal policy in *western Europe* has continued to be marked by efforts to meet the Maastricht convergence criteria for government budget deficits and gross debt. For most countries this has meant maintaining a tight policy stance in view of the fact that actual deficits and debts have been – and still are – significantly above the target values.

Concerns have been focused not only on the convergence criteria to be met to qualify for EMU, but also on ways and means to ensure continuing fiscal stability in the countries that are selected to participate in the single currency. Put simply, the perceived risk is that once a country has been admitted to EMU its budgetary discipline could weaken, while at the same time taking advantage of low interest rates for borrowing in the euro area, with possible negative spillover effects on other participating countries. A more permanent commitment to fiscal discipline is therefore seen as crucial for ensuring the internal and external stability of the future single currency.[96] A further argument is that a secure commitment to budgetary discipline will facilitate the task of the new European Central Bank to focus on price stability.

At the European Council held in Dublin in December 1996 agreement was reached on a "Stability and Growth Pact",[97] designed to ensure sustained fiscal convergence after the euro has been introduced. Basically, this pact builds on the "excessive deficit procedure" already provided for in Article 104c of the Maastricht Treaty. Proposals for automatic sanctions against countries with "excessive" deficits were not agreed upon.

The pact sets out the basic guidelines to be followed for fiscal policy, the procedure and timetable for the political decision-making process in the event that a country has an excessive deficit, and the scale of sanctions to be imposed in such a case. The salient features of the pact are as follows:

- Countries commit themselves to maintain, under normal economic circumstances, a broadly balanced budget or a budget surplus;

- A country with an excessive budget deficit, i.e. a deficit of more than 3 per cent of GDP, will be subject to sanctions, unless the deterioration is caused by exceptional circumstances, notably a severe recession. The latter is defined in the pact as an annual decline in real GDP by at least 2 per cent.

- Sanctions can be imposed if a country fails to take corrective action to bring the budget deficit below 3 per cent of GDP. These sanctions involve initially a non-interest bearing deposit which, if the deficit remains excessive after a two-year period, will be converted into a fine. The amount of the deposit and fine can vary within a range of 0.2 to 0.5 per cent of GDP, depending on the size of the budget deficit.

- The imposition of sanctions will be decided in the Economic and Financial Council (ECOFIN) on the basis of a two-thirds majority of the countries participating in EMU. The member country concerned does not have a voting right.

An immediate implication of the pact is that the process of stringent fiscal retrenchment will have to continue beyond 1997 given that most countries will still be relatively far off the new balanced budget target.

An important question is to what extent the fiscal policy rules will constrain the working of automatic stabilizers or tend to bend fiscal policy towards a pro-cyclical stance. The pact assumes that once a broadly balanced budget has been achieved, the built-in stabilizers will be able to cope with normal cyclical fluctuations without there having to be a rise in the budget deficit above the 3 per cent threshold. This, however, may not necessarily turn out to be the case. The outcome will not only depend on the depth and duration of any cyclical weakness but also on the elasticity of government net revenues with regard to fluctuations of output and incomes. Countries in which there is a high cyclical sensitivity of net revenues may therefore need to take precautionary measures either to reduce the cyclical sensitivity of the budget (e.g. by curbing the working of automatic stabilizers) and/or to run a correspondingly high structural budget surplus as a precautionary device against an excessive deficit during a cyclical downturn.[98]

Another problem is that were fines to be imposed, they would actually lead to a further deterioration in the deficit of the affected government. In addition, the timing of fiscal retrenchment imposed by the pact may not be appropriate for the cyclical position of the country concerned.

Thus, the available room for manoeuvre of fiscal policy provided by the pact may be too small to cope adequately with an economic downturn, even if the annual decline in output remains below the 2 per cent threshold. This could then either erode support for the

---

[96] T. Waigel and J. Arthuis, "Count on monetary union to succeed, on schedule", *International Herald Tribune*, 20 September 1996.

[97] Dublin European Council, 13-14 December 1996, Presidency Conclusions (http://europa.eu.int/en/record/dublin/dub-en2.htm).

[98] DIW, "Grundlinien der Wirtschaftsentwicklung", *Wochenbericht 1-2/97* (Berlin), 9 January 1997, pp. 33-37; J. Pisani-Ferry, "Fiscal policy under EMU", *The CEPII Newsletter*, No. 6 (Paris), 2nd semester 1996, pp. 1-2.

pact or increase pressures on the ECB to compensate for the failure of fiscal policy to offset recessionary tendencies. But one of the objectives of the pact was precisely to shield the new ECB from the need to conduct anti-cyclical policies.[99] By blunting the fiscal policy instrument for short-term adjustment in the individual countries, the stability pact implicitly points to a missing element in the project for monetary union, namely, an automatic fiscal transfer system with countercyclical effects between regions and/or countries, and a strong fiscal authority at the centre, as is the case in the United States.[100]

The consolidation policies pursued in *western Europe* have put major emphasis on tight expenditure controls rather than revenue enhancing measures. Weak cyclical conditions restrained the growth in net revenues in 1996. The process of budget consolidation was supported by the decline in interest rates which helped to alleviate debt service payments. This was notably the case for Italy, Portugal and Spain. In the event, the general government budget deficits in western Europe fell on average to about 4.3 per cent of GDP in 1996 compared with some 5.1 per cent in the previous year. In many countries, the various measures introduced to reduce the underlying structural budget deficits were, more or less, offset by the working of the automatic stabilizers. This can be seen from the fact that the structural budget deficit — which is the total deficit adjusted for cyclical effects — is estimated to have fallen on average by an amount corresponding to 0.9 percentage point of potential west European GDP in 1996,[101] that is, by somewhat more than the total deficit.

The dominating feature in 1996 was for general government budget deficits to decline in most countries compared with 1995. The exception was Germany, where a combination of adverse cyclical conditions and special factors led to a rise in the deficit to 3.9 per cent of GDP (table 2.6.1). In France, the deficit corresponded to 4.1 per cent of GDP in 1996, only slightly higher than the 4 per cent targeted by the government. In the United Kingdom, sustained economic expansion has led to a steady improvement in the government's financial position: borrowing fell to some 4.5 per cent of GDP in 1996 compared with a peak of 7.8 per cent in 1993.

---

[99] Ibid.

[100] L. Lindsay, "EMU: an American view", *Financial Times*, 27 November 1996.

[101] Structural budget deficits cannot be directly measured: they have to be calculated on the basis of estimates of what revenues and expenditures would be at potential output. These calculations are generally accepted to be surrounded by a relatively large margin of uncertainty.

TABLE 2.6.1

General government net lending in the ECE market economies, 1995-1997
(Per cent of GDP)

	1995	1996	1997
Western Europe	-5.1	-4.3	-2.9
4 major countries	-5.1	-4.7	-3.1
France	-4.8	-4.1	-3.0
Germany	-3.5	-3.9	-2.9
Italy	-7.1	-6.9	-3.0
United Kingdom	-5.6	-4.4	-3.5
13 smaller countries	-4.9	-3.3	-2.5
Austria	-5.9	-4.5	-3.0
Belgium	-4.1	-3.3	-2.9
Denmark	-1.6	-1.4	-0.3
Finland	-5.1	-2.6	-1.9
Greece	-9.1	-7.9	-6.5
Ireland	-2.0	-1.6	-0.9
Luxembourg	1.5	0.9	0.5
Netherlands	-4.0	-2.6	-2.5
Norway	2.9	5.9	5.0
Portugal	-5.1	-4.0	-2.9
Spain	-6.6	-4.4	-3.0
Sweden	-8.1	-3.9	-2.9
Switzerland	-1.8	-2.1	-2.4
North America	-2.2	-1.6	-1.8
Canada	-4.1	-1.8	-1.5
United States	-2.0	-1.6	-1.8
Total above	-3.6	-2.9	-2.3
Memorandum item:			
European Union	-5.2	-4.5	-3.0

Source: Member states of the European Union: European Commission, *European Economy,* Supplement A, No. 12 (Luxembourg), December 1996 and official national statistics; Norway: Royal Norwegian Ministry of Finance, *The National Budget 1997* (Oslo), 1996; Canada and United States: national statistics and *OECD Economic Outlook* (Paris), December 1996; Switzerland: Administration fédérale des finances.

A few countries already achieved budget deficits in 1996 which corresponded to the Maastricht reference value of 3 per cent of GDP or even below. This was the case for Denmark, Finland, Ireland, Luxembourg and the Netherlands. The debt criteria (60 per cent of GDP), in the strict sense, was only met by Finland, France, Luxembourg and the United Kingdom in 1996. Outside the EU, Norway clearly met both fiscal convergence criteria.

The examination as to which countries have met the various Maastricht convergence criteria, and therefore qualify for entry into the monetary union, will take place during the first half of 1998 and will be based on data for the year 1997. Accordingly, virtually all EU governments have adopted budget plans which envisage a general government net lending corresponding to 3 per cent of GDP or less in 1997. This is especially the case for France and Germany. Given the outcomes for 1996, many countries still have a sizeable budget adjustment to make, and their efforts to do so will tend to restrain the cyclical forces for recovery. Some governments are also employing one-off measures to meet the target.[102] Against the background of adverse developments in the labour market, the German government has recently raised its forecast for the budget deficit from 2.5 to 2.9 per cent of GDP in 1997. In Italy, the higher than planned budget deficit in 1996 will require further adjustment measures beyond those already adopted in the budget for 1997 if the 3 per cent target is to be reached.

The optimism of governments that the target of 3 per cent will be met in 1997 is not always shared by independent forecasters. Thus in Germany, the Council of Economic Experts (*Sachverständigenrat*) and the major economic research institutes are currently forecasting a deficit of some 3.3 to 3.5 per cent of GDP, the main reason being the expected weakness of the cyclical recovery. Similarly in France and Italy, there are alternative forecasts pointing to deficits of more than 3 per cent.

The optimism is reflected in the recent economic projections prepared by the European Commission which suggest that general government budget deficits in the EU will, indeed, decline on average to 3 per cent of GDP in 1997.[103] This decline from 4.4 per cent in 1996 mainly reflects a fall in the structural deficit, with correspondingly little change in the cyclical component. All countries, except Greece, Italy[104] and the United Kingdom, are forecast to have a budget deficit corresponding to 3 per cent of GDP or less in 1997.

Among the west European countries outside the EU, there has been a significant improvement in the government's finances in Norway since 1993. This reflects the strengthening business cycle and, notably, increases in the net cash flow from petroleum activities. Against a background of vigorous growth in 1996, fiscal policy will be tightened in 1997, contributing to a reduction in demand corresponding to 1 per cent of mainland GDP.

In the *United States*, tight expenditure control and strong growth in tax revenues led to a significant decline in the federal government's budget deficit in fiscal year 1996. It corresponded to only 1.4 per cent of GDP, the lowest ratio since 1974. The general government deficit (i.e. including state and local authority budgets) is estimated to have fallen to an average 1.6 per cent of GDP in 1996. Most forecasts point to a broadly unchanged deficit (relative to GDP) in 1997. When presenting its plans for fiscal year 1998 the administration forecast that the federal budget deficit would be eliminated by the year 2002.[105]

The successful budget consolidation policy pursued in the United States reflects to a large degree the favourable impact of strengthening economic growth on government revenues. But an important role was also played by the *timing* of fiscal retrenchment policies: the maximal restraint occurred in 1994, when the upswing was already well established. This contrasts with western Europe, where the strongest effort at fiscal retrenchment is being made in 1996-1997 at a time of relatively weak cyclical conditions (chart 2.6.4).

There was also considerable further progress in consolidating government finances in *Canada* in 1996. This was supported by strong growth in tax revenues and the favourable impact of lower interest rates on the servicing of government debt. Total government expenditure in 1996 was actually lower than in 1995. In the event, the general government deficit fell strongly in 1996, to only 1.8 per cent of GDP. This compares with 4.1 per cent in 1995 and a recent peak of 7.3 per cent in 1993.

2.7 The short-run outlook

There is a broad consensus that the moderate recovery which gradually unfolded in *western Europe* in the course of 1996 will continue in 1997. According to forecasts made at the turn of the year, real GDP will increase by about 2¼-2½ per cent in 1997.[106] Such forecasts imply that there will be no significant rebound in economic growth in the course of 1997 and that there will be hardly any significant improvement in the labour markets. Unemployment is set to remain high and broadly unchanged from its level in 1996. With considerable slack in the labour markets, ample margins of spare capacity and continuing wage restraint, expectations of inflation will remain at the very low

---

[102] In France, revenues in 1997 will be enhanced by a special transfer of FF 37.5 billion, equivalent to 0.5 per cent of GDP, from the pension fund of the telecommunications company (France Télécom), which is slated to be privatized. In Italy, the government will levy a special tax in 1997 on funds accumulated in the business sector (*fondi di quiescenza*) but which the authorities promise to pay back in 1999. The Statistical Office of the European Communities (Eurostat) has ruled that these transactions are in conformity with the existing European System of National Accounts. It is also noteworthy that in Belgium the parliament adopted three so-called "framework laws", which effectively allow the government to introduce by decree measures which will ensure that the country qualifies for membership of EMU from its inception.

[103] European Commission, *European Economy*, Supplement A, No. 12 (Luxembourg), December 1996, p. 21, table 21.

[104] For Italy, the European Commission forecasts a budget deficit corresponding to 3.3 per cent of GDP, but this did not include the revenues from the "eurotax".

[105] The United States government did not support renewed attempts to adopt a balanced budget amendment to the constitution. In fact, such a proposal failed to win sufficient support in Congress.

[106] The lower (upper) bound of this range corresponds to the growth rate for western Europe excluding (including) Turkey.

CHART 2.6.4

Fiscal impulses, economic growth and output gaps in the European Union and the United States, 1989-1997

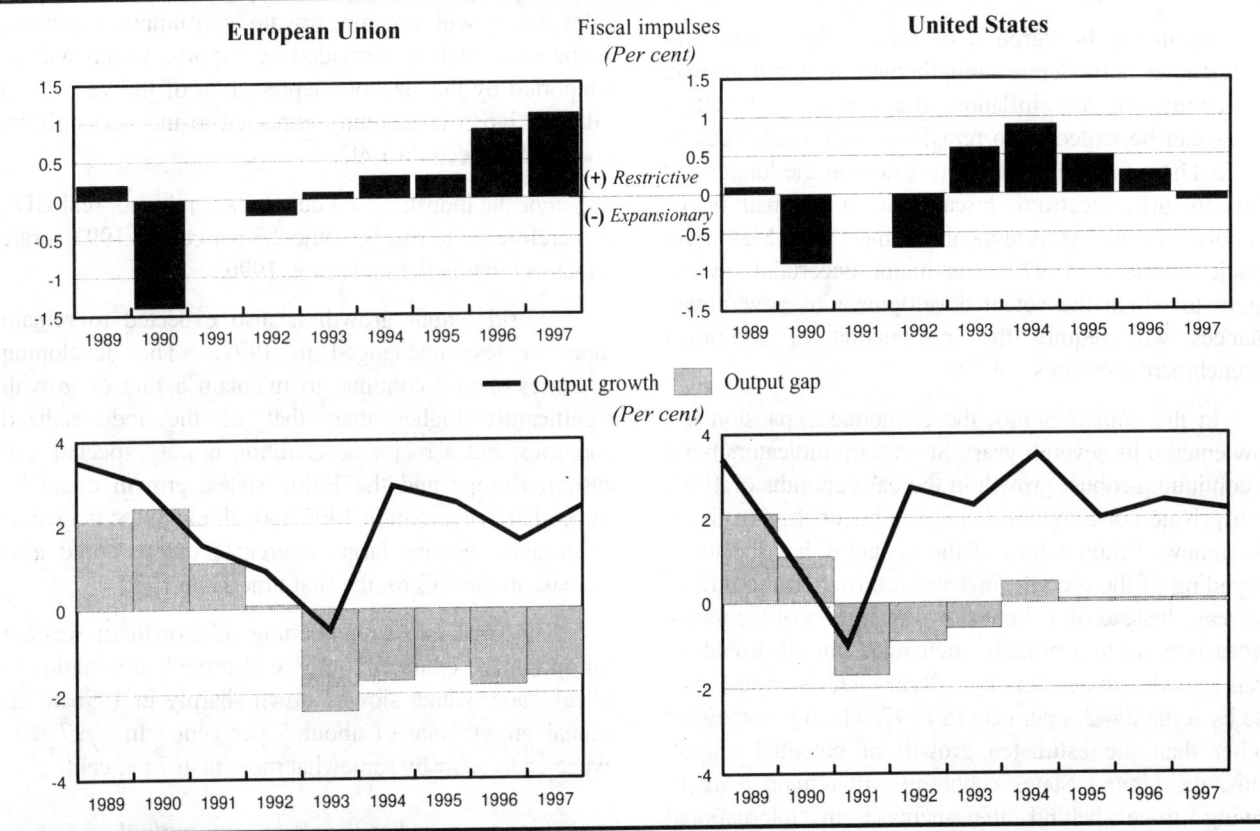

Source: OECD Economic Outlook (Paris), December 1996, annex tables 1, 11 and 31.

Note: Fiscal impulses are measured by changes in the general government structural financial balance expressed as a percentage of potential GDP. A negative (positive) sign indicates an expansionary (contractionary) effect. Economic growth is measured by annual changes in real GDP. Output gaps indicate the difference between actual and potential output as a per cent of potential GDP.

levels attained in 1996. The zeal to keep inflation low (that is, at some 1.5-2.0 per cent) will be accentuated by the efforts of monetary authorities to meet the Maastricht inflation criterion.

Export growth is likely to accelerate somewhat in 1997 and foreign demand should still be the mainstay of higher activity levels in 1997. To some degree this reflects the effect of improved price competitiveness due to the significant appreciation of the dollar. Export demand originating in the transition economies is expected to continue growing strongly in 1997. But there is also a revival of intraregional trade flows, which is mirrored in a concomitant strengthening of import growth. Changes in real net exports are therefore likely to be broadly neutral in their effect on aggregate output levels in 1997. There is likely to be only a moderate strengthening of domestic demand growth, reflecting a slight upturn in fixed investment and favourable changes in inventory accumulation. Fixed investment appears to be mainly restrained by the presence of ample margins of spare capacity and lingering uncertainty about the strength and sustainability of the recovery. Private consumption will probably continue to be rather sluggish and for the year as a whole the same growth rate as in 1996 (slightly more than 2 per cent) is forecast, which mainly reflects moderate strengthening in the course of the year. The persistent weakness of consumer demand is largely the result of weak growth in aggregate real incomes, which, in turn, reflects moderate gains in employment and wage incomes. Growth of government consumption expenditure will be dampened by ongoing fiscal retrenchment.

Output growth in France, Germany and the United Kingdom is expected to be within a range of 2¼-3 per cent in 1997, with France and Germany at the lower end of this range. A significantly lower rate of economic growth is forecast for Italy, the major reason being the draconian fiscal retrenchment exercise underway and its associated impact on domestic demand. In the United Kingdom, optimistic growth prospects have been surrounded by a larger margin of uncertainty given the difficulty of assessing the impact of the appreciation of the pound sterling on net export growth.

Among the smaller economies, growth rates are expected to be higher than in the four major economies. Robust annual GDP growth rates of 3 per cent or more in

1997 are forecast for quite a number of countries (Cyprus, Finland, Iceland, Ireland, Israel, Luxembourg, Norway, Spain and Turkey).

Against a background of already low short-term interest rates, a moderate strengthening of output growth, and continuing low inflation, the stance of monetary policy can be expected to remain broadly unchanged in 1997. The same holds for interest rates at the longer end of the maturity spectrum. Fiscal policy will remain firmly anchored to the commitment to meet the Maastricht deficit criteria in 1997. The major uncertainty is the extent to which the actual development of government finances will require the introduction of additional retrenchment measures.

In the *United States*, the economic expansion has now entered its seventh year. Short-term indicators point to continuing robust growth in the early months of 1997, with private consumption being the major driving force. The renewed momentum of the economy has led to an upgrading of the growth forecasts prepared at the turn of the year. Instead of a slight slowdown, most of the recent forecasts point to a broadly unchanged rate of growth or even a moderate acceleration. Real GDP is expected to rise by some 2½-2¾ per cent in 1997, which is somewhat higher than the estimated growth of potential output. Thus, the United States economy will remain a major driving force behind the increase in international economic activity in 1997. Inflationary pressures have remained low, despite the prolonged length of the current expansion, and expectations are for a broadly unchanged rate of inflation in 1997. The labour market will remain tight, but there are as yet no indications of significant increases in labour cost pressures.

Monetary policy has been unchanged since the beginning of 1996 but the monetary authorities have been increasingly concerned about the potential inflationary consequences of the strong (and continuing) rise in equity prices and the renewed acceleration of economic growth in the final quarter of 1996. Consequently, there was a moderate tightening of policy in late March 1997 and this may continue during the year in order to pre-empt any rekindling of inflationary expectations and pressures. The stance of fiscal policy is expected to be broadly neutral in its effect on overall economic activity in 1997.

In *Canada*, the recovery can be expected to more or less maintain the momentum gained in the second half of 1996, supported by the favourable impact of low interest rates on domestic demand and the spillover effects from the continuing robust export demand from the United States. As a result, real GDP is forecast to increase by more than 3 per cent in 1997, which represents a notable acceleration compared with 1996.

In *Japan*, forecasts are for a marked slowdown in the rate of economic growth in 1997, following a significant rebound in 1996. This deceleration mainly reflects the shift towards restrictive fiscal policy, which will dampen public investment expenditures, while higher sales taxes will restrain private consumption growth. Some offset will be provided by exports, which will be supported by the sizeable depreciation of the yen. Real GDP in Japan is currently expected to increase only by about 1.5 per cent in 1997.

For the industrialized countries combined, real GDP is therefore set to rise by some 2.3 per cent in 1997, a rate which is broadly the same as in 1996.

World output growth is also expected to remain more or less unchanged in 1997. The developing countries should continue to maintain a rate of growth significantly higher than that of the industrialized countries, but a major acceleration is not expected. In eastern Europe and the Baltic states, growth could be somewhat better than in 1996 and, although the degree of uncertainty remains large, aggregate output could also increase in the CIS for the first time since 1989.

The moderate strengthening of growth in western Europe in the course of 1997 will provide a stimulus to world trade, which slowed down sharply in 1996 to an annual growth rate of about 5 per cent. In 1997 it is expected to grow by somewhat more than 7 per cent.

Appraisal of the short-term outlook

Economic prospects have improved for western Europe in the course of 1996 and the recovery is set to continue in 1997. Despite earlier expectations, the United States economy does not appear to be in a phase of cyclical downswing but rather is going through a period of "balanced growth" with short-term fluctuations around its longer-term trend. The gap in output growth between western Europe and the United States will narrow in 1997, but the United States economy is at a much more advanced stage of the business cycle. Economic policy concerns in the United States have therefore shifted towards preventing overheating.

Inflationary pressures are expected to remain low in both western Europe and the United States. This is not surprising for western Europe, given ample margins of spare capacity and the considerable slack in the labour markets. It is, however, rather striking for the United States. The persistence of low inflation could well mark, apart from temporary factors, a significant downward shift in inflation expectations, which, in turn, can be partly related to the adverse experiences of economic agents during the high inflation period of the 1970s and 1980s. On the other hand, any marked acceleration in the rate of economic expansion of the United States economy must ultimately exhaust the available pool of labour and idle physical capital stock, with consequentially stronger upward pressures on costs and prices. A pre-emptive moderate tightening of monetary policy may therefore affirm the willingness of the authorities to keep inflation in check. Such a move may also be appropriate against the background of the considerable surge in equity prices which, ultimately, will add to inflationary pressures if not corrected in time. A tightening of United States monetary policy, to a greater extent than is currently anticipated in the financial markets, could trigger a relatively large correction of equity prices and dampen both consumer demand and business investment.

A tightening of United States monetary policy would probably also be reflected in a rise in long-term interest rates which could spill over to western Europe. There was some decoupling of long-term rates in the course of 1996 and in early 1997 but such a possibility is obviously limited given the close integration of international financial markets. Clearly, a sudden rise in United States inflation could change longer-term inflation expectations and put stronger upward pressure on world interest rates.

The pattern of real exchange rates has undergone some significant changes since the beginning of 1995. The sizeable dollar appreciation has strengthened the price competitiveness of west European firms vis-à-vis overseas producers and this will tend to support net export growth. If the current dollar exchange rate vis-à-vis major currencies (such as the deutsche mark and the French franc) were to be sustained then this would be equivalent to a stronger boost to demand than was expected by many forecasters in late 1996. On the other hand, the stronger dollar will put upward pressure on import prices in western Europe and this could alert the monetary authorities who are intent on maintaining inflation at very low levels. The dollar appreciation is tantamount to an easing of monetary conditions, and it will therefore compensate for the lack of any further reductions in official interest rates in western Europe. The question, of course, is whether the dollar appreciation will be sustained in 1997. Changes in the dollar exchange rate have always been difficult to predict in the past. Any strong reversal of the current appreciation would weaken export prospects and, if domestic demand does not pick up, business confidence in western Europe.

A major uncertainty is which and how many countries will meet the qualifications for adopting a single currency at the beginning of 1999. This will depend on economic developments in 1997 and on how the Maastricht convergence criteria, especially the fiscal ones, will be interpreted. This is largely a matter for political decision and is therefore currently subject to considerable speculation. But economic developments in the course of the first half of 1997 should make clear which countries are likely to be on course for entry into EMU, even if there is a reasonably flexible interpretation of the criteria. If actual developments run counter to the expectations prevailing in financial markets – which are already embodied in the current pattern of exchange rates and interest rates – there could be more or less large corrections with repercussions on growth prospects in the various countries.

In western Europe there is little concern that the recovery may falter in 1997. The more important issues are the structure of the recovery and the probability of it being sustained beyond 1997. Of crucial importance in this respect is the strength of business fixed investment. Fixed investment has been rather weak in western Europe since 1993. There have been significant improvements on the supply side: business profits have risen strongly relative to labour incomes, although real interest rates may still be somewhat high in several countries compared with the rate of return on financial assets. But important restraints on higher investment are the large margins of spare capacity and the modest expectations for output growth. But if fixed investment fails to take off, the current recovery is likely to peter out again into broad stagnation after 1997.

This raises the question of what room for manoeuvre exists for economic policy to support economic activity. Fiscal policy is focused on meeting the Maastricht convergence criteria and, in any case, the size of budget deficits and debt in many countries is such that there is hardly any scope for fiscal stimulus. The risk, however, is that the striving for a "point landing" on the reference value for budget deficits fixed in the Maastricht Treaty may trigger action for additional fiscal retrenchment. Such an accentuation of procyclical behaviour is likely to be counter-productive: it would further restrain the forces for growth in western Europe and probably lead to a further deterioration of government finances.

Monetary policy has been relaxed gradually over the past few years and interest rates, in general, are now quite low. But there does appear to be scope for more monetary easing without rekindling inflationary expectations, as was demonstrated by the loosening of monetary policy in the United States in 1991-1992 and in Japan. The basic problem here is that inflation targets in western Europe now appear to have centred around 1.5 per cent – as illustrated by the recent lowering of the inflation target by the Bundesbank (see section 2.6). This threatens to be too ambitious; it risks placing further restraints on growth and puts an excessive burden of adjustment on the labour market. A less ambitious inflation target would be compatible with a more satisfactory rate of output growth and allow a much needed reduction in unemployment: the experience of the United States economy, where the Federal Reserve appears to be satisfied with an inflation rate within a range of 2.5-3 per cent, suggests this is not an unreasonable expectation.

The economic policy debate over the relatively poor performance of the west European economies in recent years has focused rather narrowly on the need to enhance supply-side flexibility, especially in labour markets. There may well be a strong case for reviewing and changing many of the long-established institutions and regulations governing labour markets. But concentration on this issue is excessive insofar as it neglects the interaction of supply and demand factors in the growth process. The significant employment gains in the United States economy not only illustrate the "built-in flexibility" of the United States labour market but also the important role played by a sustained and robust growth of demand for goods and services. The notion of labour market flexibility is, moreover, rather vague and detailed studies often find that the actual degree of flexibility in economic systems perceived to be rigid is surprisingly large and has been increasing over time (see section 2.3 above). The importance of faster economic growth in facilitating structural change and greater supply-side flexibility, deserves to be brought back into the policy discussion.

Chapter 3

THE TRANSITION ECONOMIES

3.1 The general context

(i) Expectations, outcomes and outlook

(a) Expectations and outcomes

The year 1996 was a difficult one for the ECE transition economies but, on balance, positive developments and changes prevailed. The process of economic transformation deepened and strengthened in the core group of fast reforming central European countries, although on average their rate of economic growth slowed down as compared with 1995. Another positive feature was the expansion of the group of fast reformers: developments in 1996 suggest that the Baltic states have also joined this category. There was also better performance in some of the CIS countries suggesting that the transformational recession may be coming to an end in this region. There was substantial progress in macroeconomic stabilization in the Russian Federation and Ukraine, but the long awaited return to recovery and growth did not materialize. The most disturbing developments occurred in south-eastern Europe where Albania and Bulgaria were hit by serious crises; signs of instability and lack of progress in transformational reforms were obvious also in other countries in this region[107] as well as in some of the CIS countries.

Economic growth in eastern Europe as a whole decelerated in 1996 to 4 per cent, down from 5.6 per cent in 1995 (table 3.1.1). Although a slowdown occurred in a number of countries, the decline in the average growth rate was largely due to the unexpected and deep crisis in Bulgaria where GDP dropped by 10 per cent in 1996. Bearing in mind the highly uncertain character of *ex ante* forecasts for the transition countries, it can be noted that in most of the other east European economies (with the exception of Yugoslavia) outcomes were reasonably close to expectations (table 3.1.2); output growth in Poland and Slovakia exceeded the forecasts whereas in Croatia, Hungary and Slovenia it was somewhat below.[108]

Thus, the slowdown of growth in most of the countries in which it occurred in 1996 was more or less expected by policy makers. One of the main causes was the sluggish economic performance in western Europe, especially in some of the large economies (France, Germany and Italy), which contributed to a weakening of export growth from the transition countries. However, as discussed below, it also reflected some internal weaknesses in the transition countries as well as a tightening of macroeconomic policies.

The performance of the Baltic states in 1996 added to the encouraging developments among the transition countries. Economic growth strengthened (table 3.1.1) and for the first time since regaining their independence all three countries posted positive growth.

Economic activity in the Commonwealth of Independent States was uneven but the encouraging signs of recovery were more widely spread in 1996 than in 1995 (table 3.1.1). The biggest disappointments were in the two largest economies, the Russian Federation and Ukraine, which not only failed to make the turn around to recovery but actually performed much worse than expected (table 3.1.2).

As in previous years economic growth in eastern Europe has been largely based on strong industrial output performance (tables 3.1.1 and 3.1.3). In the Baltic states, however, it was notably less buoyant and the picture was rather mixed in the CIS where the aggregate results were affected by the poor performance of Russian industry. However, in most CIS countries there is a fairly close correlation between industrial performance and GDP growth (table 3.1.1). The main exceptions to these patterns were in the Baltic countries where GDP growth in 1996 was largely based on the recovery of the service sector.

---

[107] The economy of Bosnia and Herzegovina was extremely badly damaged by the war and the process of reconstruction has been very slow. One year after the Dayton Accord the economy was still barely functioning (according to some rough estimates, despite the recorded growth, industrial output in 1996 was still at some 10-15 per cent of its pre-war level) and the population relied almost exclusively on international assistance.

[108] At the time of writing this *Survey*, no official statistics were available for Albanian GDP in 1996.

TABLE 3.1.1

Basic data on the transition economies, 1994-1996
(Rates of change and shares, per cent)

	GDP (growth rates)			Industrial output (growth rates)			Inflation (per cent change, Dec./Dec.)			Unemployment rate (per cent)			Trade balance (per cent of GDP)		
	1994	1995	1996	1994	1995	1996	1994	1995	1996	1994	1995	1996	1994	1995	1996
Eastern Europe	3.9	5.6	4.0	6.7	7.6	7.2	..	..	..	13.6	12.5	11.8	-4.5	-6.5	-9.7
Albania	9.4	8.6	..	-18.6	-7.2	..	15.0	6.0	..	18.0	13.1	12.1	-23.3	-18.3	-26.7
Bosnia and Herzegovina	..	..	..	..	..	60.0[a]	94.7	-34.2	3.2	..	..	..	..	..	..
Bulgaria	1.8	2.1	-10.0*	8.5	5.4	-1.0	122.0	33.0	311.1	12.8	11.1	12.5	-2.1	-2.3	2.0
Croatia	0.6	1.7	4.4*	-2.7	0.3	3.1	-3.0	3.7	3.4	17.3	17.6	15.9	-6.7	-15.8	-17.6
Czech Republic	2.6	4.8	4.4	2.1	9.2	6.8	10.3	8.0	8.7	3.2	2.9	3.5	-2.0	-7.7	-11.5
Hungary	2.9	1.5	0.5*	9.5	4.6	3.3	21.3	28.5	20.0	10.4	10.4	10.5	-9.3	-5.9	-6.9
Poland	5.2	7.0	6.0	11.9	9.4	9.1	29.4	22.0	18.7	16.0	14.9	13.6	-4.7	-5.2	-9.6
Romania	3.9	7.1	4.1	3.3	9.4	9.8	61.9	27.7	56.8	10.9	9.5	6.3	-3.2	-6.6	-6.5
Slovakia	4.9	6.8	6.9	4.6	8.2	2.5	11.8	7.4	5.5	14.8	13.1	12.8	0.1	-1.9	-11.1
Slovenia	5.3	3.9	3.5*	6.4	2.0	1.0	18.3	8.6	8.8	14.2	14.5	14.4	-3.3	-6.3	-5.9
The FYR of Macedonia[b]	-7.2	-2.9	1.6	-10.6	-9.8	3.2	55.1	11.2	0.2	33.2	37.2	39.8	-14.4	-14.4	-22.6
Yugoslavia[b]	2.5	6.0	4.3*	1.2	3.8	6.8	8E+09	110.7	59.9	23.9	24.7	26.1	..	-8.0	-15.3
Baltic states	-0.2	1.4	3.4	-15.7	0.6	1.7	..	..	..	5.3	6.5	6.4	-9.1	-15.4	-18.6
Estonia	-2.7	2.9	3.5*	-2.0	1.9	1.1	41.8	28.8	14.9	5.1	5.0	5.6	-15.2	-19.4	-26.4
Latvia	0.6	-1.6	2.5*	-6.8	-6.3	0.7	26.1	23.3	13.2	6.5	6.6	7.2	-6.9	-11.5	-17.2
Lithuania	1.0	3.0	4.0*	-28.0	6.2	2.8	45.0	35.5	13.1	4.5	7.3	6.2	-7.6	-15.8	-15.2
CIS	-14.5	-5.7	-5.3	-21.5	-5.8	-3.4	..	..	..	4.4	5.5	6.4	7.4	7.4	6.4
Armenia	5.4	6.9	4.0	5.3	2.4	1.0	1 762.7	32.0	5.6	6.0	8.1	9.7	-20.0	-18.3	-23.2
Azerbaijan	-19.7	-12.0	1.0	-22.7	-17.2	-6.7	1 786.8	84.5	6.8	0.9	1.1	1.1	4.4	-4.6	-10.2
Belarus	-12.6	-10.0	3.0	-17.1	-11.7	3.2	1 957.3	244.2	39.1	2.1	2.7	4.0	1.1	-1.1	-4.0
Georgia	-30.0	2.4	11.0	-39.7	-9.8	7.7	9 197.5	57.4	12.6	3.8	3.4	3.2	-1.0	-8.2	-8.2
Kazakstan	-18.8	-8.9	0.5	-28.1	-8.2	0.3	1 156.8	60.4	28.6	1.0	2.1	4.1	-0.2	7.0	7.7
Kyrgyzstan	-20.1	-5.4	6.0	-28.0	-17.8	10.8	87.2	31.9	35.0	0.8	3.0	4.5	0.8	-1.9	-16.7
Republic of Moldova	-31.2	-3.0	-8.0	-27.7	-3.9	-8.5	104.6	23.8	15.1	1.0	1.4	1.5	-1.6	0.5	-10.1
Russian Federation	-12.7	-4.2	-6.0	-20.9	-3.3	-5.0	214.8	131.4	21.8	7.1	8.9	9.3	8.9	8.5	8.5
Tajikistan	-12.7	-12.4	-17.0	-25.4	-5.1	-19.8	1.1	2 382.2	40.6	1.8	1.8	2.4	0.3	29.2	15.7
Turkmenistan[c]	-18.0	-16.0	–	-24.7	-6.4	17.9	..	..	..	..	..	..	-7.7	11.1	-12.9
Ukraine	-22.9	-11.8	-10.0	-27.3	-12.0	-5.1	401.1	181.7	39.7	0.3	0.6	1.5	4.6	3.8	-3.0
Uzbekistan	-5.2	-1.2	2.0	1.6	0.1	6.0	1 281.4	144.1	..	0.3	0.3	0.4	-3.6	0.8	-2.1
Total above	-8.5	-1.6	-1.6	-14.7	-1.7	0.2	..	..	..	6.9	7.5	7.9	2.0	1.2	-0.2
Memorandum items:															
CETE-5	4.5	5.7	5.0	9.3	8.4	7.3	..	..	..	12.8	12.0	11.3	..	..	..
SETE-7	2.8	5.2	1.8	1.1	5.8	7.0	..	..	..	14.6	13.7	12.5	..	..	..
Ex-GDR Länder	9.9	5.3	2.0	13.9	5.8	4.3	3.5	2.6	1.6	13.5	14.9	15.9	..	..	..

Source: National statistics and direct communications from national statistical offices to UN/ECE secretariat (IMF and World Bank data for Albania).

Note: Aggregates are UN/ECE secretariat computations, based on previous period weights at 1992 prices and some estimates for missing components. Output measures are in real terms (at constant prices). *Industrial output* refers to gross output, not the contribution of industry to GDP. The *inflation* measure is the consumer price index except for some countries where retail prices are used (see table 3.4.1). *Unemployment* refers generally to registered unemployment at the end of the period, but to the ILO measure for Russia and some other variants elsewhere (see table 3.3.2). The *trade balance/GDP* relation refers to merchandise trade (in dollars) and GDP converted at current exchange rates; for the CIS countries it does not cover intra-CIS trade (see table 3.1.4). Aggregates shown are: *Eastern Europe* (the 12 countries below that line), with sub-aggregates *CETE-5* (central European transition economies: Czech Republic, Hungary, Poland, Slovakia, Slovenia) and *SETE-7* (south European transition economies: Albania, Bosnia and Herzegovina, Bulgaria, Croatia, Romania, The FYR of Macedonia and Yugoslavia); *Baltic states* (Estonia, Latvia, Lithuania); *CIS* (12 member countries of the Commonwealth of Independent States); and *total transition countries*.

[a] Unweighted average of public sector output growth reported by the Statistical Office of the Federation for areas under government control (87 per cent, full year) and that of Republika Srpska (34 per cent, January-October).

[b] Gross material product instead of GDP.

[c] Net material product instead of GDP.

On the demand side (see section 3.2), a notable change from 1995 was a shift between the domestic and external components of aggregate demand. In most of the transition countries where there was positive economic growth in 1996, the main impetus came from domestic demand, both private consumption and investment. Investment continued to grow at high rates in many east European countries reflecting the positive expectations of investors. In addition, as noted in sections 3.2 and 3.4, booming private consumption expenditure (most notably in central Europe) was boosted by rising real wages and incomes. Nominal incomes continued to grow at a fast pace in many of the transition economies – largely reflecting built-in inertia and slowly adjusting inflationary expectations – while actual inflation rates were declining in most countries.

TABLE 3.1.2

GDP: expectations and outcomes in the transition economies, 1995-1997
(Percentage change over the same period of the preceding year)

	1995	1996 Ex ante forecasts	1996 Jan.-June	1996 Jan.-Sept.	1996 Jan.-Dec.	1997 forecasts
Eastern Europe	5.6	5-6	3.1	3.5	4.0	3½
Albania	8.6	6	..	..	..	..
Bosnia and Herzegovina	..	..	..	..	..	..
Bulgaria	2.1	3	-6.2	..	-10.0*	-(2-4)
Croatia	1.7	6-8	1.7	3.5	4.4*	5.5
Czech Republic	4.8	4.8	4.3	4.0	4.4	4-5
Hungary	1.5	2	-	-	0.5*	2
Poland	7.0	5-6	4.0	..	6.0	5-6
Romania	7.1	4.5	..	..	4.1	-2
Slovakia	6.8	5-6	7.1	7.0	6.9	6
Slovenia	3.9	5.0	1.6	..	3.5*	4
The FYR of Macedonia^a	-2.9	2	..	..	1.6	..
Yugoslavia^a	6.0	12.5	..	..	4.3*	..
Baltic states	1.4	..	2.3	3.0	3.4	4½
Estonia	2.9	..	2.3	..	3.5*	4-5
Latvia	-1.6	0-3	1.5	2.3	2.5*	3.5
Lithuania	3.0	4.2	3.1	3.6	4.0*	5
CIS	-5.7	..	-4.6	-5.3	-5.3	..
Armenia	6.9	..	4.3	4.5	4.0	..
Azerbaijan	-12.0	..	-5.7	0.7	1.0	..
Belarus	-10.0	..	-0.2	1.0	3.0	5
Georgia	2.4	..	10.4	12.0	11.0	..
Kazakstan	-8.9	..	-0.8	1.1	0.5	2
Kyrgyzstan	-5.4	..	0.3	3.0	6.0	..
Republic of Moldova	-3.0	7.7	-4.6	..	-8.0	5-7
Russian Federation	-4.2	0/-3	-5.0	-6.0	-6.0	0-2
Tajikistan	-12.4	..	-20.2	-18.0	-17.0	..
Turkmenistan^b	-16.0	..	..	..	-	..
Ukraine	-11.8	-0.5	-8.7	-10.0	-10.0	-
Uzbekistan	-1.2	..	1.4	1.5	2.0	..
Total above	-1.6	..	-1.5	-1.8	-1.6	..
Memorandum items:						
CETE-5	5.7	..	3.6	4.2	5.0	4½
SETE-7	5.2	..	2.0	1.9	1.8	-
Ex-GDR Länder	5.3	4-6	0.7	1.7	2.0	1.0

Source: National statistics and direct communications from national statistical offices to UN/ECE secretariat (IMF and World Bank data for Albania).

Note: On regional aggregates, see the note to table 3.1.1. Forecasts are generally those of national conjunctural institutes or, if these are not available, government forecasts associated with the central budget formulation.

^a Gross material product.
^b Net material product.

TABLE 3.1.3

Gross industrial output in the transition economies, 1995-1997
(Percentage change over the same period of the preceding year)

	1995	1996 Jan.-Mar.	1996 Jan.-June	1996 Jan.-Sept.	1996 Jan.-Dec.	1997 forecasts
Eastern Europe	7.6	5.7	5.6	6.3	7.2	..
Albania	-7.2	..	..	..	..	..
Bosnia and Herzegovina	..	..	..	..	60.0^a	..
Bulgaria	5.4	-1.2	-0.4	-1.1	-1.0	..
Croatia	0.3	-0.1	0.6	2.6	3.1	..
Czech Republic	9.2	9.7	8.1	8.5	6.8	7-8
Hungary	4.6	1.8	1.1	1.5	3.3	..
Poland	9.4	8.8	8.8	9.0	9.1	..
Romania	9.4	5.7	7.8	8.8	9.8	..
Slovakia	8.2	7.2	3.3	2.8	2.5	4-4.5
Slovenia	2.0	-5.2	-3.0	-1.0	1.0	..
The FYR of Macedonia	-9.8	5.3	3.9	3.8	3.2	..
Yugoslavia	3.8	3.4	2.1	5.6	6.8	..
Baltic states	0.6	6.1	3.1	3.1	1.7	..
Estonia	1.9	-2.2	-1.5	0.6	1.1	..
Latvia	-6.3	-1.9	0.2	2.0	0.7	..
Lithuania	6.2	16.2	8.0	5.8	2.8	..
CIS	-5.8	-3.6	-2.9	-3.8	-3.4	..
Armenia	2.4	0.3	1.8	1.5	1.0	..
Azerbaijan	-17.2	-11.5	-10.2	-8.0	-6.7	..
Belarus	-11.7	-2.9	1.6	2.6	3.2	..
Georgia	-9.8	4.9	4.0	5.2	7.7	..
Kazakstan	-8.2	0.6	0.2	0.1	0.3	3
Kyrgyzstan	-17.8	4.8	13.2	11.2	10.8	..
Republic of Moldova	-3.9	4.5	8.7	-3.0	-8.5	..
Russian Federation	-3.3	-4.0	-4.0	-5.0	-5.0	-(1-2)
Tajikistan	-5.1	-24.3	-24.9	-20.2	-19.8	..
Turkmenistan	-6.4	18.1	32.0	25.9	17.9	..
Ukraine	-12.0	-3.6	-3.1	-4.3	-5.1	..
Uzbekistan	0.1	0.8	5.0	5.7	6.0	..
Total above	-1.7	-1.2	-0.8	-1.3	0.2	..
Memorandum items:						
CETE-5	8.4	7.2	6.5	6.8	7.3	..
SETE-7	5.8	3.4	4.3	5.5	7.0	..
Ex-GDR Länder	5.8	0.8	3.6	4.1	4.3	..

Source: National statistics and direct communications from national statistical offices to UN/ECE secretariat (IMF and World Bank data for Albania).

Note: On regional aggregates, see the note to table 3.1.1.

^a Unweighted average of public sector output growth reported by the Statistical Office of the Federation for areas under government control and that of Republika Srpska.

Indeed, the process of disinflation was one of the most encouraging developments in 1996: consumer price inflation decelerated, or at least remained more or less unchanged, in all transition countries with the exception of Bulgaria and Romania (table 3.1.1).[109] The most notable progress was made in the CIS countries where stabilization policies, supported by the international financial institutions, have started to produce results. In the Baltic states as well as some east European countries there were also reductions in the rate of inflation, but double-digit rates persisted in many countries. The overall progress in macroeconomic stabilization in the transition countries can to a large extent be attributed to improved efficiency in the conduct of monetary policy (see section 3.1(ii)).

The rapid expansion of domestic demand, coupled with strengthening currencies in central Europe, the Baltic states and a number of other transition countries, led to an upsurge in import demand in 1996 while export growth decelerated or stagnated (table 3.1.4). The widening of trade and current account deficits in a

---

[109] Due to the unreliability of the available data Bosnia and Herzegovina is not considered here.

TABLE 3.1.4

Foreign trade involvement of the transition economies, 1994-1996

(Merchandise trade in per cent of GDP, at current exchange rates)

	Exports			Imports		
	1994	1995	1996	1994	1995	1996
Eastern Europe	26.4	28.7	27.5	30.9	35.2	37.2
Albania	7.9	8.3	8.1	31.2	26.6	34.8
Bosnia and Herzegovina	..	..	..	..	..	..
Bulgaria	41.1	41.5	45.3	43.2	43.8	43.3
Croatia	29.9	25.7	24.2	36.5	41.5	41.8
Czech Republic	39.5	45.9	42.6	41.5	53.5	54.0
Hungary	25.8	29.4	29.6	35.0	35.4	36.4
Poland	18.6	19.4	18.6	23.3	24.6	28.2
Romania	20.4	22.2	21.6	23.6	28.8	28.1
Slovakia	48.6	49.3	46.6	48.5	51.2	57.7
Slovenia	47.6	44.8	44.9	51.0	51.1	50.8
The FYR of Macedonia	39.4	33.8	24.3	53.8	48.2	46.9
Yugoslavia	..	10.8	12.5	..	18.8	27.8
Baltic states	42.2	41.6	40.6	51.3	57.0	59.3
Estonia	56.3	50.7	48.9	71.5	70.1	75.3
Latvia	27.1	29.2	28.3	34.0	40.7	45.4
Lithuania	47.5	45.4	44.4	55.1	61.2	59.6
CIS	17.9	17.5	15.5	10.5	10.0	9.1
Armenia	8.9	8.1	9.6	29.0	26.4	32.8
Azerbaijan	22.7	13.6	11.1	18.2	18.2	21.3
Belarus	21.0	17.3	13.6	19.8	18.3	17.6
Georgia	1.9	2.9	1.8	2.8	11.1	10.0
Kazakstan	10.8	14.0	13.9	11.0	7.0	6.2
Kyrgyzstan	10.6	9.4	5.8	9.7	11.4	22.5
Republic of Moldova	11.3	16.5	15.9	12.9	16.0	26.0
Russian Federation	19.1	17.8	15.6	10.2	9.3	7.1
Tajikistan	41.0	82.5	41.7	40.8	53.3	25.9
Turkmenistan	13.2	31.7	17.7	20.8	20.6	30.6
Ukraine	12.3	15.5	15.6	7.7	11.7	18.6
Uzbekistan	14.8	16.9	20.3	18.5	16.0	22.3
Total above	21.9	22.6	20.6	19.9	21.4	20.8

Source: UN/ECE secretariat computations, based on national and CIS Statistical Committee data on trade flows (usually in dollar terms), nominal GDP levels in national currencies and current exchange rates.

Note: Foreign trade includes "new trade" among the successor states of former Czechoslovakia and the former SFR of Yugoslavia, but not intra-CIS trade. GDP shares for CIS countries are highly uncertain (because of poor national-currency GDP data, dubious exchange rates and trade values). For most transition countries, conversion of GDP at current exchange rates tends to undervalue GDP in a purchasing power sense; hence trade ratios on a purchasing power parity converted GDP would be lower. The level of the ratios is highly sensitive to exchange rate fluctuations; large swings in the ratio – as notably in Tajikistan – are likely to reflect such, often temporary, factors (e.g. flight from the currency and its reversal) rather than structural changes.

majority of transition economies in 1996 has raised concerns about the sustainability of current growth rates of output. Most east European countries, the Baltic states, and some CIS countries, are facing growing trade (table 3.1.1) and current account (table 3.6.1) deficits which are now reaching proportions that pose certain risks to macroeconomic stability (see section 3.6). The average trade deficit for eastern Europe reached 9.7 per cent of GDP in 1996 (up from 6.5 per cent in 1995) and for the Baltic states it rose from 15.4 per cent in 1995 to 18.6 per cent (table 3.1.1).

There were two principal factors behind the worsening of external balances in the east European countries. On the one hand, rising import demand is a natural component of the process of economic recovery and reconstruction which is underway in the region and in such circumstances it is not unusual to observe growth rates of imports exceeding those of exports and trade deficits increasing. At the same time, the strong import surge and the slowing down of export growth also reflect a persistent real appreciation of the foreign exchange rates of most east European and Baltic countries. So far, the current account deficits have been largely financed by net private capital inflows (although not in all countries), but if external imbalances continue to deteriorate there are bound to be increasing concerns in the financial markets about their sustainability.

The inflow of private foreign capital to the transition countries did in fact slow down in 1996 and, although these countries as a whole managed to increase substantially their foreign currency reserves in 1995, this process came to a halt in 1996. That is, the inflow of capital in 1996 just covered, on average, the current account deficits. If these trends – widening current account deficits and decelerating inflows of capital – continue in 1997, this will most probably lead to the erosion of foreign reserves and, consequently, to preventive policy responses.

The problem is not so much related to the magnitude of the external imbalances per se but to their determinants and, in particular, the declining vigour of export growth rates relative to those of imports. Although weaker demand in western Europe was a factor behind the slowdown in export growth, the more fundamental reasons appear to be related to the composition and structure of exports and more general problems of competitiveness (section 3.5).

Many of the transition countries now appear to have taken up the available "slack" for export expansion based on previously underutilized capacities. Future growth of exports will therefore have to come increasingly from new or expanding production capacities or from the restructuring and modernization of existing ones. From this perspective, the transition economies are faced with an increasing supply-side constraint due to the relatively slow pace of restructuring and this is probably one of the main fundamental reasons for growing external imbalances.

Thus (as discussed in section 3.2) the actual – and feasible – speed of restructuring of the transition economies is likely to become increasingly one of the main factors determining their future growth rates, not only in the long run but also in the short and medium term. The countries in transition will therefore increasingly have to adjust to rates of growth that match the overall process of economic restructuring in order to avoid excessive import demand and a dangerous deterioration of their external (and also, in some cases, internal) balances. Another factor contributing to the deterioration of the trade balances of some transition countries was the real appreciation of their exchange rates, a process which has been underway for several years. Policy makers are here confronted with the problem of finding a balance between maintaining the process of disinflation (which under the present exchange rate regimes of most of the countries leads to real exchange rate appreciation) and avoiding an excessive deterioration in the external balance. In the end, the resolution of this problem comes down to finding a sustainable rate of growth. As is well known, both from economic theory and from the well-documented experiences of many countries, it is not possible to have "the best of all worlds": policy makers will always have to find a feasible compromise between rates of economic growth, of inflation, and the state of domestic and external balances.

In this regard the experience of Hungary is relevant: faced with growing domestic and external imbalances, the Hungarian government introduced in 1995 a package of policies aimed at restoring macroeconomic equilibrium and creating the conditions for sustainable economic growth. After accomplishing considerable macroeconomic adjustment in 1995-1996 – which took longer and was more painful than anticipated – external and domestic balances markedly improved in 1996 and, as discussed later in this chapter, some of the important fundamentals improved significantly. The reduced pressure from public finances, rapidly rising industrial labour productivity, improved corporate profitability and export competitiveness resulting from the adjustment effort have strengthened considerably Hungary's growth prospects. This has already shown up in recent statistics: since the last quarter of 1996 both industrial output and manufacturing exports have been growing fast, investment has started to pick up and the latest surveys indicate a substantial improvement in business confidence.

The problems emerging in some of the other transition countries in 1996-1997 (mostly related to external imbalances) are in essence similar (though not identical) to those Hungary was facing in 1993-1994. Although the nature and acuteness of these problems vary across countries, the Hungarian experience may be an important and useful lesson for other countries and may point to the necessary adjustments.

Another problem which became more serious in 1996 and early 1997 was that of payment arrears in some of the CIS countries. As discussed later in this chapter, payment arrears escalated rapidly in 1996 and are now endemic in the Russian Federation. Apart from the social and political problems related to the non-payment (or to the increasing delays in payment) of wages, the accumulation of different types of arrears obviously has serious macroeconomic implications that the Russian authorities will have to face in the near future. There are signs that the problem of payment arrears is also getting worse in some of the other CIS countries.

There were no major changes in the labour markets of the transition countries in 1996. Double-digit unemployment rates prevailed in eastern Europe, whereas the reliability of the much lower official unemployment rates in the CIS countries, as well as in the Baltic states, remains statistically doubtful. Although the average rate of unemployment in eastern Europe declined slightly in 1996 (from an average 12.5 per cent in 1995 to 11.8 per cent) any dramatic improvement is unlikely either in the immediate future or the medium term.

The main problem, as with the external imbalances, is basically located in the restructuring process which is underway in the transition economies. This process implies not only a massive internal reallocation of productive resources but also a generational change in production technologies. In some cases the ongoing reindustrialization may even result in skipping one or two generations of technology as compared with that inherited from the age of central planning (telecommunications is an example of this) and reflects the technological backwardness that emerged in many industries during decades of isolation from world markets.

The process of modernization is essential for the success of economic transformation: it brings about rising productivity, generally improving productive efficiency and a better allocation of resources, and is one of the fundamental factors for sustainable growth. In the end, it is the only way to improve the welfare of the population and, eventually, to catch up with the more advanced economies of western Europe.

However, these very same new technologies will not necessarily generate jobs which are comparable in number to the previous jobs that they are replacing. Besides, a number of industries and occupations in the transition countries are likely to be phased out altogether or greatly reduced, so that new job creation – to the extent that it emerges – is likely to be concentrated in other industries and, especially, in the tertiary sectors, which in general are still underdeveloped in the transition economies, and often in different regions.

These developments, however, pose serious challenges to policy makers in the transition countries, both as regards a persistently high level of unemployment and the probable emerging mismatch between the existing regional distribution and skills endowment of the unemployed and those of the new jobs in new activities.

Developments in 1996 and early 1997 also raised concerns about the resilience and sustainability of the process of economic transformation in some of the transition countries. The primary sources of concern were the first two major setbacks in the transition process: the emergence of a massive financial crisis in Bulgaria (see section 3.1(iii)) and the collapse of large-scale fraudulent investment schemes, leading to political and social disintegration in Albania. In both cases the crisis started with large-scale market failures leading to a rapid and more general deterioration of the economic situation and to serious political crises. Whereas a peaceful political solution leading to early parliamentary elections was found in Bulgaria, the unrest in Albania went out of control and led to a virtual disintegration of some basic state institutions. Turbulent developments – although less acute in nature – have also occurred in other south European countries: there were dangerous signs of macroeconomic destabilization in Romania in early 1997; and the financial system in Yugoslavia also appears to be in deep crisis.

(b) Short-term outlook

The outlook for 1997 is generally considered to be quite uncertain in the transition economies, but it is also widely differentiated.

In eastern Europe, there is an almost inverse relation between growth performance in 1996 and the expectations for 1997: an upswing is anticipated where growth was relatively weak, whereas in countries with strong (but in some instances weakly based) growth a slowing – or even a contraction, as in Romania, for special reasons – is expected. In the more advanced east European transition countries, a moderate slowing of growth is foreseen by most observers in the *Czech Republic*, *Poland* and *Slovakia*, whereas in *Hungary*, which has gone through a severe macroeconomic restructuring manoeuvre since early 1995, a pay-off in resumed if moderate growth seems to have begun in the last quarter of 1996 and is expected to continue; a similar pick-up is also expected in *Slovenia* (table 3.1.2). Aggregate GDP growth for this group could reach some 4½ per cent, slightly down from the 5 per cent registered in 1996. Continued growth of investment demand is seen as the main driving force of domestic demand in all countries, but in the Czech Republic, Poland and Slovakia private consumption spending also continues to rise. The impact of this on growing external deficits, and in some countries the rapid growth of real wages ahead of productivity gains, are areas of concern. In Hungary and Poland, further reduction of inflation is targeted by the governments, but nonetheless it is expected to remain in double digits.

In the south European transition countries, only *Croatia* appears clearly headed for stronger growth in 1997, with a further substantial contribution expected from the resumption of tourism; and subject to the significant condition of favourable political developments, there could also be continued recovery from the depressed levels of activity due to the Balkan conflict in *Bosnia and Herzegovina*, *The FYR of Macedonia* and *Yugoslavia*, perhaps even accelerated growth. (However, for these last three countries, but especially for the first, the lack of data still precludes meaningful analysis.) In contrast, substantial falls in GDP are expected in *Bulgaria* and *Romania*, and these could possibly turn out to be more significant than is currently officially anticipated (table 3.1.2). In both countries the falls reflect very strong macroeconomic adjustment and restructuring programmes, the full dimensions of which are not yet completely clear. Falling GDP is also likely in *Albania* (although, as above, the information gap permits only guesses), but under much less policy controlled conditions and hence with less favourable implications for the medium-term outlook. A very approximative estimate is that GDP growth for this group of countries in the aggregate is likely to be close to zero if not negative. Inflation is certain to be sharply higher than in 1996 in Albania, Bulgaria and Romania (in Romania as a result of a radical freeing of prices at the beginning of the year) – and probably Yugoslavia as well – and unemployment rates are also expected to rise steeply.

In the Baltic states, after the solid pick-up in 1996, some acceleration of growth is expected in all three countries, although with rates in the 3½-5 per cent range they are still not up to the pace seen in the central group of eastern Europe in the recovery phase.

Among the CIS countries, *Ukraine* apparently is still not at the turn around point for GDP growth even on government assumptions (table 3.1.2). In *Russia*, the government (and some outside observers) expect a modest upturn finally to occur in 1997, but this is still widely doubted by economic analysts in the country who, on average, would seem to see a standstill as an achievement. Much depends on the course of policy, in particular whether the problems of financial and fiscal arrears, which are at the origin of the weakness of domestic demand, can be successfully tackled without leading to another outbreak of inflation. In both countries unemployment is likely to rise further.

Other European CIS countries also appear to have made what seem to be rather optimistic assumptions in their budget or programme formulations for 1997 (5 per cent growth in Belarus, 5-7 per cent in the Republic of Moldova. In *Belarus*, where the official claim of 3 per cent GDP growth in 1996 is already considered doubtful, expansion at a rate of 5 per cent in 1997 is not very probable, if only because it would require a large rise in energy imports from Russia which, in 1996, had already led to large payment arrears and supply cutbacks on the part of Russia's Gazprom. Moreover, no support from the multilateral organizations can be expected in the current political situation, and there are worries about resumed inflation as much of the budget deficit (over 3 per cent of GDP on the high official growth assumption) is to be financed by the central bank. In the *Republic of Moldova*, the government forecast appears to be largely based on expectations of a substantial recovery in the weather-prone agricultural sector, not a firm basis for a forecast. On the other hand, the reform policies of the new government (after the end-1996 presidential elections) appear to be quite determined, in sharp contrast to Belarus, and the country can count on some international support. Nonetheless, in both countries a modest turn around, if much smaller than the official targets, can be expected.

In *Kazakstan*, the third in size of the CIS economies, the turn around to moderate growth is expected to be reached in 1997, and in several of the smaller CIS economies in the Caucasus and in central Asia, where activity had generally turned up in the course of 1996 – in some countries already in 1994 (Armenia) or 1995 (Georgia) – the recovery is expected to continue.

(ii) Monetary policies

(a) Overview of recent developments

The emphasis of monetary policies in the transition countries is changing. In the first phase of the transition process the main focus of monetary policy was on macroeconomic stabilization: the elimination of the macroeconomic disequilibria inherited from the past and coping with the macroeconomic effects of price and trade liberalization. With progress achieved in macroeconomic stabilization, policy goals have changed in the more advanced transition countries, the emphasis shifting towards the more conventional goals of maintaining a sustainable equilibrium path for the economy while at the same time gradually lowering inflation to rates comparable to those in the developed market economies. At the same time monetary policy in the advanced reformers is becoming increasingly sophisticated. The monetary authorities are diversifying both their goals and targets and are using a wider (though still limited) range of policy instruments to pursue their monetary goals; step-by-step fine tuning is gradually replacing ad hoc measures.

The fast reformers are also paying the "price of success" in that they are having to face the problems raised by large capital inflows attracted by the opportunities in these emerging markets. Another challenging issue is the growing external imbalance in most of these countries: current account deficits are already reaching proportions that cause concern to policy makers (see section 3.6). It is likely that the management of external imbalances is going to become another focus of monetary policy in the central European transition countries.

A number of transition countries (most notably the Czech Republic, Hungary, Poland, Slovenia and Croatia) have recently experienced large monetary capital inflows (see table 3.6.18). The nature and structure of these flows has differed from country to country. In the case of Hungary, privatization revenues (from foreign investors) and other private transactions (such as borrowing by local firms from abroad) constituted the most substantial monetary inflows. In Poland the largest flows were the so-called "unclassified transactions on the current account"[110] followed by FDI and portfolio investments. In the Czech Republic the capital inflow was dominated by portfolio investment and private transactions (borrowing by local firms from abroad).[111] In Croatia and Slovenia the most important components of the flows were private transfers from abroad: remittances of incomes earned abroad by residents and/or their relatives and repatriation of deposits held abroad (or currency held outside the banking system). Initially capital inflow to Estonia was dominated by FDI, but in 1996 there was also an increase in portfolio investments and loan capital.

The inflows provoked different reactions by the monetary authorities in these countries. Among the policy responses were changes in the exchange rate regimes, effectuated in 1995-1996 by the Czech Republic, Hungary, Poland and Slovakia. By introducing floating bands for the exchange rates, the four countries chose to increase the degree of volatility and risk on the foreign exchange market, a measure aimed at reducing inflows of speculative capital.

Another measure employed by the monetary authorities in most of the countries that experienced an upsurge of capital inflows was sterilization.[112] The efficiency of sterilization has often been questioned both as regards its anti-inflationary impact and as regards its costs. The experience of the transition

---

[110] In turn, the most significant component of this item in the Polish balance of payments is the exchange of foreign currency by foreigners for the purpose of making purchases in Poland and, in fact, it constitutes a large positive balance in cross-border trade.

[111] Before the introduction of the floating band there was, for a short period of time, an upsurge of short-term (speculative) flows into the Czech Republic; afterwards they declined substantially.

[112] The conversion of monetary inflows into domestic currency results in money creation. The monetary authorities may "sterilize" this effect outright by withdrawing the extra liquidity from the money market through open market operations. In this case the cost of sterilization is absorbed by the central bank. Another way of reducing the extra liquidity is to increase reserve requirements; in this case some of the costs of sterilization are channelled to the banking system.

countries so far does not provide any unequivocal conclusions on this matter. As discussed below, in a transitional environment there may be a number of additional sources of instability of money demand apart from inflows of foreign capital. On the other hand, and especially at the present level of interest rates in the transition countries, sterilization is always costly, as the central bank (and/or the commercial banks) incur the negative interest rate differential (vis-à-vis foreign financial markets) on the amount of money withdrawn from the domestic money market.[113] An alternative to sterilization is to increase the flexibility of the foreign exchange rate and it appears that this is one of the options being tried in some countries.

In 1996, the monetary authorities in several of the central European transition countries implemented a number of adjustments in their monetary policies while at the same time continuing to pay close attention to sustaining macroeconomic equilibrium.

During the course of the year the Czech National Bank introduced some important changes in its policy. In February the fluctuation band for the koruna was increased from ±0.5 per cent to ±7.5 per cent thus boosting the role of the market in the day-to-day setting of the exchange rate. This actually resulted in an outflow of short-term capital while the central bank intervention on the foreign exchange market was practically discontinued. At the same time monetary policy was tightened considerably. During the first half of the year the intervention repo rate of the Czech National Bank was increased in four steps from 11.25 per cent to 12.4 per cent; in June the refinancing rates of the central bank also increased and in August the minimum reserve requirement was raised from 8.5 per cent to 11.5 per cent.[114] As a result of all these measures the growth of the money supply slowed down considerably in the second half of the year (affected also by the outflow of short-term capital) whereas domestic investments in koruna time deposits increased. On average, the level of real domestic monetary stocks remained broadly unchanged during the course of 1996 (chart 3.1.1).

Since March 1995 Hungary has maintained a pre-announced crawling peg exchange rate system with a ±2.25 per cent band around the central rate.[115] The monthly rate of the crawl during 1996 was 1.2 per cent, effectuated through a 0.04 per cent daily correction. In February 1996 the regulation of mandatory reserves was modified by introducing uniform reserve requirements of 12 per cent (8.5 per cent for investment banks) and the elimination of previously existing exemptions.

The operational target, defined by the Hungarian monetary authorities, is "the maintenance of an interest rate differential ... which is consistent with the pre-announced exchange rate system".[116] Accordingly, monetary policy in Hungary has been targeting directly the level of domestic interest rates and, hence, money demand (rather than money supply). One of the delicate goals of this policy was to find a sustainable balance between stimulating domestic saving and discouraging excessive speculative capital flows.[117]

Intensive private capital inflows into Hungary continued in 1996 and they were partly sterilized by the central bank through open market interventions; however, the persistent strong demand for forints resulted in a real expansion of domestic monetary stocks, mostly in the form of personal and corporate interest-bearing forint deposits (see chart 3.1.1).

The Polish monetary authorities followed a very cautious and gradualist monetary policy in 1995-1996. The exchange rate regime in Poland is a crawling band similar to that in Hungary, but with a wider fluctuation band (±7 per cent since May 1995). Throughout 1996 the pre-announced rate of devaluation of the zloty amounted to 1.0 per cent a month (also effectuated on a daily basis). The National Bank of Poland has been pursuing a policy of gradually lowering refinancing rates in line with disinflation. In January 1996 the two key interest rates were reduced by 2 percentage points: the rediscount rate to 23 per cent and the Lombard rate to 26 per cent. This was followed in June by a further reduction of 1 percentage point in each rate.

Towards the end of the year, however, the monetary authorities reversed course and tightened monetary policy in response to signs of overheating in the economy and, especially, to the growing demand for credit by the corporate sector.[118] In November the National Bank of Poland raised its repo rates (while leaving the key rediscount and Lombard rates unchanged) which led to a general rise of interest rates. In addition, the mandatory reserve requirements for the commercial banks were also increased in February 1997.

---

[113] The technical implementation of sterilization requires that the monetary authorities have at their disposal sufficient quantities of securities (e.g. government bonds or treasury bills) to perform open market operations. Both in the Czech Republic and Poland, due to the lack of sufficient government securities, the monetary authorities had to issue special central bank bills to perform these operations.

[114] Czech National Bank, *Report on Monetary Development in the Czech Republic for the Period January-September 1996* (Prague), p. 27.

[115] The new exchange rate system was introduced in the framework of the 1995 stabilization package.

[116] National Bank of Hungary, *Recent Economic Developments in Hungary* (Budapest), September 1996, p. 22.

[117] National Bank of Hungary, op. cit., p. 24.

[118] Thus, while total money supply increased by 29 per cent in 1996 (December over December), total lending to the corporate sector during the same period grew by 42.7 per cent. Narodowy Bank Polski, *Informacja Wstepna*, No. 12 (Warsaw), 1996, p. 2.

CHART 3.1.1
Indices of real money supply and composition of nominal broad money in selected transition economies, 1995-1996

— Index of real money supply, domestic currency[a] (January 1995=100, left scale)[b]
--- Share of M1 in nominal broad money (per cent, right scale)
— Share of domestic currency in nominal broad money (per cent, right scale)

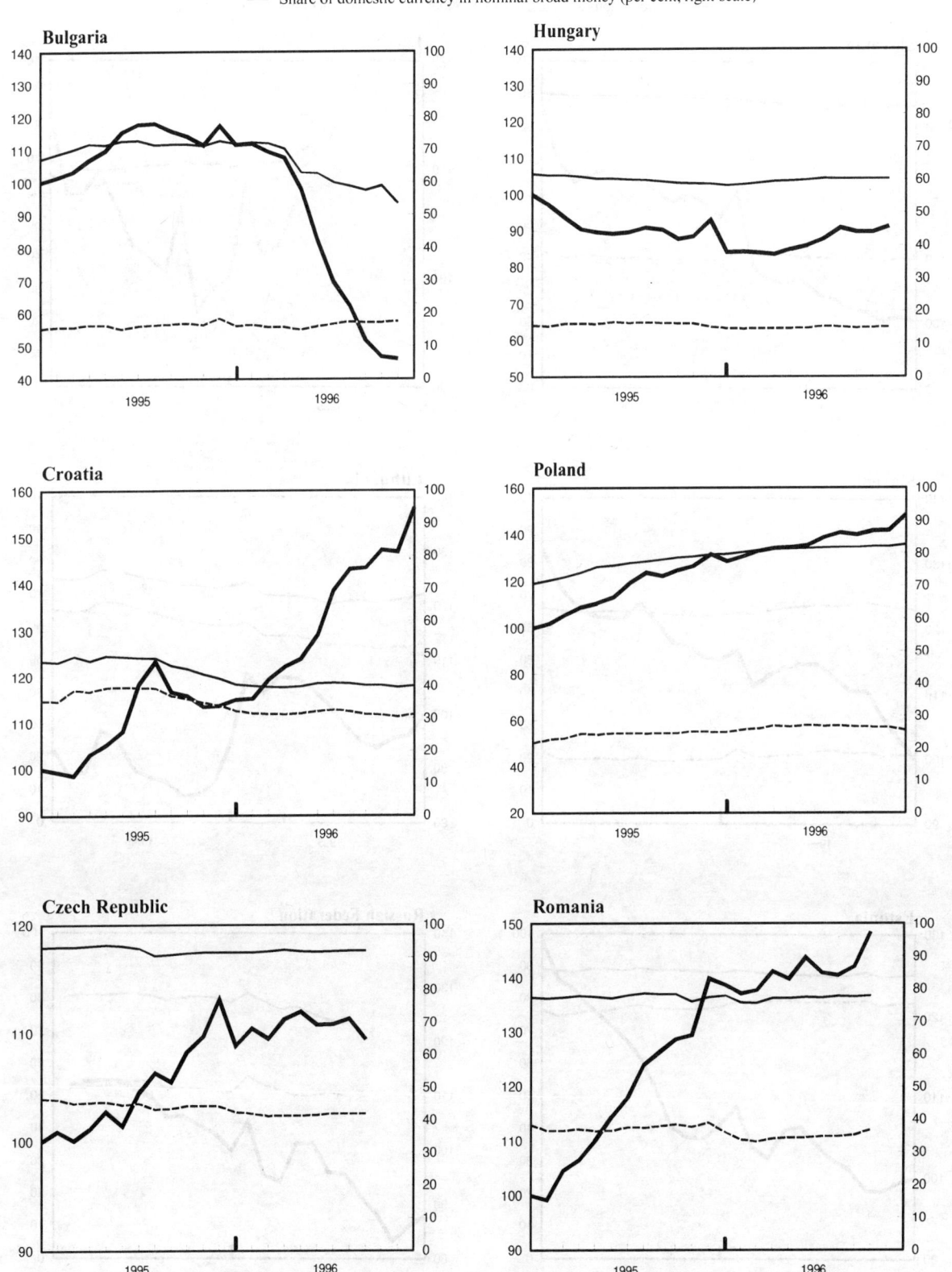

(For source and notes see end of chart.)

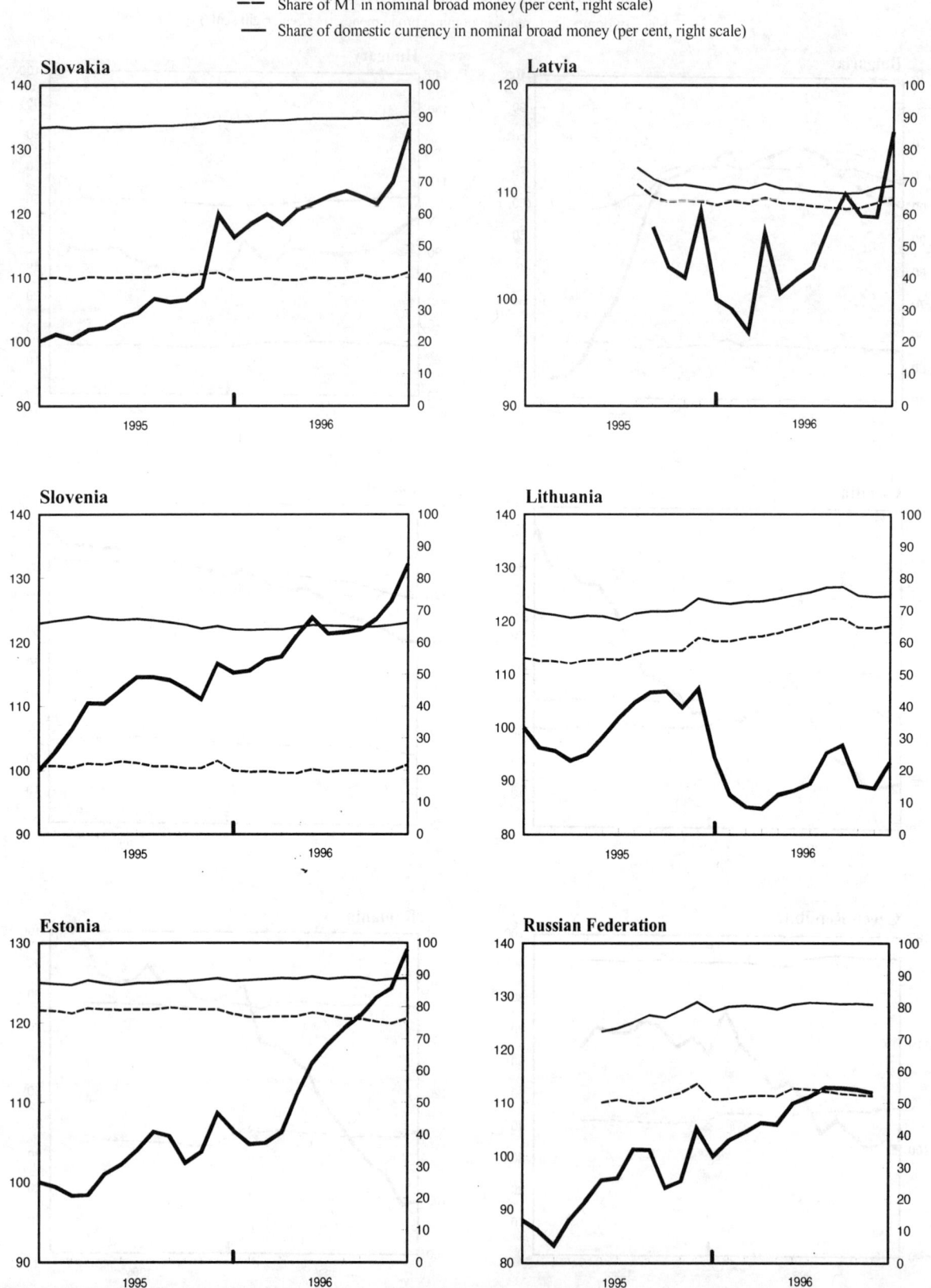

CHART 3.1.1 (concluded)
Indices of real money supply and composition of nominal broad money in selected transition economies, 1995-1996

Source: National statistics and direct communications from national statistical offices to UN/ECE secretariat; IMF, *International Financial Statistics* (Washington, D.C.), February 1997; Government of the Russian Federation/Working Centre for Economic Reform, *Russian Economic Trends*, November 1996.

a Nominal stocks deflated by CPI.
b For Latvia and Russian Federation, January 1996=100.

Slovakia has been one of the most successful among the transition countries in lowering inflation and this has been largely attributed to a carefully designed and implemented monetary policy. However, the deteriorating external balance prompted the National Bank of Slovakia to tighten its monetary policy in 1996. In July the central bank raised its Lombard rate from 13 to 15 per cent and increased the mandatory reserve ratio for domestic currency time deposits from 3 to 9 per cent. As of August, the fluctuation band of the koruna on the foreign exchange market was also increased from 3 to 5 per cent.

Monetary developments in Croatia and Slovenia in 1995-1996 broadly followed a similar pattern. Both countries maintain a flexible exchange rate system and have experienced substantial capital inflows; currency substitution is traditionally high. Despite the risk associated with these features both countries have made significant progress in disinflation. The monetary authorities in both countries have sought to sterilize the monetary inflows but their approach was somewhat different from that of the other central European economies: in both cases much greater emphasis has been placed on reserve requirements and on regulating the conversion of foreign exchange stocks into domestic currency. In addition, at an earlier phase in the transition the Bank of Slovenia had also engaged in open market interventions (a policy which was subsequently abandoned) whereas these were negligible in the case of Croatia.

The monetary authorities in Croatia and Slovenia have maintained relatively stringent reserve requirements and regulations on foreign exchange deposits, which constitute a large portion of the broad money supply in both countries (chart 3.1.1). Thus, in Slovenia, household sight deposits in foreign currencies at commercial banks must be backed 100 per cent by liquid foreign exchange deposits held by the banks abroad, the minimum reserve requirement for foreign exchange time deposits of up to 3 months being 75 per cent.[119] Reserve requirements are also rather high in Croatia.

After nearly two years of disinflation and seeming stability, the macroeconomic situation in Romania rapidly deteriorated in the early months of 1997: within less than a month, in January-February, the nominal exchange rate of the leu depreciated by more than 100 per cent; consequently, inflation started to pick up rapidly. This was, however, the result of macroeconomic pressures and imbalances which had been accumulating during the last two years and, especially, in 1996 as a result of lax and incoherent policies.

During the last two years, and especially in 1996, the authorities fostered monetary expansion and boosted incomes in the pursuit of demand-driven growth (see sections 3.2 and 3.4); in addition, large amounts of subsidized credit were allocated to the agricultural sector. On the other hand, distortive interventions on the foreign exchange market[120] led to an artificial boosting of the official exchange rate of the leu, but at the same time a parallel market emerged, quoting an exchange rate at some 15-20 per cent below the official rate. Administrative price controls also led to significant price distortions. The external balances of Romania also deteriorated, and towards the end of 1996 macroeconomic imbalances were reaching such dangerous proportions that a major macroeconomic adjustment was required.[121]

In January 1997 a full-fledged foreign exchange market was re-established in Romania and the exchange rate was liberalized (together with a broader programme of price liberalization and in the context of a major macrostabilization effort by the Romanian authorities). At the same time the national bank undertook drastic measures to tighten monetary control: central bank refinancing was cut and refinancing rates were raised to unprecedently high levels in an attempt to arrest monetary expansion and boost confidence in the currency.[122] It remains to be seen whether and to what extent the new Romanian stabilization programme will achieve its goals; in any case, the painful but necessary macroeconomic consolidation will require a persistent policy effort.

Among the Baltic states, Estonia and Lithuania maintain a currency board which limits substantially the freedom to conduct an independent monetary policy (see box 3.1.1).[123] Latvia has maintained a fixed exchange rate regime and, accordingly, the focus of monetary policy there has been on interest rate management. The Bank of Latvia intervenes on the money market by setting its refinancing rate and by open market operations. It also intervenes on the foreign exchange market at the fixed rate.

---

[119] OECD, *1996-1997 Economic Review – Slovenia* (Paris), 1997.

[120] In March 1996 the National Bank of Romania withdrew the licences of most participants on the interbank foreign exchange market, leaving only four large banks to operate in the market and thus creating conditions for its easy manipulation. Following this measure, the IMF froze disbursements of the previously agreed stand-by loan to Romania. In August the authorities initiated an even more controversial measure, compelling some 100 large companies to settle their energy bills in foreign exchange, a measure which would have effectively confiscated the foreign exchange earnings of these companies (this decision was revoked soon afterwards).

[121] For example, Romania's total fiscal deficit (including the quasi-fiscal deficit resulting from subsidized credits and other implicit subsidies) in 1996 is estimated at some 13 per cent of GDP. Statement by Prime Minister V. Ciorbea, *Financial Times*, 18 February 1997.

[122] At the weekly auction on 20 February the one-week refinance rate surged to 226 per cent and a week later, on 27 February, to 427 per cent. CS First Boston, *Emerging Economics Research – Europe*, 5 March 1997.

[123] Under a currency board policy instruments are constrained but not completely impotent. For example, the monetary authorities can affect monetary circulation through reserve requirements, prudential regulations (such as capital adequacy and liquidity requirements), and banking supervision.

BOX 3.1.1

Currency boards

A currency board is a central bank operating under a specific set of exchange rate and monetary arrangements. Under a currency board regime the exchange rate of the domestic currency is fixed to a stable reserve currency (and this is regarded as a long-term obligation) and the monetary authorities undertake to maintain domestic base money (the sum of currency notes in circulation and commercial banks' reserves at the central bank) at a level equivalent to the official foreign exchange reserves. Consequently, changes in the level of foreign exchange reserves are automatically reflected in changes in the level of high-powered domestic money.

Due to the full backing of the monetary base, domestic currency can be exchanged without limitation for foreign currency. However, this does not apply directly to deposits denominated in domestic currency: deposited money must be first withdrawn and then the cash can be converted into foreign exchange.

Compared with other exchange rate and monetary regimes, the currency board is the least sophisticated as under it the central bank has no direct control over any policy instrument (table 3.1.6). Most importantly, with a currency board the monetary authorities have no discretionary power to create base money and thus the central bank de facto has no sovereignty to pursue an independent monetary policy.

A number of important implications and restrictions stem from the loss of monetary sovereignty. One is that fiscal deficits cannot be financed by direct borrowing from the central bank: the government can finance a deficit only to the extent that it can sell debt financing securities on the market. Another serious implication is that the central bank cannot perform open market operations and cannot act as a lender of last resort. Thus a currency board implies a tight financial discipline at both the macro- and micro-level.

However, the restrictions of a currency board are at the same time its virtues. On the positive side, apart from the unlimited convertibility of domestic cash money, a currency board introduces full transparency and predictability of the monetary and exchange rate regimes which allows economic agents to commit themselves to longer planning horizons. The tight fiscal and monetary discipline contribute to macroeconomic stabilization as domestic prices gradually converge to prices in the reserve currency country. Monetary and financial discipline also boost public confidence in financial institutions.

Although a currency board sets rigid controls over the level of domestic high-powered money, it does not directly restrict the growth of other monetary aggregates (such as M1) which are linked to base money through money multipliers. For a given level of base money an increasing level of bank deposits denominated in domestic currency (resulting from increased public confidence in the banks) will lead to an expansion of M1 and thus provide more credit to the economy.

Historically currency boards were the typical monetary regimes in former colonies throughout the world and after the Second World War they gradually started to disappear. For some time a currency board was regarded as just a rudimentary form of central banking closely related to the colonial system and which was dying away together with the colonial system itself.

The rebirth of the concept was marked by the reintroduction of a currency board in Hong Kong in 1983 (after a currency crisis) and, especially, after the establishment of a currency board in Argentina in 1991. In the case of Argentina, the currency board was introduced as a central policy instrument in a full-scale stabilization programme, after several previous stabilization efforts had failed.

After the demise of communism several European transition countries turned to the currency board as a policy tool to stabilize their economies. In Europe a currency board was first introduced by Estonia in 1992 together with the launching of its new national currency. Lithuania established one in April 1994 with the explicit goal of macroeconomic stabilization through monetary and fiscal discipline.[1] Following the Dayton Accord in 1996, a currency board is to be introduced also in Bosnia and Herzegovina.

In Bulgaria the proposal to set up a currency board, as part of a strict monetary and fiscal programme to stabilize the economy, was made by the IMF in discussions with the Bulgarian government during the latter half of 1996. Although the interim government is empowered to open negotiations with the international financial institutions, the eventual introduction of the currency board is expected to take place only after the parliamentary elections in mid-April.

---

[1] At the time of writing this *Survey* there was a debate in Lithuania regarding the eventual replacement of the currency board by the conventional instruments of national monetary policy. Having re-established credibility in monetary policy and the exchange rate, the new coalition government is expected to maintain a stable exchange rate but to use the wider scope for monetary policy to offset the possibly destabilizing effect of large foreign capital inflows which are likely to be attracted by the government's new privatization programme.

In all three countries, following a relatively stable macroeconomic performance in recent years, interest rates have been steadily declining. In Latvia this process was led by the central bank: during 1996 in nine consecutive steps, the Bank of Latvia reduced its refinancing rate from 24 per cent in January to 9.5 per cent in December.[124] In January-February 1997 it was reduced further to 8 per cent. The monetary authorities in Estonia and Lithuania do not intervene on the money markets directly; thus the decline in interest rates in these two countries (which was similar to that in Latvia) reflected their newly established monetary equilibrium.

At the same time, there are a number of dissimilarities in the monetary performance of the three Baltic states (chart 3.1.1). Developments in Estonia in 1996 were affected by continuing large inflows of foreign capital which resulted in an increase of domestic monetary stocks. Real monetary stocks also grew in Latvia but this was mainly caused by domestic money creation. In contrast monetary stocks in Lithuania fell dramatically at the beginning of 1996 as a result of withdrawals of deposits in the wake of the major banking crisis that shook the country at the end of 1995.

Policy makers in Lithuania have been debating recently the efficiency of the currency board arrangement as well as the current pegging of the litas to the dollar. It is envisaged to gradually change the existing monetary arrangements in Lithuania with the eventual goal of dismantling the currency board (but not before 1998) and re-establish the monetary functions of the Bank of Lithuania. The plans also include a switch from the current peg of the litas to the dollar to either a basket of currencies or to the euro.

Recent monetary performance in Russia has been marked by a number of contradictory developments. In February 1996 Russia embarked on an ambitious stabilization programme supported by a three-year IMF agreement on an Extended Fund Facility (EFF) loan of $10.1 billion. The programme's targets included, *inter alia*, a substantial reduction of inflation, increasing public confidence in the rouble, and a lowering of the budget deficit. Policy was largely based on tightening the money supply. Changes in exchange rate policy were also introduced: the fixed (and adjustable every six months) fluctuation band of the rouble was replaced in mid-1996 with a crawling band with a pre-announced rate of the crawl.[125]

During 1996 the Russian monetary authorities followed closely the monetary targets set in the programme. The Central Bank of Russia tightened considerably its control over the money supply by substantially reducing credit to the government and refinancing of commercial banks in line with the quantitative monetary targets.[126] In addition, the central bank intervened on a number of occasions to defend the floating band of the rouble and maintain the credibility of the exchange rate regime. These measures obviously boosted public confidence in the Russian stabilization policies and contributed to the notable deceleration of inflation.

However, monetary developments in Russia are complex and sometimes perplexing. Despite the notable stabilization of the rouble, public confidence in the domestic currency (and in domestic financial institutions) is still not very high, an indication of which is the massive capital flight.[127] The continuing capital drain – which is of a scale and persistence that are unique among the transition countries – is in fact one of the causes of the severe demonetization of the Russian economy and this sets it apart from the rest of the transition countries (see table 3.1.5).

Monetary austerity in 1996 added to the negative implications of demonetization and provoked a number of side effects. Austerity kept real interest rates in Russia very high (despite the fall in their nominal level) and contributed to the continuation of the transformational recession in 1996.[128] In turn, it also increased the costs to the government of financing its budget deficit. Also, due to the liquidity shortage in the banking system, the government found it difficult to place securities on the money market. Consequently, in August 1996, some of the restrictions on the domestic money market were relaxed, allowing easier access of foreigners to certain categories of government securities. While this provided relief to the problems of servicing the fiscal deficit, the interest costs to the budget were very high.[129] The incoming

---

[124] Bank of Latvia, *Monetary Policy Highlights – 1996* (Bank of Latvia, internet website).

[125] The de facto average monthly nominal depreciation of the rouble in the second half of 1996 amounted to 1.5 per cent (Government of the Russian Federation/Working Centre for Economic Reform, *Russian Economic Trends, Monthly Update* (Moscow), 21 February 1997, p. 4). The target set for 1997 is 9 per cent annual nominal depreciation which is equivalent to monthly rate of the crawl of 0.7 per cent.

[126] These targets were specified in the stabilization programme as ceilings on the net credit exposure of the Central Bank of Russia.

[127] According to some estimates, some $60-$80 billion left Russia between 1993 and 1996; in just the first half of 1996 capital flight amounted to about $20 billion. Government of the Russian Federation/ Working Centre for Economic Reform, *Russian Economic Trends*, Vol. 5, No. 3 (Moscow), 1996, pp. 41, 100.

[128] During the second half of 1996 the real yield on GKOs (government treasury bills denominated in roubles), which is in fact the floor on interest rates in the Russian economy, was estimated at some 35-40 per cent per annum. Ibid.

[129] Under the crawling band (defended by the central bank) and the high level of domestic interest rates, there exists a very high interest rate differential which makes GKOs attractive to foreign capital but costly to the budget. The liberalization of the securities market induced substantial inflows of portfolio capital into Russia: by January 1997 (less than six months after the opening of the market) foreigners already held

TABLE 3.1.5

Monetization in selected transition countries: share of monetary aggregates[a] in GDP, 1994-1996

(Per cent)

	$M1^b$			Broad money, domestic currency[c]			Total broad money[d]		
	1994	1995	1996[e]	1994	1995	1996[e]	1994	1995	1996[e]
Albania	..	20.3	..	..	30.6	..	..	37.5	..
Bulgaria	10.9	9.4	6.8	43.3	40.9	25.7	63.4	57.5	41.0
Croatia	6.6	8.2	9.9	..	10.3	12.6	..	22.1	31.3
Czech Republic	36.0	33.0	31.7	69.2	68.2	70.0	74.9	73.8	76.1
Hungary	20.6	16.8	15.1	33.3	28.4	26.9	53.5	47.2	45.0
Poland	7.2	7.6	9.2	22.0	23.8	29.1	30.8	31.3	35.6
Romania	5.8	6.9	6.8	10.1	14.3	15.1	13.9	18.5	19.6
Slovakia	25.2	24.1	25.6	51.8	52.2	57.2	59.5	59.8	64.0
Slovenia	7.2	8.0	8.0	20.9	25.1	26.6	33.5	37.8	41.0
Yugoslavia[f]	8.1	7.2	..	9.8	9.5	..	42.2	35.1	..
Estonia	17.6	17.1	18.3	19.9	18.9	21.3	21.7	21.5	23.7
Latvia	15.0	15.7	12.9	..	..	13.8	29.5	23.8	20.2
Lithuania	12.2	11.8	11.1	15.8	14.6	12.9	21.6	21.1	17.4
Russian Federation	8.0	6.9	7.4	10.1	9.4	11.3	14.5	13.6	14.1

Source: National statistics and direct communications from national statistical offices to UN/ECE secretariat; IMF, *International Financial Statistics* (Washington, D.C.), various issues; Government of the Russian Federation/Working Centre for Economic Reform, *Russian Economic Trends* (Moscow), November 1996.

[a] Averages of monthly or quarterly figures.

[b] Currency in circulation plus demand deposits.

[c] M1 plus time deposits in domestic currency.

[d] Broad money in domestic currency plus foreign currency deposits.

[e] For 1996 the monetary aggregates refer to January-September for the Czech Republic; January-October for Romania; January-November for Bulgaria, Hungary and the Russian Federation; the full year for Croatia, Estonia, Latvia, Lithuania, Poland, Slovakia and Slovenia. GDP data for 1996 are estimates.

[f] For Yugoslavia, gross material product; 1995, estimates.

portfolio capital also raises some concern as it is likely to be short term and speculative in nature. As discussed below, an eventual change in the market environment may reverse money demand which, in turn, may have highly destabilizing macroeconomic effects.

One of the most distortive side effects of demonetization and monetary austerity was the mushrooming of payment arrears. In 1996 the growth of non-payments in Russia was endemic and is widely considered to be the most serious financial problem of the economy.[130] Even on conservative estimates, the stock of arrears in Russia at the end of 1996 amounted to almost twice the stock of total nominal broad money (domestic currency) in the Russian economy.[131] This is an indication of a serious degree of suppressed money demand which is hardly instrumental to the normal functioning of the economy.[132] And, as a matter of fact, the vacuum in formal monetary exchange has led to an outburst of money surrogates (mostly promissory notes), which are flooding the Russian economy and creating even more disruption to the normal monetary circulation.[133]

Obviously, this is not a sustainable situation over the medium and longer run and adjustments will have to be made to restore monetary equilibrium. But, great care will be needed in remonetizing the economy, as the process will carry the potential danger of an inflationary outburst.

---

about one third of the GKO market or some $13 billion of debt instruments. Government of the Russian Federation/Working Centre for Economic Reform, *Russian Economic Trends, Monthly Update*, 21 February 1997 (Moscow), p. 4.

[130] In spite of the extensive public debate on payment arrears in Russia, the quantitative magnitude of the problem remains confused by the lack of reliable statistics. According to Goskomstat, at the beginning of December 1996 the total amount of arrears in the Russian economy was R552 trillion (Goskomstat Rossii, *Sotsial'no-ekonomicheskoe polozhenie Rossii* (Moscow), 12 December 1996). Other estimates put the figure as high as R1,100 trillion (*Izvestiya* (Moscow), 30 January 1997).

[131] Non-payment in Russia has taken a variety of forms. Undoubtedly, the most widely publicized and politically most sensitive is the non-payment of wages. However, in spite of their major social and political impact, wage arrears are only a relatively small share of the total stock of non-payments. According to Goskomstat estimates quoted above, in December 1996 wage arrears accounted for only 9 per cent of total arrears.

[132] This is a somewhat paradoxical development given the huge monetary overhang which existed in Russia before the start of economic transformation.

[133] See Yu. Latynina, "Den'gi po-rossiiski", *Izvestiya* (Moscow), 30 January 1997.

(b) Some issues in monetary policy: coping with unstable money demand

The introduction and conduct of a consistent and coherent monetary policy has been one of the centrepieces of the stabilization effort in the transition countries. The success in achieving macroeconomic equilibrium and disinflation by targeting monetary goals critically depends on the actual causal links between macroeconomic variables in the economy and on the actual impact of the monetary policy instruments used.

Recent macroeconomic developments in some transition countries have revealed new and specific features of the transitional environment which pose serious challenges to the monetary authorities and to policy makers. On the one hand, some of the countries more advanced in the transition process faced unexpected and large inflows of foreign capital which created additional problems in the financial markets in these countries. On the other hand, recent developments in some of the south-east European countries, such as the financial crisis in Bulgaria (discussed in more detail in section 3.1(iii)), the collapse of financial pyramids in Albania and the macroeconomic destabilization in Romania, indicate the fragility of macroeconomic equilibrium in the transition economies and its inherent vulnerability to financial disturbances. These developments point once again to the imperative of carefully matching the design of monetary policy to the specifics of the transitional environment which often fail to meet the assumptions of textbook economic models.

While mainstream economic theory suggests the existence of a causal connection between the quantity of money and inflation, the actual relationship and the transmission mechanism involved are by no means straightforward.

One of the conventional operational assumptions both in theory and in practice has been that of the constant velocity of money circulation which, at a given level of real output, translates into a direct proportionality between the quantity of money in circulation and the rate of inflation. Although theoretically convenient, this assumption has been questioned by a number of empirical studies even in mature market economies. It is all the more questionable in the highly volatile macroeconomic environment of the economies in transition. Indeed, a previous issue of this *Survey* found quite a weak correlation (in some cases non-existent and in others inverse) between changes in the money supply and in the consumer price index during the early phase of the transition process.[134]

A possible interpretation of this instability in the relation between money stocks and inflation in the transition countries is the highly unstable demand for money in a transitional environment. Recent changes in real monetary aggregates (chart 3.1.1) clearly demonstrate the variability of money demand in the transition countries. One obvious reason for these substantial movements in the real money supply in some of the transition countries in recent years is the inflow of foreign capital. Inflows (outflows) of capital may result in increases (decreases) in the level of domestic monetary stocks.[135] However, while capital movements may give a clue to some of the observed trends they are not sufficient to explain the large variation of monetary aggregates seen in the transition countries. Moreover, the rates of change of real money stocks in some countries over relatively short time spans are clearly incompatible with the conjecture of a constant income velocity of money circulation.

There are a number of factors that contribute to the instability of money demand in the transition countries. First of all, due to the underdeveloped (and virtually non-existent at the outset of transition) capital markets, the function of money as a store of wealth is rather important. But also, due to the scarcity or lack of alternative or sufficiently attractive investment opportunities, a large share of personal (and even corporate) savings is held in different forms of money balances. The most important monetary components which perform the functions of savings and/or store of wealth are the interest-bearing, domestic and foreign currency deposit accounts in the banking system.[136] Apart from the underdevelopment of capital markets, the share of wealth stored in such monetary assets presumably reflects the confidence of the public in the banking system. On the other hand, the composition of the broad money stock is largely influenced by tradition, investors' preferences and the opportunity costs of holding wealth in other forms or of investing in other types of assets.

One specific implication of this situation is the fact that the broad money stock is quite heterogeneous and the demand functions for its subcomponents may be quite different in nature. In fact there exists strong

---

[134] UN/ECE, *Economic Survey of Europe in 1994-1995*, pp. 169-170.

[135] Not all capital flows affect the level of monetary stocks. For example in-kind FDI has no monetary effect whatsoever; capital inflows which are not converted into domestic currency do not affect domestic monetary stocks but only broad money. The most sensitive capital flows are those that are converted into domestic currency such as portfolio investment, credit from abroad, which is spent in the country, and speculative short-term capital.

[136] Unfortunately, the available statistical information does not allow a precise assessment of the actual magnitude of this phenomenon but there are good reasons to believe that the share of total wealth stored in the form of money balances in the transition economies is on average much higher than in the case of mature market economies.

complementarity in the demand for different subcomponents, similar to that in a closed demand system subject to budget constraints. For a given budget constraint (total wealth) investors may rearrange their portfolios on the basis of the expected returns to different monetary investments, implying the existence of direct and cross-price elasticities. A rearrangement of portfolios may also result from changes in investors' preferences regarding the functions of investment and store of wealth.

Another implication, which follows directly from the above, is that money demand and, hence, velocity may be rather sensitive to changes in the own rate of interest on money. This is precisely the area which needs careful scrutiny as the own interest rate on money is by and large one of the main monetary policy instruments. The controversy arises because the monetary authorities tend to treat the own interest rate on money mostly as a tool to control money supply, whereas at the same time it has a very strong impact on money demand. For example, high real domestic interest rates (which policy usually aims at reducing money supply) may lead at the same time to rising demand for domestic currency assets for investment purposes. This may trigger reverse currency substitution and a rearrangement of the investment portfolios of domestic investors. In addition, the emergence of interest rate differentials with respect to international financial markets will tend to attract inflows of speculative capital (which is another sign of unstable money demand). The capital inflow, in turn, leads to expansion of the domestic money supply, a result which is exactly the opposite to what the monetary authorities might have aimed at by maintaining high real interest rates. If the monetary authorities try to stick to predefined monetary targets, they will need to sterilize the capital inflows which, however, may be costly. On the other hand, sudden and abrupt reversals in interest rate policy are likely to induce reverse changes in money demand which, in turn, may result in destabilizing shocks to the economy (see section 3.1(iii)).

The extent to which unstable and changing money demand directly translates into inflationary pressures depends on the depth and the structure of financial markets as well as on the existence of mechanisms linking asset prices to those of goods and services. Changing demand for specific types of monetary assets and the rearrangement of investors' portfolios naturally lead to changes in asset prices. To the extent that the pricing of the assets affected by changing money demand is exogenous with respect to the prices of goods and services (as are, for example, – at least in the short run – the prices of bonds and real estate), there may be little effect on inflation in the short run.

There is, however, one important exception: domestic assets denominated in foreign exchange (which, as noted, have a significant weight in investors' portfolios in many transition countries). In this case, depending on the exchange rate regime, there may be endogenous links between the formation of the prices of monetary assets and those of goods and services. Indeed, to the extent that the exchange rate is allowed to float (e.g. in a completely flexible exchange rate regime or where the exchange rate is pegged within a band) changes in the demand for assets denominated in foreign currency will induce changes in the exchange rate and this, in turn, will translate into inflationary or deflationary pressure. The degree of proliferation of such effects will depend on the existence of built-in stabilizers in the economy capable of effectuating a corrective feedback to bring the system back into equilibrium. Again, as indicated by the experience of some transition economies, such safeguards may not be in place, thus creating the danger that the destabilization of money demand might lead to a run on the currency and a financial crisis.

In the circumstances facing a transition economy, the monetary authorities may in fact be faced with contradictory goals and outcomes. They often need to strike a delicate balance, especially in coordinating exchange rate and interest rate policies, so as to avoid destabilizing disturbances caused by abrupt changes in money demand. This may not be an easy task in a situation where capital markets are underdeveloped and when the available policy instruments are limited.

Some of the conjectures outlined above are supported by data on the composition of the nominal stocks in the transition countries (chart 3.1.1) as well as in statistics on the monetization of the transition economies (table 3.1.5). The latter reveal important cross-country differences both in terms of the nature and dynamics of money demand and in terms of the stance of monetary policies.

One apparent source of variation is the exchange rate regime which implies different levels of discretion over the tools of monetary policy and, indeed, different policy goals under different exchange rate regimes (see table 3.1.6).[137] The countries that apply a fixed exchange

---

[137] The ECE transition countries are applying a wide variety of exchange rate regimes at the beginning of 1997. Among the east European countries the Czech Republic and Slovakia adhere to a fixed exchange rate (with a floating band since 1995); since 1995 Hungary and Poland have maintained a crawling band (before that the regime was a crawling peg); the Yugoslav regime is a de facto adjustable peg, whereas the currency is allowed to float in Albania, Bulgaria, Croatia, Romania and Slovenia. The Baltic states stick to fixed exchange rate regimes (and among them Estonia and Lithuania have introduced currency boards). From mid-1995 Russia switched from a floating exchange rate to a crawling band whereas most of the other newly independent CIS countries adhere to flexible exchange rate regimes. In February 1997, Belarus announced a decision to switch to a regime of a fixed exchange rate with a band.

TABLE 3.1.6

Constraints on monetary policy instruments under different exchange rate and monetary regimes

Exchange rate/monetary regimes	Types of policy instruments		
	Pre-announced commitments of the monetary authorities on	Policy-controlled instruments (under the discretion of the monetary authorities)	Residual (no or limited discretion of the monetary authorities)
Flexible exchange rate	–	Money supply, interest rate	Exchange rate
Crawling peg; fixed exchange rate	Exchange rate	Interest rate	Money supply
Currency board	Exchange rate, money supply	–	Interest rate

Source: UN/ECE secretariat.
Note: Stylized picture.

rate regime (the Czech Republic, Slovakia and the Baltic states) on average display a higher proportion of M1 in GDP (table 3.1.5) and, in most cases, a lower share of foreign exchange deposits in broad money (which is defined by the area between the domestic currency curve and the 100 per cent line in the country figures in chart 3.1.1)[138] both of which reflect the lower foreign exchange risk under a fixed exchange rate.[139]

In fact, the share of domestic currency in the broad money supply (chart 3.1.1) can also be regarded as an indication of the degree of confidence in the national currency.[140] Apart from the Czech Republic, Slovakia and Estonia, confidence in the currency, by this measure, was also increasing in 1995-1996 in Poland and Lithuania and remained more or less unchanged in Hungary and Slovenia. In relative terms, currency substitution is still quite high in Croatia[141] and Slovenia, which partly reflects a historic tradition, whereas it increased in Latvia (after the 1995 banking crisis) and, especially, in Bulgaria (as a result of the deep financial crisis which hit the country in 1996).

The area between the M1 and the domestic currency curves in chart 3.1.1 measures the share of interest-bearing monetary assets denominated in domestic currency in the total money stock and indicates the relative importance of these types of assets in investors' portfolios. The cross-country variation in the shares of such assets reflects, on the one hand, the rate of return on these assets and, on the other, the degree of diversification of the banking products available on the financial markets of the different countries.

On average, the degree of monetization (measured by the quantity of money per unit of output) in most of the transition economies – with the exception of the Czech Republic and Slovakia – remains rather low (table 3.1.5).[142] There are also puzzling differences both in the level of monetization across countries and in its rates of change in recent years. Besides, there appears to be no systematic correlation between the level of monetization and the rates of change of inflation: deceleration of inflation was compatible both with increasing and with decreasing levels of monetization in different countries; the same held for accelerating inflation.

This is another indication of the continuing monetary turbulence in the transition countries which highlights the importance of cautious and consistent macroeconomic policies. Obviously, it is very hard to judge the overall success of monetary policy in such an environment, but common sense suggests that a policy can be regarded as successful if it is sustainable in the medium and long run in terms of outcomes and costs and if it leads to the strategic goal of achieving a stable macroeconomic equilibrium.

(iii) The crisis in Bulgaria

Although considered a laggard in the process of systemic reform, until 1996 Bulgaria's economic performance was following the more or less typical path of the transition economies in starting with a period of depression and inflation followed by a gradual recovery of output and disinflation. The crisis which hit the country in 1996 – a drop of GDP by some 10 per cent and

---

[138] This can also be regarded as a measure of the degree of currency substitution.

[139] Actually, under a fixed exchange rate regime the liquid component of the money stock (M1) can also be considered as a part of the investors' portfolios as it guarantees the store of value of the money balances to the extent that it is preserved in the reserve currency or currency basket.

[140] Note that this proposition is only valid if at the same time public confidence in the banking system is sufficiently high and if there is no capital flight (i.e. if financial assets owned by residents stay in the banking system).

[141] The Croatian data shown in the chart only refer to "new" foreign exchange deposits (made after 1992) and do not include the old foreign exchange accounts (from the era of the SFR of Yugoslavia) which were effectively frozen in 1990. At end-1996 the total amount of frozen "old" accounts was roughly one third of the total value of "new" ones.

[142] In 1995 in the EU, average shares of the corresponding monetary aggregates (measured as the arithmetic average for the EU member states except Luxembourg) were: M1/GDP = 24.3 per cent; Broad money/GDP = 69.9 per cent. Moreover, the EU averages have been quite stable over time: in 1990 they were 25.4 per cent and 63.6 per cent, respectively. UN/ECE secretariat calculations, based on IMF, *International Financial Statistics* (Washington, D.C.), February 1997 and UN/ECE, *Economic Survey of Europe in 1995-1996*, p. 168.

year-on-year inflation exceeding 300 per cent – was a major deviation from the prevailing pattern of developments in the other transition countries. Although this is the first case of a major reversal in the transition process, it does sound a warning that the process of economic transformation may still be rather fragile and vulnerable to upsets and that stabilization and recovery are not necessarily irreversible.

Bulgaria's painful recent experience deserves closer attention in order to determine whether it is a unique phenomenon among the transition economies or whether the underlying factors of the crisis could emerge in other countries as well.

(a) The genesis of the crisis

Bulgaria started the process of economic transformation from a much more unfavourable starting point as compared with other central and east European countries; its disadvantages included an exceptionally high level of dependence on CMEA (especially Soviet) markets; an unsustainable level of foreign debt (which subsequently led to a default in 1990); severe domestic macroeconomic imbalances; remoteness from the major west European markets and greater vulnerability to negative externalities (such as United Nations sanctions on Yugoslavia). Because of its considerable dependence on Soviet markets the structure of the Bulgarian economy was probably more similar to the economies of some of the ex-Soviet republics than to those of other central and east European countries. Consequently, the impact of the breakup of the CMEA on the Bulgarian economy resembled the shock to the successor states of the USSR from its disintegration. However, while the successor states of the Soviet Union started their new-found independence practically debt free (Russia inherited the entire Soviet debt) and could start the process of economic transformation with clean balance sheets and currency reforms, its severe debt burden limited considerably the policy alternatives open to Bulgaria at the outset of transition.[143]

Economic transformation started in Bulgaria in 1991 when a stabilization programme was launched together with wide ranging price liberalization, the opening up of the market to private economic agents, and the liberalization of foreign trade. A new constitution and a number of legislative acts paved the way for the creation of new institutions. However, political instability, a lack of public consensus behind the reforms, and stop-go policy measures led to a marked deceleration in the pace of reform in subsequent years. In general, the first phase of economic transformation in Bulgaria was characterized by a slow rate of reform and erratic economic policies. Ambivalent attitudes towards privatization caused major delays in its implementation and by 1996 only a handful of large state owned firms had been privatized. The state owned sector still contributed about 85 per cent of industrial output in 1996. Nominally, international assistance to Bulgaria was not negligible, but the results were disappointing as most of the programmes it supported stalled in the course of their implementation.[144] With hindsight, the amount of external assistance (both financial and technical) required to have a significant impact on Bulgaria's inherited problems was probably much greater than was actually offered.

The loss of over 50 per cent of Bulgaria's total exports over the course of 1989-1991, mostly from manufacturing industry, meant that a large number of industrial state owned firms started to experience serious financial problems. As is now much clearer, the loss of markets was not temporary but permanent: a number of industrial enterprises were designed to supply only the Soviet market and subsequently proved to be unfit for restructuring and survival in market conditions. Most of these firms will therefore probably have to be closed down. This is a fairly general phenomenon among transition countries; however, the magnitude and the acuteness of the problem appear to have been more severe in Bulgaria than in any other central and east European country.

Until 1996 policy makers in Bulgaria failed to confront the complexity and gravity of this problem. Of course, from their point of view, they were faced with a daunting task and subject to serious political constraints, not the least of which arose from the sheer magnitude of the adjustment effort required of the population. It amounted to no less than the closing down of large parts of Bulgarian industry plus active restructuring and privatization of what remained. However, the reluctance to address these major restructuring problems resulted in their complete neglect.

Moreover, no effort was made until 1996 to impose hard budget constraints on the operation of state owned firms: the authorities tolerated weak financial discipline and the accumulation of payment arrears. In spite of the adoption of new bankruptcy legislation in 1994, no large state owned firm was declared bankrupt until the summer of 1996. Instead, loss-making firms were allowed to continue operation as going concerns and this had a

---

[143] In 1991 (the first year in which market based exchange rates appeared) Bulgaria's gross foreign debt amounted to 140 per cent of GDP (UN/ECE database).

[144] In the period 1991-1996 Bulgaria signed four stand-by agreements with the IMF but none of them was successfully finalized because of subsequent failures to meet the conditions attached to the loans. During the same period 11 World Bank supported projects were initiated for a total of $488 million; however, only $96.4 million was actually disbursed and only one project was successfully finalized (*Pari* (Sofia), 23 February 1997). Additional assistance was received from the EU and G-24.

contaminating effect on the behaviour of all economic agents. Weak financial discipline became endemic and the general lack of hard budget constraints discouraged firms from restructuring. The delay of privatization created a power vacuum and a lack of proper governance of state owned firms. This encouraged rent seeking and corrupt practices by company managers and state officials which, according to some studies, reached an unprecedented level.[145]

Economic policy played a pervasive role in this process and instead of enforcing hard budget constraints actually eroded financial discipline in the economy. A series of unconditional financial bail outs was particularly destructive.[146] This rather costly policy approach created a vicious circle: the problem of bad loans remained unsolved and a wave of new bad credits emerged. The role of the commercial banks in this process was also counter-productive due to distorted incentives and improper supervision.[147] The banks' practices in this period were characterized by imprudent and risky lending, corrupt behaviour (including insider lending on a large scale), widespread violations of banking regulations, and distortive administrative intervention in the day-to-day operation of state owned banks.

(b) The dimensions of the crisis

The Bulgarian crisis has its origins in the inefficient and unviable state owned sector of the economy but due to soft budget constraints and repeated bail outs it spread to the public finances, the banking system and the financial markets.

Public finances

As a result of the expensive financial rescue operations, Bulgarian domestic public debt increased

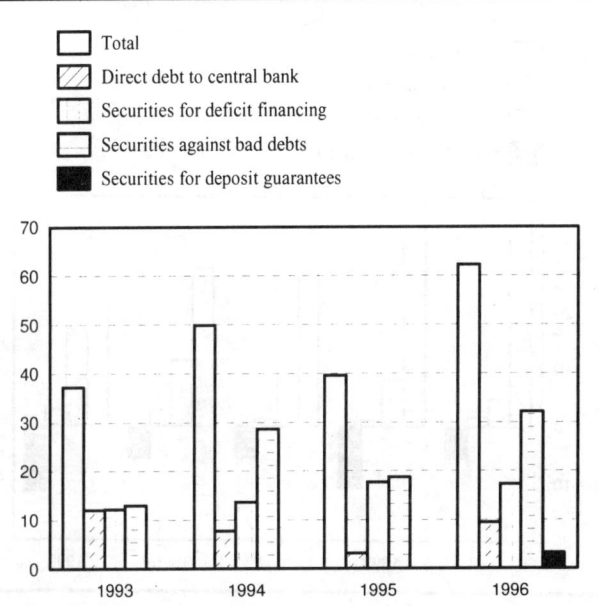

CHART 3.1.2

Structure of domestic public debt in Bulgaria, 1993-1996
(Per cent of GDP)

Source: Ministry of Finance/Bulgarian National Bank, *Government Debt Management* (Sofia), various issues.

rapidly within the space of a few years (see chart 3.1.2). This, in turn, had further negative repercussions. The ever increasing public sector borrowing persistently crowded out business investment while snowballing interest payments on the debt started to crowd out all non-interest budget expenditure (chart 3.1.3). Although the resumption of servicing of the foreign debt in 1994 added to the level of foreign interest payments, it was the interest paid on the domestic public debt, which reached almost 17 per cent of GDP in 1996, that was most damaging.

In such a situation the authorities could not avoid chronic large budget deficits, which were almost completely monetized, in spite of the fact that existing legislation required the government to finance its deficits primarily through the issue of government securities.[148] With the escalation of the financial crisis in 1996, which drained the banks of liquid resources, direct borrowing from the central bank increased substantially (see chart 3.1.2). On several occasions the central bank also had to intervene as a primary buyer of deficit financing securities, with the same macroeconomic impact.

---

[145] R. Avramov and K. Guenov, "The rebirth of capitalism in Bulgaria", *Bank Review 4* (Sofia), 1994, pp. 3-24.

[146] In 1991, the Bulgarian government announced its decision to convert into public debt 25 per cent of the non-performing bad loans made before 1990 and that eventually those remaining would be dealt with in a similar manner. The bail out was completely unconditional and did not require any change in behaviour, thus creating the overwhelming expectation of an "all-forgiving" policy on the part of the state. The outcome was that practically all state owned firms gradually stopped servicing their old bank credits and this led to the snowballing of bad loans and a sharp deterioration in the banking system. In December 1993 a special law to deal with non-performing bank credits made to enterprises before 1990 enabled all old non-performing loans in the banks' portfolios to be replaced with government securities. As in the previous case, this law only dealt with the stock aspect of the bad loans problem and did not address at all the flow aspect: no preventive measures were taken to stop the emergence of new bad loans. In 1994-1995 the authorities introduced a series of measures to bail out two large financially troubled state owned banks. And finally, in 1996, another controversial law was passed by the parliament which amounted to an unconditional state guarantee of deposits in failed banks.

[147] For an analysis of incentives in the Bulgarian banking system see OECD, *1996-1997 Economic Review – Bulgaria* (Paris), 1997.

[148] The Law on the Bulgarian National Bank adopted in 1991 provided that the central bank could make only temporary short-term loans (cash advances with maturity of up to three months) to the government. However, this provision was specified vaguely and there were ways to circumvent it. Besides, at the end of 1996, the parliament voted a law compelling the central bank to extend a special long-term loan to the government to cover the soaring budget deficit, thus contradicting the spirit of the Law on the Bulgarian National Bank.

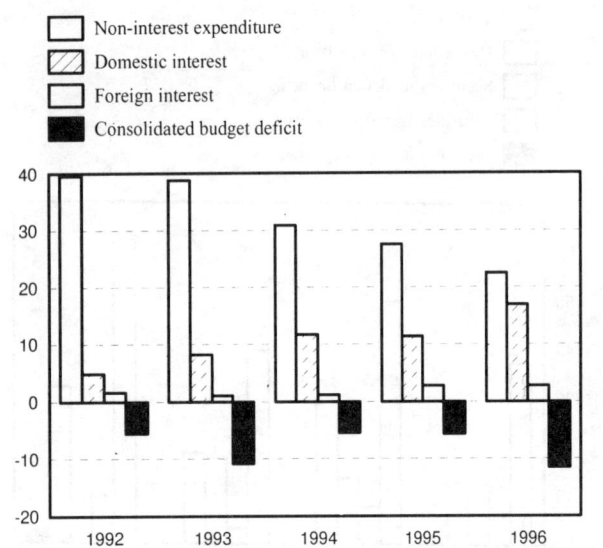

CHART 3.1.3

Budget expenditure and budget deficit in Bulgaria, 1992-1996
(Per cent of GDP)

Source: National statistics.

Another channel of monetizing the fiscal deficit was through the Lombard refinancing of commercial banks by the central bank. It is usually assumed that deficit financing through the issue of government securities is not inflationary as it does not lead to new money creation. However, the existing regulations in Bulgaria allowed the banks to use the bonds as collateral for refinancing from the central bank and, when they did so, this generated new central bank credit. Thus any net expansion of Lombard credits extended against deficit financing paper was equivalent to new net lending by the central bank to the government.

A third channel of monetization consisted of central bank operations with securities issued against bad debt. Initially these securities were placed in the portfolios of the commercial banks included in the bail out operations. Subsequently, however, the commercial banks were able to use them as collateral for central bank refinancing with results similar to those described above. Moreover, on a number of occasions the central bank bought such securities from financially troubled banks in exchange for liquid resources, thus resulting in net money creation.

Even considered in isolation, the resulting fiscal crisis in Bulgaria was very grave. But it was also part of several vicious circles which gave rise to an exponential growth of financial imbalances and increasing macroeconomic instability, and by 1996 the fiscal situation was practically out of control due to the rapid escalation of domestic public debt.

The banking system

The banking crisis also escalated as a result of soft budget constraints which gave rise to the emergence and rapid growth of a new wave of bad loans. At the same time the crisis was exacerbated by weak banking supervision and improper banking practices (including corruption) which were tolerated by the authorities.

The confidence of the public in the banking sector was a key factor in the relative stability of the financial system until 1995, but the erosion of confidence in 1995-1996 played a crucial role in the crisis of 1996. Due to the absence of a wider choice of investment opportunities, most of the savings of the population before 1996 were deposited in the banking system, thereby providing the basis for financing the budget deficit and for the expansion of the activities of the commercial banks.

Over time the negative results of protracted financial indiscipline accumulated in the banking system and it was only a matter of time when the deterioration in the quality of the banks' assets would become apparent.[149] In fact, despite the persistent worsening of their loan portfolios, the banks were largely kept afloat by the inflow of new household deposits, a process very similar to that underlying the operation of financial pyramids.

Public confidence in the banks started to decline in late 1995 when some of them began to experience liquidity problems as a result of the unsustainable volume of bad loans. The panic was initially confined to the banks most affected by liquidity problems, but it gradually escalated with the subsequent closure of several banks and towards mid-1996 it developed into a full-scale run on the entire banking system with massive withdrawals of deposits.[150] Inconsistent policies in dealing with deposit insurance also contributed to the worsening of the banking crisis.[151]

---

[149] At the end of 1995 the standard (performing) loans in the Bulgarian banking system amounted to just 25.9 per cent of all loans extended by commercial banks. Bulgarian National Bank, *Annual Report 1995* (Sofia), 1996.

[150] In the period October 1995 to September 1996 the cumulative withdrawals of domestic currency time deposits amounted to 82 billion lev (equivalent to $840 million at the spot exchange rate) and the cumulative withdrawals of foreign exchange deposits were equivalent to $670 million. Moreover, as the banking crisis was coupled with a simultaneous run on the currency, there was massive capital flight. Apart from the withdrawals of foreign exchange deposits, most of the domestic currency assets withdrawn from the banks were also converted into foreign currency which then remained outside the banking system.

[151] Before 1996 there was no formal deposit insurance in Bulgaria, except for the State Savings Bank which inherited from the past a 100 per cent state guarantee on deposits. However, having no experience with banking crises, the population presumably implicitly anticipated similar guarantees on all deposits in commercial banks. The first formal deposit insurance scheme was put into operation by the central bank in December 1995 (when the central bank itself was faced with deteriorating liquidity

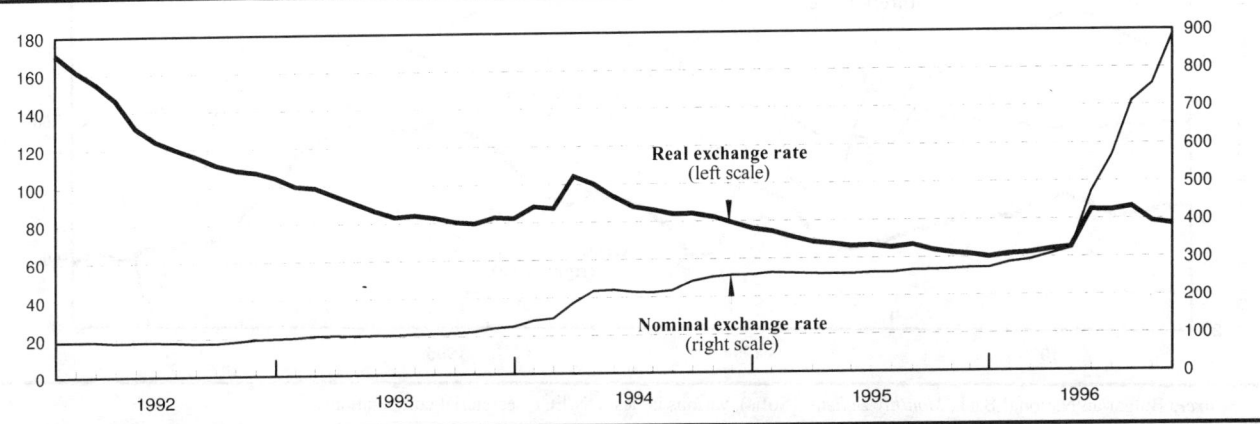

CHART 3.1.4

Exchange rate[a] indices of Bulgaria, 1992-1996
(January 1993=100)

Source: Bulgarian National Bank, Monthly Bulletin (Sofia), various issues; UN/ECE secretariat computations.

[a] Bulgarian levs per dollar.

The deposit drain reinforced the liquidity problems of the commercial banks. By the end of 1996, 15 commercial banks (both state owned and private) were placed under conservatorship by the central bank and bankruptcy procedures were initiated against several of them. Furthermore, the collapse of the banking system aggravated the fiscal crisis and created another vicious circle.

Due to the liquidity crisis in the banking system, most new government security issues by mid-1996 were undersubscribed and remained partially unsold. In turn, the Ministry of Finance, which was also experiencing a cash shortage, started to pay the interest due on outstanding securities through new issues, thus reducing the cash supply to the banks and further worsening their liquidity problems. Consequently, the Ministry of Finance was forced to apply more and more frequently for cash advances from the central bank which often intervened as the first buyer of government securities. In addition, the central bank increased the unsecured refinancing of the commercial banks. Finally, a special law was passed in December, obliging the central bank to extend a substantial one-time direct credit to finance the budget deficit. As noted already, the effect of these measures was direct monetization of the fiscal and quasi-fiscal deficits which in turn reinforced the general financial crisis.

of the banking system); however, it provided for only a very limited protection of the deposits. The announcement of such a scheme was perceived by the public as another sign of instability in the banking system and thus it had the perverse effect of adding to the banking panic. The revision of the scheme in March 1996 (offering more generous protection) did little to stop the run on deposits. Parliament finally voted in June a law providing a 100 per cent guarantee on household deposits and 50 per cent on corporate deposits in failed banks. However, by that time public confidence was already too severely damaged and even this measure was unable to check the run on deposits.

Foreign debt and the exchange rate

The escalation of the Bulgarian crisis was also influenced by the foreign debt burden. In June 1994 Bulgaria reached a restructuring and rescheduling agreement with the London Club creditor banks in which most of the outstanding debt was transformed into Brady bonds and a small portion of the debt was bought back at a discount. Consequently, Bulgaria had to resume immediately the servicing of its debt. The resumption of debt service, however, was problematic due to the limited debt servicing capacity of the economy.[152] The balance of payments position of Bulgaria remained highly unfavourable: there was no sustained inflow of (long-term) foreign capital and the debt maturity structure after the London Club agreement implied a very tight repayment schedule.[153]

The deteriorating situation in the banking system and in the public finances, coupled with the foreign debt burden, severely affected the financial markets and especially the foreign exchange market. Bulgaria was hit by two currency crises, in 1994 and in 1996, and the second was especially severe (chart 3.1.4). On both occasions, however, but especially in 1996, the run on the currency was amplified by the lack of coherent macroeconomic policies.

[152] Relative to the size of the economy and to the level of merchandise exports Bulgaria's gross foreign debt is not much higher than, say, the gross debt of Hungary during the first phase of transition. However, the maturity structure of the debt (long-term debt prevailed in Hungary while medium- and short-term obligations were significant in the case of Bulgaria), the debt servicing capacities of the two countries, and the level of foreign reserves differed substantially.

[153] The average level of annual debt service in the period 1995-1999 amounts to some $1 billion, which is a heavy burden equivalent to 20-25 per cent of export earnings in 1995-1996.

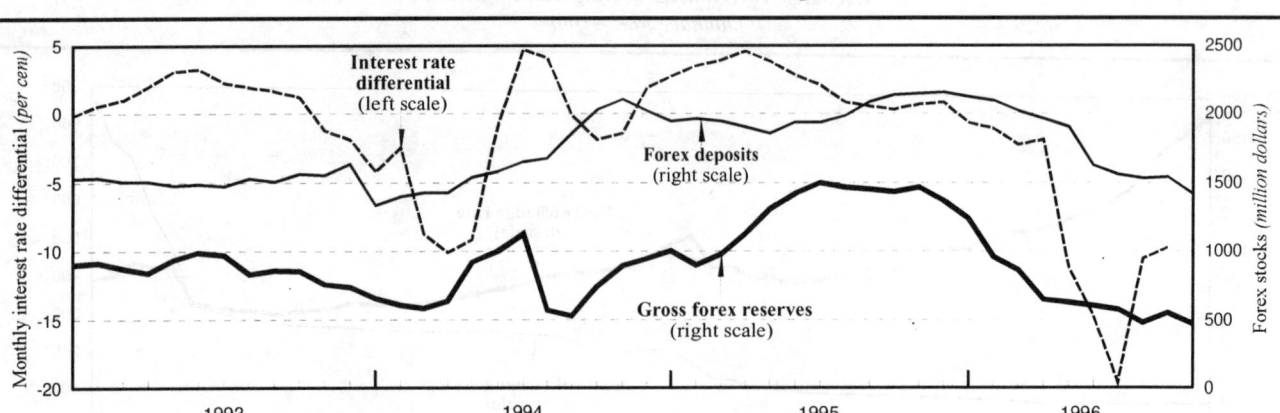

CHART 3.1.5

Interest rate differential[a] and forex stocks in Bulgaria, 1993-1996

Source: Bulgarian National Bank, *Monthly Bulletin* (Sofia), various issues; UN/ECE secretariat computations.

a *Ex post* dollar return on one-month time deposits denominated in domestic currency.

At the outset of the transition, Bulgaria opted for a flexible exchange rate regime because of its limited foreign exchange reserves. The declared policy was a "managed float", with central bank intervention aimed mainly at smoothing short-run movements in the exchange rate,[154] but targeted levels or corridors for the nominal rate were never pre-announced. The philosophy of the stabilization effort was based on control of the money supply which, as noted above, was not successful. In reality, the most important policy tool of the central bank was the basic interest rate. Due to the high level of dependence of the commercial banks on central bank refinancing, the pricing policy of the entire banking sector was linked to the basic interest rate, and almost all interest rate contracts in Bulgaria were defined in percentage points relative to the basic rate.

Throughout 1992 and most of 1993 the monetary authorities maintained a policy of "nominal exchange rate stability", implicitly using the exchange rate as a nominal anchor for the inflation target. The interest rate was kept relatively high to maintain a sizeable differential between the rates on domestic and foreign exchange assets (chart 3.1.5).[155] In the second half of 1993 the central bank reduced the basic interest rate, presumably in response to lower inflation and, possibly, pressure from the fiscal authorities. This reduction eliminated the interest rate differential and destabilized money demand. At the same time, it was also inconsistent with another change in monetary policy. On the eve of the London Club agreement, when debt service was about to be resumed, the central bank was under pressure to strengthen foreign exchange reserves in order to meet the up-front payments on the deal: it therefore discontinued its intervention on the foreign exchange market in support of the domestic currency. The weakened money demand and the pressure on the exchange rate resulted in a run on the currency which culminated in March 1994. The crisis ended in April with the signing of a stand-by agreement with the IMF providing financial assistance in preparation for the London Club agreement.[156]

However, the 1994 crisis was confined to the foreign exchange market. Two factors played an important role in arresting the crisis. First, the banking system was still relatively stable (at least, it was perceived as such by the public and enjoyed its confidence) and the currency crisis did not lead to a run on the banks. The crisis resulted mainly in currency substitution (a reverse switch to foreign exchange deposits) but the assets remained in the banking system. And second, the assistance provided by the international financial institutions temporarily eased the pressure on the foreign exchange market.

In 1995-1996 events proceeded in a similar fashion, but this time the general economic situation was much more unfavourable. Bulgaria had started servicing its foreign debt and the external financial assistance was exhausted. The authorities were unable to raise long-term capital to support the servicing of the debt: foreign direct investment was not forthcoming; no significant steps were taken to launch privatization; and there was no attempt to issue new sovereign bonds. In these circumstances, and in view of the needs for debt servicing, the risky strategy was adopted of raising short-term capital inflows as bridging finance until a new IMF and World Bank financial package, expected in 1995,

---

[154] In reality, the central bank intervened on a number of occasions to defend the exchange rate.

[155] The interest rate differential was calculated as the *ex post* dollar return on one-month time deposits denominated in domestic currency.

[156] This was complemented by further assistance from the World Bank and the EU.

could be agreed. The policy instruments were basically the same as in 1992-1993 (although there was a different policy goal): high nominal interest rates and positive interest rate differentials in favour of domestic assets were maintained for almost the whole of 1995.

In the short term the results were favourable: foreign capital flowed in and foreign exchange reserves increased (chart 3.1.5). The accompanying stability of the nominal exchange rate and appreciation of the real exchange rate (chart 3.1.4) had a favourable effect on disinflation and permitted a modest growth of output.

Towards the end of 1995, however, the situation started to deteriorate. During the year the government had failed to reach an agreement with the international financial institutions[157] and thus no new balance of payment support was forthcoming. At the same time the fiscal crisis, due to the escalation of domestic debt, started to squeeze the budget because of the high interest rate. As a result of pressure from the fiscal authorities, the central bank began lowering interest rates in 1995, repeating the same destabilizing policy as two years earlier, although this time the reduction of interest rates was more pronounced. The reaction of the markets was almost identical to that in 1993-1994 but stronger: destabilization of money demand, demonetization and capital outflow. As regards the impact of short-term capital movements, the Bulgarian experience bears some resemblance to the recent Mexican crisis. As observed during financial crises in other countries, when foreign capital starts to flow out, domestic capital tends to follow.[158] Up to a point the central bank defended the domestic currency by intervening on the foreign exchange market but the main outcome was a depletion of reserves (chart 3.1.5).

The fragile state of the banking system was responsible for another important difference in the development of events. The run on the currency in 1996 was coupled with a simultaneous run on the banking system resulting in massive capital flight and the failure of a number of commercial banks. At first the central bank attempted to rescue the troubled banks by providing them with fresh liquidity, mostly uncollateralized refinancing.[159] However, with the escalation of the crisis towards mid-1996 the first bankruptcy procedures were initiated against some banks.

Policy implications

The lack of policy coherence resulted in more vicious circles. When the monetary authorities were faced with several emergencies in 1996, they reverted to their main policy tool, the interest rate, raising it to unprecedented levels in a vain attempt to stop the slide.[160] But the fiscal result was the opposite of what the authorities were aiming at: after the central bank increased the basic interest rate the costs to the budget also rose; consequently, the central bank had to cope with financing the rising fiscal deficit which was done through monetization and led to galloping inflation. In turn, injections of liquidity ended up on the foreign exchange market, further eroding the exchange rate. The combined effect of these vicious circles, coupled with the resignation of the government and a subsequent political crisis, gave rise to hyperinflationary hikes in January-February 1997.

At the moment of writing this *Survey*, the policy discussion in Bulgaria on ways out of the crisis is focused around the proposal to introduce a currency board (see box 3.1.1). A strong policy commitment to financial discipline, as embodied in a currency board, is considered essential for restoring public confidence in financial institutions and in economic policy in general. At the same time it is thought that a fixed exchange rate will contribute to arresting the hyperinflation and to a lowering of interest rates which, in turn, will reduce the burden of interest payments on public expenditure.

While rapid stabilization is probably a necessary condition for resolving a crisis of this type, it should be remembered that the fundamental roots of the crisis are microeconomic in origin. Successful structural reforms are also necessary for attaining a stable macroeconomic environment and a basis for sustained economic growth.

(c) The lessons of the Bulgarian crisis

The Bulgarian economic crisis underlines the possible fragility of the process of transition from plan to market in a number of countries and points to some of the unexpected hurdles to be overcome. Although the Bulgarian experience is unique as regards the particular combination of several adverse developments, including inadequate policy responses, the individual elements of the crisis are typical in a transitional environment and can be identified in a number of other transition countries.

The primary source of the Bulgarian financial crisis was the large, unrestructured, loss-making state owned industrial sector and the main policy related problem was the failure by the authorities to address this issue at an early stage by restructuring, privatization, imposing hard

---

[157] A precondition for new financial assistance was the initiation of a major restructuring programme that the Socialist government was reluctant to undertake at that time.

[158] For a discussion of this issue see J. Sachs, A. Tornell, and A. Velasco, "Financial crises in emerging markets", *Brookings Papers on Economic Activity*, 1 (Washington, D.C.), 1996, pp. 147-215.

[159] In fact the central bank was compelled initially to act in this manner as no bankruptcy legislation was in place at the time and technically it had no option but to continue supporting the ailing banks. Such legislation was adopted later in 1996.

[160] In September 1996 the central bank raised the basic interest rate to 25 per cent a month.

budget constraints and a faster rate of liquidation of the biggest loss makers. A policy of "muddling through" and expensive, unconditional bail outs not only failed to resolve the problems but, on the contrary, aggravated them considerably.

In this respect, the Bulgarian crisis provides a warning to policy makers in other transition economies. Although most other east European transition countries are more advanced in the process of restructuring and privatizing the industrial sector, all of them, albeit to different degrees, face the problem of restructuring or closing down loss-making firms which are still owned or controlled by the state. Indeed, the closing down of unviable, large state owned enterprises is a very difficult policy problem in any country, but the Bulgarian experience indicates that keeping loss-making state firms afloat may entail very high fiscal costs. Bad loans have also been a major problem throughout central and eastern Europe and governments in other countries have engaged in financial bail outs of banks and enterprises as well (see section 3.2). Again, the negative developments in Bulgaria underline the importance of a careful and well thought out approach to dealing with these issues. One major lesson from the Bulgarian experience is that any of these problems, if left unchecked or not addressed by adequate policies, may give rise to a major financial crisis.

The macroeconomic aspects of the Bulgarian crisis also indicate the vital necessity of coherent and consistent macroeconomic policies in a transitional environment. While this may seem a statement of the obvious, the fact is that in a fragile environment with missing or weak institutions, underdeveloped markets, and inexperienced agents, the transition economy may be especially vulnerable to shocks, so that even relatively small disturbances may result in major economic disruptions. In this regard, the destabilizing impact of monetary policy on money demand in Bulgaria is a case in point. As demonstrated by the evolution of events on two occasions – in 1994 and in 1996 – policy changes that were perceived by the monetary authorities as minor but appropriate to the current state of the economy were interpreted by the markets as critical and led to major disturbances with a highly destabilizing effect on the economy. Moreover, destabilizing policy actions tend to be asymmetric: once the damage has been done, simply reversing policy is not enough to undo it.

The Bulgarian crisis indeed provides evidence on the actual impact of policy measures and policy mistakes in a highly fragile and vulnerable transition environment. The escalation of the financial crisis in Bulgaria can be seen as a textbook example of the dangers of moral hazard. The lack of commitment by the authorities to enforcing hard budget constraints and the resort to repeated bail outs led to a complete erosion of financial discipline not only among the financially troubled firms but throughout the whole economy. Thus, inadequate policies amplified and accelerated the financial crisis, which in turn provided an incentive to shy away from the active restructuring of state owned firms. The issue of coordination (or lack of coordination) between monetary and fiscal policy in the period preceding and during the crisis is another textbook example of how fiscal pressures to ease monetary policy (and the submission of monetary policy to such pressures) may end up with the opposite fiscal result to that intended and with an increase rather than a reduction in macroeconomic disequilibria. Bulgaria's painful experience with short-term capital flows is not new, but it underlines the vulnerability of a fragile transition economy to external shocks; here again, policy tended to worsen rather than improve the situation.

Economic policy undoubtedly contributed to the escalation of the Bulgarian crisis, but another problem arising in the transition from a planned to a market economy was also important and the Bulgarian experience in this respect may also be relevant to other transition countries.

As already noted, the primary sources of financial instability in Bulgaria were the losses generated in the state owned sector of industry. Within this sector there is a specific group of enterprises which are typically chronic loss makers with large accumulations of bad debt, and which have little or no possibility of ever becoming viable in a market environment.[161] As such they are hardly worth restructuring, and therefore have very little or no chance of being privatized. They also provide examples of cases where hard budget constraints are unlikely to be sufficient to encourage them towards profitability as their inefficiency is mostly technologically determined.[162] However, although many such enterprises are technically insolvent, they continue to operate for political or social reasons.

Such state owned enterprises exist in all transition countries, although their number and their share in the economy differ from country to country. Most of them are typical products of the socialist era: as a rule they tend to be industrial giants, initially designed according to the needs of intra-CMEA specialization, equipped

---

[161] It can be argued that the losses generated by such enterprises should be allocated directly to general government expenditure and be added to the general fiscal deficit: as it will not be possible for them to redeem their implied liabilities in the future, they will have to be met from other sources of public revenue.

[162] The only way to improve the productive efficiency of such an enterprise is through upgrading the production technology which requires new investment. Some of them, however, may be unfit for upgrading: for example, the costs of upgrading may compare unfavourably with the costs of a greenfield investment in a new production site of the same capacity but with new technology.

with obsolete physical assets, and heavily overstaffed. They are usually concentrated in mining, in heavy industrial branches, or in industries in which a country was specialized within the CMEA.

The conventional policy wisdom about such enterprises is that they are to be closed and better sooner than later. However, it should be pointed out that even this most radical of restructuring measures is not cost free. Such enterprises are usually heavily indebted and this debt may be much larger than the market value of their assets. If creditors are to be compensated under a liquidation scheme the state may incur further expenditure. The liquidation procedure itself can also be quite costly and protracted. And, probably most important, the closure of a large state owned industrial enterprise is likely to result in a large and possibly permanent rise in unemployment (at least in the medium term)[163] implying a further burden on the general government budget.

There may also be serious political constraints on restructuring policy. Large industrial enterprises often have significant political and lobbying power which may be used to inhibit radical policies; policy decisions may be constrained also by social considerations, for example, when such enterprises are the major regional employers.[164] In any case, there appears to be a number of important and inherent constraints on the rate at which many unviable, loss-making enterprises can be closed down. In the meantime these state owned enterprises will continue to be a source of persistent financial imbalance in the economy, the magnitude of which will depend on the size of the loss-making sector. Their eventual liquidation, as already mentioned, may reduce but not eliminate this imbalance: residual fiscal burdens will remain due to the rise in unemployment which may be long lasting.

The magnitude and severity of problems stemming from this situation crucially depend on the size of the loss-making sector and thus on the structure of the economy inherited from the past.[165] The speed and decisiveness of policy reform are also important but, as already noted, they may be restricted by political constraints and by the very size of the unviable sector of the economy. If the resulting financial and fiscal imbalances are sufficiently large, the available economic policy options to cope with them may be seriously limited.[166] To the degree that the problem exists in all transition economies, it affects their macroeconomic performance and can be regarded as one of the sources of potential instability and persistent inflation.[167]

The logic of this analysis suggests that probably the best chance of breaking this vicious circle in the medium term is through new investment. Some current loss-making enterprises – if modernized through new investment – may become profitable. New investment in other sectors of the economy would reduce the fiscal imbalance to the extent that such investment creates new jobs. Also, high rates of economic growth may lead to net new job creation (alleviating some of the fiscal pressures) and would increase the resources available to the authorities to sterilize the remaining adverse effects. The trouble is that if a country is already facing serious problems of this type such a scenario is not likely in the medium term. The reason is that the inherent instability which is built into such an economy is in fact an anti-growth factor and a deterrent to new private investment.

The chances of breaking out of this vicious circle therefore depend largely on the assistance programmes of the international financial institutions and other forms of official aid. In this regard it might be appropriate to reconsider once again the idea of a new multinational policy initiative specifically designed to promote investment and growth in those transition countries which are most affected by such inherited structural problems.

Most of the structural problems discussed above are not new: both policy makers and the international financial institutions were aware of them and, indeed, the latter had been urging the Bulgarian authorities to address them for a long time. However, Bulgaria's painful experience has revealed, perhaps for the first time, how deeply rooted and severe these problems are as well as the dangers they contain both for macroeconomic stability and the sustainability of the transition process. There is more evidence now that the magnitude of the transformational problems in some countries may be so great that they simply do not possess the necessary resources to solve them on their own, even if their

---

[163] Simply making the workforce of a loss-making enterprise redundant does not imply any net new job creation. Some of the workers may eventually find new jobs but, if no new jobs emerge in the economy as a whole, this will be at the expense of others. A reduction of the number of unemployed will take place only as a result of net new job creation, and that may take some time to bring about.

[164] The existence of such constraints is confirmed by a number of empirical studies; see for example, W. Carlin, J. Van Reenen and T. Wolfe, "Enterprise restructuring in the transition: an analytical survey of case study evidence from central and eastern Europe", EBRD, *Working Paper*, No. 14 (London), July 1994.

[165] This is in fact one of the key factors in determining whether the economy starts the transition to a market economy in "favourable" or "unfavourable" initial conditions.

[166] Under certain conditions there may not be any policies to neutralize and counterbalance the problem. For example, if the cumulative losses grow faster than the nominal GDP, this will force the economy into an inherently unstable performance path: this is equivalent to a trajectory in which nominal public debt is growing faster than nominal GDP.

[167] For example, the primary sources of the growing non-payments in some of the successor states of the former Soviet Union (including Russia) are existing loss-making enterprises. The endemic growth of non-payments is a sign of inherent financial and fiscal imbalances in these countries.

policies are right. Besides, the needed restructuring and adjustment cannot take place overnight; it is a process which will require a sustained effort over a long period both by policy makers in the country and by the donors. The key to successful transformation in the lagging transition countries may therefore lie in a new symbiosis of a more generous and long-term programme of international assistance coupled with a coherent and well-defined long-term policy commitment by the national authorities.

3.2 Output and demand

(i) Output

(a) Main developments in output

The development of output in the ECE transition economies in 1996 was uneven. Positive growth continued in eastern Europe (with the notable exception of Bulgaria) but it lost some of its previous vigour, most of the east European economies growing at lower rates than in 1995 (table 3.1.1). For the first time, however, there was positive, albeit modest, economic growth in all three Baltic states. Output performance in the Commonwealth of Independent States was mixed: an increasing number of countries showed signs of recovery in 1996, but for the CIS as a whole output continued to decline, primarily because of the failure of the Russian Federation to make the expected return to positive growth.

The slowing down of the recovery in eastern Europe, and the continuation of the recession in much of the CIS region, unveiled some new impediments to the difficult transition from a command to a market economy The deterioration of performance in eastern Europe resulted from a combination of supply and demand factors. While growing domestic demand favourably affects the process of recovery in the region it is obviously not sufficient as an engine of growth, especially for the smaller economies. On the other hand, the east European countries, the Baltic states, and a number of the CIS countries are increasingly dependent on western Europe as their major export market. A downturn in the growth of west European demand – such as occurred in 1996 – may therefore seriously upset (probably with multiplier effects) output in the transition countries. Moreover, the supply side of the economy in most of these countries is still too feeble to compete on an equal footing in international markets, especially in a situation of widespread sluggish demand.

In spite of the progress made with privatization, the process of market oriented enterprise restructuring is far from over in the transition countries. Unrestructured, large state owned enterprises still exist in most of them; and, as discussed later, apart from being a burden on their economies, the existence of such enterprises creates an unhealthy environment which leads to increased risks to macroeconomic stability and the prospects for growth.

The core group of fast reforming *central European* and *Baltic countries*, on average, performed relatively well in 1996. Poland, the original leader in bold economic reforms, is now the leader among the transition countries – and in Europe as a whole – in terms of economic growth. In 1996 it became the first of the transition countries to surpass the pre-transition level of GDP of 1989 (appendix table B.1).

The Polish success is all the more noteworthy as it seems to be based on both sound and consistently improving macroeconomic fundamentals and a coherent pattern of structural reform.[168] After a relatively modest start in the first half of 1996 (GDP grew by some 4 per cent during the first two quarters), growth in Poland markedly accelerated in the second half of the year and especially during the fourth quarter when it was boosted by strong domestic demand. Throughout the year – and indeed since the recovery began in 1992 – the robust performance of the Polish economy was backed by a booming industrial output (chart 3.2.1). Construction also rose strongly in 1996 (table 3.2.3).

Since the launching of the stabilization programme in March 1995, the Hungarian economy has undergone a major macroeconomic adjustment. The main goal of the programme – establishing the conditions for balanced economic growth – required a reduction of the domestic and external macroeconomic imbalances and, especially, of the fiscal and current account deficits, which had become a threat to macroeconomic stability. This painful adjustment, which resulted in a substantial cut of domestic demand (and which was probably greater than initially expected), started to bear fruit towards mid-1996: after slowing down for a year, growth started to accelerate (notably in industry – see chart 3.2.1) and the Hungarian economy finished the year on a high note. Manufacturing output grew especially strongly in the exporting branches (exports of manufactured products increased by 13.5 per cent) and particularly in the new industries supported by inflows of FDI.

By most measures the tough stabilization effort was successful: the fiscal and current account deficits declined substantially in 1996, manufacturing exports swelled, and the rate of inflation slowed down. Hungary now seems set for a much faster rate of growth than has occurred during the last few years, a take-off that has been long awaited, both domestically and abroad.

---

[168] It should be noted that the figure for the apparent trade deficit in the case of Poland (table 3.1.1) is a somewhat biased measure due to the existence of large (unrecorded) cross-border trade flows in which Poland enjoys a significant surplus, which partly offsets the "official" trade deficit (see table 3.6.2 where these flows appear as "non-classified items" in the current account).

CHART 3.2.1

Indices of gross industrial output in selected transition countries, 1995-1996

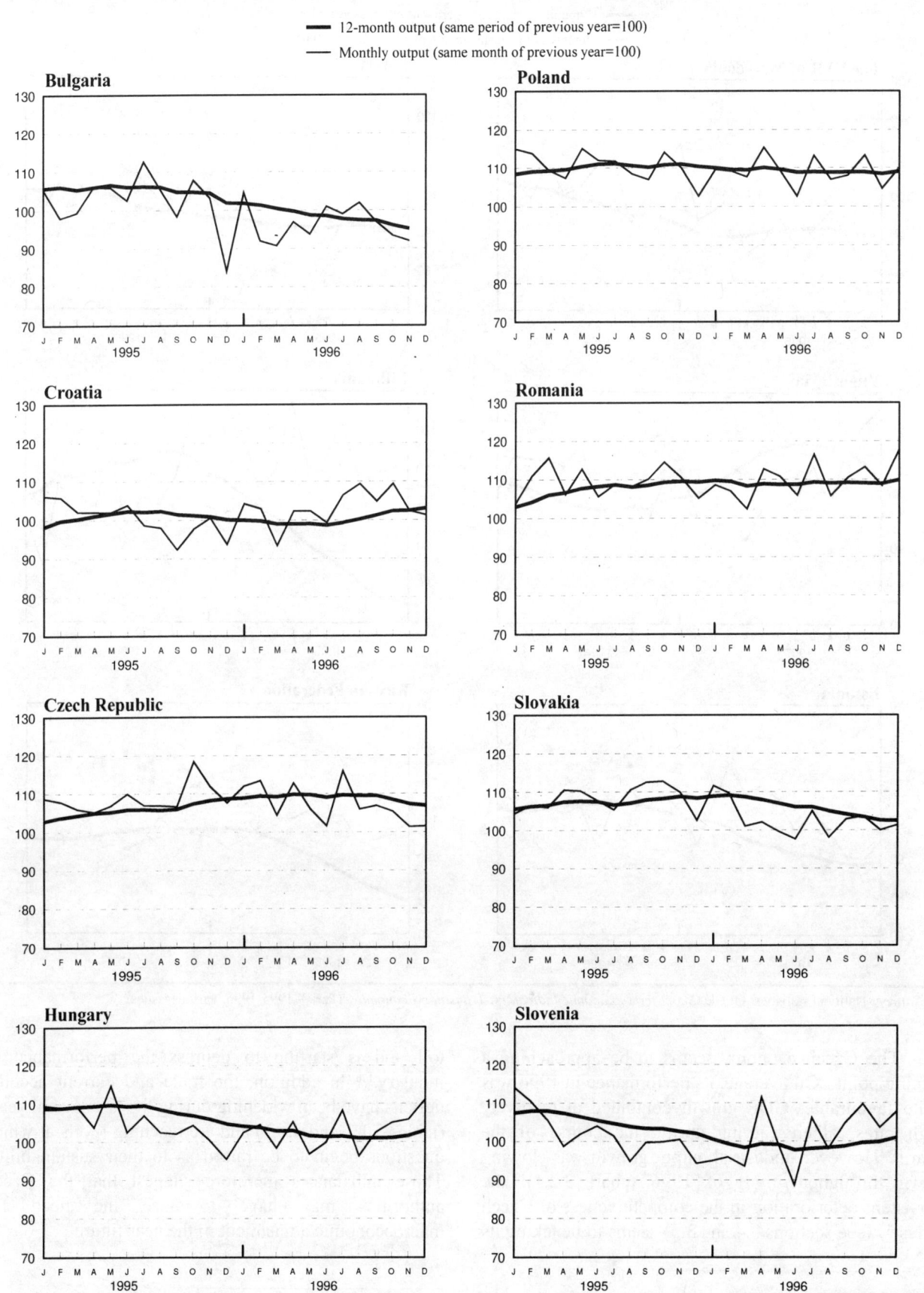

(For source and notes see end of chart.)

CHART 3.2.1 (concluded)
Indices of gross industrial output in selected transition countries, 1995-1996

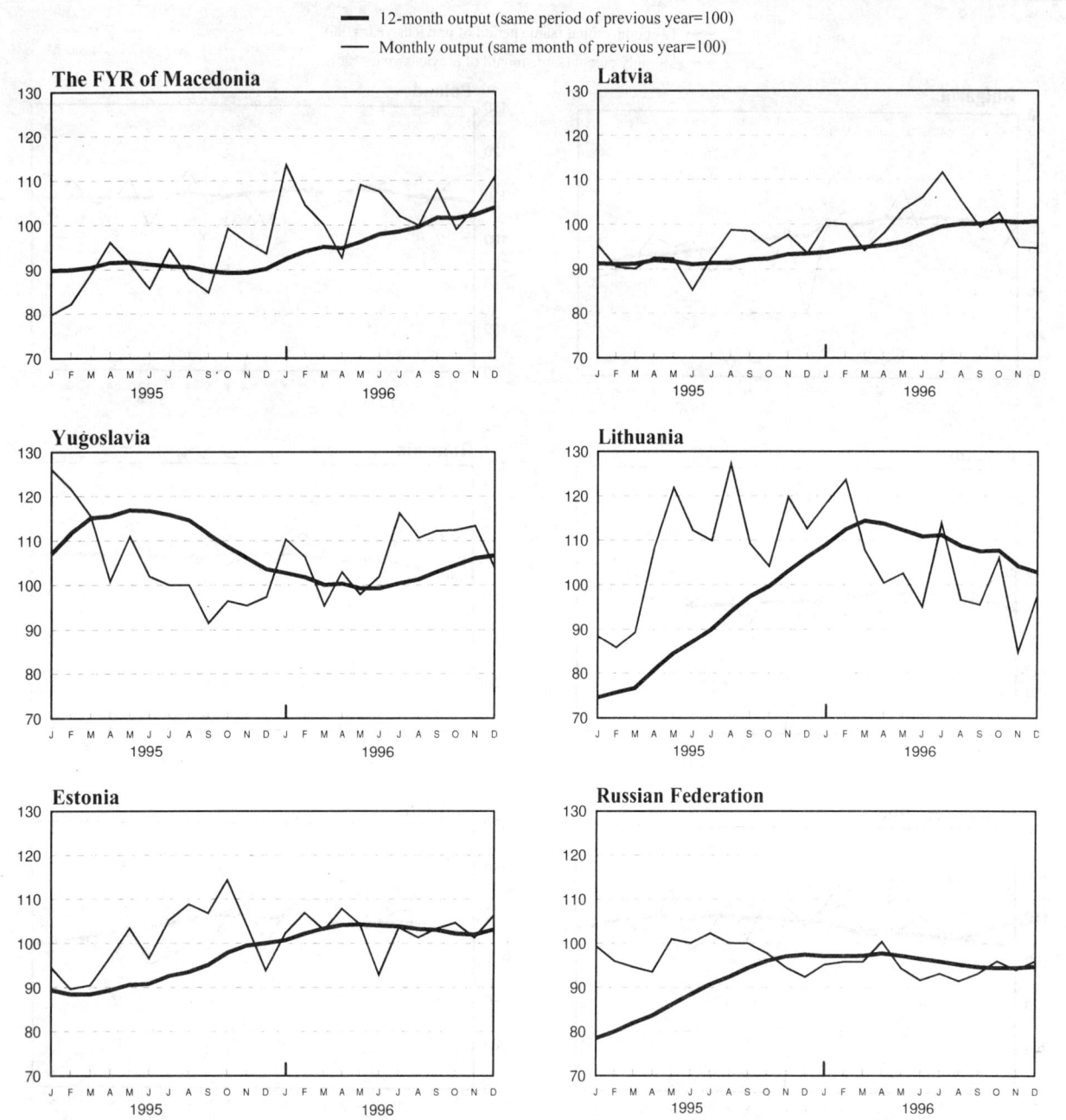

Source: National statistics; OECD, *Short-term Economic Indicators: Transition Economies* (Paris), 1995-1996, various issues.

The Czech economy seems to be approaching a turning point. On average, its performance in 1996 was quite favourable: GDP growth continued at relatively high rates, albeit varying during the course of the year.[169] However, industrial output growth was slowing down throughout most of 1996 (chart 3.2.1): the persistent deterioration in the competitiveness of Czech industry (see sections 3.4 and 3.5) seems to be taking its toll and is starting to depress the performance of industry.[170] In addition, the trade and current account deficits have been widening during the last several years (tables 3.1.1 and 3.6.1) and are reaching levels at which questions begin to be raised as to their sustainability. These imbalances therefore suggest that the Czech authorities may have to face the need for macroeconomic adjustment in the near future.

---

[169] GDP growth rate in the first and second quarters of 1996 was 4.6 per cent, then it fell to 3.6 per cent in the third quarter but rose again to 4.7 per cent in the last quarter of the year.

[170] According to preliminary figures, industrial output actually declined during the first months of 1997.

Slovakia followed a fairly similar pattern of macroeconomic economic performance in 1996, but some of the underlying forces were different. The rate of growth of GDP in Slovakia in 1996 (6.9 per cent) was in fact the highest among the east European and the Baltic states, but for the most part it resulted from the expansion of non-manufacturing activity and, especially, of government financed infrastructure projects; industrial output growth actually decelerated markedly from its 1995 rate (chart 3.2.1) and according to preliminary estimates the volume of exports of goods and services even declined. As in the Czech Republic, Slovakia ran sizeable trade and current account deficits and competitiveness was deteriorating. Changes in macroeconomic policy may thus be necessary in Slovakia as well.

In Slovenia, economic performance was mixed in 1996. After a strong recovery in 1994 which continued until mid-1995, Slovenian industry plunged into a recession which continued until almost the last quarter of 1996 (chart 3.2.1). Slovenia is particularly dependent on exports of manufactures to Germany and this decline can be largely attributed to the downturn in the growth of German import demand in this period. In addition, a longer-term trend of real appreciation of the currency has negatively affected the competitiveness of exports. In the second half of 1996 industrial output started to recover and this influenced overall economic performance. Domestic output in 1996 was boosted by large investment projects in infrastructure and services, which contributed to the relatively good outcome for the year as a whole in terms of GDP growth.

There was a marked recovery for the first time in Croatia in 1996 although it also reflects a somewhat mixed performance. The main driving force was the strong recovery of the tourist industry thanks to the return of foreign holiday makers following the end of the war in Bosnia and Herzegovina. There was also some recovery of industrial output (chart 3.2.1) but Croatian competitiveness was adversely affected by an overvalued currency.

Economic performance in both Croatia and Slovenia was positively influenced by the end of United Nations sanctions and the resumption of both countries' economic relations with Yugoslavia and the other successor states of the former SFR of Yugoslavia.

However, in general, instability continues to prevail throughout the rest of *south-eastern Europe*. In 1996 the first serious transition crisis occurred in Bulgaria (see 3.1(iii)) and there were severe economic setbacks in several other countries in the region.

The financial crisis in Bulgaria had profound repercussions on the performance of the real economy and output deteriorated throughout the year. The collapse of the banking system and the vertiginous rise in interest rates resulted in a virtual halt of credit activity in the second half of 1996. The large real depreciation of the exchange rate, which resulted from the run on the currency, adversely affected domestic producers who depended on imported intermediate goods and other inputs. All this, coupled with shrinking domestic demand, resulted in a severe depression of economic activity and serious disruptions to supply. Indeed, the decline in output in 1996 – a drop of some 10 per cent in GDP, according to preliminary estimates – is similar to the initial transformation shock experienced by the transition economies at the outset of reforms.[171] In addition, Bulgaria was hit by grave grain shortages in 1996 – due to a poor harvest and mismanagement of grain reserves – which, in turn, led to widespread shortages of bread.

In Romania, economic performance in 1996 seems to contrast quite favourably with developments in neighbouring Bulgaria. GDP grew at a relatively high rate, strong industrial growth continued for the second consecutive year, and the financial system operated without significant, observable strain. However, output growth in Romania in the last two years does not appear to have been based on sound macroeconomic fundamentals. Expansionary monetary and incomes policies and increasing subsidies to various sectors of the economy in 1995 and 1996 boosted final domestic demand and output growth in the short run, but this was achieved at the cost of serious and growing macroeconomic imbalances: a widening current account deficit (which, in contrast to central European countries, was not matched by private capital inflows and thus led to a fast growth of foreign indebtedness and official borrowing), a growing fiscal deficit, and the artificial suppression of prices and the exchange rate through administrative controls.

The new government which took office in December 1996 has proclaimed a radical change in policy aimed at restoring macroeconomic equilibrium and implementing deep structural reforms. The necessary adjustment, which requires liberalization of the economy and an end to the unbalanced expansionary policy, is likely to result – if carried out – in a marked reduction of domestic demand and output. This cooling down, however, seems to be indispensable if the overheated Romanian economy is to head towards a soft landing instead of the crash that occurred in Bulgaria.

After a deep and prolonged recession, there were signs of recovery in the economy of The FYR of Macedonia in 1996: both GDP and industrial output

---

[171] The reported decline in gross industrial output was much smaller (table 3.1.1); however, the decline in industrial value added may be much larger.

increased – albeit rather modestly – after six years of continuous decline. However, the impact of the gradual improvement of political relations in the region on economic performance in The FYR of Macedonia was somewhat mixed. On the one hand, the lifting of the trade embargo imposed by Greece removed serious impediments to transportation and trade and gave an impetus to economic recovery. On the other, the lifting of United Nations sanctions on Yugoslavia deprived local agents of the lucrative business of (illegal) re-routing of sizeable trade flows to neighbouring Serbia.

The economy of The FYR of Macedonia still appears to be quite depressed (an indication of which is the extremely high level of unemployment, which is the highest of all the transition countries for which unemployment statistics are available). Tight stabilization policies in recent years, while successfully curbing inflation, also appear to have contributed to the low level of economic activity. At this stage of the transition, a new balance of economic policy allowing for more flexibility might be more beneficial for the country.

The reliability of the statistical information about recent economic performance in Yugoslavia is somewhat questionable. Most of the available figures indicate that a recovery is underway, although its strength, as officially reported, may be overstated. After a long period of stagnation, Yugoslav industrial output started to pick up in the second half of 1996, mostly driven by strong domestic demand, and this contributed to the relatively high rate of GDP growth.

Moderate recovery continued in the *Baltic states* in 1996. In all three countries there was significant progress in macroeconomic stabilization; however, this was achieved at some cost in output growth: so far none of the Baltic states has shown much sign of dynamism in output performance. Another reason for relatively slow growth (and one which distinguishes the Baltic states from the faster growing central European economies) is that domestic demand has generally been sluggish and so far has not been a major determinant of growth. At the same time there has been a widening of trade and current account deficits in all three countries.

In 1996 industrial output was virtually stagnant in Estonia and Latvia, and in Lithuania its growth decelerated sharply after a strong performance in 1995 (chart 3.2.1). At the same time other sectors of economic activity were buoyant. Growth in Estonia was boosted by strong expansion of the service sector (especially of financial services and tourism) and of construction (table 3.2.3); some services and utilities also grew solidly in Latvia and Lithuania. The latter two economies seem to have largely recovered from the banking crises of 1995. Latvia also benefited from the revival of the port of Ventspils and of its transit trade (Ventspils is the only terminus for exports of Russian crude oil to Europe). Macroeconomic performance strengthened considerably thanks to sound fiscal and monetary policies.

The diverging economic trends which emerged in the *CIS countries* in 1995 continued in 1996. On average, economic performance in the CIS region improved in 1996: in seven countries there was positive GDP growth compared with just two in 1995, but this was concentrated mainly in the smaller CIS countries. Counter to hopes and some expectations, output in the two largest economies – the Russian Federation and Ukraine – did not begin to recover in 1996. Moreover, in Russia the decline in GDP (6 per cent) was even larger than in 1995. Considering the high level of dependence of the other CIS economies on Russia, this continued slump undoubtedly affected their economic performance. Nevertheless, there were encouraging signs of recovery, or near recovery, in an increasing number of CIS countries and these may indicate that the end of the transition recession in the CIS may be approaching.

Overall economic performance in the Russian Federation in 1996 was greatly affected by the continuing fall of industrial output (by some 5 per cent in 1996) across the board, including even the leading sectors based on primary resources (table 3.2.7). On the positive side, while output performance deteriorated on average in industrial enterprises, and especially in the group of large and medium-sized enterprises, small enterprises and joint ventures reported positive growth in 1996.[172] Falls in output also occurred in other major sectors such as agriculture and construction as well as in services (table 3.2.3).

Continued monetary and fiscal austerity – in line with the February 1996 stabilization programme – may have added to the continuing recession in Russia. Aggregate domestic demand still appears to be severely depressed: there was no revival of investment in 1996 and final consumption fell in spite of the general rise of real incomes in 1996 (the causes of this somewhat paradoxical performance are discussed later). The continuing drain of financial resources through massive capital flight (see section 3.1(ii)) is another brake on recovery: it has been debilitating the Russian economy and has reduced its potential for future growth, at least in the short run.

---

[172] During the first three quarters of 1996 the output of large and medium-sized industrial enterprises declined by 8 per cent (from the same period of 1995); it grew in the group of small enterprises and joint ventures by 13 per cent; however, the latter group is still rather tiny: in 1996 it produced only some 15 per cent of total Russian industrial output. Government of the Russian Federation/Working Centre for Economic Reform, *Russian Economic Trends*, Vol. 5, No. 3 (Moscow), 1996, pp. 79-80.

TABLE 3.2.1

Share of major sectors in GDP[a] in eastern Europe and the Baltic states, 1991-1996

(Per cent of GDP, at current prices)

	1991	1992	1993	1994	1995	1996[b]		1991	1992	1993	1994	1995	1996[b]
Albania							**Slovakia**						
Agriculture	43.7	54.1	55.8	55.5	..	..	Agriculture	5.7	6.2	4.9	7.4	6.3	5.7
Industry	31.6	17.0	13.8	12.6	..	..	Industry	52.7	37.3	32.0	32.1	32.1	28.6
Construction	6.4	7.6	8.8	9.5	..	..	Construction	7.4	7.0	4.9	5.1	5.2	5.2
Services	18.3	21.3	21.5	22.5	..	..	Services	34.2	49.5	58.1	55.3	56.4	60.5
Bulgaria							**Slovenia**						
Agriculture	14.5	12.0	10.6	12.3	13.9	7.7	Agriculture	5.7	5.8	5.1	5.4	5.0	5.2
Industry	32.9	34.4	29.2	27.0	28.5	31.0	Industry	39.9	35.9	33.4	33.8	32.1	32.6
Construction	4.4	6.0	5.8	5.1	5.2	3.6	Construction	4.1	4.3	4.7	4.8	5.1	5.6
Services	48.2	47.5	54.4	55.6	52.5	57.7	Services	50.3	54.0	56.8	56.0	57.9	56.7
Croatia							**The FYR of Macedonia**						
Agriculture	10.0	11.3	10.6	11.1	..	..	Agriculture	14.0	19.0	11.9	..	..	..
Industry	25.5	27.4	28.1	25.9	..	..	Industry	31.0	34.1	31.5	..	..	..
Construction	4.6	2.7	2.2	2.1	..	..	Construction	5.7	4.2	6.1	..	..	..
Services	59.9	58.6	59.1	60.9	..	..	Services	49.3	42.7	50.6	..	..	..
Czech Republic													
Agriculture	6.0	6.1	6.5	5.8	5.2	5.1							
Industry	47.4	42.9	37.0	34.8	34.1	34.5							
Construction	6.8	5.3	5.2	5.9	7.4	6.6							
Services	39.8	45.8	51.3	53.5	53.4	53.8							
Hungary							**Estonia**						
Agriculture	8.5	7.2	6.6	6.7	7.2	..	Agriculture	21.8	14.2	11.0	10.1	8.1	7.4
Industry	29.0	27.3	26.2	25.3	26.9	..	Industry	35.3	31.2	24.5	24.2	22.9	23.5
Construction	5.4	5.9	5.3	5.1	4.9	..	Construction	7.9	5.2	6.6	6.4	5.3	5.7
Services	57.1	59.6	61.9	62.8	61.0	..	Services	34.9	49.4	57.9	59.3	63.7	63.4
Poland							**Latvia**						
Agriculture	6.4	7.1	7.2	7.1	7.3	..	Agriculture	23.2	17.6	11.8	9.5	9.8	8.7
Industry	35.8	35.5	35.7	36.2	36.0	..	Industry	38.2	29.8	30.7	25.3	25.3	26.0
Construction	8.8	8.1	7.1	6.4	5.8	..	Construction	5.8	5.0	4.3	6.0	7.7	6.9
Services	49.1	49.3	50.0	50.3	50.9	..	Services	32.9	47.5	53.2	59.3	57.2	58.4
Romania							**Lithuania**						
Agriculture	19.7	18.5	21.6	20.6	20.5	19.8	Agriculture	20.2	11.6	11.0	7.3	9.3	13.3
Industry	39.5	37.3	34.9	37.6	35.7	37.3	Industry	45.3	39.4	30.4	25.8	29.0	27.7
Construction	4.5	4.7	5.4	6.8	6.8	6.9	Construction	5.0	9.3	7.8	8.7	6.7	7.0
Services	36.3	39.5	38.1	35.0	36.9	36.0	Services	29.6	39.7	50.8	58.2	55.0	52.0

Source: National statistics and direct communications from national statistical offices to UN/ECE secretariat.

[a] Percentage share of total value added.

[b] January-June for Bulgaria and Estonia; January-September for the Czech Republic, Lithuania and Slovakia.

According to preliminary estimates, in the first months of 1997 industrial output and the volume of retail sales in Russia grew — albeit modestly — in both January and February (compared with the same period of 1996), the first time this has happened in almost a decade.[173] It remains to be seen whether these positive developments will strengthen in the course of the year.

Progress was made with macroeconomic stabilization in Ukraine in 1996, but not in terms of output: GDP continued to decline at a double-digit rate for the sixth consecutive year. Moreover, the slide appears to have continued in the early months of 1997: preliminary estimates indicate a further shrinking of both GDP and industrial output. The process of enterprise restructuring and privatization in Ukraine has encountered serious delays and the unrestructured industrial sector lacks the dynamism which is essential for a robust recovery. In 1996 the Ukrainian authorities made only the first step towards deep economic reform; a second important step is planned in a reform package which has been negotiated in the framework of an eventual IMF programme which envisages wide ranging economic liberalization, as well as structural and fiscal reforms.[174] At the time of writing, this agreement has not been finalized.

---

[173] Statement by Prime Minister V. Chernomyrdin, quoted by Interfax News Agency, 7 March 1997. The reliability of these preliminary figures, however, has been questioned for methodological reasons. See *Financial Times*, 25 March 1997 and *Finansovye Izvestiya* (Moscow), 13 March 1997.

[174] Ministry of Economy of Ukraine/European Centre for Macroeconomic Analysis of Ukraine, *Ukrainian Economic Trends, Monthly Update* (Kiev), November 1996.

TABLE 3.2.2

Share of major sectors in GDP[a] in the CIS countries, 1991-1996

(Per cent of GDP, at current prices)

	1991	1992	1993	1994	1995	1996[b]
Armenia						
Agriculture	24.8	30.3	51.0	43.5	..	..
Industry	37.5	32.7	22.9	29.2	..	..
Construction	11.8	6.1	4.1	6.7	..	..
Services	25.9	30.8	22.0	20.7	..	..
Azerbaijan						
Agriculture	32.4	27.7	26.7	32.7	26.7	..
Industry	25.0	31.3	24.5	16.7	21.7	..
Construction	6.7	7.8	7.4	7.5	2.6	..
Services	35.9	33.2	41.4	43.2	49.1	..
Belarus						
Agriculture	20.9	23.4	16.9	14.1	12.7	..
Industry	41.2	37.8	29.3	28.2	27.0	..
Construction	7.8	7.2	7.6	6.1	5.8	..
Services	30.1	31.6	46.2	51.6	54.6	..
Georgia						
Agriculture	28.7	55.5	76.9	34.2	46.7	35.0
Industry	28.7	12.8	7.1	25.4	17.0	17.6
Construction	8.4	7.0	0.8	6.5	3.6	3.4
Services	34.3	24.6	15.1	33.9	32.7	43.9
Kazakstan						
Agriculture	29.0	29.1	18.3	14.4	12.9	13.5
Industry	27.7	30.9	30.6	26.7	26.4	22.5
Construction	7.7	7.4	9.2	9.4	6.4	3.8
Services	35.5	32.6	41.8	49.5	54.3	60.2
Kyrgyzstan						
Agriculture	36.9	39.0	40.0	39.9	43.1	32.3
Industry	29.3	35.1	27.2	22.3	16.9	23.4
Construction	6.6	4.1	5.5	3.5	7.0	5.1
Services	27.2	21.8	27.3	34.3	32.9	39.2
Republic of Moldova						
Agriculture	34.2	37.5	..	..	..	..
Industry	25.4	29.8	..	..	..	..
Construction	5.5	4.9	..	..	..	..
Services	34.9	27.8	..	..	..	..
Russian Federation						
Agriculture	14.0	7.3	8.2	6.5	7.6	7.5
Industry	39.3	35.0	35.0	33.3	33.5	30.7
Construction	9.4	6.3	7.9	9.1	8.6	9.0
Services	37.3	51.4	48.9	51.1	50.3	52.7
Tajikistan						
Agriculture	..	25.9	23.2	19.0	21.7	..
Industry	..	34.8	36.2	34.6	35.3	..
Construction	..	8.6	9.9	12.0	7.9	..
Services	..	30.6	30.7	34.4	35.2	..
Ukraine						
Agriculture	22.3	19.6	20.0	15.3	14.2	..
Industry	41.9	41.8	27.6	36.7	36.0	..
Construction	7.9	7.1	6.4	7.8	9.7	..
Services	27.9	31.6	45.9	40.2	40.1	..
Uzbekistan						
Agriculture	37.2	35.4	27.9	34.5	28.5	..
Industry	26.5	26.6	22.4	17.0	16.4	..
Construction	10.4	9.5	9.0	7.2	7.8	..
Services	25.9	28.4	40.8	41.3	47.3	..

Source: National statistics and direct communications from national statistical offices to UN/ECE secretariat.

[a] Percentage share of total value added.

[b] January-June for Georgia and Kyrgyzstan.

Among the other CIS countries, 1996 marked a notable turn around to positive growth reported from Belarus, Kyrgyzstan and Uzbekistan; in Azerbaijan, Kazakstan and Turkmenistan the fall in output seems close to bottoming out. In Armenia and Georgia the output growth in 1995 has been maintained, but the decline has continued in the Republic of Moldova and Tajikistan. The Moldovan economy, which is heavily dependent on agriculture, was badly affected by poor weather in 1996.[175] A more detailed account of economic developments in the central Asian CIS countries is given in chapter 4.

(b) Sectoral developments and industrial restructuring

The process of economic restructuring which is underway in the transition countries has profound implications for economic performance. The process – by and large driven by the emerging market forces – implies substantial shifts of factors and resources among different sectors of economic activity. However, after an initial period of dramatic changes during the first few years of transition, the restructuring process has tended to slow down somewhat in recent years (tables 3.2.1 and 3.2.2). There are a number of reasons for this apparent deceleration.

It should be underlined that the strikingly rapid structural changes that occurred at the start of the transition were mostly of a passive nature: they simply reflected the different speeds at which different economic activities were shrinking. With the start of the recovery in aggregate output, the very nature of the dynamics of structural change has altered. Shifts in the composition of output are beginning to reflect, to an increasing degree, the more active nature of the process of restructuring: the emergence of new economic agents as well as the different rates of growth of different economic activities.

Understandably, active restructuring is a fundamental economic issue for the economies in transition and by its nature is much slower than the earlier process when much capacity was simply being destroyed. The current stage involves defining the new place of these economies in the international division of labour and reflects their emerging comparative advantages in an increasingly interdependent world economy. The directions of economic restructuring in the transition countries are being shaped in the first place by the direction of the flows of new investment and, frequently, by the flows of FDI. However, it should be added that the speed of restructuring has been impeded by the lack, or the underdeveloped state, of capital markets and by the generally low degree of factor mobility in the transition countries.

---

[175] Government of Moldova/European Expertise Service, *Moldovan Economic Trends, Quarterly Issue* (Chisinau), February 1997, pp. 9-10.

TABLE 3.2.3
Gross output of agriculture and construction in the transition countries, 1991-1996
(Annual percentage change)

	Agriculture						Construction					
	1991	1992	1993	1994	1995	1996	1991	1992	1993	1994	1995	1996
Albania	-24.0	-15.0	..	..	..	..	..	..	..	..	..	..
Bulgaria	-0.3	-6.1	-18.3	6.8	15.4	-13.0	-59.3	1.5	-10.5	-6.7	2.5	-18.0^a
Croatia^b	-7.0	-13.0	5.0	-3.0	1.0	..	-25.0	-32.0	-11.0	-5.0	-4.0	9.0
Czech Republic	-8.9	-12.1	-2.3	-6.0	5.0	-0.2	-27.5	19.7	-7.5	7.5	8.5	4.8
Hungary	4.0	-25.7	-9.2	9.7	1.9	5.0	-12.6	1.5	2.6	12.4	-17.6	-1.1
Poland	-1.6	-12.7	1.3	-7.8	10.7	-4.0	11.4	6.6	8.0	0.5	8.1	6.4
Romania	0.8	-13.3	10.2	0.2	4.5	1.8	-26.5	4.6	11.5	29.0	13.2	7.0
Slovakia	-7.4	-13.9	-8.0	2.6	0.1	1.9	-30.5	6.2	-32.3	-6.8	2.5	3.7
Slovenia	-0.2	-5.5	-3.5	6.4	3.6	..	-25.7	-11.7	-18.2	-0.1	0.9	-2.5
The FYR of Macedonia^b	17.9	–	-20.3	8.2	5.0	-4.0*	..	5.1	-5.8	-10.6	-2.3	5.1^a
Yugoslavia^b	10.0	-18.0	-3.0	6.0	4.0	-3.0^a	-9.2	-21.6	-24.4	-10.9	-2.0	-10.0^a
Estonia	-21.8	-7.7	-6.9	-11.0	-2.4	-2.5^c	..	..	20.3	-4.2	4.0	19.2^c
Latvia	-2.1	-15.1	-13.4	-16.9	-3.4	-11.0	-43.8	-50.3	-49.5	11.6	16.5	1.0^a
Lithuania	-4.6	-23.8	-7.3	-18.1	1.8	..	..	..	-38.9	0.9	-0.9	-33.5^a
Armenia	–	-13.0	24.0	3.0	5.0	7.0	-35.0	-92.0	-24.0	-35.0	..	..
Azerbaijan	–	-25.0	-15.0	-13.0	-7.0	5.0	-14.0	-30.0	-27.0	-36.0	-43.0	..
Belarus	-5.0	-9.0	4.0	-14.0	-5.0	2.0	4.0	-29.0	-15.0	-11.0	-31.0	..
Georgia	-36.0	-13.0	-12.0	11.0	13.0	15.0	-62.0	-68.0	-54.0	-14.0	-3.0	..
Kazakstan	-10.0	1.0	-5.0	-20.0	-23.8	4.8	0.5	-47.0	-39.0	-15.0	-37.0	..
Kyrgyzstan	-10.0	-5.0	-10.0	-18.0	-9.0	3.0	-13.0	-35.0	-31.0	-42.0	..	..
Republic of Moldova	-10.0	-16.0	10.0	-25.0	3.0	-10.0	-9.0	-26.0	-44.0	-51.0	-17.0	..
Russian Federation	-5.0	-9.0	-4.0	-12.0	-8.0	-7.0	-2.0	-36.0	-8.0	-24.0	-9.0	-11.0
Tajikistan	-4.0	-27.0	-4.0	-25.0	-28.0	-15.0	..	..	..	..	..	..
Turkmenistan	-4.0	-9.0	8.0	-11.0	-18.0	-2.0	11.0	20.0	45.0	..	..	..
Ukraine	-13.0	-8.0	2.0	-16.0	-4.0	-8.0	-8.0	-37.0	-10.0	-23.0	-35.0	-36.0^a
Uzbekistan	-1.0	-6.0	1.0	-8.0	-3.0	-7.0	5.0	-32.0	-5.0	-22.0	4.0	..

Source: National statistics and direct communications from national statistical offices to UN/ECE secretariat.

^{a} January-September.

^{b} Construction, effective working time.

^{c} Half year.

In addition, there were other factors that affected the structure of output in recent years and which have led to a slowing down or even a reversing of some earlier changes. The most important of these was the relatively fast recovery of gross industrial output in eastern Europe, often at rates exceeding the rates of growth of GDP. In most east European and Baltic countries the share of industry in GDP seems to have stabilized since 1994 and in some of them has even begun to rise following a relatively strong performance in the last few years (table 3.2.1).

On the other hand, performance in other sectors such as agriculture and construction has not followed a uniform pattern and generally was quite mixed, even during years of strong GDP growth (table 3.2.3).

In 1996, grain crops in many east European countries (both in the south and in the north) were affected by bad weather. In the Russian Federation the grain crop was bigger than in 1995 but it was still one of the worst in the past 30 years.[176] The Russian output of other important agricultural products such as meat and milk decreased in 1996.[177] On the other hand, agricultural output increased in a number of CIS countries (table 3.2.3) and this had a positive impact on overall economic growth in these countries.

The agricultural sector in the transition countries is suffering from low efficiency and a markedly lower level of technology than in western Europe. Low labour costs have not always been sufficient to offset the gap in competitiveness and most of the countries are still heavily subsidizing their farmers and applying various measures of trade protection. In this regard, the agricultural sector in those fast reforming transition countries which are striving to join the EU will soon be facing new pressures from the need to adjust to EU regulations and conditionality. As to the CIS countries, progress in restructuring of agriculture has generally been much slower than in other areas of reform.

---

[176] Russia produced 69.3 million tonnes of grain in 1996 (compared with 63.4 million tonnes in 1995). However, three quarters of Russia's farms ran at a loss in 1996. Interfax News Agency, 21 January 1997 and OMRI, *Daily Digest*, 22 January 1997.

[177] During the first three quarters of 1996 the production of meat in Russia fell by 13 per cent and the production of milk by 8 per cent, compared with the same period of the previous year. Government of the Russian Federation/Working Centre for Economic Reform, *Russian Economic Trends*, Vol. 5, No. 3 (Moscow), 1996, p. 84.

TABLE 3.2.4

Official estimates of the private sector's contribution to GDP in selected transition countries, 1991-1996

(Percentages)

	1991	1992	1993	1994	1995	1996
Bulgaria	27	38	41	42	48	..
Czech Republic	17	28	45	56	66	75[a]
Hungary	30	47	55	60	65	75
Poland	42	47	52	53	58	..
Romania	24	26	32	39	45	52
Slovakia[b]	..	22	26	58	65	77[a]
Yugoslavia	..	..	..	33	..	..
Latvia	..	..	..	55	..	..
Lithuania	16	37	57	62	65	..
Kyrgyzstan	26	28	39	43	..	..
Republic of Moldova	..	..	..	..	..	..
Russian Federation	..	14	21	62	70	..

Source: National statistics and direct communications from national statistical offices to UN/ECE secretariat.

[a] January-September.

[b] Employment including cooperative sector and entrepreneurs.

The private sector is also expanding in most transition countries (table 3.2.4).[178] In fact some countries (most notably Hungary and the Czech Republic) have virtually completed the process of privatization. In Hungary, with very few exceptions, most "privatizable" companies have already been privatized and all large state owned commercial banks, with the exception of one, were privatized in 1995-1996.[179] The situation is similar in the Czech Republic, where the successful completion of the mass privatization programme[180] instigated the imitation of similar voucher programmes in a number of other east European, Baltic and CIS countries. These programmes (in most cases underway or nearing completion) have accelerated considerably the transfer of ownership in the transition countries.

One positive outcome of the process of restructuring in the transition countries is the significant gain in labour productivity, especially in industry, which can be observed in many transition countries in recent years (table 3.2.5). In most central and eastern European countries this process has been underway for several years already and the gains in industrial labour productivity as a rule have exceeded substantially the growth of industrial output. This suggests that, in general, recovery has been accompanied by enterprise restructuring and adjustment and, in particular, by the shedding of excess labour. The latest available figures indicate that the Baltic states and an increasing number of CIS countries (with some notable exceptions) are joining this group (table 3.2.5). However, in 1996 there was a slowdown in productivity growth in some east European countries, most likely because of weaker output and export growth and reflecting some losses in international competitiveness.

The recovery of industry has followed very different branch patterns of growth in the various transition economies (tables 3.2.6 and 3.2.7). Even in the fast reforming countries where, on average, there was a robust recovery of industrial output, growth has not been uniform across the board. In addition, the branch pattern of recovery has varied with different phases of the transition.

In 1994-1995, export-led industrial output growth prevailed, helped by the recovery in western Europe, the main export market for most east European manufacturers. During this period central and east European producers benefited from growing western demand for metals, chemicals and other manufactured inputs and intermediates (mostly resource-intensive, standardized, and bulk products). They also benefited from the availability of idle and underutilized capacities which could be put into operation quickly to satisfy the growing external demand. Indeed, this can be traced in the data on capacity utilization in manufacturing industry for those transition countries for which such data exist (table 3.2.8).

With the slowdown in the pace of recovery in western Europe in 1996, this element of demand-driven industrial growth lost momentum in the transition countries. Another factor that is increasingly curbing the growth of manufacturing exports from many of the transition countries is the persistent real appreciation of their currencies which has negatively affected their international competitiveness (see section 3.5). This has contributed to the stagnation, or in some cases recession, in 1996 in some of the export oriented manufacturing branches which had led the industrial recovery in previous years (table 3.2.6).

The continued recovery in central Europe, however, contributed to a strengthening of the domestic markets of the transition countries. The growing confidence of investors and consumers has been boosting domestic demand in a number of central European countries, a tendency which became especially pronounced in 1996. Indeed, as discussed below, expanding domestic demand – both final consumption and fixed investment – was the main factor of growth for a number of transition countries in 1996. This, in turn, affected the performance of individual industries and the pattern of industrial restructuring.

---

[178] It should be noted that the reported shares of the private sector in the different countries should be treated with caution because of a lack of uniform and universally agreed methodology for its estimation.

[179] The only remaining large state owned bank – the National Commercial and Credit Bank (K&H) – is due to be privatized in 1997. However, there also remain a number of large state owned enterprises which require further restructuring before being privatized.

[180] The state still holds sizeable stakes in a number of large Czech companies and banks but it intends to sell them in the near future.

TABLE 3.2.5

Measured labour productivity in the transition countries, 1991-1996
(Annual percentage change)

	Total labour productivity [a]						Labour productivity in industry [b]					
	1991	1992	1993	1994	1995	1996[c]	1991	1992	1993	1994	1995	1996[d]
Albania	-26.7	19.0	14.6	-1.4	11.4	..	..	..	..	..	..	..
Bulgaria	1.4	1.0	0.1	1.2	0.5	-9.6[e]	-5.2	-3.1	-2.8	12.6	6.6	6.2
Croatia	-12.7	0.1	2.6	3.0	3.0	6.5	-13.4	-0.8	-6.4	1.8	13.5	..
Czech Republic	-9.3	-3.9	0.7	1.8	2.1	2.7	-21.4	-0.2	-0.5	7.9	8.6	7.6
Hungary	-2.3	6.8	4.7	5.3	3.4	1.1[e]	-7.2	-0.2	16.3	14.7	10.5	3.9
Poland	-1.2	7.2	6.4	4.2	5.1	..	-4.2	16.6	9.7	12.8	6.0	9.9
Romania	-12.5	-5.9	5.5	4.5	12.9	..	-18.7	-10.0	10.4	8.6	16.2	13.8
Slovakia	-2.4	-7.5	-1.3	6.0	5.3	6.3	-10.7	-3.6	-5.0	8.0	8.3	2.5
Slovenia	-0.3	1.3	5.1	7.2	4.2	2.3[f]	-2.3	-3.2	6.8	11.2	5.8	7.6
The FYR of Macedonia	-9.5	-9.1	-9.2	-2.0	7.3	..	-9.9	-9.7	-9.4	-5.0	..	..
Yugoslavia	-8.9	-25.4	-28.8	4.6	7.5	..	-11.4	-17.1	-35.7	4.0	6.5	8.5
Estonia	-11.9	-8.2	3.8	-1.5	4.5	..	..	-28.1	-12.0	..	..	..
Latvia	-9.6	-32.3	-8.0	4.0	-0.3	4.1	4.8	-28.6	-19.8	5.7	4.0	2.4
Lithuania	-15.2	-32.5	-27.3	-7.2	5.0	2.2	-5.5	-25.9	-23.2	-12.7	14.9	8.3
Armenia	-11.1	-38.4	-6.7	8.4	7.8	7.1	-0.3	-41.4	0.3	7.5	20.0	4.1
Azerbaijan	-5.2	-19.3	-22.9	-17.9	-11.6	1.1	7.3	-18.6	3.1	-20.0	-12.0	-1.1
Belarus	1.3	-7.2	-9.4	-10.2	-4.1	7.7	0.7	-3.5	-7.5	-13.1	-0.9	8.2
Georgia	-12.4	-24.2	-33.0	-28.2	3.5	10.5[f]	..	..	..	..	..	..
Kazakstan	-4.2	-9.6	-10.3	-14.5	-8.4	2.7	-2.3	-10.4	-1.9	-21.8	1.3	4.1
Kyrgyzstan	-8.1	-17.7	-7.6	-18.4	-5.2	3.7	7.4	-23.7	-14.8	-21.5	-3.4	20.3
Republic of Moldova	-17.5	-28.4	-0.2	-30.9	-2.5	-3.1[f]	-4.3	-25.5	69.9	-23.6	12.0	-0.9
Russian Federation	-3.1	-12.4	-7.1	-9.7	-1.2	-5.0[e]	-6.3	-13.8	-12.0	-11.4	4.5	-2.8
Tajikistan	-10.0	-28.7	-14.7	-10.3	-12.8	..	-2.0	-22.2	..	..	-0.1	-15.1
Turkmenistan	-6.4	35.6	5.4	-19.1	-16.4	..	9.5	-12.1	-5.9	-22.6	-6.4	14.9
Ukraine	-7.2	-8.1	-12.2	-19.9	-14.4	-9.2	-4.0	-1.8	-3.0	-19.0	-10.6	3.0
Uzbekistan	-5.0	-10.5	-2.2	-4.0	-1.5	2.1	0.5	-5.8	1.9	12.3	-1.2	11.9

Source: National statistics and direct communications from national statistical offices to UN/ECE secretariat.

[a] GDP or NMP per one employed person, see tables 3.1.1 and 3.3.1; based on annual average employment.

[b] Gross industrial output per one employed person in industry.

[c] January-September except where otherwise mentioned.

[d] In Bulgaria, Hungary, Latvia, Poland, Slovakia, Ukraine and Yugoslavia estimated on the basis of the change in industrial employment in January-September 1996 over the corresponding period in 1995.

[e] Full year.

[f] January-June.

With domestic demand picking up, the engineering industry also started to recover in a number of countries, most notably in the Czech Republic, Hungary, Poland, Romania and Slovakia; there was also positive growth in some branches of manufactured consumer products (table 3.2.6). However, as discussed in the next section, domestic consumers are becoming more demanding and diversified in their tastes, while competition on the domestic markets of the transition economies is rapidly increasing. Recent experience shows that domestic producers still often tend to lose against competition from the more powerful and more technologically advanced multinationals and other foreign suppliers: this has been one element in the growth of import demand and the widening of trade deficits.

Thus local manufacturers have not always been able – due to inferior product quality and generally low competitiveness – to take full advantage of the boom in domestic demand. Manufacturers throughout eastern Europe are still predominantly locked into the niche of moderately priced, low- to medium-quality, traditional products.[181] Although there is still a demand for such products, it is declining, both domestically and abroad, and does not provide the basis for a viable strategy in the medium term. But the key to success in moving into new product areas is investment in manufacturing; the creation of an appropriate investment climate is therefore a high priority for policy makers. In any case, much still needs to be done in terms of modernization and restructuring before east European manufacturers are able to compete on a more or less equal footing with the leading producers on the world markets for manufactures.

---

[181] There are some exceptions regarding the emergence of highly competitive new east European manufactured products, in most cases the result of FDI and cooperation with multinational conglomerates. For example, the booming new (or modernized) automobile and auto parts industries of the Czech Republic, Hungary and Poland have resulted in the establishment of eastern manufacturers in both domestic and foreign markets.

TABLE 3.2.6

Growth of industrial output by branch in eastern Europe and the Baltic states, 1994-1996

(Annual percentage change)

	Fuels	Energy	Ferrous metals	Non-ferrous metals	Engineering	Chemicals	Building materials, wood products	Light industry	Food processing	Other	Total industry
Bulgaria											
1994	-3.4	-7.1	27.5	12.7	0.3	36.4	11.9	8.5	-1.3	-28.1	8.5
1995	6.9	9.9	8.9	-6.4	3.5	17.0	4.3	-9.6	1.5	-15.8	5.4
1996[a b]	13.3	9.4	-19.6	4.3	0.3	3.8	-0.4	-6.5	-7.5	-19.9	-1.1
Croatia											
1994	-9.0	-7.6[c]	..	..[d]	-18.1	3.9	..	-7.2[e]	4.5	..	-2.7
1995	19.2	4.9[c]	-36.0	13.0	14.6[f]	1.1	-0.4[g]	-9.9[h]	2.3	..	0.3
1996	-9.1	25.4[c]	-33.3	12.7	-8.4[f]	-1.6	28.9[g]	-12.8[h]	0.8	..	3.1
Czech Republic[i]											
1994	0.6	-2.8	2.8	..[d]	-4.8	3.8	..	5.6	-2.2	..	-0.1
1995	-1.4	3.4	12.3	..[d]	9.2	-1.1	..	0.1	4.0	..	7.1
1996	5.0	2.5	-2.6	..[d]	12.6	4.2	..	-5.5	3.9	..	6.8
Hungary											
1994	-16.3	1.8	20.6	..[d]	20.4	5.9	3.5	4.3	5.6	..	9.5
1995	-13.3	1.8	5.5	1.7	21.0	-1.2	3.5[j]	-4.8	1.6	..	4.6
1996[k]	2.8	5.0	-3.1	0.1	16.2	-0.7	-6.1[j]	-3.0	0.2	..	3.3
Poland											
1994	6.0	4.7	16.7	14.6	11.1	17.3	..	13.4	12.8	..	11.9
1995	-0.9	0.9	14.8	4.8	25.8	13.1	..	11.8	9.3	..	9.4
1996	1.1	-0.1	1.0	5.9	15.9	5.7	..	12.7	13.9	..	9.1
Romania											
1994	1.3	1.1	14.2	..[d]	25.6	3.0	..	16.9	12.4	13.0	3.3
1995	-0.6	2.1	17.5	..[d]	25.2	9.3	..	6.1	2.1	..	9.4
1996[a]	-2.1	5.3	-0.6	..[d]	29.1	-3.9	..	-5.6	-3.4	..	8.8
Slovakia[i]											
1994	18.9	2.9	..	..	-4.1	5.5	2.8	-4.0	-8.1	-3.6	2.9
1995	6.1	1.7	..	..	39.3	10.8	7.8	-4.1	-2.1	9.0	7.5
1996	-3.9	5.1	..	..	12.5	2.7	..	-2.0	5.7	11.0	2.9
Slovenia											
1994	-6.8[l]	..	3.1	17.4	6.5	18.6	19.0	-6.3	3.9	-4.3	6.4
1995	39.8[l]	..	-5.4	0.5	3.6	7.5	-0.5	2.6	1.6	0.8	2.0
1996[k]	-30.2[l]	..	-3.1	-3.4	-16.5	4.2	14.1	-0.3	2.0	..	0.5
The FYR of Macedonia											
1994	–	6.5	-43.3	-12.8	..	16.6	-9.8	-6.4[m]	-1.0	..	-10.6
1995	5.2[n]	3.0	3.0	-3.0	-34.7[f]	2.0	-28.1[j]	-24.9[m]	–	..	-9.8
1996	..	..	..	..	..	..	..	..	..	..	..
Yugoslavia											
1994	–	0.1	2.5	10.5	-9.2	12.6	-8.6[j]	2.6	2.1	-1.0	1.2
1995	7.5	..	13.5	19.0	-7.1	17.1	1.9[j]	-16.8	12.5	..	3.8
1996[o]	..	3.0	43.0	22.0	3.0	21.0	-7.0[j]	8.0	–	14.0	6.8
Estonia											
1994	-1.1	-2.4	24.7	3.1	7.4	44.2	21.2	-5.5	-10.4	-8.5	-2.0
1995	-4.4	-2.0	13.2	-9.0	11.2	8.3	..	9.9	-3.3	-4.5	1.9
1996[l]	7.9	11.5	..	-9.2	-16.7	2.2	..	5.5	-10.4	..	1.1
Latvia											
1994	23.7	-1.7	..	..	-20.1	-32.4	..	-22.8	-7.4	..	-6.8
1995	-17.3	-0.1	..	..	-11.2	-5.3	..	-14.0	-4.5	..	-6.3
1996[a]	15.2	1.1	..	..	-12.0	-6.0	..	-1.6	7.2	..	2.4
Lithuania											
1994	..	..	..	..	..	..	..	..	..	..	..
1995	-21.5	26.2	4.5	-5.0	-7.8	-4.2	-1.1	-5.0	-3.3	..	6.2
1996[o]	1.4	16.3	-17.1	-5.7	-6.2	14.3	8.5	4.5	-3.2	..	3.0

Source: National statistics and direct communications from national statistical offices to UN/ECE secretariat.

Note: The sectoral breakdown is not strictly comparable among the countries.

a January-September.
b Industrial enterprises of the public sector only.
c Electricity only.
d Included in ferrous metals.
e Textiles and clothing only.
f Machine building only.
g Building materials only.
h Textiles, clothing, footwear.
i Enterprises with more than 25 employees for 1994; for 1995 enterprises with more than 100 employees.
j Wood products only.
k January-November.
l Oil extraction and natural gas.
m January-July.
n Coal production.
o January-October.

TABLE 3.2.7
Growth of industrial output by branch in the CIS, 1994-1996
(Annual percentage change)

	Fuels	Energy	Ferrous metals	Non-ferrous metals	Engineering	Chemicals	Building materials, wood products	Light industry	Food processing	Other	Total industry
Armenia											
1994	..	-3.5	78.6	54.0	-9.3	22.7	-10.7	-21.0	17.3	..	5.3
1995	..	-11.3	9.1	54.6	-2.6	15.1	64.1	-7.4	-15.2	..	2.4
1996	..	8.0	..	21.3	-17.4	-10.6	5.7	-13.5	5.6	..	1.0
Azerbaijan											
1994	-8.7	-9.7	-74.1	-66.8	-27.7	-33.5	-35.4	-23.9	-24.5	..	-22.7
1995	-7.8	-6.2	-64.9	-41.7	-35.0	-17.3	-37.4	-14.7	-35.7	..	-17.2
1996	-3.1	-2.9	-50.0	-70.0	-8.6	9.1	7.7	-36.4	-10.7	..	-6.7
Belarus											
1994	-45.2	-11.3	-12.5	74.8	-14.1	-16.8	-28.8	-23.0	-5.5	..	-17.1
1995	12.4	-16.2	-0.8	-28.0	-20.8	10.7	-21.2	-34.5	-12.3	..	-11.7
1996	-5.4	-1.6	22.3	2.9	1.6	4.6	..	13.2	2.9	..	3.2
Kazakstan											
1994	-13.8	-14.4	-30.0	-25.7	-43.2	-40.9	-55.1	-55.9	-25.9	..	-28.1
1995	-11.7	-2.7	23.3	8.5	-21.1	5.7	-34.9	-41.5	-20.2	..	-8.2
1996	1.6	-8.9	-25.4	2.8	-9.4	-28.8	-32.5	..	29.7	..	0.3
Kyrgyzstan											
1994	-31.7	7.5	-38.9	-0.2	-54.3	-41.6	-38.9	-36.8	-31.5	..	-28.0
1995	-25.9	0.5	117.0	-16.6	-16.4	-15.5	-18.6	-34.9	-35.7	..	-17.8
1996	-10.5	0.4	..	6.9	-6.7	-9.1	33.3	–	50.0	..	10.8
Republic of Moldova											
1994	..	-18.3			-39.5	-41.6	-1.2	-47.7	-31.1	..	-27.7
1995	..	-5.9	..	-3.4	-43.8	-20.5	-22.1	15.5	..	-3.9
1996	..	5.6	..	..	-24.4	33.3	-9.4	..	-9.7	..	-8.5
Russian Federation											
1994	-10.2	-8.8	-17.3	-8.9	-30.8	-24.5	-27.3	-46.0	-17.5	..	-20.9
1995	-1.1	-1.0	8.6	2.8	-9.5	7.7	-7.4	-31.3	-7.6	..	-3.3
1996	-2.7	-2.4	-2.6	-5.0	-11.1	-2.0	-24.4	-25.0	-8.6	..	-5.0
Tajikistan											
1994	-35.4	-9.9		-7.4	-38.5	-32.0	-37.6	-37.9	-30.4	..	-25.4
1995	-19.0	6.4	..	1.6	-26.5	6.4	-73.1	12.1	-22.5	..	-5.1
1996	-7.7	-2.1	..	-12.5	13.0	-21.1	20.0	-25.7	-22.6	..	-19.8
Turkmenistan											
1994	-40.0	-14.9	-36.2	24.1	-10.2	-46.2	-23.9	-8.2	5.5	..	-24.7
1995	-8.5	-3.3	166.1	29.7	19.8	23.3	-22.3	1.7	-8.1	..	-6.4
1996	9.1	6.3	-93.2	..	155.9	-0.8	-3.2	61.2	-18.6	..	17.9
Ukraine											
1994	-17.2	-12.5	-28.8	..	-38.4	-25.5	-37.4	-47.2	-18.5	..	-27.3
1995	-10.5	-5.4	-6.5	..	-24.8	-9.0	-28.0	-29.1	-13.2	..	-12.0
1996	-5.2	-7.0	14.7	..	-14.3	-4.4	..	-2.9	-5.8	..	-5.1
Uzbekistan											
1994	2.5	-3.6	-32.9	-7.2	9.0	-22.2	-12.1	9.1	3.9	..	1.6
1995	-0.8	1.3	-11.4	-7.1	19.3	11.6	-13.3	-1.8	-12.2	..	0.1
1996	1.3	0.1	30.0	27.4	39.6	15.8	5.6	5.9	29.1	..	6.0

Source: CIS Interstate Statistical Committee; Russian Federation Goskomstat.

The above conclusions apply *a fortiori* to producers in the Commonwealth of Independent States. Indeed, total industrial output in the CIS as a whole in 1996 amounted to just under 50 per cent of its 1989 level, after seven years of continuous decline, whereas industrial output in eastern Europe in 1996 had reached 74 per cent of its 1989 level and has been growing since 1994 (appendix table B.4).[182] The recovery of industrial output in some of the CIS countries is marginal given the enormous falls which occurred in some of them in previous years (see table 3.2.7 and appendix table B.4).[183]

---

[182] The overall decline in the CIS is very much influenced by developments in Russia's industrial sector. Some CIS countries fared better: for example Uzbekistan's industrial output did not go into a major recession and has been growing since 1993 (appendix table B.4). It should also be noted that the fall in industrial output in the Baltic states was even greater than in the CIS: in 1996 it was still at only 38 per cent of its 1989 level (appendix table B.4).

[183] In this regard, the large positive growth rates in some CIS industrial branches in recent years mostly reflect recovery from extremely low levels and do not necessarily reflect large increments in absolute terms.

TABLE 3.2.8

Capacity utilization in manufacturing industry in selected transition countries, 1992-1996

(Per cent)

	1992	1993	1994	1995				1996			
				QI	QII	QIII	QIV	QI	QII	QIII	QIV
Bulgaria	58	56	59	60	60	61	63	63	60	59	64
Czech Republic	76	74	78	78	81	80	84	..	..	..	..
Hungary	68	71	75	76	76	..	79	77	77	80	..
Poland	..	63	66	68	69	69	71	69	71	70	..
Slovakia	..	72	77	74	75	74	76	78	79	77	78
Estonia	..	54	57	55	57	59	59	53	58	60	..
Latvia	..	46^a	48	49	51	50	51	52	52	55	56
Lithuania	..	50^b	50	44	44	44	45	46	47	49	50
Belarus	..	..	..	55	51	46	50	47	46	50	51
Russian Federation	..	62^a	52	51	36	33	46	44	45	39	44

Source: OECD, *Short-term Economic Indicators: Transition Economies* (Paris), various issues.

Note: Averages for 1992, 1993 and 1994 are computed from originally reported quarterly figures.

^{a} QII-QIV.

^{b} QIII-QIV.

The pattern of Russia's industrial performance is indicative of some of the trends prevailing in other CIS countries as well. Apart from metals and some resource-intensive manufactures in 1995, no single industrial branch in Russia registered output growth in the last three years. Moreover, the decline of industrial output in 1996 resumed across the board, affecting all mining and manufacturing branches, including the production of fuels, Russia's most important export item (table 3.2.7).[184] The situation in some Russian manufacturing branches is especially severe: in 1996 the output of light industry dropped by 25 per cent; that of building materials by 24 per cent; of engineering by 11.1 per cent, and food processing by 8.6 per cent. In all these sectors the decline has been underway for several years in a row. The chemical industry (including petrochemicals) – which is a potential source of growth in Russia – also suffered a further setback in 1996 and is reportedly working at a mere 34 per cent of capacity.[185]

The profitability of Russian industrial enterprises – and their financial state in general – deteriorated further after the launching of the austerity programme in February 1996, and this, in turn, had a negative macroeconomic impact. Thus, over 40 per cent of the 6500 industrial companies finished 1996 with serious losses;[186] consequently, the taxable income of the corporate sector in the year shrank substantially and generated much less revenue for the state budget than expected.[187] In fact this was one of the main underlying factors for the considerable exacerbation of the crisis in Russian public finances in 1996.

The bad financial state of some enterprises and the lack of, or delay in, restructuring is not only a problem of the Russian economy. As discussed in section 3.1(iii), the continued operation of large, loss-making state owned enterprises was one of the underlying factors of the financial crisis in Bulgaria. However, the problem exists – although its magnitude may differ – in most transition economies and it is a potential threat to their macroeconomic stability. Such enterprises often present policy makers with real dilemmas: closing them may be next to impossible due to political and social considerations, while keeping them afloat with public subsidies is extremely costly and poses a threat to the financial stability of the country.

The authorities in a number of transition countries have recently engaged in a new round of financial rescue operations to bail out financially troubled enterprises.[188]

---

[184] For example Russia produced in 1996 some 293 million tonnes of crude oil, 2 per cent less than in 1995. Reported by Interfax News Agency on 21 January 1997.

[185] Statement by Russia's Deputy Minister of Industry, M. Dvortsin, as reported by Interfax News Agency on 14 February 1997.

[186] Ibid.

[187] Thus, the level of *nominal* taxable corporate income in industry, construction and transportation during the first three quarters of 1996 declined by 46 per cent from the same period of 1995. As a share of GDP, taxable corporate income in January-September 1996 was a meagre 4.5 per cent whereas in the same period of 1995 it had been 12.7 per cent. See A. Frenkel, "Prognoz osnovnyh pokazatelei sotsial'no-ekonomicheskogo razvitiya Rossii na 1997 g.", *Voprosy statistiki*, 1 (Moscow), 1997, pp. 44-54.

[188] In 1993, Hungary also implemented a major bail out of financially troubled state owned enterprises (manufacturing firms in heavy industry and food processing, as well as agricultural cooperatives and the Hungarian Railways company). Their debts were purchased by the state and later a large proportion of them was transferred to the Hungarian Investment and Development Bank with the goal of eventual recovery. While this operation helped to rehabilitate the creditor banks, it did very little in terms of enterprise restructuring: only a very small portion of the restructured enterprise debt was actually recovered and the bail out had

In 1995 the Romanian government selected 151 big loss-making, state owned enterprises (responsible for 49 per cent of the total losses in the public sector of the economy) in a programme of financial surveillance. The enterprises were required to elaborate plans for restructuring and financial recovery which included, *inter alia,* the possibilities of rescheduling, reducing and writing off their outstanding debts (with the final costs being borne by the state). At the beginning of 1997 the overall results of the programme remained highly questionable: the very high public cost already absorbed by the programme had not brought about notable changes in enterprise performance and it is likely that the programme will be abandoned.[189]

In February 1997 the Slovak government proposed a plan for enterprise "revitalization" which is aimed at the financial rescue of "strategic" enterprises. The programme envisages debt relief – subject to the approval of a special commission – for a number of heavily indebted Slovak firms. The total amount of debt relief (which is due to be implemented through rescheduling or write-offs) is estimated at some 70 billion koruna ($2.1 billion), which is roughly equivalent to the total debt of these enterprises to the four largest commercial banks and to the state budget. At the time of writing, the programme is still subject to approval by the Slovak parliament.

The Russian government has also initiated in early 1997 a plan for rescuing financially troubled enterprises. The programme envisages a restructuring of enterprise debt to the federal budget which would take the form of debt-for-equity swaps by transferring company shares to the State Property Committee in exchange for the enterprise debt (effectively equivalent to partial renationalization of the companies). According to the provisions of the programme, part of the debt of an enterprise can be rescheduled in certain cases but the government can sell the acquired shares in the event that current liabilities to the state are not met.[190]

While these types of measures may provide some temporary relief, they tend to only postpone the necessity of facing up to the actual restructuring problems in all their severity. Besides, in the final analysis, financial rescue implies rising levels of public debt and, in order to avoid potentially worse crises, it may be salutary to recall in this regard Bulgaria's woeful experience with this type of policy.

In spite of the progress with structural reforms in Poland, it is estimated that there are still around 300 "strategic" enterprises and several industrial sectors that have remained largely untouched by privatization and restructuring efforts and there are fears that they may turn into "an enclave of non-market economy".[191] Among these enterprises the large and inefficient coal mining sector is a special case and is likely to remain a problem in the medium term.[192] The Polish authorities have yet to adopt a comprehensive policy for dealing with these difficult policy issues.[193]

The restructuring of the coal mining sector itself – which suffers from low productivity throughout the whole of eastern Europe and the CIS region – is a major problem in a number of transition countries (as it was a couple of decades ago in most of the industrialized world and continues to be so in some developed market economies). In fact, much adjustment has been made, often with the active involvement of the World Bank: between 1990 and 1996 more than 100 coal mines were closed in transition countries, involving the loss of more than 440 thousand jobs.[194] It is expected that by the year 2000 between 376,000 and 400,000 additional jobs will disappear from the coalmining industries of central and eastern Europe.[195] However, whatever arguments in terms of increased economic efficiency, they are often overwhelmed by political constraints and there are definite limits to the speed at which this sensitive sector can be restructured.

These examples are indicative of some of the major restructuring problems that policy makers are still facing – and are likely to continue to face for some time – in the transition countries. There are no easy solutions to them: they will require determined and persistent efforts on the part of the governments and further sacrifices on the part of the population of these countries.

---

almost no positive effect on enterprise performance. See I. Ábel and L. Szakadát, "Bank restructuring in Hungary", a paper presented at the Workshop on Bank and Enterprise Restructuring in Central and Eastern Europe, Athens, 28 February-1 March 1997.

[189] L. Croitoru, "Enterprise restructuring in Romania", a paper presented at the Workshop on Bank and Enterprise ...

[190] Reported in OMRI, *Daily Digest*, 6 March 1997.

[191] Statement by M. Belka, then economic advisor to the Polish President and subsequently Minister of Finance, in *Polish Banking & Economic Review*, Special Edition (Warsaw), November 1996.

[192] For example, at the end of 1995 the average labour costs per tonne of coal in Poland amounted to $17, compared with $5-$6 in the United States and Canada. In 1995, the coal mining sector recorded losses amounting to 413 million zloty and coal mines owed to the state 8.3 billion zloty in arrears of tax and social security contributions. See OECD, *Economic Surveys, Poland 1997* (Paris), p. 80.

[193] The story of the prolonged efforts to close down the Gdansk shipyard is indicative of the political difficulties of implementing such unpopular and painful measures.

[194] For example, during this period, 15 mines were closed and 60,000 jobs were lost in the Czech Republic; in Hungary the numbers were 20 mines and 20,000 lost jobs; in Romania 12 mines and 60,000 lost jobs; in the Russian Federation 37 mines and 147,000 lost jobs; in Poland 9 mines were closed, 13 were in liquidation, and a total of 150,000 jobs were lost. UN/ECE, *Highlights*, No. 30, February 1997.

[195] Ibid.

(ii) Demand

(a) Main developments in demand

In 1996, demand-side developments were relatively dynamic – albeit varying in nature – in most of the ECE transition countries. In most of central Europe and the Baltic countries (with the notable exception of Hungary) there were some conspicuous shifts in the composition of final demand: export growth slowed down while at the same time domestic demand (both consumption and investment) grew strongly, partly offsetting the negative impact of net exports and taking the lead as the main demand-side determinant of growth. As discussed in section 3.5, the loss of momentum in export performance reflects both a weakening of demand in the main trading partners of the transition countries and deterioration or stagnation in the latter's export competitiveness.

A similar, leading role of domestic demand occurred in all those CIS countries which had positive GDP growth in 1996, although in some of them investment remained depressed. In terms of the pattern of demand, the Russian Federation was an exception in having a large and expanding trade surplus. In south-eastern Europe and in the rest of the CIS, developments on the demand side were mixed and sometimes varied in the course of the year.

In principle, the contribution of the components of final demand to economic growth can be analysed on the basis of aggregate national accounts data (tables 3.2.9 and 3.2.10). Unfortunately, full national accounts are still not produced in all transition countries and where they are, they are usually available with a substantial time lag. However, the partial data which were available at the time of writing this *Survey* confirm some of the statements outlined above regarding shifts in the composition of demand. Thus, for the countries for which national accounts in real terms are available (the Czech Republic, Latvia, Romania and Slovakia), the preliminary and partial statistical data confirm the leading role of domestic demand as a determinant of growth in 1996.

Nominal national accounts (also partial and preliminary) are available for a larger number of transition countries (table 3.2.10). These indicate (with the exception of Bulgaria and the Russian Federation) a rise in the shares in 1996 of the domestic components of final demand and a sizeable negative contribution from net trade. In a number of cases (the Czech Republic, Slovakia and the Baltic states) these changes were quite pronounced.

Because of the incompleteness of national accounts data on the changes in final demand (table 3.2.11 and 3.2.13), it is useful to refer also to some indirect indicators of demand, such as changes in retail trade and real wages (table 3.2.12), and in the volume of investment outlays (table 3.2.13).

Although the driving forces behind the observed changes in demand in 1996 varied among the transition countries, there are some similarities. In the Czech Republic, Poland, Slovakia (table 3.2.13) and probably also in Slovenia, the main source of increased demand in 1996 came from booming investment: in all four countries it grew at double-digit rates. In the Czech Republic the investment drive was mostly in the non-financial business sector with the biggest increases in manufacturing, transport and communications.[196] In Poland, the boom strengthened a trend which had been underway for several years: business investment outlays increased substantially across the board in most economic branches with the private sector leading the process.[197] Most of the exceptionally high growth of investment in Slovakia came from large publicly-financed investment projects, mostly in infrastructure and utilities (construction of roads and the nuclear power station at Mochovice).[198] However, with the completion of these projects such a high rate of increase in investment is unlikely to continue. According to preliminary estimates, Slovenian investment also reached an unusually high peak[199] in 1996, with the bulk of new investment going into road construction, transport and communications.

At the same time, in most of these countries, personal consumption was also on the rise. The primary source for soaring consumption was the accelerated growth of real wages (table 3.2.12) and the strengthening of consumer confidence. One sign of increasing consumer confidence in Poland, for example, was an upsurge in household borrowing for purchases of consumer goods and durables, which led to substantial growth in household indebtedness.[200] Added to the acceleration in the growth of wages following the relaxation of wage regulation, this was one of the factors that prompted the tightening of monetary policy towards the end of the year (see section 3.1(ii)). On a negative note, there is growing concern that the consumption boom is leading to reduced savings and could compromise future investment growth.

---

[196] Czech National Bank, *Report on Monetary Development in the Czech Republic for the Period January-September 1996* (Prague), p. 12.

[197] Central Statistical Office, *Poland Quarterly Statistics*, No. 3 (Warsaw), December 1996, pp. 15-16.

[198] CS First Boston, *Economic Research – Europe*, 9 December 1996.

[199] Some preliminary estimates put the rate of real growth of investment outlays in 1996 at 74.6 per cent but there may be distortions in this estimate due to methodological problems. Institute of Macroeconomic Analysis and Development, *Slovenian Economic Mirror* (Ljubljana), January 1997, p. 15.

[200] Central Statistical Office, *Poland Quarterly Statistics*, No. 3 (Warsaw), December 1996, pp. 5, 13.

TABLE 3.2.9

Contribution of final demand components to real GDP growth in selected transition countries, 1992-1996

(Percentage points)

	1992	1993	1994	1995	1996		1992	1993	1994	1995	1996
Bulgaria						*Estonia*[a]					
Consumption	-2.7	-3.1	-3.8	-2.3	..	Consumption	..	-3.9	3.3	4.2	..
Fixed investment	-1.2	-2.8	0.1	1.2	..	Fixed investment	..	2.0	2.5	0.6	..
Changes in stocks	-1.2	-1.2	-1.8	-0.9	..	Changes in stocks	..	-2.2	-0.3	5.2	..
Net trade	-2.3	6.0	6.9	-4.9	..	Net trade	..	-4.5	-9.5	-1.7	..
Exports	..	..	..	..	..	Exports	..	14.0	19.4	..	..
Imports	..	..	..	..	..	Imports	..	-18.5	-28.9	..	..
Statistical discrepancy	0.1	-0.3	0.4	8.9	..	Statistical discrepancy	..	–	1.4	-7.2	..
GDP	-7.3	-1.5	1.8	2.1	..	GDP	..	-8.6	-2.7	1.2	..
Czech Republic						*Latvia*					
Consumption	6.0	1.5	2.5	1.2	3.8	Consumption	-23.4	-3.3	1.5	-0.9	1.2
Fixed investment	2.1	-2.1	4.4	4.1	3.9	Fixed investment	-3.7	-2.2	0.1	1.7	0.9
Changes in stocks	-5.8	1.7	-0.3	4.3	2.2	Changes in stocks	-10.7	-23.5	4.8	1.0	..
Net trade	-5.2	-1.0	-4.3	-6.8	-5.5	Net trade	2.9	14.1	-5.8	-3.0	..
Exports	3.4	4.3	0.1	3.4	3.2	Exports	6.8	-18.0	-6.2	..	..
Imports	-8.6	-5.3	-4.4	-10.2	-8.8	Imports	-3.9	32.1	0.4	..	..
Statistical discrepancy	-3.4	-1.0	0.4	2.0	–	Statistical discrepancy	–	–	0.1	-0.4	..
GDP	-6.4	-0.9	2.6	4.8	4.4	GDP	-34.9	-14.9	0.6	-1.6	2.5
Hungary						*Belarus*					
Consumption	0.5	4.6	-2.0	-5.1	..	Consumption	-7.0	-2.9	-7.4	-11.3	..
Fixed investment	-0.6	0.4	2.7	-1.0	..	Fixed investment	-4.3	-3.3	-3.5	-5.2	..
Changes in stocks	-3.6	5.0	1.7	2.7	..	Changes in stocks	-0.7	-0.1	-6.5	-0.5	..
Net trade	0.6	-10.6	0.5	4.9	..	Net trade	3.4	-3.9	2.3	2.6	..
Exports	0.7	-3.5	4.3	4.6	..	Exports	-15.8	-12.9	–	-1.0	..
Imports	-0.1	-7.1	-3.7	0.3	..	Imports	19.3	9.0	2.2	3.7	..
Statistical discrepancy	–	–	–	–	..	Statistical discrepancy	-1.2	-0.5	2.5	4.2	..
GDP	-3.1	-0.6	2.9	1.5	..	GDP	-9.6	-10.6	-12.6	-10.1	..
Poland						*Kyrgyzstan*					
Consumption	2.7	4.0	3.1	3.2	..	Consumption	-11.7	-10.7	-19.4	-14.3	..
Fixed investment	0.5	0.6	2.0	4.1	..	Fixed investment	-5.1	-3.1	-3.8	8.3	..
Changes in stocks	-3.4	2.2	-0.1	1.8	..	Changes in stocks	6.7	-6.8	-0.3	0.9	..
Net trade	2.8	-3.0	0.6	-1.4	..	Net trade	-4.3	5.2	5.1	-4.4	..
Exports	3.3	1.0	4.3	..	..	Exports	-25.9	8.5	4.4	-14.6	..
Imports	-0.5	-4.0	-3.7	..	..	Imports	21.6	-3.3	0.6	10.2	..
Statistical discrepancy	–	–	-0.3	..	..	Statistical discrepancy	0.5	-0.1	-1.7	4.1	..
GDP	2.6	3.8	5.2	7.0	..	GDP	-13.9	-15.5	-20.1	-5.4	..
Romania						*Republic of Moldova*					
Consumption	-4.4	1.1	3.5	10.4	3.1	Consumption	-23.6	-1.1	-29.3	..	..
Fixed investment	1.7	1.6	4.2	2.0	1.3	Fixed investment	-6.5	-8.5	-9.3	..	..
Changes in stocks	-3.5	-1.4	-3.0	-0.8	-0.2	Changes in stocks	-0.6	-7.4	-10.4	..	..
Net trade	-1.1	0.9	3.0	-4.1	-0.1	Net trade	3.3	6.1	17.9	..	..
Exports	0.5	2.0	3.7	7.3	-0.7	Exports	-2.5	17.7	-28.9	..	..
Imports	-1.6	-1.1	-0.7	-11.4	0.6	Imports	5.8	-11.6	46.8	..	..
Statistical discrepancy	-1.4	-0.6	-3.8	-0.4	–	Statistical discrepancy	-1.7	9.7	-0.1	..	..
GDP	-8.8	1.5	3.9	7.1	4.1	GDP	-29.1	-1.2	-31.2	-3.0	..
Slovakia						*Russian Federation*					
Consumption	-1.3	-1.2	-2.7	2.1	8.4	Consumption	-2.3	-0.5	-0.7	-1.4	-2.7
Fixed investment	-1.5	-1.4	-1.7	1.7	9.7	Fixed investment	-14.5	-6.2	-5.1	-1.2	-2.9
Changes in stocks	-3.1	-0.7	-1.7	5.8	2.4	Changes in stocks	-3.8	-4.0	-3.5	-1.3	-0.9
Net trade	-1.6	0.4	11.1	-3.5	-12.7	Net trade	9.9	3.3	-1.2	0.3	5.3
Exports	17.9	-0.1	8.7	2.4	-1.0	Exports	-13.3	-1.3	3.3	5.8	-5.4
Imports	-19.5	0.5	2.4	-5.9	-11.7	Imports	23.2	4.6	-4.5	-5.5	10.7
Statistical discrepancy	0.9	-0.9	-0.1	0.7	-0.9	Statistical discrepancy	-3.8	-1.3	-2.3	-0.5	-4.8
GDP	-6.5	-3.9	4.9	6.8	6.9	GDP	-14.5	-8.7	-12.7	-4.2	-6.0
Slovenia						*Ukraine*					
Consumption	-2.3	8.6	3.2	4.7	..	Consumption	-4.7	-15.7	-7.3	2.3	..
Fixed investment	-2.8	2.2	2.5	4.5	..	Fixed investment	-3.0	-5.7	-6.2	-6.5	..
Changes in stocks	2.2	1.6	-0.1	2.0	..	Changes in stocks	-0.3	–	–	0.1	..
Net trade	-2.5	-9.5	-0.4	-7.3	..	Net trade	-1.0	3.1	-2.1	1.0	..
Exports	-18.3	0.4	6.5	0.5	..	Exports	-11.8	-1.6	1.6	5.9	..
Imports	15.8	-9.9	-6.9	-7.8	..	Imports	11.0	4.7	-3.7	-4.9	..
Statistical discrepancy	–	–	–	–	..	Statistical discrepancy	-0.9	4.1	-7.3	-8.7	..
GDP	-5.5	2.8	5.3	3.9	..	GDP	-9.9	-14.2	-22.9	-11.8	..

Source: National statistics and direct communications from national statistical offices to UN/ECE secretariat.

[a] January-June in 1995.

TABLE 3.2.10

Composition of final demand in current prices in selected transition countries, 1994-1996
(Percentage of GDP)

	Consumption			Fixed investment			Changes in stocks			Net trade		
	1994	1995	1996	1994	1995	1996	1994	1995	1996	1994	1995	1996
Bulgaria	91.2	85.8	88.9[a]	13.8	14.2	12.7[a]	-4.4	0.2	-3.4[a]	-0.6	-0.2	2.0[a]
Czech Republic	79.3	76.7	73.5[b]	28.9	31.0	27.7[b]	-7.2	-3.2	6.5[b]	1.0	-4.5	-7.6[b]
Hungary	84.3	79.4	..	20.1	19.3	..	2.1	3.5	..	-6.5	-2.2	..
Poland	83.1	81.5	..	16.2	17.1	..	-0.3	1.2	..	1.0	0.2	..
Romania	77.3	80.8	80.2	20.3	21.9	23.3	4.5	2.4	2.1	-2.1	-5.2	-5.5
Slovakia	72.3	69.2	73.1	29.5	29.2	36.6	-6.2	-0.7	1.6	5.5	1.8	-11.1
Slovenia	77.3	78.1	77.2	19.6	21.2	22.9	0.5	2.0	2.0	2.6	-1.3	-2.1
The FYR of Macedonia	88.8	..	..	17.0	..	..	-0.7	..	..	-5.1	..	..
Estonia	81.3	80.5	82.1[a]	26.0	25.0	24.6[a]	2.6	1.9	3.0[a]	-10.7	-8.7	-10.4[a]
Latvia	78.8	82.3	80.4	14.9	16.6	16.8	4.2	3.1	11.6	2.1	-2.1	-8.8
Lithuania	90.6	89.3	87.9[c]	20.6	21.4	18.9[c]	-5.2	1.2	-2.6[c]	-6.0	-11.9	-10.9[c]
Armenia	105.8	..	..	20.2	..	..	3.2	..	..	-33.8	..	..
Azerbaijan	104.0	89.9	..	27.3	12.3	..	-11.4	7.4	..	-15.9	-11.4	..
Belarus	84.4	79.6	..	29.2	25.1	..	-0.3	–	..	-13.3	-4.7	..
Georgia	115.7	114.2	..	0.9	0.8	..	0.4	0.3	..	-17.0	-15.3	..
Kazakstan	83.3	77.9	..	19.0	18.8	..	2.4	5.5	..	-4.7	-2.3	..
Kyrgyzstan	97.0	89.9	..	12.1	22.0	..	-3.4	-2.4	..	-6.0	-8.0	..
Republic of Moldova	71.1	58.8[a]	..	9.3	7.4[a]	..	25.1	36.6[a]	..	-6.0	-3.4[a]	..
Russian Federation	69.1	67.6	69.1	21.5	19.8	20.5	3.7	3.3	2.7	4.6	4.0	5.3
Ukraine	67.8	69.8	..	23.5	20.0	..	11.6	14.1	..	-3.2	-4.0	..
Uzbekistan	83.9	..	..	20.5	..	..	-0.1	..	..	-4.3	..	..

Source: National statistics and direct communications from national statistical offices to UN/ECE secretariat.

[a] First half.
[b] January-September.
[c] First quarter.

TABLE 3.2.11

Real consumption in selected transition countries, 1993-1996
(Annual percentage change)

	Private consumption expenditure[a]				Government consumption expenditure[b]				Total			
	1993	1994	1995	1996[c]	1993	1994	1995	1996[c]	1993	1994	1995	1996[c]
Bulgaria	-0.8	-2.6	-1.6	..	-12.5	-11.9	-8.2	..	-3.6	-4.5	-2.9	..
Czech Republic	2.9	5.3	4.8	5.3	-0.1	-2.3	-7.1	1.0	2.0	3.1	1.6	4.3
Hungary	3.4	0.2	-7.4	-3.0	9.8	-7.4	-3.1	-1.0	5.4	-2.3	-6.1	-2.4
Poland	5.8	4.5	4.4	..	3.5	2.2	2.9	..	5.1	3.9	4.5	..
Romania	0.9	2.2	19.0	6.2	2.7	11.0	-7.6	-7.2	1.3	4.2	12.5	3.5
Slovakia	-1.4	-0.1	3.4	7.2	-2.1	-10.6	1.6	24.2	-1.6	-3.5	2.9	12.1
Slovenia	13.6	4.6	7.0	..	5.3	2.1	2.5	..	11.4	4.0	5.9	..
Estonia[d]	-10.2	3.0	3.3	..	14.2	8.2	10.9	..	-5.3	4.3	5.2	..
Latvia	-7.4	3.2	-3.0	..	1.6	-0.9	3.1	..	-4.9	2.0	-1.2	..
Azerbaijan	-24.9	-22.2	-0.3	..	1.2	-5.6	-16.6	..	-22.5	-19.7	-2.8	..
Belarus	-1.5	-13.4	-21.8	..	-10.5	-3.0	-1.8	..	-4.3	-10.4	-15.4	..
Kyrgyzstan	-9.1	-19.6	-19.9	..	-19.7	-22.0	4.4	..	-11.6	-20.1	-14.9	..
Republic of Moldova	-1.4	-36.9	..	..	-1.9	-52.9	..	..	-1.6	-42.2	..	..
Russian Federation	1.2	-0.7	3.7	-5.7	-6.4	-2.8	1.7	-1.4	-1.0	-1.3	-2.3	-4.5
Ukraine	-27.3	-8.8	9.4	..	0.2	-12.2	-8.6	..	-19.6	-9.7	2.8	..

Source: National statistics and direct communications from national statistical offices to UN/ECE secretariat.

[a] Expenditures incurred by households and non-profit institutions serving households.
[b] Expenditures incurred by the general government on both individual consumption of goods and services and collective consumption of services.
[c] January-September for Czech Republic and Hungary.
[d] 1995 data refer to January-June.

TABLE 3.2.12

Retail trade and real net wages in the transition countries, 1993-1996

(Percentage change over same period of preceding year)

	Retail trade[a]				Real net wages[b]			
	1993	1994	1995	1996	1993	1994	1995	1996
Bulgaria	-3.6	4.2	2.7	-9.3	..	-22.5	-4.1	-19.4
Croatia	-28.1	13.2	12.5	14.8[c]	-0.5	14.4	40.2	6.3[c]
Czech Republic	-2.1	5.5	4.8	11.4[c]	3.7	7.7	7.7	8.0
Hungary	2.8	-6.1	-8.6	-1.6[d]	-2.3	5.4	-11.3	-6.7[c]
Poland	7.0	3.0	2.3	8.4[c]	-2.9	0.5	3.1	6.2[c]
Romania	-7.7	8.4	29.0	7.9[c]	-16.7	0.4	12.6	9.7
Slovakia	10.0	1.6	2.0	7.4[c]	-3.9	3.7	4.7	8.2
Slovenia	3.3	5.2	3.1	3.9[c]	14.8	7.1	5.2	3.9[c]
The FYR of Macedonia	-9.5	-10.3	-1.8	20.7[c]	28.9	-10.2	-4.3	-1.4[c]
Yugoslavia	-26.7	64.7	11.1	..	..	..	..	..
Estonia	42.0	27.4	10.4	2.0[e]	3.9	10.2	6.2	0.4[c]
Latvia	-14.0	23.9	-1.0	-11.0[f]	5.0	12.0	-0.4	-7.2
Lithuania	-37.0	-16.1[g]	1.5[g]	7.0[h]	..	13.6	2.8	1.2
Armenia	-28.0	-1.4	54.0	12.5	-77.5	-43.5	23.2	33.6
Azerbaijan	-48.1	-24.8	-1.0	11.8	-38.4	-57.9	-19.1	12.5
Belarus	-14.2	-10.6	-23.0	23.3	-6.5	-30.9	-4.9	4.6
Georgia	-70.0	-25.0	155.0	11.8	..	..	..	..
Kazakstan	-17.0	-53.0	-14.4	38.5	-36.8	-35.9	1.8	3.6
Kyrgyzstan	-10.1	-4.4	-6.0	1.9	-53.3	-28.0	10.5	2.3
Republic of Moldova	-38.0	-41.7	-3.3	-5.6	-65.2	-40.6	3.1	4.8
Russian Federation	1.9	0.1	-7.2	-4.0	3.2	-7.8	-25.9	4.7
Tajikistan	-21.7	-28.8	-79.9	-40.1	-73.3	-52.0	-73.3	-25.3
Turkmenistan	-5.5	-56.6	..	..	..	..	..	..
Ukraine	-35.0	-13.6	-13.9	-11.4	-47.7	-11.1	19.3	4.4
Uzbekistan	27.5	-9.6	-4.3	21.0	85.7	-52.0	..	..

Source: National statistics.

[a] Retail trade for the CIS refers to total retail trade (including unregistered trade) except Azerbaijan and Belarus (1993), Georgia (1993-1995), Kazakstan (1993-1994), Republic of Moldova, Tajikistan, Turkmenistan and Ukraine (1993-1996), for which the coverage is incomplete.

[b] Nominal wages deflated by consumer price indices. For Bulgaria, the Czech Republic, Slovakia, the Baltic states and CIS countries relating to gross wages.

[c] January-September.

[d] January-October.

[e] January-November.

[f] Goods only.

[g] State, consumers' cooperatives and large private companies (with five and more employees).

[h] Including trade in small private companies and market places.

Another source of concern in 1996 was the slowing down in the pace of export growth (in Slovakia, exports of goods and services even declined in real terms – see table 3.2.9). At the same time, booming domestic consumption and investment, as well as appreciating currencies in real terms, have led to rapid growth of imports and to widening trade and current account deficits.

In Croatia there was an unexpectedly large increase in domestic demand in 1996 (mostly private consumption) thanks to the return of foreign tourists. Consumption was also boosted by higher real wages, which continued to rise after very strong growth during the previous two years (table 3.2.12). However, investment demand remained depressed and the trade balance worsened due to weak export performance (both investment and exports probably declined in value in 1996).

External balances also deteriorated in the Baltic states but this was not so much due to booming domestic demand (which was somewhat sluggish in all three countries) but to deteriorating trade performance, exports lagging behind the growth in imports. Preliminary data suggest that, as a result of fiscal tightening, private consumption stagnated or grew only marginally in Latvia in 1996[201] and this led to a sizeable increase in stocks (table 3.2.10); at the same time there was a modest growth in investment. There seems to have been a moderate growth of both private consumption and investment in Estonia.

The pattern of demand in Hungary in 1996 was distinctly different from that in the other central European

---

[201] As shown in table 3.2.12, there was a significant drop in the reported volume of retail trade but the coverage of these data is not comprehensive.

TABLE 3.2.13

Investment in selected transition countries, 1993-1996
(Annual percentage change)

	Gross capital formation				Gross fixed capital formation				Investment outlays			
	1993	1994	1995	1996	1993	1994	1995	1996	1993	1994	1995	1996
Bulgaria	-20.2	-10.1	64.7	..	-17.5	1.1	8.8	..	..	..	..	..
Czech Republic	-2.2	22.1	38.4	23.4[a]	-7.7	17.3	14.1	17.9[a]	..	..	..	..
Hungary	32.3	19.8	6.3	..	2.0	12.5	-4.3	-9.0[a]	2.5	12.3	1.2	-4.0
Poland	14.9	9.0	27.6	..	2.9	9.2	18.5	..	2.3	8.2	17.1	26.5
Romania	0.7	4.8	4.5	4.2	8.3	20.7	8.6	5.3	8.4	26.3	10.7	4.5
Slovakia	-7.3	-12.3	33.0	42.8	-4.1	-5.1	5.8	33.3	..	..	8.2	..
Slovenia	21.3	11.9	29.6	..	11.9	12.6	20.8	..	..	..	14.3	..
Estonia	-0.9	8.0	19.2[b]	..	10.0	10.2	2.3[b]	..	..	..	..	..
Latvia	-76.7	53.2	19.3	..	-15.8	0.8	12.6	..	-40.8	7.0	8.5	..
Armenia	..	..	..	..	..	..	..	..	-24.0	-35.0	..	..
Azerbaijan	..	..	..	..	..	..	..	..	-39.0	89.0	-18.0	74.0
Belarus	-12.5	-33.5	-26.1	..	-15.4	-17.2	-27.0	..	-15.0	-11.0	-31.0	-10.0
Georgia	..	..	..	..	-49.1	..	..	..	-62.0	-0.5	2.0	19.0
Kazakstan	..	..	..	..	..	..	..	..	-39.0	-15.0	-37.0	-35.0
Kyrgyzstan	-50.3	-35.0	..	..	-21.8	-28.9	..	..	-31.0	-42.0	82.0	18.0
Republic of Moldova	-23.5	-40.7	..	..	-41.3	-76.5	..	..	-44.0	-51.0	-17.0	-15.0
Russian Federation	-29.4	-31.8	-12.4	-20.1	-25.8	-26.0	-7.5	-18.5	-12.0	-24.0	-10.0	-18.0
Tajikistan	..	..	..	..	..	..	..	..	..	-43.0	-25.0	..
Turkmenistan	..	..	..	..	..	..	..	..	45.0	..	..	..
Ukraine	-24.6	-27.1	-37.1	..	-30.5	-41.1	-56.7	..	-10.0	-23.0	-35.0	-20.0
Uzbekistan	..	..	..	..	..	..	..	..	-5.0	-22.0	4.0	7.0

Source: National statistics and direct communications from national statistical offices to UN/ECE secretariat.

[a] January-September.

[b] January-June.

transition countries. Following the austerity programme of 1995, all the components of aggregate final domestic demand continued to fall for the second consecutive year (tables 3.2.11, 3.2.12 and 3.2.13).[202] The negative impact of the stabilization programme on domestic demand now appears to have been stronger than originally expected. For example, real wages in 1996 fell more than twice as much as envisaged by the government at the beginning of the year.[203] However, this macroeconomic adjustment is starting to pay off: domestic and external imbalances were reduced considerably, manufacturing exports performed quite well and, as noted earlier, the economy appears set to embark on a path of self-sustained growth.

In Romania domestic demand changed considerably in the course of 1996, although the average annual figures indicate strong growth of private consumption and investment (tables 3.2.11 and 3.2.13). In the pre-election period (general elections were held in Romania in November) and, especially, in the first half of the year, monetary and incomes policies were expansionary and there was loose control over public finances. This loosening of policy emerged in 1995, in an attempt to boost domestic demand and economic growth, but its populist character intensified considerably in 1996. The growing macroeconomic imbalances prompted a deceleration in the rate of expansion in the second half of the year and the stance of policies was reversed after the November elections. The new government announced its intention to restore macroeconomic equilibrium by severe tightening of monetary and fiscal policies, in order to avoid a financial crisis, and to accelerate structural reforms. This type of adjustment inevitably requires substantial curbs on domestic demand. In fact this process was already underway in Romania in the second half of 1996: while the volume of retail trade grew at double-digit rates during the first two quarters, its growth markedly decelerated in the third quarter and actually turned negative in the last quarter of the year, partly reflecting the erosion of incomes due to the rising inflation.[204] A similar decline appears to have hit investment activity during the last quarter of 1996.[205] If

---

[202] According to preliminary estimates, during the first three quarters of 1996 real private consumption expenditures in Hungary declined by 3 per cent and real gross fixed capital formation dropped by 9 per cent from the same period of 1995. Hungarian Central Statistical Office, *Gross Domestic Product, Third Quarter* (Budapest), 1996.

[203] During the first three quarters of 1996 real net wages declined by 6.7 per cent (table 3.2.12). Preliminary estimates for 1996 as a whole indicate a drop of more than 5 per cent. National Bank of Hungary, *Recent Economic Developments in Hungary* (Budapest), February 1997, p. 8.

[204] UN/ECE secretariat estimates, based on Romanian national statistics.

[205] While real investment outlays in Romania grew during the first three quarters of 1996 at an annualized rate of some 8 per cent, the

the government continues with its new austerity programme, the decline of domestic demand – both in private and in government consumption as well as in investment – can be expected to continue in 1997.

The deep financial crisis in Bulgaria reverberated throughout the whole economy and, *inter alia*, led to notable changes in aggregate demand. The first effects of the crisis were on the supply side of the economy where, as already discussed, it resulted in widespread disruptions. However, the accelerating rate of inflation in the second half of the year, and especially during the first months of 1997, led to a rapid erosion of real incomes. This, in turn, resulted in a considerable reduction of personal consumption: thus, according to preliminary estimates, the volume of retail trade in the last quarter of 1996 was some 30 per cent lower than in the same period of the previous year.[206] The shrinking of private consumption probably accelerated in the first quarter of 1997 in the wake of the hyperinflation which occurred in January and February. Although full-year national accounts data were not available at the time of writing this *Survey*, all of the other components of final demand in Bulgaria were negatively affected by the crisis in a similar manner.[207] The severe constraints on domestic demand, which are likely to continue in 1997, will be a deterrent to a rapid post-crisis recovery in Bulgaria.

In the Commonwealth of Independent States, an important change in 1996 was the apparent recovery of domestic demand in some CIS countries: at least, the volume of retail trade grew in double-digit rates in Armenia, Azerbaijan, Belarus, Georgia, Kazakstan and Uzbekistan (table 3.2.12). There was also a notable recovery in real wages throughout this region. Investment has been slower to recover, but there was a marked turn around in Azerbaijan, Georgia and Uzbekistan and significant growth in Kyrgyzstan (table 3.2.13).

However, the evolution of demand in the two largest economies – the Russian Federation and Ukraine – was characterized by some puzzling features. One was the apparent coexistence of two seemingly contradictory trends: namely, growth in real wages and a contraction of the volume of retail trade (table 3.2.12). At the same time, there was a large drop in real investment outlays in both countries (table 3.2.13), and other information suggests that the rise in real incomes was not transformed into increased (formal) savings.[208]

The available information on some recent developments in the Russian Federation suggests several possible interpretations.[209] One plausible explanation of the paradox is the continuation and even intensification of capital flight from Russia:[210] a growing share of domestically earned incomes was probably exchanged for foreign currency which then was either transferred out of the country or held outside the formal banking system.[211] Another possible interpretation (which is not an alternative but rather a supplement to the above) is that the official statistics report wages on an accrual basis, i.e. the level of wages that were *due* to be paid rather than those that were *actually* paid.[212] With the aggravation of the payments crisis in Russia in 1996 (see section 3.1(ii)), an increasing share of wages was not paid on time and the volume of wage arrears increased throughout the year.[213] Thus the reported growth in real wages may not accurately reflect the actual increase in the purchasing power of income earners and, hence, would not have translated into higher consumer demand. As discussed in the next section, the wide income differentiation in Russia (although slightly reduced in 1996, at least on an accrual basis) also contributes to the depressed consumer demand in the country.

Investment in the Russian Federation continued to decline at a double-digit rate in 1996 (table 3.2.13). This reflects both the continuing low confidence of investors and a reduction in state funds allocated for

---

reported average rate of growth for the year as a whole was only 4.5 per cent (table 3.2.13), implying falling investment during the last quarter of the year.

[206] UN/ECE secretariat estimates, based on Bulgarian national statistics.

[207] One indirect indication is the contraction of imports in the course of 1996 and, accordingly, the trade balance: while Bulgaria's trade balance was negative during the first three quarters, the preliminary figures for the year as a whole indicate a small surplus (table 3.1.1).

[208] According to official Russian estimates, the savings rate in Russia in 1996 remained unchanged from 1995 at around 23 per cent. Government of the Russian Federation/Working Centre for Economic Reform, *Russian Economic Trends*, Vol. 5, No. 3 (Moscow), 1996, p. 70.

[209] No comparable Ukrainian statistics were available but at least some of the issues noted for Russia probably apply to Ukraine as well.

[210] Regarded here in the broader sense as flight from the official financial system. Money hoarding, or keeping cash "under the mattress" (almost exclusively in the form of foreign exchange), is a form of residential capital flight which, however, has the same monetary effect on the financial system as the actual flight of capital across the border.

[211] For example, in January 1997 Russians bought $5.2 billion of foreign currency which accounted for 24 per cent of Russians' total income that month. The population's foreign currency purchases increased from 14 per cent of total income in 1995 to 18.5 per cent in 1996. Goskomstat estimates as reported by OMRI, *Daily Digest*, 4 March 1997.

[212] Government of the Russian Federation/Working Centre for Economic Reform, *Russian Economic Trends*, Vol. 5, No. 3 (Moscow), 1996, p. 57.

[213] As reported by the Goskomstat, the amount of wage arrears of the group of enterprises burdened with wage arrears in December 1996 amounted to 272 per cent of the monthly wage fund due to be paid, up from 153 per cent in January 1996 (Goskomstat Rossii, *Sotsial'no-ekonomicheskoe polozhenie Rossii* (Moscow), 12 December 1996). According to a recent survey, only 30 per cent of wages in Russia were paid on time and in full in 1996, down from 45 per cent in 1995 (reported in OMRI, *Daily Digest*, 13 December 1996).

investment, a result of the tough stabilization programme. The very strong Russian export performance in 1996 – for the third consecutive year – failed to boost GDP growth because of the offset from severely depressed domestic demand. The growth in Russian exports thus reflects to a certain degree a diversion of resources from the depressed domestic market.

(b) Demand-side factors of growth in the transition

The widely diverging patterns of economic performance and the different rates of recovery in the east European countries raise a number of questions – both conceptual and empirical – regarding the determinants of growth during the transition from plan to market. Identifying the principal factors of growth and revealing the specific mechanisms and causal relations that feature in the transition process is of key importance to policy makers in these countries, since high and self-sustained rates of growth would be an indicator of success in the process of economic transformation. However, so far the level of understanding of the particularities of recovery in a transition economy is not very high and many puzzles remain. An indication of this state of affairs is the generally low degree of reliability of economic forecasts for the countries in transition.

A previous issue of this *Survey* focused on the supply-side determinants of growth in the transition;[214] in what follows, a brief review is made of some of the demand-side factors that are likely to affect the process of recovery and growth in the transition economies. It is not intended to conduct an exhaustive discussion of the topic; but simply to outline some of the important and specific developments on the demand side that have played, and are likely to continue to play, a role in shaping the dynamics of economic performance in these countries.

Shifts in aggregate demand

The process of economic transformation has led to major structural changes in aggregate demand in the former centrally planned economies. Decades of adherence to planning dogmas had led to serious distortions in the composition of aggregate demand with an overaccumulation of fixed capital with very low rates of return and repressed levels of consumption and shortages of consumer goods. The process of economic liberalization leading to a market based formation of demand brought about significant changes in its structure, notably a fall in the share of gross capital formation and an increase in the share of consumption).

Another fundamental change in the pattern of demand, which resulted from the end of the Cold War, was the large decline of military demand. The virtual elimination of a wide range of military orders made a large amount of production capacity redundant and a large number of employees in the defence industry unemployed. This effect was most pronounced in Russia but it was present in all the former Warsaw Pact countries.

These changing patterns of demand (equivalent to permanent shifts in, and changes in the slope of, the aggregate demand curve) have had a fundamental impact on economic performance in the transition countries. Obviously, the more orthodox the economic system of a country under communism, the more pronounced have been the subsequent shifts in aggregate demand and their repercussions on performance.

The logic of these changes suggests that the transition shifts in aggregate demand are likely to have led to a general fall in the level of domestic demand and, in particular, of demand for domestically produced goods; thus, they had a negative impact on growth during the transition. Although, as discussed below, the causal relations were different, the negative impacts hit both the manufacturing sectors producing capital goods (and construction) as well as those supplying consumer goods. In fact this is one of the fundamental features of the transition depression observed throughout the region in the first years after the start of reforms.[215]

The adjustment of the investment ratio from disproportionately high to more "normal" levels[216] took place as the result of simultaneous shifts in both the business and public components of aggregate investment demand. Business investment – very much in line with conventional cyclical development – contracted substantially during the transition recession (the latter's duration varied across countries and still continues in some) but then started to pick up quickly in the recovery phase. However, it follows from the discussion above that in the long run the relative share of business investment is likely to remain lower than in the pre-transition period, even in a phase of steady growth.

Public investment also fell considerably during the recession, a reflection not only of the general decline in economic activity but also of the withdrawal of the state

---

[214] UN/ECE, *Economic Survey of Europe in 1995-1996*, pp. 67-72.

[215] See also J. Kornai, "Transformational recession: the main causes", *Journal of Comparative Economics*, Vol. 19, No. 1 (Orlando, FL), August 1994, pp. 39-63.

[216] For example the share of gross capital formation in GDP for the CIS countries as a whole dropped from 34 per cent in 1991 to 28 per cent in 1995 and continued to decline to around 20 per cent in 1996 (CIS Statistical Committee, *CIS Statistical Bulletin*, 1(191) (Moscow), 1997, pp. 75, 77). But it should be added that not all former centrally planned countries displayed disproportionately high investment ratios at the outset of the transition. For example this ratio had already started to decline in Hungary and Poland in the 1980s. Consequently, the necessary transitional adjustment was smaller in these countries.

from a range of activities, as well as the fiscal constraints of stabilization programmes. In the face of the rising budget deficits in many transition countries, public investment expenditure was usually among the first items to be subject to spending cuts and its subsequent recovery during the upturn in activity has been much less pronounced than that of private investment.

Changes in the patterns of investment demand

The decline of investment was a shock to the transition economies which, in some cases, led to a virtual collapse of some domestic suppliers of capital goods as well as construction companies. Likewise, it was a shock also to traditional suppliers of capital goods in other transition countries. Indeed, the sectors producing investment goods have been among the biggest losers from the painful adjustment of aggregate demand in the course of transition.[217]

Moreover, when recovery got underway and investment started to grow rapidly in some of the transition countries, this did not always result in an equi-proportionate recovery of the demand for domestically produced investment goods. The main problem lies in the actual technological level of the capital goods sectors in the transition economies. After decades of isolation from world market competition, as well as the COCOM strategic embargo on technology transfer, there is a large gap between the technology levels of eastern Europe and those of the more developed market economies, a gap which is especially pronounced in the more sophisticated, R&D and technology-intensive manufacturing branches.

The process of economic restructuring and industrial modernization of the transition countries, however, may sometimes call for a different set of technologies (rather than a replication or elaboration of existing techniques) and this can lead to a specific type of new investment demand in which the fastest growing segment is that of sophisticated equipment with a high technological and know-how content.[218] This type of demand (often associated with foreign direct investment) can hardly be met by domestic suppliers which results in the acceleration of import demand and widening of trade deficits.[219]

The technological backwardness of the capital goods sectors is a structural problem for most transition economies, especially as catching up with western technology in these particular sectors is all the more difficult due to their high capital intensity. Moreover, as world demand for investment goods is increasingly shifting towards high technology products, east European producers cannot rely on growing export markets for their output. Thus the future of these industries in the transition economies will largely depend on their ability to modernize their production capacities and/or their success in attracting FDI. In any case, it is doubtful whether this process can be successful across the board; at least some of the existing production capacities of heavy industry will have to be phased out.

At the same time, it needs to be taken into account that the process of systemic transformation also encompasses a delayed transition to a post-industrial economy in many of the former centrally planned economies. This transition by itself is also rather painful, particularly as the former systems of central planning had resulted in overindustrialization and underdevelopment of the tertiary sector. However, while many existing industries were based on mass industrial production with large-scale employment concentrated in perhaps one or a few locations, the newly emerging activities in the service sector, while providing new job opportunities, do so under quite different conditions. Apart from requiring specific skills (which are not always to be found among traditional blue collar workers), the new service activities rarely offer geographical concentrations of large-scale employment and so often there are locational mismatches between new job opportunities and those that have been lost. This makes both the structural and the frictional unemployment problems especially severe in the transition economies.

To sum up, the adjustment of fixed investment to the new market environment is likely to result in a permanent reduction of investment ratios from their pre-transition levels, at least, in the medium and long run. In addition, due to the inherent supply-side constraints, even booming domestic investment demand is unlikely to

---

[217] For example, in the period 1990-1995, the gross output of the engineering branches in Russia shrank by a total of 60 per cent; in Ukraine by 58 per cent; in Bulgaria between 1989 and 1993 by 66 per cent (Russian Federation Goskomstat, *Russia in Figures, 1996* (Moscow), 1996; Ministerstvo Statystyky Ukrainy, *Statistichnii shchorichnik Ukraini za 1995 rik* (Kiev), 1996; National Statistical Institute, *Statistical Yearbook 1994* (Sofia), 1995). This decline was not so pronounced in the central European transition countries: in Poland it stopped in 1992-1993; in the Czech Republic and Hungary in 1994; in Slovakia there was a substantial turn around in 1995 (see table 3.2.6).

[218] The rapid expansion of modern car (and car parts) manufacturing in central Europe is a case in point. For example, the new Volkswagen-Skoda assembly plant in Mlada Boleslav is considered to be among the most modern car assembly facilities in the world. Also, some of the world's leading car manufacturers are using the new facilities in central Europe to experiment with new, highly automated production lines and new organizational techniques that are often opposed by unionized west European labour.

[219] For example, it is estimated that Czech domestic demand for engineering goods in the period 1993-1996 expanded by 43 per cent whereas the total sales of the Czech engineering sector increased in the same period by only 28 per cent. Although domestic output grew substantially it failed to match the even faster growing demand for engineering products; the gap in demand was filled by foreign suppliers. *Czech Business and Trade*, No. 11 (Prague), 1996.

TABLE 3.2.14

Machinery and equipment share in total investment of selected transition countries, 1991-1996

(Per cent)

	1991	1992	1993	1994	1995	1996
Bulgaria	47	46	36	46	44	..
Bulgaria[a]	..	..	..	49	48	50[b]
Croatia	..	51	50	49	48	..
Czech Republic	46	46	49	50	49[b]	..
Hungary	46	48	48	46	49	44[b]
Poland	30	38	38	41	47	..
Poland[c]	..	..	..	51	57	59[d]
Romania	39	40	43	52	45	44
Slovakia[e]	51	..	48	44	49	..
Slovenia	54	51	50	51	..	..
Estonia	..	..	58	52	52	39
Latvia[e]	33	34	37	41	52	..
Lithuania	25	23	37	33	41	..
Belarus[f]	47	33	38	44	45	..
Republic of Moldova	52	40	31	31	35	..
Russian Federation	32	25	24	20	..	..
Ukraine	35	34	28	43	44	..

Source: UN/ECE secretariat Common Database; national statistics and direct communications from national statistical offices.

[a] Share in total investment outlays of public sector only.
[b] January-June.
[c] Based on an incomplete sample.
[d] January-September.
[e] At constant prices.
[f] Including surveying and geological work.

induce matching growth in the domestic capital goods industries, at least in the short run, but instead will generate additional import demand.

At some point, the adjustment of investment demand from pre- to post-transition levels will come to an end and from then investment activity will only be determined by the expectations and confidence of investors as to future returns, as well as the other factors such as risk, cost of capital, etc., which normally enter the investment function in a market economy. As regards this new market based type of investment activity, there is already a dramatic differentiation among the transition countries. On the one hand, in the more advanced reforming countries, there has been a strong recovery of investment in the last three to five years and in some of them (Hungary, the Czech Republic, Slovenia, Estonia and Poland) there has also been a large inflow of foreign direct investment. At the same time the continuing downward trend of investment in most of the CIS countries, as well as the lack of a stable recovery in some of the south-east European countries, seems to indicate something more than just the problems of "transformational adjustment". There does appear to be a more serious lack of progress in these countries in establishing the necessary economic and institutional environment required to encourage investment: macroeconomic and political stability, transparency and predictability of laws and their enforcement, in effect, the minimum set of institutions required to support efficient, market based economic activity.

The delays in creating the appropriate investment climate are deterrents to economic recovery but, more important, they seriously undermine the medium- and long-term growth prospects in the affected countries. Business investment, and especially investment in machinery and equipment (see table 3.2.14), is crucial for the modernization of the transition economies and for improving the competitiveness of their producers on both international and domestic markets. To the extent that some transition countries are now lagging behind others in new investment, they are likely to lag further behind in terms of growth and development levels in the future unless exceptional efforts are made to catch up. Encouraging investment and improving the investment climate thus need to be high on the agenda of policy makers especially in those transition economies where, so far, investment has not picked up; at the same time, probably greater emphasis needs to be given to these issues in the assistance programmes of the international financial institutions and by other suppliers of official and multinational assistance.

Changes in private consumption patterns

The changes in private consumption patterns during the transition process have been no less dramatic than those in aggregate investment. The liberalization of trade provided domestic consumers with access to the wide diversity of consumer goods available on the world markets and which, before 1989, used to be beyond the reach of most local consumers. Typically this resulted (at least in the first phase of the transition) in extensive substitution of higher quality western goods or cheap imports from developing countries for domestic products or products previously supplied within the CMEA trade arrangements. The supply-side disturbances in the transition countries and the disruption of their mutual trade exerted an additional pressure to buy western products. Although the demand for local and traditional consumer products eventually partly recovered, the opening up of the transition countries in general has resulted in a permanent reduction in the demand for locally produced consumer goods as well as in the demand for those produced in other transition countries. This transitional adjustment in private consumption (as in the case of investment) delivered a shock to local producers, especially to those who could not switch to alternative markets. As a result the consumer goods sectors of the transition economy were also adversely affected by the transitional shift in aggregate demand but, on average, to a smaller extent than the capital goods sector. Besides, the adverse effects varied considerably among industries and countries.

The liberalization of prices and international trade, as well as the elimination of most subsidies, led to significant changes in relative prices which instigated further shifts in the patterns of private consumption demand. These changes were reinforced by the fall in the average level of real income during the period of transformational recession and which still continues in some countries (especially in the CIS). This initial adjustment resulted in a general decline in the level of private consumption demand.

The impact of income differentiation

The changes in the distribution of income and wealth that have taken place in the transition countries since the collapse of communism have also had a profound impact on private consumption demand. Centrally planned economies were, on the whole, highly egalitarian societies: a number of empirical studies have shown that, on average, these countries had the lowest levels of measured income inequality as compared with all other country groups in the world.[220] In terms of consumer demand, this even distribution translated into relatively homogeneous consumption patterns with a significant proportion of consumers clustered into what used to be a "middle-income" category. However, pre-transition data on income inequality should be treated with some caution due to the price distortions and supply shortages, but also because of the lack of correlation between money income and access to resources under the communist regime (where the latter was largely determined by position in the political hierarchy).

The release of market forces and of the previously dormant and suppressed entrepreneurial spirit, *inter alia*, brought this apparent egalitarianism to an end. Indeed, the changes in wealth and income distribution are among the most dramatic changes that have taken place in the transition countries in recent years. It seems likely that never before in peacetime human history has such a profound redistribution of measured income and wealth taken place within such a short period of time.[221]

The most striking redistribution of measured income and wealth in recent years has occurred in some of the successor states of the Soviet Union. Thus, the ratios between the top and the bottom decile shares of income in the CIS countries in 1991 were in the range of 2-5; already in 1994 these ratios had jumped into the range of 7-15; in Russia alone this ratio in 1994 reached the level of 15.1 but decreased slightly in 1995-1996. The ratio between the top and the bottom quintile shares of income in Russia increased from 3.3 in 1990 to 8.5 in 1995.[222] Income differentiation increased systematically and substantially in Bulgaria as well: the quintile ratio grew from 2.7 in 1988 to 6.5 in 1995. The changes in measured income distribution in most of the other east European transition countries were not so pronounced: in Poland the quintile ratio peaked in 1994 at 6.3 (from 3.6 in 1988) but later declined to 4.8 in 1995; in the Czech Republic this ratio increased from 2.6 in 1988 to 3.9 in 1994; in Hungary from 3.3 in 1988 to 3.9 in 1993; and in Romania from 3.3 in 1989 to 4.3 in 1994.[223] As a result, there emerged in most transition countries a relatively small layer of wealthy individuals, usually involved in entrepreneurial activity, and a growing number of poor people, whereas the middle-income groups have been diminishing in size (with the possible exception of some central European countries).[224]

The growing visible income inequality has undoubtedly led to important shifts in market effective consumer demand in the transition countries: the previously homogeneous demand patterns have become increasingly heterogeneous. And, obviously, the greater the change in the degree of income differentiation, the greater will be the shifts in consumption patterns. On the one hand, the *nouveaux riches* have already switched, thanks to the openness of the economies, to increased consumption of the consumer luxuries available on the world market. On the other hand, increasingly large cohorts of impoverished people can barely afford the basic necessities. An indication of this is the increasing

---

[220] See, for example, K. Deininger and L. Squire, "A new data set measuring income inequality", *The World Bank Economic Review*, Vol. 10, No. 3 (Washington, D.C.), 1996, pp. 565-591, which is based on a comprehensive World Bank database covering 108 countries over a period of several decades.

[221] The quantitative measurement of changes in the distribution of income and wealth is a notoriously difficult task, especially when it covers a period which includes an abrupt change of regime such as the transition from plan to market. Without going into the technical problems, a word of caution is needed about the accuracy of all such quantitative measurements (for details and discussion on methodological issues see B. Milanovic, "Income, inequality and poverty during the transition: a survey of the evidence", *MOCT-MOST*, No. 6 (Bologna), 1996, pp. 131-147 and K. Deininger and L. Squire, loc. cit.). Besides, most studies so far have focused on income distribution and almost no publicly available, systematic evidence on wealth redistribution during the transition exists. As to income distribution, the discussion here refers to two of the most widely used measures: 1) the Gini coefficient, which

quantifies the area between the actual income distribution line (the Lorenz curve) and the line of absolute income equality; and 2) the ratios between the top quintile's share of income to the bottom quintile's share (within a breakdown of income earners into five descending quintiles). Similarly, a ratio between the top and the bottom deciles is used if income earners are grouped into 10 descending deciles.

[222] CIS Statistical Committee, *CIS in 1996* (Moscow), 1997, p. 23; Russian Federation Goskomstat, *Uroven' zhizni naseleniya Rossii* (Moscow), 1996, p. 80.

[223] World Bank, *Economic Growth Project Data Sets* (World Bank, internet website); Natsionalen Statisticheski Institut, *Bulgaria '95 Sotsialno-Ikonomichesko Razvitie* (Sofia), 1996, p. 94; J. Vecernik, "Changing earnings distribution in the Czech Republic: survey evidence from 1988-1994", *Economics of Transition*, Vol. 3, No. 3 (Oxford), 1995, pp. 355-371; OECD, *Economic Surveys: Poland 1997* (Paris), 1997, pp. 86-87.

[224] See B. Milanovic, loc. cit.

CHART 3.2.2

Change in income inequality during the transition and consumption in 1993-1995 for selected transition countries

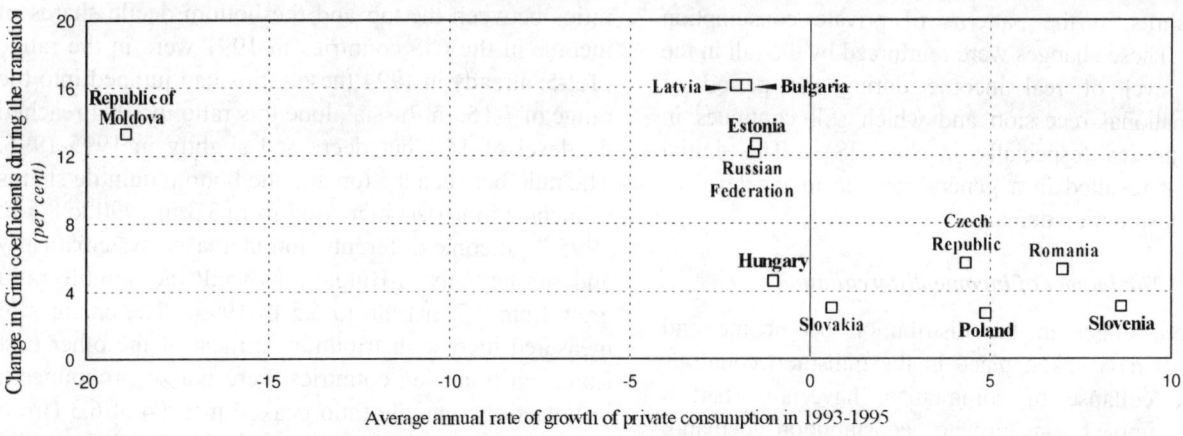

Source: World Bank, *Economic Growth Project Data Sets* (Washington, D.C.); Natsionalen Statisticheski Institut, *Bulgaria '95 Sotsialno-Ikonomichesko Razvitie* (Sofia), 1996, p. 94; J. Vecernik, "Changing earnings distribution in the Czech Republic: survey evidence from 1988-1994", *Economics of Transition*, Vol. 3, No. 3 (Oxford), 1995, pp. 355-371; OECD, *Economic Surveys, Poland 1997* (Paris), 1997, pp. 86-87; UNICEF, *Children at Risk in Central and Eastern Europe: Risks and Promises*, Regional Report, No. 4, 1996, statistical annex; P. Cornelius and B. Weder, "Economic transformation and income distribution: some evidence from the Baltic countries", *IMF Staff Papers*, 43(3) (Washington, D.C.), 1996, pp. 587-604; national statistics; UN/ECE secretariat computations.

Note: The changes in the Gini coefficients refer to the following periods: 1989 to 1995 for Bulgaria, Estonia and Poland; 1989 to 1994 for the Czech Republic, Latvia, the Republic of Moldova, Romania and Slovakia; 1989 to 1993 for Hungary and Slovenia; 1990 to 1995 for Russia.

share of household income spent on food which is observable in all CIS countries and in some east European transition countries.[225]

Growing income differentiation (especially when it is coupled with economic decline or stagnation) pushes an increasing number of people into categories of low absolute levels of income (and into poverty) and this has a depressing effect on aggregate consumption demand. Even if the average per capita level of real income in a country remains unchanged, increasing inequality of incomes is likely to result in declining aggregate consumption demand and, especially, of demand for mass consumption goods. In the first place, this is due to differences in the marginal propensities to save and to consume in the high and in the low income groups (the rich are likely to save an increasing share of their increased income). But, in addition, there will be an effect due to differences in the income elasticities of demand for different types of goods: the income gains in the high income groups are likely to affect only demand for specific types of goods (mostly luxuries) and will not offset the reduced demand for mass consumption products in the lower income categories.

It is thus reasonable to expect a negative correlation between changes in income inequality and in consumption. Recent data for some of the transition countries tends to confirm such a proposition, as can be seen in chart 3.2.2 which plots the rates of growth of total consumption against changes in the Gini coefficient in selected transition countries. Thus, the growing income inequality during the transition (at least in its initial phase) has probably been a negative factor in economic growth in this period and is continuing to be so in countries with rising income gaps between the wealthy and the poor.[226]

Besides, it appears that the degree and speed of changes in the visible distribution of wealth and income have further implications as regards the social and

---

[225] For example, the share of household income spent on food increased between 1991 and 1995 in some CIS countries as follows: in Azerbaijan from 57 to 74 per cent; in Belarus from 36 to 62 per cent; in Kazakstan from 43 to 56 per cent; in Kyrgyzstan from 43 to 59 per cent; in the Republic of Moldova from 37 to 47 per cent; in Russia from 38 to 52 per cent (but dropped to 50 per cent in 1996); in Uzbekistan from 49 to 69 per cent (CIS Statistical Committee, *CIS in 1996* (Moscow), 1997, pp. 133-134). In Bulgaria this share increased from 38 per cent in 1989 to 46 per cent in 1995 (Natsionalen Statisticheski Institut, *Bulgaria '95 Sotsialno-Ikonomichesko Razvitie* (Sofia), 1996, p. 95). The share of household income spent on food is high in other transition countries as well: in the first half of 1996 it was estimated at 55 per cent in Lithuania (Lithuanian Department of Statistics, *Survey of Lithuanian Economy*, No. 2 (Vilnius), 1996, p. 21). It should be noted that the available empirical information on household expenditure is based on household budget surveys; the latter are usually biased towards lower income categories and thus the above estimates may also be somewhat biased.

[226] The relationship between growth and income inequality is one of the most controversial issues in economics, both in theory and in empirical studies. Probably the first comprehensive studies on these issues were those of Kuznets (see for example, S. Kuznets, "Economic growth and income inequality", *American Economic Review*, 45 (Nashville), 1955, pp. 1-28). Kuznets hypothesized different relations between growth and inequality at different stages of economic development; however, growth itself was considered exogenous. In later studies it was proposed that there might exist endogenous links between growth and social change and, in particular, income inequality (for a discussion see K. Deininger and L. Squire, loc. cit.). The evidence from recent developments in the transition countries tends to support the latter conjecture.

political support for the reform process. A smaller degree of income differentiation, especially when coupled with economic growth and the growth of income in all or most income categories, seems to be essential for the emergence (or preservation) of a significantly large segment of the population which will remain supportive of the reform process. Such a feature is observable in the more successful reformers of central Europe and this is another characteristic that distinguishes them from the CIS countries and from the countries of south-eastern Europe. Pushing large numbers of the population into the cohorts of the poor, as has happened in the latter countries, creates resistance to reform: the losers from transition – or rather those who perceive they may be – are unlikely to support policies that will lead to a deterioration in their social status and drive them into poverty.

In any case, growing real incomes for a majority of the population has been one of the important demand-side factors for growth in some of the transition countries of central and eastern Europe, especially in 1995-1996. It has stimulated at least a partial recovery in some of the depressed domestic economic sectors as well as encouraging the growth of supply of new products and services.

The impact of trade liberalization

The liberalization of trade in the transition countries opened new windows of opportunity for participating in the international division of labour and for exploiting their comparative advantages. At the same time, greater openness has made their economies more susceptible to changes in external demand. Thus on the one hand, provided supply is sufficiently responsive, the transition countries now have the opportunity to respond to increased external demand for their products. On the other, their economies are now more vulnerable to external shocks.

The experience of the first phase of the transition process provided strong evidence as to the importance of exports as a factor for stimulating recovery and growth. The exposure of the previously planned economies to external competition has also revealed comparative advantages of these countries which had been previously obscured by the administrative trade arrangements within the CMEA. In some cases liberalization unveiled a hidden export potential which had not been used before; in other cases it resulted in a permanent shrinking of exports.

However, it is important to stress that the degree to which each country was capable of capturing the new opportunities, or to which it was negatively affected by the erosion of foreign markets, depended to a significant extent on the course of reforms. Indeed, external demand can only be met if supply is responsive, i.e. if producers behave as market type agents in a competitive environment and if the existing legislative and regulatory framework supports this type of behaviour. Conversely, non-market behaviour by economic agents and institutional hurdles are an effective barrier to export growth even if there is a potential external demand.

In this regard, the recent export performance of the transition countries varies considerably: in some cases there have been spectacular growth rates of exports, indicating the emergence of virtuous circles of successful market reforms and capitalization on the countries' comparative advantages; in other countries, exports have stagnated, often below pre-transition levels, which points to delays in systemic reforms and, probably, economic underperformance.[227] Of course, in the countries worst affected from the loss of traditional markets the necessary adjustment is much greater and much more painful and thus it might need a longer time to be implemented even if the political will is there and if the policies are right.

Chart 3.2.3 presents a comparative picture of exports and GDP growth in some transition countries in 1993-1995.[228] Despite an obviously large variance the graph tends to support the proposition of a positive correlation between GDP and export growth, but there are some specific features in the areas of positive and negative economic growth. The chart suggests that in the phase of recovery there is a positive correlation between the rates of growth of exports and of GDP, while during the phase of recession the relation is somewhat more ambiguous. Indeed, in some of the transition countries the period of economic decline has been consistent with the growth, in some cases notable, of exports. This paradoxical performance has mostly occurred in successor states of previously larger entities (the Soviet Union and the SFR of Yugoslavia) and may be due to a process of gradual redirection of trade flows to new markets in a situation of highly depressed domestic markets.[229]

---

[227] For example, in 1995 total Polish exports in current dollar terms were 65 per cent higher than in 1987; total Czech and Slovak exports (excluding mutual trade) increased by 61 per cent over Czechoslovak exports in 1987, Hungarian exports increased by 34 per cent. Conversely, in 1995 total Bulgarian exports were 35 per cent below their 1987 level and Romanian exports were still 12 per cent down; see appendix table B.10.

[228] Exports are in current dollar terms due to the lack of data in constant domestic prices for most countries. However, as the same measure applies to all countries, provided there are no large differences in the changes in dollar export prices, this should not result in a notable bias of comparative performance. The export data for Albania and the Baltic states are for 1994-1995 only.

[229] It should be noted that the trade data for the first years after the breakup are not very reliable and this may be a source of some distortion.

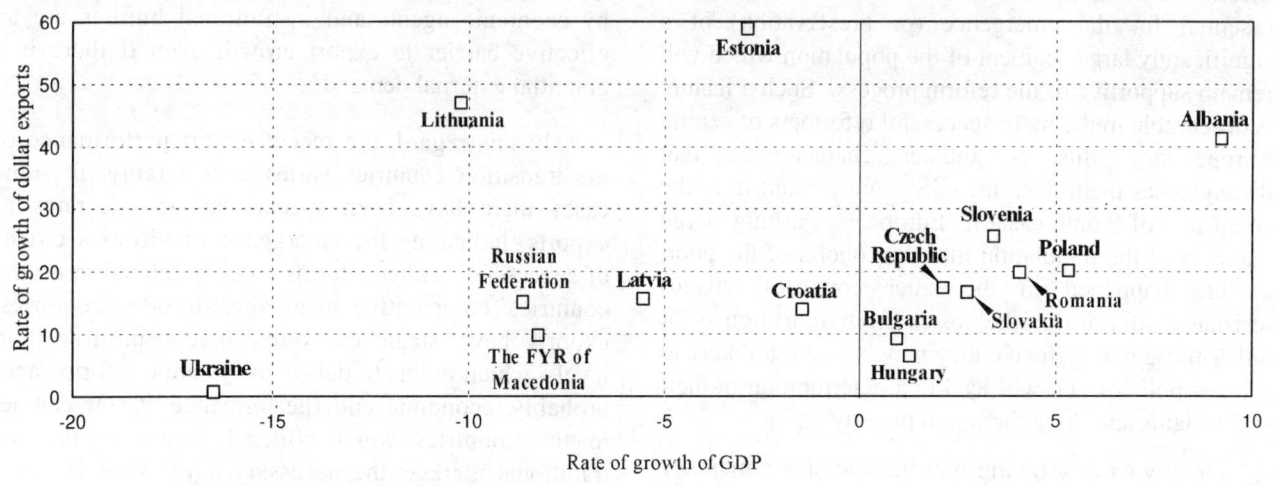

CHART 3.2.3

Dollar exports and GDP in selected transition countries, 1993-1995
(Average annual growth rates)

Source: National statistics; UN/ECE secretariat computations.

It is yet to be seen whether and to what extent the export boom observable in most transition countries will be sustainable in the medium and the long run. In 1994-1995 part of the export growth can clearly be attributed to cyclical factors. The transition countries managed to accommodate a growing external demand, especially in western Europe, due to the existence of some idle capacities (mainly in unsophisticated, homogeneous and bulky products such as raw materials, metals, chemicals and some other intermediate goods). Indeed, this was one of the factors behind the recovery in eastern Europe in that period.

However, there are inherent limits to this type of growth related both to supply and demand. The existing production facilities in these sectors are in general obsolescent and there are technological limitations to the expansion of output; in some cases their productive efficiency is questionable due to the persistence of certain price distortions (such as subsidized energy). On the other hand, the unexpected deceleration of growth in western Europe in 1996 indicates that the transition countries cannot rely on high rates of expansion of foreign demand on a long-term basis.

Although their existing comparative advantages will continue to be important to the transition countries for some years to come, they cannot be regarded as providing a secure basis for growth in the medium and longer run. In a world of increasing international competition, only broadly based, competitive and high quality exports can support high rates of self-sustained growth. A prerequisite for achieving this is the deep reconstruction and modernization of the supply side in the transition countries.

Besides, the central European transition countries are basically labour scarce economies in the medium term, but with relatively high, and somewhat outdated, human capital stocks. It is difficult to see how they can maintain a long-term future in labour-intensive, relatively unskilled, and internationally standardized products. So over the medium term they have to shift to high skill, knowledge-intensive products. That means a shift to a strategy designed to augment and improve their human capital stocks.

3.3 Labour markets

(i) Overview

There was no general improvement in the labour market situation in the transition countries as a whole in 1996. Nevertheless, in some of them the improvements which started in 1994-1995, continued in 1996, and in a few even strengthened. This uneven performance, already apparent during 1994 and 1995, became much more marked in 1996 given the large differences in output performance, the speed of restructuring, and the state of the overall reform process among the countries.

In *eastern Europe*, the level of employment in the region as a whole probably stagnated or even further declined during the first three quarters of 1996 largely reflecting the slowdown in overall economic activity. On the other hand, the level of east European unemployment, which had peaked at the end of the first quarter of 1994, continued to decline in 1996. This was mainly due, however, to sharp declines in Poland and Romania which largely offset the increases in most of the other countries. Although unemployment rates fell slightly further, partly due to growing numbers of discouraged unemployed people leaving the labour force in some countries, they still remained high. At the end of 1996, the average east European unemployment rate was 11.8 per cent, down from 12.5 per cent a year earlier. In the *Baltic states,* the available quarterly data suggest that employment started to increase in the first three quarters of 1996 in Lithuania but continued to fall more rapidly in Latvia. The rate of unemployment in the three Baltic states was broadly unchanged, averaging 6.4 per cent at the end of 1996. In the *CIS countries,* employment has continued to fall although at much slower rates than might have been expected given the general economic situation. In most of them this is due to the general stance of macroeconomic and reform policies, which still do not encourage the release of redundant workers. Official "registered unemployment" has remained extremely low (some 4 per cent or less of the labour force, except in Armenia), even lower than the fall in employment would imply, due to shortcomings in the official statistics.

(ii) Employment

(a) Changes in total employment in 1996

Available quarterly data[230] indicate that in the first three quarters of 1996, during which there was a slowing down of output growth, demand for labour probably stagnated or even fell slightly in *eastern Europe* as a whole (table 3.3.1). It was only in the Czech Republic, Poland and Slovakia (and probably in Romania) that net job creation, albeit less than in 1995, continued in 1996, a reflection of their better output performance compared with other countries in the region. In the other countries where quarterly data are available, total employment continued to decline in 1996. In Hungary and Yugoslavia there was some slowdown in the rate of decline, but in Slovenia the shakeout of labour accelerated, particularly in industry which was strongly affected by a deterioration in external demand. Employment also continued to deteriorate considerably in The FYR of Macedonia.

Among the Baltic countries, total employment continued to fall in the first three quarters of 1996 in Latvia,[231] but it started to recover in Lithuania. This small recovery in total employment in Lithuania, despite a large fall in industrial employment, reflects GDP growth of 3 and 4 per cent, respectively, in 1995 and 1996.

In the first three quarters of 1996, employment continued to fall in the *CIS countries*,[232] except in Georgia and Turkmenistan. In most countries employment fell at similar rates to those in 1995 and, again, by much less than the fall in output (by 1.2 per cent against a nearly 5.3 per cent decline in output), thus implying a further increase in excess employment in many of these countries.

(b) Excess employment

During the early stages of the transition process, the demand for labour in *eastern Europe* fell together with output, but by a much smaller amount. During 1990-1993, employment fell by some 16 per cent in eastern Europe, while output (real GDP) declined by 23 per cent (appendix tables B.1 and B.5). In 1994, employment stopped falling when output increased by nearly 4 per cent in the region as a whole. Although the economic upturn continued in 1995 with output increasing by 5.6 per cent, employment fell slightly in the region as a whole. Employment gains continued in 1995 only in Bulgaria, the Czech Republic and Poland, and appeared for the first time in Slovakia. Thus, during 1994-1995 output in eastern Europe recovered by 9.7 per cent, but employment shrank further, albeit slightly.

This slow response of labour demand to changes in output, both during the recession and the recovery, was of course reflected in measured productivity. The latter has followed a U-shaped path of development which, although resembling a typical cyclical performance, was nevertheless markedly pronounced in the east European economies, particularly in industry. This implies a considerable degree of excess employment in all the former centrally planned economies which continued, and in most cases

---

[230] Assessing recent changes in employment in the transition economies is particularly difficult as quarterly employment data are still not available for some countries. Furthermore, in those countries where they are available, they are not comparable with annual data which have a much broader coverage. The coverage of quarterly data is not only different from that of annual data but also varies among countries. For example, quarterly data exclude enterprises with less than 25 employees in the Czech Republic and Slovakia, less than 10 employees in Hungary, less than 5 employees in Poland and less than 3 employees in Slovenia. Thus, by excluding the new and rapidly expanding small private enterprises, quarterly data, to varying degrees in individual countries, may overestimate the decline (or underestimate the rise) in employment. For example, in the Czech Republic, during the first three quarters of 1996, employment fell by 2.6 per cent according to the quarterly data based on enterprises with 25 and more employees. In contrast, according to the labour force survey (which is used for table 3.3.1), with a much larger coverage of enterprises, there was an increase in employment by more than 1 per cent during the same period. Czech Statistical Office, *Indicators of the Economic and Social Development of the Czech Republic*, No. 3 (Prague), 1996, p. A13.

[231] There are no quarterly data for Estonia.

[232] The 3 per cent increase in employment in Ukraine in 1995 was probably a result of the fuller coverage of individual farmers by the first labour force survey conducted in Ukraine in October 1995. Earlier estimates, based on the old series, indicated a nearly 2 per cent fall in employment which seems more consistent with a nearly 12 per cent decline in GDP in 1995.

increased, during the early stages of transition. The presence of excess employment permitted enterprises to increase output without increasing their workforce during the recovery phase. In other words, the recent recovery in many east European countries has sharply increased the labour productivity of those already employed rather than creating a net rise in new jobs. Furthermore, in countries where enterprise restructuring has gained momentum, labour shedding in industry has continued during the recovery although at notably slower rates than during the recession. This has raised labour productivity in industry much faster than in the rest of the economy, thus dampening the growth in unit labour costs and permitting a significant increase in the unit profits of industry.[233] Nevertheless, the surge of productivity in industry was not due only to the combination of output growth and labour shedding. There was also a non-cyclical, "embodied" gain related to the restructuring of output, the introduction of more modern management, and upgrading of the physical and human capital stock.[234]

This combination of cyclical and non-cyclical elements in productivity growth varied greatly among countries. For example, in Hungary, industrial labour productivity in 1996 was more than 40 per cent above its 1990 level while industrial output was still some 9 per cent below its 1990 level. In contrast, in the Czech Republic, despite a sharp recovery in output (nearly 17 per cent) during 1994-1996, industrial labour productivity was still slightly below its level in 1990, suggesting a much slower adjustment of labour and, therefore, the existence of a still larger margin of excess employment. In fact, industrial employment in the Czech Republic in 1996 was only one fifth below its level in 1990, compared with more than one third in Hungary. A similar, albeit somewhat less pronounced, picture also emerges in Slovak industry. In contrast to both Hungary and the Czech Republic, in Poland, where the output recovery not only started earlier but also continued more vigorously, industrial labour productivity in 1996 was some two thirds above its 1990 level with a much smaller reduction in employment (20 per cent) than in Hungary (37 per cent). This suggests not only larger gains from the removal of past inefficiencies and from "embodied" productivity in Poland than in Hungary, but also a more profound restructuring of enterprises than in the Czech Republic.

The most serious degree of excess employment exists in the *CIS economies*, where real GDP for the region as a whole fell at a rate more than four times faster than employment between 1989 and 1996. In the Russian Federation, real GDP in 1996 was nearly 45 per cent below its 1989 level while the net loss of jobs in the total economy was 12 per cent.[235] Thus the chronic "labour hoarding" of the Russian economy during the Soviet era has actually increased during the transitional recession. There are various reasons for this intensification of disguised unemployment. First, in the absence of a comprehensive social safety net, many workers have preferred to remain on the payrolls of enterprises without being paid, or being paid extremely low wages, rather than register as unemployed, in order to benefit from some of the social provisions of the workplace. Secondly the authorities are reluctant to encourage large-scale redundancies by implementing bankruptcy provisions in order to avoid a surge in open unemployment which would exacerbate the social costs of transition. The latter have already increased considerably due to a marked rise in the numbers of people living in poverty, in suicide rates and other indicators of social stress, as well as in various forms of criminality. Thirdly, managers caught between weak aggregate demand and excess employment in their enterprises[236] use various methods to alleviate labour costs. They often resort to production stoppages[237] and lay-offs to reduce the average wage; such stoppages have the same effect as redundancy on unit labour costs but without the employer having to pay the "severance pay". Other ways of dealing with excess employment are unpaid or partially paid involuntary leave.[238] The most common rate for such partial payment is the minimum wage, which is considerably below the

---

[233] See section 3.4 for a discussion of unit costs and productivity in industry.

[234] See UN/ECE, *Economic Survey of Europe in 1995-1996*, chap. 3.2, for a discussion of productivity changes since the start of the reforms.

[235] The decline in measured productivity is somewhat less pronounced in industry where most of the job losses were concentrated. In Russia, between 1989 and 1996, even though industrial output fell 53 per cent, employment declined by nearly 30 per cent.

[236] According to the most recent *Russian Labour Flexibility Survey 1996*, "no less than 45.3 per cent of managements said they could produce the same level with fewer workers, with 55.5 per cent of firms with more than 500 workers stating that they could do so. The overall figure was 10 percentage points more than in 1995". G. Standing, *Reviving Dead Souls: Russian Unemployment and Enterprise Restructuring*, ILO (Geneva), January 1997, p. 14.

[237] "According to the *Russian Labour Flexibility Survey 1996*, on average, firms had *stopped production completely* for 2.9 weeks in 1995 for economic reasons, and 2.3 weeks in the first five months of 1996. They had *partially stopped production* for 3.6 weeks in 1995 and 2.6 weeks in the first five months of 1996. For those that had stopped partially, on average 34.1 per cent of workers were directly affected. *All these figures were substantially higher than recorded in 1993 and 1994.* If one assumed a working year of 48 weeks, then the figures imply that in 1996 about 7.7 per cent of working time was lost due to total stoppages and 1.9 per cent due to partial stoppages. Using the same measures for 1994 and 1995, one can conclude that this form of concealed unemployment was substantially worse in 1996". G. Standing, ..., p. 16.

[238] In September 1996, the number of persons involuntarily working part-time or on compulsory temporary leave was nearly 5 million, or 7 per cent of the work force. Russian Federation Goskomstat, *Sotsialno-ekonomicheskoe Polozhenie Rossii, 1996 g.*, No. 10 (Moscow), November 1996, p. 180.

TABLE 3.3.1

Total employment in the transition countries, 1990-1996

(Annual average percentage change)

	Total employment					Employment in industry			
	1990-1995[b]	1994	1995	1995[a] QI-QIII	1996[a] QI-QIII	1990-1995[b c]	1994	1995	1996[a] QI-QIII
Eastern Europe[d]	-16.0	0.1	-0.5	1.1	0.3	-30.2	-3.4	-1.7	..
Albania	-21.3	11.0	-2.5	-0.2	-1.3	..	..	..	..
Bosnia and Herzegovina	..	..	..	..	..	..	..	..	..
Bulgaria	-24.1	0.6	2.1	..	..	-43.1	-3.7	-1.1	-5.3
Croatia	-26.1	-2.3	-1.3	..	-2.8	-40.5	-4.4	-11.6	..
Czech Republic	-7.2	0.8	2.6	2.7	1.1	-22.9	-5.3	0.6	-0.6
Hungary[e]	-27.4	-2.2	-1.8	-0.9	-0.6	-37.5	-4.5	-5.4	-1.2
Poland	-13.3	1.0	1.8	1.2	0.8	-23.8	-0.8	3.2	-0.8
Romania[e]	-13.3	-0.5	-5.2	..	..	-34.6	-4.9	-5.8	-3.4
Slovakia[e]	-14.6	-1.0	2.0	2.1	0.7	-28.4	-3.2	-0.1	0.3
Slovenia	-20.7	-1.8	-0.3	-0.4	-0.8	-35.9	-4.3	-3.6	-6.2
The FYR of Macedonia	-26.1	-5.3	-9.5	-7.7	-6.7	-35.6	-5.9	-13.2	-8.9
Yugoslavia	-14.7	-2.0	-1.4	-2.3	-1.9	-19.4	-2.7	-2.5	-1.8
Baltic states	-15.5	-4.1	-1.6	-1.9	..	..	..	..	..
Estonia	-19.8	-1.2	-1.5	-1.8	..	..	..	..	..
Latvia	-15.5	-3.2	-1.3	-1.4	-1.8	-41.7	-11.8	-9.9	-1.6
Lithuania	-13.6	-5.8	-1.9	-1.7	1.2	-37.2	-17.5	-7.6	-5.1
Total above	-16.0	-0.2	-0.5	..	..	..	..	..	..
CIS	-9.9	-3.2	-1.4	-2.1	-1.2	-24.1	..	..	..
Armenia	-7.3	-2.8	-0.8	-1.5	-2.5	-38.8	-2.1	-14.6	-3.0
Azerbaijan	-2.5	-2.2	-0.5	-0.4	-0.4	-25.0	-3.4	-5.9	-5.7
Belarus	-15.3	-2.7	-6.2	-5.9	-6.2	-23.7	-4.6	-10.9	-4.6
Georgia	-35.9	-2.5	-1.1	..	1.4	-54.6	-8.3	-8.6	..
Kazakstan	-12.3	-5.0	-0.5	-2.2	-1.6	-29.3	-8.0	-9.4	-3.6
Kyrgyzstan	-5.6	-2.1	-0.2	-0.1	-0.7	-38.7	-8.2	-14.9	-7.9
Republic of Moldova	-5.1	-0.4	-0.5	-0.4	-1.8	..	-5.3	-14.2	-7.7
Russian Federation	-12.1	-3.3	-3.0	-3.2	-0.8	-24.7	-10.7	-7.5	-2.3
Tajikistan	-1.4	-2.7	0.5	..	..	-27.1	..	-5.0	-5.5
Turkmenistan	12.1	1.4	0.5	0.6	0.4	-0.2	-2.7	–	2.6
Ukraine	-6.7	-3.8	3.0[f]	..	-0.9	-20.8	-8.8	-7.7	..
Uzbekistan	7.2	-1.3	0.3	1.0	-0.6	-6.8	-9.5	1.3	-5.3
Total above	-11.7	-2.4	-1.1	..	..	..	..	..	..
Memorandum items:									
CETE-5	-15.1	0.2	1.4	1.5	0.4	-26.7	-2.8	0.8	-0.5
SETE-7	-17.5	-0.1	-3.1	..	..	..	..	..	..

Source: National statistics and direct communications from national statistical offices to UN/ECE secretariat.

a Percentage change over the same period of preceding year.

b Cumulative change over the period.

c For the CIS and the Baltic states, 1991-1995.

d Quarterly aggregates include only Albania, Czech Republic, Hungary, Poland, Slovakia, Slovenia, The FYR of Macedonia and Yugoslavia.

e End of year.

f The increase in total employment in 1995 was probably a result of the fuller coverage of individual farmers on the basis of the first labour force survey conducted in October 1995 (see text).

---

subsistence level.[239] Enterprises also promote short-time working and encourage prolonged maternity leave which can last up to two years.[240]

Certainly, Russia is not the only CIS country where there is significant excess employment. For example, in the Republic of Moldova, some 8 per cent of the labour force was on compulsory unpaid leave during the second quarter of 1996 and another 2 per cent of the labour force was working part-time in mid 1996.[241]

---

[239] In December 1996, the minimum wage was equivalent to only 20 per cent of the official subsistence level and less than 8 per cent of the nominal average wage in the total economy. Russian Federation Goskomstat, *Sotsialno-ekonomicheskoe Polozhenie Rossii, 1996 g.*, No. 12 (Moscow), January 1997, p. 189.

[240] Taking into account the labour input lost due to partial and complete stoppages of production, the number of workers on administrative leave, and the full-time equivalent of labour input lost because of enforced short-time working, suppressed unemployment in Russian industry in 1996 is estimated at over a third of the workforce. This excludes any unreal maternity leave and unpaid employment. G. Standing, ..., p. 20.

[241] Government of Moldova/European Expertise Service, *Moldovan Economic Trends* (Chisinau), August 1996, p. 21.

Excess employment in the Baltic states probably declined during the recent recovery, but it still remains significant given the much larger fall in output (some 50 per cent) compared with that for employment (about 15 per cent) between 1989 and 1996. However, statistics implying excess employment have so far been published only in Estonia: these showed that some 7 per cent of the labour force was on part-time work in the third quarter of 1996.[242] The situation is probably worse in Latvia and Lithuania where the transitional recessions were deeper and longer.

(iii) Unemployment

(a) Levels and rates of registered unemployment in 1996

The total number of unemployed in the transition countries as a whole was 14.4 million at the end of 1996, some 365,000 persons more compared with a year earlier (table 3.3.2).

In *eastern Europe*, unemployment peaked in early 1994 at about 7.5 million people and since then has declined slowly to 6.1 million at the end of 1996, 7 per cent less than a year earlier and 19 per cent below its peak in March 1994. However, while in 1995 the declines were widespread, in 1996 they were largely concentrated in Poland and Romania. Unemployment also continued to fall in 1996, albeit at much lower rates, in Albania and Hungary but elsewhere it remained broadly unchanged (Slovakia, Slovenia) or continued to rise. Moreover, for the first time since 1994, there were increases in the number of people unemployed in the second half of 1996, especially in Bulgaria and the Czech Republic (in the latter rising above its peak level in 1994).

Given the rise in the number of unemployed in many countries, together with further contractions in the labour force in some of them, *rates of unemployment* remained high. In December 1996, the average unemployment rate was 11.8 per cent, with most countries in a range of between 10.5 per cent (Hungary) and just below 16 per cent (Croatia).[243] The main exceptions were the Czech Republic, where the rate increased but remained very low at 3.5 per cent,[244] and Romania where it was just above 6 per cent, more than 3 percentage points lower than a year earlier.[245] During 1996, unemployment rates continued to increase in The FYR of Macedonia and Yugoslavia, reaching 40 and 26 per cent respectively.[246]

In the *Baltic states*, both the level and the rates of unemployment in 1996 increased in Estonia and Latvia, but declined considerably in Lithuania. In December 1996, unemployment rates ranged between 5.6 per cent in Estonia and 7.2 per cent in Latvia.

In most of the *CIS countries*, the number of unemployed persons continued to rise during 1996 (by some 12 per cent), reaching some 8 million for the region as a whole at the end of the year. However, officially registered unemployment still remains artificially low, particularly in those countries where the transitional recession continued for the fifth year running. In December 1996, it varied between 0.4 per cent (Uzbekistan) and 4.5 per cent (Kyrgyzstan), the exception being Armenia where it was nearly 10 per cent.

These low rates of registered unemployment in the CIS countries and, to a lesser extent, in the Baltic states, both in comparison with the steep output decline[247] and with east European rates, reflect both the excess employment discussed above and the omission from the *registered* unemployed of large numbers of the jobless who are willing to work. This is revealed by the much higher rates of unemployment which emerge from the labour force surveys that are now intermittently conducted in a number of these countries.[248]

---

[242] Direct communication from the Statistical Office of Estonia.

[243] In Croatia, despite the continued increase in the number of unemployed persons, the unemployment rate reported for December 1996 was nearly 2 percentage points lower than in 1995. This fall reflects a revised estimate of the labour force based on a fuller coverage of the private sector, and hence a statistical discontinuity.

[244] Apart from active labour market policies, slow micro-level restructuring and relatively strong growth in private sector employment, the significant increase in the non-participation rate at the start of the reforms (due to early retirements and increased numbers on "disability" pensions), and the favourable conditions prior to transition in the structure of employment, including abundant labour skills, all contributed to low unemployment in the Czech Republic. For a more comprehensive discussion of the reasons for the low Czech unemployment rate see *World Economic and Social Survey 1994* (United Nations publication, Sales No. E.94.II.C1), pp. 197-198; OECD, *Review of the Labour Market in the Czech Republic* (Paris), 1995; IMF Staff Country Report No. 95/103, *Explaining the Gap Between Slovak and Czech Unemployment Rates, Slovak Republic Selected Issues* (Washington, D.C.), October 1995.

[245] According to the Romanian authorities, the sharp reduction of unemployment in 1996 is attributed to continued expansion in the private sector and to a special government project to cut unemployment by various means, including job-creating programmes. *BBC Summary of World Broadcasts*, EEW-0450, 29 August 1996.

[246] The exceptionally high rates in these two republics of the former SFR of Yugoslavia, however, partly reflect deficiencies in the measurement of total employment and therefore of the labour force. The data reported on employment in The FYR of Macedonia and Yugoslavia cover only the social sector in agriculture. See UN/ECE, *Economic Survey of Europe in 1995-1996*, p. 86, footnote 158.

[247] Real GDP in 1996 was around 50 per cent or more below its level in 1989 in the majority of the CIS and the two Baltic states (Latvia and Lithuania); see appendix table B.1.

[248] Labour force surveys (LFS) are now conducted in all the east European countries except Albania, Bosnia and Herzegovina, Croatia and The FYR of Macedonia. Among the CIS and the Baltic states, only Estonia, Lithuania, Russia and Ukraine conduct a regular LFS (the latter does it annually).

TABLE 3.3.2

Registered unemployment in the transition countries, 1994-1996
(Thousands and per cent of labour force, end of period)

	Unemployment (thousands)						Unemployment rate (per cent)					
				1996						1996		
	1994	1995	QI	QII	QIII	QIV	1994	1995	QI	QII	QIII	QIV
Eastern Europe	7 190	6 583	6 811	6 210	6 007	6 114	13.6	12.5	13.0	11.9	11.6	11.8
Albania	262	171	165	162	161	158	18.0	13.1	12.2	12.4	12.3	12.1
Bosnia and Herzegovina	..	..	..	..	..	..	..	..	..	..	..	..
Bulgaria	488	424	434	380	402	479	12.8	11.1	11.4	10.0	10.5	12.5
Croatia	248	249	264	253	262	269	17.3	17.6	15.2	14.7	15.1	15.9
Czech Republic	167	153	159	144	169	186	3.2	2.9	3.0	2.7	3.2	3.5
Hungary	520	496	528	482	501	478	10.4	10.4	11.6	10.6	11.0	10.5
Poland	2 838	2 629	2 726	2 508	2 341	2 360	16.0	14.9	15.4	14.3	13.5	13.6
Romania	1 224	998	1 031	798	665	658	10.9	9.5	9.8	7.6	6.3	6.3
Slovakia	372	330	343	311	314	330	14.8	13.1	13.3	12.1	12.2	12.8
Slovenia	124	127	121	114	118	124	14.2	14.5	13.9	13.3	13.7	14.4
The FYR of Macedonia[a]	196	229	234	237	239	245	33.2	37.2	38.1	38.7	38.9	39.8
Yugoslavia[a]	751	777	806	821	835	827	23.9	24.7	25.3	25.8	26.2	26.1
Baltic states	197	245	272	246	238	237	5.3	6.5	7.4	6.7	6.4	6.4
Estonia[b]	35	34	40	35	36	37	5.1	5.0	5.9	5.2	5.3	5.6
Latvia	84	83	88	89	89	91	6.5	6.6	7.0	7.0	7.0	7.2
Lithuania	78	128	144	122	113	109	4.5	7.3	8.3	7.0	6.4	6.2
Total above	7 387	6 828	7 083	6 456	6 245	6 354	13.0	12.3	12.7	11.6	11.2	11.5
CIS	6 011	7 185	7 450	7 724	7 842	8 027	4.4	5.5	5.9	6.1	6.2	6.4
Armenia	92	132	139	149	153	159	6.0	8.1	8.7	9.3	9.8	9.7
Azerbaijan	24	28	29	30	31	32	0.9	1.1	1.1	1.1	1.1	1.1
Belarus	101	131	167	176	185	183	2.1	2.7	3.5	3.7	4.1	4.0
Georgia	76	61	61	47	52	58	3.8	3.4	3.3	2.6	3.0	3.2
Kazakstan	70	140	206	244	269	282	1.0	2.1	3.1	3.7	4.0	4.1
Kyrgyzstan	13	50	72	80	79	77	0.8	3.0	4.2	4.6	4.7	4.5
Republic of Moldova	21	25	27	26	26	23	1.0	1.4	1.6	1.5	1.6	1.5
Russian Federation[c]	5 478	6 431	6 476	6 665	6 700	6 788	7.1	8.9	8.9	9.2	9.2	9.3
Tajikistan	32	35	42	52	55	46	1.8	1.8	2.3	2.5	2.8	2.4
Turkmenistan	..	..	..	..	..	..	..	..	..	..	..	..
Ukraine	82	127	200	221	258	351	0.3	0.6	0.9	1.0	1.1	1.5
Uzbekistan	22	25	31	34	34	28	0.3	0.3	0.4	0.4	0.4	0.4
Total above	13 398	14 013	14 533	14 180	14 087	14 378	6.9	7.5	8.0	7.8	7.7	7.9
Memorandum items:												
CETE-5	4 020	3 735	3 877	3 559	3 443	3 478	12.8	12.0	12.5	11.5	11.2	11.3
SETE-7	3 170	2 848	2 934	2 651	2 564	2 636	14.6	13.7	13.8	12.5	12.1	12.5
Russian Federation[d]	1 637	2 327	2 676	2 605	2 470	2 506	2.1	3.2	3.7	3.6	3.4	3.4

Source: National statistics and direct communications from national statistical offices to UN/ECE secretariat.

[a] The data reported on employment cover only the social sector in agriculture, hence unemployment rates are biased upwards (see text).

[b] Registered unemployed job seekers which has a broader definition than the registered unemployed with regards to age and past employment record.

[c] Based on monthly Russian Federation Goskomstat estimates according to the ILO definition, i.e. including all persons not having employment but actively seeking and available for work. Data for 1995 and 1996 were revised by Goskomstat.

[d] Registered unemployment.

(b) Comparison of registered and labour force survey unemployment

A comparison of unemployment rates based on labour force surveys with those derived from registered unemployment figures clearly indicates that the latter considerably understate the incidence of open unemployment in both the CIS and the Baltic states (table 3.3.3). In the Russian Federation, for example, registration statistics reported a 3.4 per cent unemployment rate (2.5 million *registered* unemployed) in September 1996. However, actual open unemployment, on the basis of the ILO definition, was estimated at 6.7 million or 9.2 per cent of the labour force, which implies that only 37 per cent of the jobless turned to the employment offices when looking for work. In Ukraine, the labour force survey conducted in October 1996 showed an open unemployment rate of 7.6 per cent, 5 times higher than the officially registered rate for the same month (1.5 per cent). Although labour force survey data are not available for the other CIS countries, the situation is probably similar to that in the Russian Federation or even worse.[249]

---

[249] In Georgia, for example, an informal survey found that 15.5 per cent of adults described themselves as unemployed in the second quarter of 1996

TABLE 3.3.3

Comparative measures of unemployment in selected transition countries, QI-QIII 1996
(Per cent of labour force)

	1996		
	QI	QII	QIII
Bulgaria			
Labour force survey unemployment	15.3	13.5	13.7
Registered unemployment	11.4	10.0	10.5
Czech Republic			
Labour force survey unemployment	3.4	3.2	3.5
Registered unemployment	3.0	2.8	2.9
Hungary			
Labour force survey unemployment	9.8	9.2	9.3
Registered unemployment	11.5	10.8	10.9
Poland			
Labour force survey unemployment	14.0	12.4	11.6
Registered unemployment	15.5	14.7	13.8
Romania			
Labour force survey unemployment	9.3	5.9	5.9
Registered unemployment	9.8	8.2	6.6
Slovakia			
Labour force survey unemployment	11.7	10.7	10.5
Registered unemployment	13.5	12.4	12.3
Slovenia			
Labour force survey unemployment[a]	..	7.3	..
Registered unemployment	..	13.7	..
Estonia			
Labour force survey unemployment	..	11.2	10.0
Registered unemployment	..	5.2	5.3
Lithuania			
Labour force survey unemployment	..	14.2	15.6
Registered unemployment	..	7.4	6.6
Russian Federation			
Labour force survey unemployment	8.9	9.2	9.2
Registered unemployment	3.7	3.6	3.4
Ukraine			
Labour force survey unemployment[b]	..	..	7.6
Registered unemployment	..	..	1.5

Source: National statistics and direct communications from national statistical offices to UN/ECE secretariat.

[a] The labour force survey is conducted once a year in May.

[b] The labour force survey is conducted once a year in October.

Similarly, in Estonia and Lithuania, labour force surveys conducted in the third quarter of 1996 show unemployment rates of 10 per cent and 15.6 per cent respectively, as compared with 5.3 and 6.6 per cent on the basis of registered unemployment. Moreover, as can be seen in table 3.3.3, the registered statistics not only severely distort the incidence of unemployment in these countries but may also show a different trend. Indeed, in the Russian Federation, and particularly in Lithuania,

compared with the 2.6 per cent registered unemployment rate. TACIS European Expertise Service, *Georgian Economic Trends, Quarterly Review* (Tbilisi), Second Quarter 1996, p. 69. A representative survey of the Republic of Moldova, excluding Transnistria, which took place in June 1995, indicated that the real unemployment rate was close to 11 per cent compared with the registered rate of less than 2 per cent. CIS Statistical Committee, *Statisticheskii Bulleten*, No. 3 (Moscow), February 1996, p. 107.

where the differences between the two measures are markedly large, data based on registrations continued to show some improvement during 1996 when actual unemployment measured by the survey was steadily increasing. The reverse is true for Estonia.

In contrast, in the majority of east European countries the survey based estimates of unemployment rates tend to give similar results to those based on registration data, the main exceptions being Bulgaria and Slovenia.[250]

(c) Changes in the labour force

The labour force declined between 1990 and 1995 in all the transition countries shown in table 3.3.4,[251] except in Poland where it increased by 2.5 per cent. Moreover, Poland is the only country where activity rates in 1995 were the same as in 1990. In all the other transition countries they were below their 1990 levels, particularly in Albania and Hungary (by about one fifth). The largest contraction in the labour force among the countries shown in table 3.3.4 was in Hungary (some 17 per cent), which was almost entirely due to the fall in activity rates. In fact, among the east European countries, the working-age population fell only in Bulgaria and remained broadly stable in Hungary. In all the others it increased and partly offset the effect of falling activity rates on the size of the labour force, particularly in Albania, the Czech Republic and Slovakia.

In the three CIS and particularly in the Baltic countries, in contrast to the majority of east European countries, the working-age population fell and contributed to the decline in the labour force. However, the fall in labour force in these countries was relatively less (except in Estonia and Belarus) than in some of the east European

---

[250] Slovenia is exceptional in that the registered unemployment rate is much higher (by some 6 percentage points) than the rate based on labour force surveys. This can be explained by the much broader definition of unemployed applied by the employment offices which counts as unemployed each person who comes into the office and seeks help, declaring himself to be unemployed. "In reality, about 20 per cent of the registered unemployed have part-time jobs and 15 per cent, for various reasons, are not actively seeking employment". Institute of Macroeconomic Analysis, *Analysis of Economic Trends in 1994 and Projections for 1995* (Ljubljana), 1995, pp. 13-14. A similar situation seems also to prevail in Poland where registered unemployment rates are higher by more than 2 percentage points than the survey rate; this is due to the fact that a certain number of the registered unemployed are actually employed in marginal activities and register to find another job or just to collect benefits while not seeking work. *Wiadomosci Statystyczne*, No. 8 (Warsaw), 1995, pp. 14-15.

[251] The estimates in table 3.3.4 are intended to convey orders of magnitude and to show changes over time; they should not be used to compare levels among countries. The activity rates are calculated by dividing the labour force (employment plus unemployment) by the working-age population. The working-age population is defined as the 15-64 years age group for all countries. Therefore, the rates shown in table 3.3.4 can differ considerably from the officially published rates based on different age groups. However, this should not affect their changes over time significantly since changes in school leaving and retirement ages tend to be infrequent.

TABLE 3.3.4

Changes in the labour force in selected transition countries, 1990-1995
(Percentages)

	Activity rate[a]		Total change in labour force[b] 1990-1995	Change in working-age population 1990-1995	Change in labour force due to change in activity rates[c] 1990-1995
	1990	1995			
Albania	78	63	-13.5	6.6	-18.9
Bulgaria	70	64	-9.9	-1.7	-8.4
Czech Republic	79	74	-4.1	3.0	-7.0
Hungary	79	65	-17.3	-0.3	-17.1
Poland	69	69	2.5	2.7	-0.2
Romania	72	68	-4.5	0.9	-5.4
Slovakia	74	70	-1.1	4.8	-5.6
Slovenia	74	65	-9.7	3.1	-12.4
Estonia	76	67	-13.7	-2.3	-11.7
Latvia	79	75	-9.8	-4.8	-5.3
Lithuania	75	72	-5.4	-0.9	-4.5
Belarus	76	68	-12.2	-1.4	-10.9
Russian Federation	76	74	-3.8	-0.8	-3.0
Ukraine	74	70	-5.8	-1.3	-4.5

Source: UN/ECE secretariat estimates, based on national statistics.

Note: The "interaction effect", the joint effect of changes in both population and activity rates, is ignored in the above calculations. However, given the state of the data (particularly those for the labour force, which are based on only recorded employment and registered unemployment) and uncertain estimates of the working-age population, these calculations only provide orders of magnitude and they should be treated as such (see text).

[a] Activity rate = labour force ÷ working-age population (15-64 years).
[b] Labour force = employment + registered unemployment.
[c] $[(WAP_{1995} \times AR_{1990}) - (LF_{1995})] \div (WAP_{1995} \times AR_{1990})$; where WAP is the working age population (15-64 years); AR is the activity rate; LF is the labour force.

countries due to the relatively smaller decline in their activity rates. Nevertheless, this relatively favourable evolution of the labour force despite declining working-age population and activity rates in these countries is probably due to the inflated figures for employment (due to the excess employment discussed above).

(d) Structure of unemployment by age, sex and duration

Not only do the registered open unemployment rates in many east European countries resemble the rates prevailing in western Europe,[252] but also the structure of their unemployment shares similar features with that in western Europe, namely, relatively high female rates and shares, very high youth unemployment rates, and a large and growing share of long-term unemployed reaching nearly half of all those unemployed.

In all the transition countries for which the relevant statistics are available, *female unemployment* rates continued to be higher than male rates in 1996, except in Hungary[253] (table 3.3.5). Moreover, in some countries (particularly in Poland and Slovakia) the difference between the two rates became even more pronounced than in 1995, and in Slovenia the female rate exceeded the male rate for the first time due to a fall in the latter. Not only the rate but also the *share of women in total unemployment* in most countries remained very high in 1996. In September 1996, it was only in Hungary and Slovenia that the female share was slightly less than 50 per cent (table 3.3.6). Among the east European countries the female share was highest in the Czech Republic and Poland where nearly 60 per cent of all the unemployed were women; and in the latter it increased by nearly 4 percentage points between 1995 and 1996, at the same time as total unemployment fell sharply. In the Baltic states, the higher share of women in total unemployment was broadly unchanged in 1996, varying between some 52 per cent in Latvia and 70 per cent in Estonia, the highest rate of all the countries show in table 3.3.6. In Russia, the share remained above 60 per cent.

---

[252] At the end of 1996 the registered unemployment rate in the CEFTA countries was 11.3 per cent (table 3.3.2), only 1 percentage point above the west European average, and 1.2 percentage points above the average of the four major west European countries (table 2.3.4)

[253] The main reason for this is that in Hungary most of the decline in employment was concentrated in agriculture, heavy industry and construction, in which female workers were only a small share of the labour force. Female employment in Hungary is concentrated mostly in trade, catering, education, health and public administration, in which employment either has increased or was least hit by labour redundancy. Another reason is the provision of extended child-care leave which diminishes the female workforce. E. Ehrlich and G. Révész, *Human Resources and Social Stability during Transition in Hungary*, International Centre for Economic Growth and Institute for World Economics of the Hungarian Academy of Sciences (San Francisco), 1995.

TABLE 3.3.5

Registered unemployment rates by sex in selected transition countries, 1995-1996
(Per cent of labour force)

	September 1995			September 1996		
	Total	Male	Female	Total	Male	Female
Czech Republic	3.0	2.1	3.9	3.2	2.4	4.1
Hungary	10.3	11.2	9.1	11.0	11.6	10.3
Poland	15.0	13.3	17.0	13.5	11.4	15.9
Romania	9.2	7.8	10.9	6.3	5.4	7.4
Slovakia	13.2	11.8	14.9	12.2	10.5	14.2
Slovenia	14.2	14.3	14.0	13.7	13.5	14.0
Latvia	6.0	5.5	6.6	7.0	6.4	7.7
Lithuania	6.3	5.5	7.1	6.4	5.8	7.0
Russian Federation	2.9	2.1	3.8	3.4	2.4	4.6

Source: National statistics and direct communications to UN/ECE secretariat.

TABLE 3.3.6

Share of women, youth and long-term unemployed in total unemployment in selected transition countries, 1995-1996
(Percentages)

	Women		Youth[a]		Long-term[b]	
	September		September		QIII	
	1995	1996	1995	1996	1995	1996
Bulgaria	54.8	54.7	22.5	22.6	65.8	60.0
Croatia	51.5	49.8	34.1	32.9	53.8[c]	51.3[c]
Czech Republic	61.9	59.8	30.3	31.0	31.1	30.6
Hungary	43.8	44.7	27.5	26.0	51.3	49.7
Poland	55.8	59.5	37.5	33.0	42.2	40.4
Romania	56.1	53.8	45.8	47.8	47.0	51.7
Slovakia	52.4	54.5	30.4	29.5	54.4	51.7
Slovenia	46.9	48.0	33.1	29.8	52.9	52.1
Estonia	68.9	69.9	20.5	18.2	29.0	31.0
Latvia	52.7	51.5	20.2	17.8	24.6[c]	29.5[c]
Lithuania	56.4	55.5	19.2	18.1	24.0	32.0
Russian Federation	63.5	62.5	36.2	35.2	30.3	32.9

Source: National statistics and direct communications from national statistical offices to UN/ECE secretariat.

[a] Persons less than 25 years old.

[b] Persons who have been unemployed for more than 12 months. Data refer to labour force survey results in the third quarter, except in the Russian Federation (first quarter) and Slovenia (second quarter).

[c] Registered unemployment.

Youth unemployment remains one of the most pressing social problems in all the transition countries. The available data suggest that young people under 25 are disproportionately affected by unemployment. In Romania, where the registered unemployment rate (6.3 per cent) is one of the lowest among the transition countries, the share of young people in total unemployment was nearly 50 per cent in September 1996 (table 3.3.6). In the other countries of eastern Europe, the young accounted for between one quarter and one third of total unemployment. It was less than one fifth in the Baltic states and reached 35 per cent in the Russian Federation. According to labour force surveys, the *youth unemployment rate* in most of the east European countries was around two times (in Romania nearly three times) higher than the average rate in the third quarter of 1996.[254]

Given the increasingly stagnant nature of the unemployment pool, partly due to people changing their employment without entering the ranks of the unemployed (job-to-job transfers) and partly because of the redundancy of skills due to a sharp change in the structure of output, the numbers of *long-term unemployed* (those out of work for more than one year) increased rapidly during the early phase of transition. One of the most salient recent features of unemployment in east European countries is that despite a relatively strong recovery in output since 1994 and a fall in total unemployment in 1995, the number of jobless people remaining out of work for more than one year increased rapidly during 1994-1995. Labour force surveys, however, indicate that between the third quarters of 1995 and 1996 (table 3.3.6), there was a slight decline in the share of the long-term unemployed in total unemployment in all the east European countries except Romania, where it increased from 47 per cent to nearly 52 per cent. Notwithstanding this decline, the proportion of long-term unemployment remained very high in eastern Europe (at some 50 per cent or more of the total in most countries) in the third quarter of 1996; the exceptions were the Czech Republic and Poland where relatively lower shares of long-term unemployed fell further (to some 30 and 40 per cent respectively), probably due to the growing personal service sector where skill requirements are relatively less important. In the Baltic states and in the Russian Federation, the share of the long-term unemployed continued to increase and in the third quarter of 1996 was around one third of total unemployment; but in all of them it was still lower than the average share for eastern Europe which may, *inter alia*, imply a higher proportion of "discouraged" unemployed people leaving the labour force than in eastern Europe.

Long-term joblessness is not only economically wasteful, painful for individuals and a potential source of social instability, but it also creates serious difficulties for the labour market clearing mechanism. Not only the experience of western Europe in the 1980s but also the recent experience of the advanced transition countries have shown that the reduction of unemployment during recoveries is much more difficult when there is a high incidence of long-term unemployment.

In addition to high and rising numbers of long-term unemployed, the process of enterprise restructuring still has a long way to go and structural adjustment is far from

---

[254] OECD, *Short-term Economic Indicators, Transition Economies*, No. 1 (Paris), January 1997.

complete in most transition economies. Single-digit unemployment rates are therefore unlikely to be achieved in the near future, and further increases in unemployment rates are likely especially in those countries where reforms are less advanced.

3.4 Costs and prices

(i) Overview

Inflation in the transition countries continued to fall, on average, during 1996. The rate measured by the change in the consumer price index between December 1995 and December 1996, was 20 per cent or less in the majority of countries; in eight of them it was in single digits. However, in contrast to 1995, the outcome in many east European countries fell short of expectations, mainly because productivity growth was not only less than forecast but also slower than the rate of increase in wages. Import price pressures were subdued, most of the east European currencies remaining stable or even appreciating against the deutsche mark. Furthermore, world market prices for manufactured goods were also fairly stable and primary commodity prices (except energy which increased by 15.5 per cent) fell by nearly 9 per cent. Thus, even though the dollar appreciated against most currencies (by some 5 per cent against the deutsche mark) in 1996 compared with 1995, changes in import unit values were negligible.[255] The attempt to maintain or increase gross operating profits, despite rapidly increasing labour costs (except in a few cases), ceased to put upward pressure on producer prices in many of the countries, and particularly in the export oriented industries, because of weak demand and competitive pressures.

This combination of downward pressure on producer prices and upward pressure on labour costs could have a dampening effect on fixed investment in domestic industry, which still has a long way to go to recover fully from the transitional recession. Slow micro-level restructuring, partly due to stronger labour pressures and social concerns in the absence of comprehensive social safety nets, as well as other factors such as continued liberalization of administered prices, pre-election loosening of fiscal policy in some countries, etc., have all helped to maintain underlying inflationary pressures which were already increasing due to the strong growth of real wages since 1994. This real wage growth was one of the main factors behind the heating up of the economy in some east European countries, while in others, more moderate growth in output was not accompanied by offsetting moderation in employment, thus leading to increases in real unit labour costs. While inflation rates in most east European countries fell less than expected in 1996, there were actually a number of countries where they accelerated.

In the Baltic countries and most of the CIS, in contrast, the reduction of inflation was significant in 1996, and in many cases much more than expected, mainly due to the continuation of tight macroeconomic policies, relatively stable exchange rates, and much improved productivity.

(ii) Consumer prices, performance and prospects

Consumer price inflation in the transition countries as a whole continued to fall during 1996. Over the 12 months to December 1996, the increase in prices averaged 20 per cent or less in 16 out of 25 countries (table 3.4.1).[256] However, both the performance vis-à-vis earlier expectations and the underlying causes of inflation varied greatly among the individual countries.

In *eastern Europe* within-year inflation in 1995 had moderated in all countries except Hungary and Yugoslavia. In 1996, in contrast, it accelerated in a number of countries, and especially in Albania, Romania and Bulgaria. The acceleration was most severe in *Bulgaria* (from 33 per cent in 1995 to 311 per cent in 1996) due to the loosening of macroeconomic policies in the wake of the banking and liquidity crisis in the early months of the year and the subsequent collapse of the exchange rate. From the end of May, the lev depreciated strongly throughout the rest of the year, reaching 487 lev per dollar at the end of December compared with 70.7 lev at the end of 1995.[257] In January and February 1997, the monthly inflation rate jumped to 43.8 per cent and 243 per cent respectively. Amid these hyperinflationary monthly rates and a further collapse in output, the government resigned; early elections are set for April and, at the time of writing, the interim government and the IMF are negotiating a draft letter of intent and are preparing for the introduction of a currency board by June 1997.

In *Romania*, after some two years of significant effort to lower inflation, monetary and incomes policies were loosened in late 1995 in the run-up to the November 1996 parliamentary elections. After the surge in aggregate demand in the last quarter of 1995 and the early months of 1996, and the run on the leu, the government tried to stabilize the exchange rate through direct intervention in order to avoid an eventual surge in import prices (particularly of energy) rather than taking the more painful measures to restore monetary and

---

[255] For more details on world commodity prices and import unit values see chap. 2.4 and section 3.5, respectively.

[256] The 25 countries exclude Turkmenistan and Uzbekistan for which there are no monthly data.

[257] For a discussion of the Bulgarian crisis, see section 3.1(iii).

TABLE 3.4.1

Consumer prices in the transition countries, 1995-1996

(Percentage change)

	Annual average 1996 All items	1995 All items	December over previous December 1996			
			All items	Food	Non-food goods	Services
Albania	11.1[a]	6.0	17.3[b]	..	..	..
Bosnia and Herzegovina	-21.2	-34.2	3.2	-11.5	2.0	41.7
Bulgaria	123.1	33.0	311.1	303.7	329.1	306.7
Croatia[c]	3.6	3.7	3.4	4.9	1.0	8.7
Czech Republic	8.9	8.0	8.7	7.9	7.5	11.3
Hungary	23.6	28.5	20.0	17.7	20.9	22.3
Poland	19.8	22.0	18.7	19.1	17.7	19.4
Romania	38.8	27.7	56.8	55.2	60.3	53.6
Slovakia	6.0	7.4	5.5	3.4	6.6	5.1
Slovenia[c]	9.7	8.6	8.8	12.9	7.0	12.9
The FYR of Macedonia[c]	4.1	11.2	0.2	-4.7	0.8	2.9
Yugoslavia	90.5	110.7	59.9	50.3	71.5	114.3
Estonia	23.1	28.8	14.9	13.1	13.0	16.8
Latvia[c]	17.7	23.3	13.2	7.7	17.4	18.6
Lithuania	24.7	35.5	13.1	13.6	..	..
Armenia	18.7	32.0	5.6	2.3	1.9	30.3
Azerbaijan	19.8	84.5	6.8	0.2	9.5	104.3
Belarus	52.7	244.2	39.1	43.4	30.3	29.6
Georgia	39.0	57.4	12.6	..	..	..
Kazakstan	39.1	60.4	28.6	16.4	7.4	139.3
Kyrgyzstan	30.3	31.9	35.0	39.1	20.0	46.6
Republic of Moldova	23.5	23.8	15.1	11.7	14.6	29.8
Russian Federation	47.8	131.4	21.8	17.7	17.8	48.4
Tajikistan	422.4	2 382.2	40.6	34.9	41.2	80.0
Turkmenistan	..	..	..	..	..	..
Ukraine	80.2	181.7	39.7	17.4	18.8	112.7
Uzbekistan	..	93.7[d]	..	..	..	..

Source: UN/ECE secretariat estimates, based on national statistics.

[a] January-October 1996 over the corresponding period in 1995.
[b] October 1996 over December 1995.
[c] Retail price index. For Croatia the food price index is from the cost of living index.
[d] September 1995 over December 1994.

fiscal stability. Inflation accelerated substantially after April, and because of the government's intervention on the foreign exchange market the loan agreements with the IMF were suspended. Within-year inflation reached 56.8 per cent in 1996, more than twice the rate in the previous year and almost triple the initial target of 20 per cent agreed by the government and the IMF. However, despite a decline in exports, output, particularly in the non-agricultural sectors, continued to grow strongly during the first half of 1996 thanks to the rapid rise in domestic consumption, fuelled by rising real wages and pensions, before they started to weaken due to accelerated inflation. After the change of government which followed the November elections, a restrictive macroeconomic policy stance was announced. But although the new government has stressed its commitment to the IMF requirements (*inter alia*, full price and exchange rate liberalization, containment of the government's deficit, a faster rate of privatization and industrial restructuring, subsidy cuts, etc.), the need to preserve social stability may still inhibit immediate action on all aspects of the reforms at once.[258] Nevertheless, on 1 January 1997, both energy and food prices were liberated and this was immediately followed by a doubling of the price of fuel and a tripling of those for bread and a number of other staple foods. By February, the leu had been devalued by more than 100 per cent as part of a strategy to eliminate the parallel foreign exchange market and, in particular, to switch resources from consumption to net exports. These measures will give another boost to the consumer price index in the coming months, but in order to break the wage-price spiral, wage indexation, introduced by the previous government during the pre-election campaign, was also abandoned in January 1997. The government's "programme of national economic salvation" foresees a 2 per cent contraction in output and an annualized rate of inflation of under 30 per cent in the second half of 1997, so that the within-year inflation rate for 1997 as a whole should not exceed 90 per cent.

---

[258] For further details on monetary policy see section 3.1(ii).

Inflation, although at much lower rates, also accelerated in *Albania* during 1996. The 10-month cumulative rate jumped from 2.9 per cent in October 1995 to 17.3 per cent in October 1996, mainly because of a loosening of fiscal and wage policies. Inflation also rose, although slightly, in the *Czech Republic* during 1996, the within-year rate reaching 8.7 per cent compared with 8 per cent in 1995. Reflecting to some extent the slow pace of micro-level restructuring and the influence of very low unemployment rates on wage demands, real wages growth[259] (8 per cent, year-on-year) continued to exceed by far the increase in productivity[260] (some 3 per cent). In the first half of 1996 retail sales volume increased by some 10 per cent compared with about 5 per cent during the same period in 1995. The seasonal decline in food prices during the summer, however, reduced the pressure for higher consumer prices during the third quarter. The price index temporarily jumped in July (by 1 per cent) due to a 15 per cent increase in energy prices and a 25 per cent increase in controlled rents. In the fourth quarter of 1996 the annualized rate of inflation fell to 5.3 from 7.4 per cent in the same period of the previous year. Nevertheless, without wage restraint and a faster rate of restructuring, the rate of disinflation in the Czech economy may further lose momentum and even be reversed in the short term.[261]

In *Slovenia*, inflation also started to climb slowly during the first half of 1996. Very high real interest rates, while limiting private consumption growth, created liquidity problems in many sectors which, together with a strong appreciation of the tolar in real terms, slowed down production growth and led to an acceleration in unit labour costs during the first half, which had already been rising for some time due to the slow rates of restructuring and privatization. The relaxation of incomes policies[262] in early 1996 in the run-up to the December 1996 elections also fuelled higher inflation, as did the rise in various government controlled prices, particularly for transport.[263] Nevertheless, year-on-year monthly rates of inflation fell back to single digits in the last quarter despite another large increase in public service charges in December (equivalent to more than half the total rate of change in the CPI). Hence, for the year as a whole, the rate of inflation in Slovenia remained broadly unchanged from the 1995 rate, at 8.8 per cent.

As in the Czech Republic, there was some overheating of the economy in Poland and Slovakia in 1996 but in both countries fiscal policy remained consistent and monetary policy remained tight, particularly after the first quarter. Furthermore, exchange rates remained relatively stable (strongly appreciating in real terms). Nevertheless, in *Poland* disinflation was again slower than the government's target (the within-year rate in 1996 was 18.7 per cent against a target of 17 per cent) despite relatively cheaper imports[264] and faster productivity growth. The shortfall was due to the continued adjustment of administered prices, a bad cereal harvest which pushed up food prices,[265] and buoyant consumer spending,[266] which was supported by strong growth in real household incomes and a large increase in consumer credit,[267] particularly early in the year. The government is aiming for an end-year inflation rate of 13 per cent in 1997. However, ongoing price adjustments, growing export difficulties and increasing wage pressures could make this target difficult to achieve, particularly during an election year, even though the 1997 budget is again centred on an anti-inflationary stance.[268]

---

[259] As measured by average gross wages deflated by the CPI.

[260] Real growth in GDP deflated by the growth in total employment.

[261] Some 100,000 Czech railway workers staged a five-day strike in February 1997 to protest the loss-making Czech railways' restructuring plan which, *inter alia*, foresees large redundancies as part of the effort to reduce its significant debt. The government, in order to end this labour unrest and to prevent it from spreading to other sectors, proposed delaying the issue to end-May when new proposals would be drawn up in consultation with the rail unions. This was the major, but not the only one, of several public sector disputes that the Czech government has faced recently while trying to restrain public sector wage growth. During 1996 there were also various disputes concerning education and health sector workers.

[262] Labour unions staged several strikes in 1996 demanding wage increases of up to 40 per cent and to protest proposed government cuts in health care benefits and sick pay.

[263] Increase in prices under direct government control contributed 27 per cent to the change in CPI in 1996. Their contribution was 30 per cent and 26 per cent, respectively, in 1994 and 1995. Institute of Macroeconomic Analysis and Development, *Slovenian Economic Mirror*, Vol. III, No. 1 (Ljubljana), January 1997, p. 8.

[264] Due to a sharp increase in non-resident short-term capital inflows, the national bank let the zloty appreciate by 2.5 per cent within its band in December 1995. In January 1996, import duties were reduced by 2 percentage points and the temporary import surcharge was cut from 5 to 3 per cent. Also in early January, the zloty's crawling peg was reduced from 1.2 to 1 per cent per month.

[265] Third quarter cereal prices were some 60-70 per cent higher than in 1995 and this increase spilled over to livestock prices.

[266] Retail trade volume increased by 8.4 per cent in 1996 (first three quarters) compared with 2.3 per cent in 1995 (table 3.2.12).

[267] In 1996, the expansion of total domestic credit accounted for nearly 80 per cent of the total growth in M2. This represented a significant change compared with 1995, when the monetary expansion was almost fully accounted for by net capital inflows. CS First Boston, "Poland: an economic success story with political question marks", *Emerging Economics Research – Europe*, 20 February 1997, p. 4. Within total domestic credit, consumer credit doubled, albeit from a very low base, *Financial Times*, 24 February 1997.

[268] The 1997 budget, influenced by the desire to achieve all the Maastricht criteria in the medium term, targets the budget deficit at 2.8 per cent of GDP.

Slovakia, by contrast, in spite of achieving eastern Europe's highest output growth rate in 1996, also had one of the lowest inflation rates (5.5 per cent down from 7.4 per cent). A further tightening of monetary policy, a stable exchange rate, and balanced fiscal management all contributed to this favourable performance. Nevertheless, there was a sharp slowdown in productivity gains in industry and a significant deterioration in the trade deficit;[269] thus, mounting pressure on the exchange rate poses a risk to the sustainability of this high growth/low inflation combination.

In *Hungary*, inflation had accelerated after the introduction of the stabilization programme in March 1995,[270] but it came down firmly during 1996: in spite of strong underlying cost-push factors such as increases in administered prices and pre-announced changes in the crawling peg exchange rate, tight and coordinated fiscal and monetary policies finally exerted their full impact. In addition, the gradual strengthening of the exchange rate and a reduction of the 8 per cent import surcharge by 1 percentage point in July and October (which is intended to be phased out by July 1997) attenuated the import price pressure on domestic inflation. Thus, in spite of a weaker than foreseen output performance and sharply reduced growth in industrial labour productivity, the inflation target of 20 per cent (for December 1996 over December 1995) was reached. The 1997 budget calculations are based on an average annual rate of 18 per cent (or 15 per cent for December 1997 over December 1996), a target which would not be easy to achieve if there were any fiscal relaxation in the run-up to the general elections in 1998. Furthermore, there are more consumption taxes and utility price increases to come, some of which were postponed in 1996, as well as higher energy prices which are seen as an essential test of the government's commitment to the modernization of the energy sector. In addition, the crawling peg devaluation of the forint is set to continue in 1997 and will remain at the rate of 1.2 per cent per month until the second quarter of 1997.

In *Croatia*, retail price inflation in 1996 remained low and decelerated from 3.7 per cent in 1995 to 3.4 per cent in 1996. In *The FYR of Macedonia* retail prices actually fell during the spring and summer and at the end of 1996 were only 0.2 per cent above their end-1995 levels. In *Yugoslavia*, the rate of consumer price inflation has continued to moderate, particularly after the summer months, but at some 60 per cent for the year as a whole it was the second highest (after Bulgaria) of all the 25 transition countries for which consumer price indices are available.

In all three *Baltic countries* inflation fell for the fourth consecutive year. Within-year rates were within the 13-15 per cent range, half the rates prevailing in 1995. Disinflation was particularly strong in *Lithuania* where the annualized rate fell from nearly 27 per cent in the second half of 1995 to 6.5 per cent in the same period of 1996, a rate significantly below those in many of the more advanced transition countries of eastern Europe. This relatively strong deceleration in the inflation rate in Lithuania, despite a much stronger wage inflation than in the other two Baltic states, was mainly due to continued restraint on fiscal expenditure and a strong appreciation of the litas against the deutsche mark due to its link to the dollar. In *Latvia*, monetary and fiscal policies in 1996 continued to be tight; real wages in the total economy, compared with 1995, actually declined for most of the year; and the increase in unit labour costs slowed down further. In *Estonia*, however, disinflation lost some momentum in 1996 as continued price liberalization put upward pressure on the consumer price index during the first half of 1996. Nonetheless, price rises abated during the rest of the year: the annualized monthly rates in the second half of the year fell to single digits, down from 37.5 per cent on average in the first quarter.

Disinflation in 1996 was most significant in the *CIS countries*. Annual average rates of change, which were in four digits in 1994 and three digits in 1995 in the majority of the countries (appendix table B.7 and chart 3.4.1), fell to within the range of 18-40 per cent in 1996 in more than half of them. Given the sharp deceleration during the course of the year, the fall in the within-year inflation rate in 1996 was even more notable: in two countries the rate was down to single digits (Armenia and Azerbaijan) and the highest was 41 per cent (Tajikistan where, nevertheless, it was down from more than 2000 per cent in 1995). Disinflation was particularly strong in the third quarter when in five countries (Armenia, Azerbaijan, Georgia, Kyrgyzstan and Tajikistan) prices actually fell below their average levels in the second quarter. Within-year inflation was higher in 1996 than in 1995 only in Kyrgyzstan.

This general and strong disinflation in the CIS countries is the result of their own macroeconomic policy efforts (particularly with monetary policy) to achieve the targets and fulfil the conditions set by the international financial institutions for the disbursement of agreed loans, as well as a more effective decoupling of national monetary systems within the CIS. The spillover effect of the strong and persistent fall in inflation since the beginning of 1995 in the Russian Federation, which is still the largest single trade partner of most of them, has also contributed significantly.

---

[269] Section 3.5, table 3.5.3.

[270] This programme included, *inter alia*, a large devaluation of the forint (initially 9 per cent followed by a crawling peg of 1.9 per cent per month) and an import surcharge (8 per cent) which, given the high import content of consumption, was the major reason behind the acceleration of the CPI during the spring of 1995.

The Transition Economies

CHART 3.4.1

Consumer prices in the transition countries, 1995-1996

(Percentage change over the same month of preceding year)

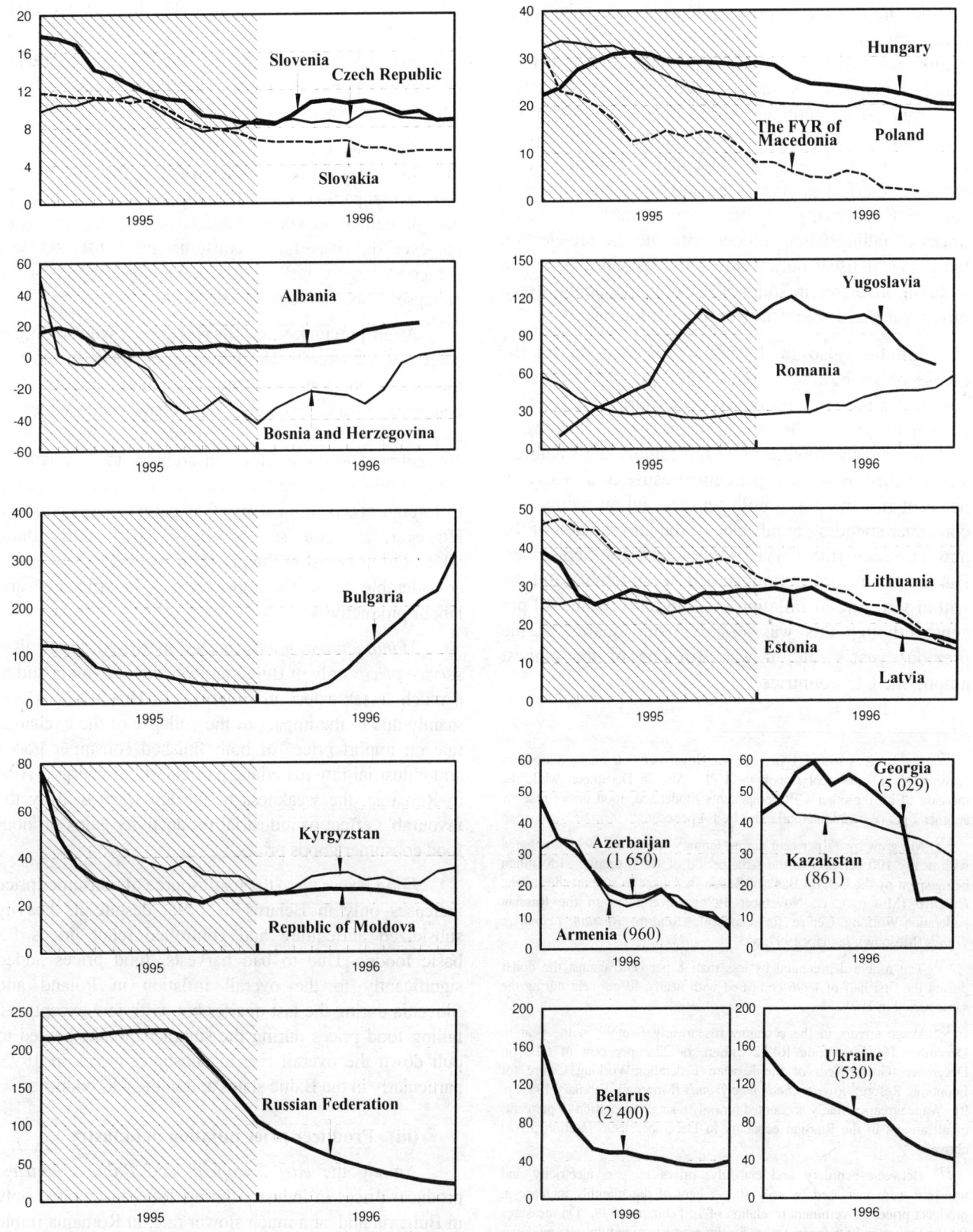

Source: National statistics.

Note: Figures in parentheses for some of the CIS countries are the highest rates reached during 1995. Tajikistan is excluded as it still had four-digit year-on-year rates during 1996. Turkmenistan and Uzbekistan are excluded due to lack of data. Countries are grouped according to scale within the subregions.

In *Russia*, in 1996, the rate of increase in consumer prices continued on the steady downward trend which had started in early 1995. The monthly rate of change fell from nearly 18 per cent in January 1995 to virtual price stability in the third quarter of 1996. Despite some acceleration in the monthly rate in the last quarter, due mainly to seasonal factors,[271] the within-year rate in 1996 was 21.8 per cent, less than one sixth of the rate in 1995. Monetary policy continued to be tight.[272] Furthermore, the effect of the relatively strong rouble on import price inflation, particularly in the first half,[273] combined with the adverse effect of large wage and pension arrears[274] on consumer demand, partly offset the upward pressure on prices of falling labour productivity and the pre-election loosening of fiscal policy during the second quarter. In addition, increases in some of the principal administered prices were held below the overall inflation rate.[275]

Inflation also moderated in *Ukraine* during the course of 1996, albeit after surging in the first quarter due to the pre-winter subsidy cycle during the closing months of 1995. After a significant fiscal tightening in early spring, the monthly rate fell rapidly until August, when it rose by some 6 per cent because of a nearly 30 per cent rise in energy utility prices and an increase in consumer spending in advance of the introduction of the new currency (the hryvna) announced for September. But the downward trend resumed in September and the within-year rate of inflation in 1996 fell to some 40 per cent; although this was less than one quarter of the previous year's rate, it was still one of the highest among the CIS countries.

---

[271] Food prices rose 2.1 per cent in November – a major reason for November's 1.9 per cent rise in the CPI. Also in December, while the increase in all the other CPI components moderated, food prices rose by another 2 per cent and the total index by 1.4 per cent.

[272] M2 grew by 27 per cent during January-November 1996, compared with nearly 100 per cent in the same period of 1995, despite $1.8 billion being spent by the Central Bank of Russia in August to support the rouble. *Izvestiya* (Moscow), 12 November 1996; Government of the Russian Federation/Working Centre for Economic Reform, *Russian Economic Trends* (Moscow), January 1997, p. 3.

[273] The rouble depreciated by less than 8 per cent against the dollar during the first half of 1996 compared with nearly 40 per cent during the same period in 1995.

[274] Wage arrears in the economy rose nearly fivefold in the year to December 1996, reaching R47.2 trillion, or 22.2 per cent of GDP in December (Government of the Russian Federation/Working Centre for Economic Reform, *Russian Economic Trends* (Moscow), February 1997, p. 9). Wage arrears actually accounted for only a small proportion (9 per cent) of all arrears in the Russian economy in December 1996 (section 3.1(ii) above).

[275] Between February and end-July, prices of gas, electricity and transport were increased by only 80 per cent of the monthly increase in producer prices (government resolution of 12 February 1996, "On measures for the limitation of price increases for the products of natural monopolies", *Rossiiskaya gazeta* (Moscow), 20 February 1996). Since August these prices have been allowed to increase at a rate not exceeding the change in producer prices in the preceding month (government resolution of 17 July 1996, *Rossiiskaya ...*, 1 August 1996). These resolutions will remain valid during the first half of 1997 (*Rossiiskaya ...*, 11 February 1997).

In *Belarus* the speed of disinflation was even faster, the within-year rate of inflation falling from nearly 250 per cent in 1995 to less than 40 per cent in 1996. However, in contrast to Russia, Ukraine and most of the other CIS countries, inflation in Belarus was essentially repressed through the use of central government price subsidies and cheap credits (particularly to agriculture and construction) and de facto price controls at both the wholesale and retail levels.

The lowest inflation rates among the CIS countries in 1996 were in those countries where tight policy stances were accompanied by strong output and/or productivity growth, namely in *Armenia* and *Georgia*. In *Azerbaijan*, however, the sharp fall in consumer price inflation (to 7 per cent) largely reflects the deflationary effect of the collapse in household incomes.

As in previous years, *service prices* were the main source of the overall rise in consumer prices in most of the transition countries and particularly in the CIS countries. To a large extent, this probably reflects the fact that administered price adjustments have been concentrated in this sector. In addition to its relatively lower productivity, the service sector is also less exposed to foreign competition than other parts of the economy. However, in some service sectors, particularly those which did not exist in the pre-reform period, there is still considerable scope for improvements in efficiency and labour productivity.

Manufactured goods prices increased faster than *service prices* only in Bulgaria, Romania, Slovakia and to a much lesser extent in Belarus. In Bulgaria this was mainly due to the impact of the collapse of the exchange rate on import prices of both finished consumer goods and industrial raw materials, particularly of energy. Also in Romania, the weakness of the leu washed away the favourable effect of industrial productivity gains on non-food consumer goods prices.

Food prices were the main source of consumer price increases only in Belarus in 1996, despite its heavily subsidized agriculture and controlled retail prices for basic foods. Due to bad harvests, food prices added significantly to the overall inflation in Poland and Slovenia during the last quarter but, with few exceptions, falling food prices during the summer months helped to pull down the overall rise in consumer prices in 1996, particularly in the Baltic states and in the CIS countries.

(iii) Producer price inflation in industry

Among the *east European* and *Baltic countries*, producer prices in industry during 1996 accelerated only in Bulgaria and, at a much slower rate, in Romania (table 3.4.2 and chart 3.4.2). To a large extent this reflected the effect of exchange rate depreciation on material input prices, particularly energy. In all the other countries of the region, industrial producer price inflation decelerated

TABLE 3.4.2

Producer prices in industry[a] in the transition countries, 1995-1996

(Percentage change)

	Annual average		December over December	
	1995	1996	1995	1996
Albania	..	..	..	..
Bosnia and Herzegovina	68.7	-4.8	18.4	-4.6
Bulgaria	48.9	129.7	33.7	338.1
Croatia	0.8	1.3	1.6	1.5
Czech Republic	7.8	4.9	7.4	4.6
Hungary	28.5	22.3	31.3	20.6
Poland	26.0	13.2	19.0	10.7
Romania	35.3	50.0	32.2	60.4
Slovakia	9.1	4.0	7.1	4.7
Slovenia	12.4	6.7	7.5	5.9
The FYR of Macedonia	4.7	–	2.2	-0.6
Yugoslavia	75.7	89.1	112.0	51.3
Estonia	25.6	14.7	21.8	9.8
Latvia	12.0	13.8	15.9	7.9
Lithuania	28.8	16.5	20.4	12.3
Armenia	187.8	36.7	70.5	17.6
Azerbaijan	1 340.1	70.6	452.2	20.9
Belarus	538.6	37.6	123.6	34.7
Georgia	..	..	..	..
Kazakstan	173.6	24.7	42.6	18.3
Kyrgyzstan	34.0	44.7	25.0	58.9
Republic of Moldova	52.2	30.2	46.3	19.5
Russian Federation	237.6	50.7	175.3	25.5
Tajikistan	351.7	341.9	633.8	82.3
Turkmenistan	..	..	..	..
Ukraine	450.8	54.6	177.4	26.1
Uzbekistan	792.5	135.7[b]	279.6	67.1[c]

Source: UN/ECE secretariat estimates, based on national statistics.

[a] Industry = mining + manufacturing + utilities.

[b] January-November 1996 over the corresponding period in 1995.

[c] November 1996 over December 1995.

and in most of them by even more than consumer prices, a reflection of the dampening effect of relatively strong exchange rates on unit material input costs (except energy) as well as smaller gross operating profits in industry due to weakened export demand. On the other hand, significantly higher real wages and the consequent increase in consumer demand tended to put upward pressure on prices at the retail level. A notable example of this was in the Czech Republic where increases in consumer prices accelerated while those for producer prices fell significantly.

In many of the *CIS countries*, producer price inflation also decelerated more rapidly than consumer prices in 1996, in some of the countries for the first time since the beginning of the reforms, but in the majority the increase in producer prices was still larger than that for consumer prices.[276] In general, this suggests that weak demand rather than improved efficiency was the main factor behind the disinflation in some of the CIS countries in 1996. Furthermore, a higher rate of inflation at the producer level was largely due to falling output combined with excess employment, the presence of monopolies, and higher distribution costs in industry.

(iv) Wages and unit labour costs in industry

Wage disinflation, which started in most of the east European countries in 1993 and in the CIS and Baltic countries in 1994, continued through 1996. However, in most of the *east European* and *Baltic countries* the annual rate of change in *average nominal gross wages in industry* in 1996 (compared with 1995) remained high and again increased faster than industrial producer prices. *Real product wages*[277] fell only in Bulgaria, Yugoslavia and, for the second consecutive year, in Hungary albeit at a much slower rate than in 1995. They rose even faster than in 1995 in the other four CEFTA countries and Lithuania. Meanwhile, the growth of industrial production[278] in the majority of countries slowed down in 1996 (table 3.1.1) and, in some, much faster than the decline in employment (table 3.3.1). Thus, *productivity* growth, which had improved strongly in 1994-1995 weakened in 1996, except in Poland (largely thanks to continued strong output growth), Slovenia (in spite of marginal output growth), and Yugoslavia (a combination of improved output, particularly during the second half of the year, and continued labour shedding). Nevertheless, despite the deceleration, productivity growth remained quite high in a number of countries, notably in Romania (nearly 14 per cent: the highest rate of change in the region) and, to a lesser extent, in the Czech Republic and Lithuania (particularly in the first half of the year in both).

The rise in *nominal unit labour costs*,[279] which had slowed down considerably in 1995, continued to decelerate in some of the east European and Baltic countries in 1996. The improvement was generally smaller than in 1995, although there was a notable deceleration in Poland and particularly in Slovenia. In Poland this was due to a combination of continued strong output growth and a marked deceleration in wage increases, while in Slovenia it was mainly due to increased labour shedding in industry (6 per cent). In contrast, the rise in nominal unit labour costs accelerated considerably in Bulgaria (from 50 per cent in 1995 to more than 90 per cent in 1996) and, at much lower rates, in Hungary and Slovakia. There was also some acceleration in the Czech Republic, the result of stable but high wage growth and smaller productivity gains.

---

[276] There are no data for producer prices in Georgia and Turkmenistan.

[277] Nominal gross wages in industry deflated by the change in the producer price index. This may also be seen as an indicator of the change in gross operating profits.

[278] Measured by the annual average rate of change in gross industrial output.

[279] Nominal wage growth deflated by the change in productivity.

CHART 3.4.2
Consumer and industrial producer prices in the transition countries, 1995-1996
(Indices, January 1995=100)

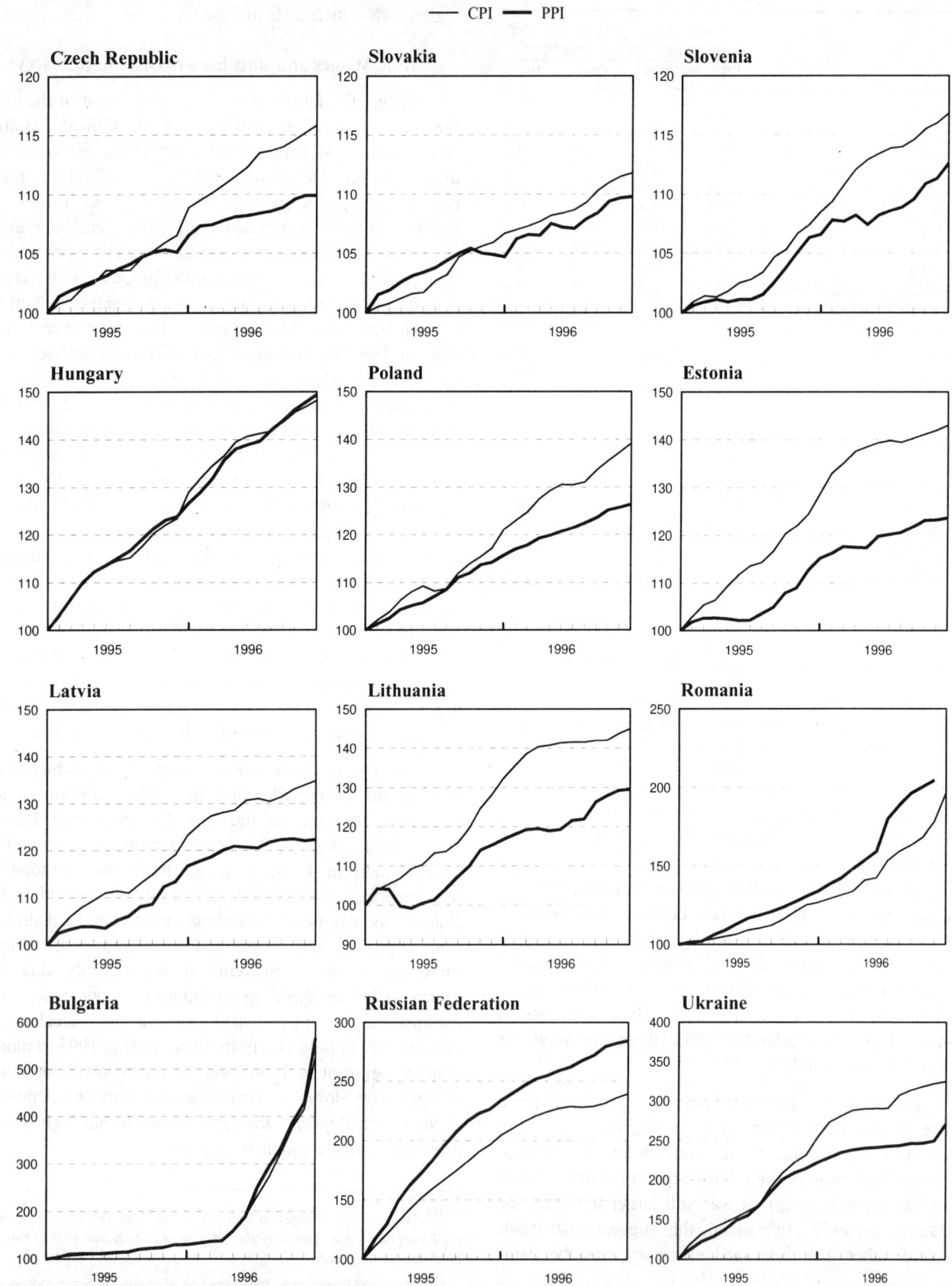

Source: National statistics.
Note: Countries are grouped according to scale.

TABLE 3.4.3

Wages and unit labour costs in industry[a] in the transition economies, 1995-1996
(Annual average percentage change)

	Nominal wages[b]		Real product wages[c]		Labour productivity[d]		Unit labour costs[e]		Real unit labour costs[f]	
	1995	1996	1995	1996	1995	1996[g]	1995	1996[g]	1995	1996[g]
Albania	29.3	..	..	..	..	..	..	..	..	..
Bosnia and Herzegovina	620.5	293.4	327.1	313.3	..	..	..	..	..	..
Bulgaria	59.8	105.6	7.3	-10.5	6.6	6.2	49.9	93.6	0.7	-15.7
Croatia	45.7	11.7	44.6	10.3	13.5	..	28.4	..	27.4	..
Czech Republic	18.1	17.7	9.5	12.2	8.6	7.6	8.8	9.4	0.9	4.3
Hungary	21.2	21.4	-5.7	-0.7	10.5	3.9	9.7	16.9	-14.6	-4.4
Poland	31.9	26.3	4.7	11.6	6.0	9.9	24.4	14.9	-1.2	1.5
Romania	54.2	53.4	14.0	2.3	16.2	13.8	32.8	34.8	-1.8	-10.1
Slovakia	15.2	14.7	5.5	10.3	8.3	2.5	6.4	11.9	-2.6	7.5
Slovenia	17.0	13.1	4.0	6.0	5.8	7.6	10.5	5.1	-1.7	-1.5
The FYR of Macedonia	10.4	2.7	5.4	..	..	..	..	..	..	..
Yugoslavia	109.1	82.9	19.0	-3.3	6.5	8.5	96.3	68.5	11.7	-10.9
Estonia	36.3	19.5	8.5	4.2	..	..	..	..	..	..
Latvia	28.1	20.3	14.4	5.7	4.0	2.4	23.2	17.5	10.0	3.3
Lithuania	39.4	29.4	8.2	11.1	14.9	8.3	21.3	19.5	-5.8	2.5
Armenia	214.0	48.8	9.1	8.1[h]	20.0	4.1	161.7	42.9	-9.1	3.8[h]
Azerbaijan	298.4	44.8	-72.3	-15.2	-12.0	-1.1	352.8	46.4	-68.6	-14.3
Belarus	668.3	60.5	20.3	16.7	-0.9	8.2	675.1	48.3	21.4	7.9
Georgia	..	..	..	..	..	..	..	..	..	..
Kazakstan	186.7	36.2	4.8	9.2	1.3	4.1	182.9	30.8	3.4	4.9
Kyrgyzstan	60.1	26.9	19.5	-12.3	-3.4	20.3	65.7	5.4	23.7	-27.1
Republic of Moldova	30.8	33.0	-14.1	2.2	12.0	-0.9	16.7	34.2	-23.3	3.1
Russian Federation	125.6	48.0	-33.2	-1.8	4.5	-2.8	115.8	52.3	-36.1	1.0
Tajikistan	109.5	265.5	-53.6	-17.3	-0.1	-15.1	109.7	330.5	-53.6	-2.6
Turkmenistan	639.5	763.7	..	..	-6.4	14.9	690.0	651.7	..	..
Ukraine	425.1	20.6	-4.7	-22.0	-3.9	3.0	446.4	17.1	-0.8	-24.3
Uzbekistan	289.2	99.1	-56.4	-15.5[h]	-1.2	11.9	293.7	77.9	-55.9	-24.5[h]

Source: UN/ECE secretariat estimates, based on national statistics and direct communications from national statistical offices.

[a] Same definition of industry as table 3.4.2.

[b] Average gross wages in industry except in Croatia, The FYR of Macedonia and Yugoslavia: net wages in total economy; in Albania, Bosnia and Herzegovina and all the CIS countries: gross wages including social assistance in total economy.

[c] Nominal wages deflated by producer price index.

[d] Gross industrial output deflated by industrial employment.

[e] Nominal wages deflated by productivity.

[f] Real product wages deflated by productivity.

[g] Except in Bulgaria, Hungary, Latvia, Poland, Slovakia, Ukraine and Yugoslavia, estimated on the basis of the change in industrial employment in January-September 1996 over the corresponding period in 1995.

[h] January-November 1996 over the corresponding period in 1995.

In contrast to 1994-1995, unit labour costs increased faster than producer prices in the majority of the east European countries in 1996. *Real unit labour costs*[280] increased particularly strongly in Slovakia (7.5 per cent) which, given the significant degree of producer price disinflation and the stable exchange rate, implies a sharp squeeze on industrial unit profits particularly in export oriented industries. Real unit labour costs also increased, albeit at slower rates, in the Czech Republic (more than 4 per cent) and Poland (1.5 per cent). In Poland they had been declining from the beginning of 1993 until 1995, and together with the dampening effect on unit material costs of a relatively strong zloty, were a major support to investment growth sustained by internal financing. In the Czech Republic the faster growth in real unit labour costs in 1996 reflects the sharp increase in real wages and the slowdown in productivity growth, whereas in Poland it was due to wage inflation pulling ahead of price increases and reduced profit margins, especially in export industries.

The rate of increase in *nominal gross wages* in most of the CIS countries continued to fall sharply in 1996 and, in contrast to the majority of east European and Baltic countries, nominal wages rose more slowly than producer prices. *Real product wages* increased only in Armenia, Belarus, Kazakstan and the Republic of Moldova, and except in the latter, the rise was due to a more rapid

---

[280] Unit labour costs deflated by the change in industrial producer prices. In other words, real product wages deflated by the change in productivity.

deceleration of inflation than in 1995. Real product wages continued their downward trend in the majority of the CIS countries in 1996, mainly because of accelerated price disinflation. The main exception was Belarus where the increase in real product wages in 1995 and 1996 reflected the very different policies pursued in Belarus as compared with the majority of the other CIS countries.

Industrial labour productivity improved in 1996 in most of the CIS countries, and in some significantly. The main exception to this generally favourable tendency was Tajikistan (a fall of some 15 per cent) and, to a much lesser extent, Azerbaijan, the Republic of Moldova and Russia. In the latter, the relatively strong rouble combined with extremely high real interest rates not only choked off inflation but also economic activity. Industrial output continued to fall faster than employment, reflecting, *inter alia*, the stance of economic policies which do not discourage large enterprises from keeping redundant workers on their payrolls.[281]

Given the relative changes in wages and productivity, *unit labour costs* in most of the CIS countries rose more slowly in 1996 than in 1995, although the rate of increase still remains much higher than in most of the east European and Baltic countries. Unit labour costs in some of the CIS countries also continued to increase more slowly than producer prices (Azerbaijan, Kyrgyzstan, Ukraine, Uzbekistan and, to a much lesser extent, Tajikistan). Exchange rates in the CIS countries during 1996 were generally stable and in some cases even appreciated (and strongly so in real terms). Unit profits in industry, which have been improving strongly since 1993, therefore probably increased still further with the main exception of Belarus. In Russia, attempts to maintain or increase gross operating profits together with falling productivity, have been the main source of domestic pressure for higher producer prices since the reforms got underway.

3.5 Foreign trade

(i) Overall trade results

The highly dynamic *export* development of transition economies in 1995 was not sustained through 1996. The value growth of the ECE transition countries' total exports (in current dollars) of about 5 per cent in 1996 was markedly lower than the 19 per cent average annual growth in 1994-1995. The growth of exports in value terms slowed most notably in the east European economies, which in aggregate reported an increase of just 1 per cent from 1995 (table 3.5.1), whereas the Baltic and CIS economies, including the Russian Federation, maintained stronger growth at 16 and 10 per cent, respectively. The main factors behind the slowing export performance were a further weakening of western demand, but also domestic economic policies, in some cases, and a lag in structural reforms, in general, in many of these countries. Governments' concentration on anti-inflationary measures led to a real appreciation of most of the domestic currencies which affected export competitiveness and squeezed exporters' margins. But the transition economies' competitiveness was even more affected by the continued rapid growth of unit labour costs, in particular in the industrial sector. Industrial real wage growth in recent years has outpaced productivity growth in most east European and Baltic countries, and the gap between these growth rates in general widened in 1996; this gap also widened in many CIS countries where the decline in real wages was less pronounced than the decline in productivity (section 3.4).

Import growth in current value terms also slowed in the transition countries, but less substantially. For all ECE economies in transition the rise was 12 per cent, compared to the average annual rate of 16 per cent in 1994-1995. Imports grew by 11 per cent in the east European countries, by 24 per cent in the Baltic states and by 12 per cent in the CIS, if the Russian Federation is included. The growth of east European imports mainly reflects the continued robust upturn in domestic demand, which has replaced net exports as the main support of economic growth in the region; to a smaller degree this also holds for the Baltic countries. In the Russian Federation and in a few other CIS countries, however, net exports were the only component of aggregate demand which was growing. The general real appreciation of domestic currencies made imports more competitive on the local markets. This was also enhanced by a further liberalization of trade regimes, including cuts in the average import tariff rates (by some 1-2 percentage points in many east European countries) and in import surcharges earlier introduced as temporary measures in order to restrict imports (in Hungary, Poland, Slovakia, etc.), and some one-time measures like zero-tariff quotas for some foodstuffs or the temporary lifting of import duties for selected goods.[282]

These developments in foreign trade flows have led to considerably weakened *trade balances* of the ECE transition economies in general, in particular in eastern Europe. External positions deteriorated notably also in the Baltic countries and in the CIS region. Only the Russian Federation posted a further growth in its surplus, as a result of declining imports and resumed fast growth in exports in the second half of 1996 (table 3.5.2).

---

[281] For a discussion of excess employment in the transition countries, see section 3.3(ii).

[282] A jump in Slovakia's imports in December can partly be attributed to such a measure: the temporary lifting of the tariff on small cars was about to expire on 1 January 1997. Romania's government used such interventions quite often in the course of 1996.

TABLE 3.5.1

Foreign trade of the European transition countries by direction, 1994-1996

(Value in billion dollars, growth rates in per cent)[a]

Country or country group[b]	Exports				Imports			
	Value 1995[c]	Growth rates			Value 1995[c]	Growth rates		
		1994	1995	1996[d]		1994	1995	1996[d]
Eastern Europe, to and from:								
World	92.4	16.4	25.0	1.1	112.3	9.9	31.3	11.4
Transition economies	24.5	5.2	24.2	6.5	28.4	-1.6	30.3	9.4
Soviet Union/successor states[e]	6.6	8.5	27.5	9.5	12.2	-6.2	26.9	8.8
Eastern Europe[f]	11.7	6.9	27.0	5.2	10.6	3.3	34.1	9.5
Developed market economies	60.2	24.5	25.3	-0.9	75.3	16.6	31.3	11.5
Developing countries	7.8	2.6	24.9	-0.1	8.5	0.1	33.3	16.4
Baltic states, to and from:								
World	5.9	12.9	35.5	16.1	8.0	27.1	41.7	24.0
Transition economies	2.9	-2.9	18.5	22.2	3.4	0.7	22.9	12.9
CIS	2.1	-6.4	19.1	19.5	2.5	-9.1	16.1	2.2
Baltic states	0.6	35.2	22.9	41.5	0.5	24.4	58.1	35.9
Developed market economies	2.8	47.8	59.9	8.4	4.4	75.4	64.2	30.7
Developing countries	0.1	-2.8	20.4	52.7	0.2	-6.5	-21.7	70.1
Russian Federation, to and from:								
World	63.7	8.4	29.5	8.0	33.2	5.2	17.0	-5.4
Transition economies	14.8	-14.9	36.3	21.9	6.3	-15.8	38.1	-12.7
Eastern Europe[f]	8.0	-19.5[g]	42.4	15.7	3.5	-16.6[g]	55.5	-21.4
Baltic states	2.3	26.4	31.5	16.3	1.1	-17.4	57.4	-39.8
Developed market economies	38.6	21.7	19.5	3.2	23.1	20.2	15.9	-7.8
Developing countries	10.3	0.4	20.1	5.8	3.8	-4.5	-12.2	21.5
Other CIS countries, to and from:								
World	13.7	13.1	43.8	17.1	11.3	17.0	34.3	62.6

Source: National statistics and direct communications from national statistical offices to UN/ECE secretariat; for the Russian Federation, State Customs Committee data; CIS Interstate Statistical Committee data.

[a] Growth rates are calculated on values expressed in dollars. Values for 1995 and growth rates for 1994 and 1995 include the "new" foreign trade (trade among successor states of former Czechoslovakia and SFR of Yugoslavia).

[b] "Eastern Europe" refers to Albania, Bulgaria, Croatia, the Czech Republic, Hungary, Poland, Romania, Slovakia and Slovenia. For lack of adequate data, the trade of Bosnia and Herzegovina, The former Yugoslav Republic of Macedonia and Yugoslavia (Serbia and Montenegro) are not covered. The partner country grouping follows the practice until recently prevalent in the national statistical sources, which differs from the breakdown usually employed in United Nations publications. Thus, "transition economies", which covers the ex-socialist trade partners, includes Cuba and the former SFR of Yugoslavia, in addition to the east European countries, the former Soviet Union and the Asian centrally planned economies. "Developed market economies" excludes Turkey and includes Australia, New Zealand and South Africa.

[c] Values for 1995 revised according to the new methodology of foreign trade reporting in the Czech Republic and Lithuania. Growth rates through 1995 according to the previous reporting system.

[d] Aggregated growth rates for eastern Europe calculated using preliminary full-year data for Bulgaria and Poland.

[e] Data from six reporting countries only (Bulgaria, Czech Republic, Hungary, Poland, Romania and Slovakia).

[f] Excluding the former SFR of Yugoslavia.

[g] Trade with all former CMEA members (i.e. including Cuba, Mongolia and Viet Nam).

The data for 1996 are still incomplete, and those that are available are sometimes very uncertain,[283] but the main changes are reasonably clear and are presented in more detail in sections 3.5(ii)-(v) below.

The considerable deceleration in the value growth of both exports and imports reflected dollar exchange rate movements against other major hard currencies in 1995-1996. The changes in world market prices including fast growing energy prices and overall decline in the prices for industrial raw materials and intermediates also seriously affected trade performance of these countries – caused lower dollar value for exports from many east European countries and improved substantially export earnings in Russia and some other net fuel exporters among the CIS countries. Hence, in terms of *volume*, for which the data, however, are much more uncertain, the deceleration in trade growth was generally less pronounced and in a few cases (for instance, Russia's exports) there was none.

The very limited data available for the first months of 1997 point to a continuation of the trends seen in 1996. In January-February 1997, the gap between exports and imports widened further in the CEFTA and Baltic countries, while at least two south European transition countries (Croatia and Romania) recorded some improvement in their trade balances. The Russian

---

[283] On the persistence of statistical problems in transition economies, see UN/ECE, *Economic Bulletin for Europe*, Vol. 48 (1996), chap. 3.

TABLE 3.5.2

Trade balances of the European transition countries, 1992-1996

(Billion dollars)

	1992	1993	1994	1995^a	1996^b
Eastern Europe					
World	-6.3	-12.9	-11.9	-19.9	-31.6
Transition economies	-2.0	-3.0	-2.0	-3.9	-5.0
Developed market economies	-4.7	-10.2	-9.7	-15.2	-24.4
Developing countries	0.4	0.3	-0.2	-0.8	-2.2
Baltic states					
World	..	-0.3	-0.9	-2.1	-3.1
Transition economies	..	0.1	–	-0.5	-0.2
CIS	..	-0.2	-0.1	-0.4	–
Developed market economies	..	-0.3	-0.8	-1.6	-2.7
Developing countries	..	-0.1	-0.1	-0.1	-0.2
Russian Federation^c					
World	5.4	17.5	24.8	30.6	37.4
Transition economies	4.3	5.7	7.3	8.5	12.6
Eastern Europe^d	2.7^e	4.6^e	4.2	4.5	6.5
Developed market economies	1.5	10.2	13.0	15.5	18.6
Developing countries	-0.4	1.6	4.5	6.5	6.3

Source: National statistics and direct communications from national statistical offices to UN/ECE secretariat.

^{a} 1995 trade balances are derived from export and import data reported by the Czech Republic and Lithuania according to the new methodology.

^{b} Trade balance for eastern Europe includes preliminary full-year figures for Bulgaria and Poland.

^{c} For the Russian Federation: Goskomstat data for 1992-1993; State Customs Committee data for 1994-1996. The two series are not fully comparable. Data for 1995 revised. All data exclude intra-CIS trade.

^{d} Excludes the former SFR of Yugoslavia.

^{e} Russian Federation balance with all former CMEA members (i.e. including Cuba, Mongolia and Viet Nam).

Federation, too, seems to have recorded a further rise in its trade surplus in spite of some levelling of world market fuel prices. Exports remain sluggish in most east European countries as western import demand, and in particular that of the German market, has not yet shown any revival and the countries themselves have not yet implemented any more effective measures leading to improved export competitiveness. Most continue to adhere to anti-inflationary programmes which generally include exchange rate policies that lead to real currency appreciation and shrinking export margins. Some governments enhanced their export promotion programmes as from the beginning of 1997 including tax preferences and state sponsored credit facilities (the Czech Republic, Slovakia, etc.), but these measures are hardly sufficient.

Export growth in the transition economies is expected to remain weak in the first half of 1997, but is likely to pick up somewhat later in the year if western import demand resumes its growth as expected. For countries relying on oil and industrial raw material exports, the prices of which are expected to remain more or less steady on the world markets, export earnings may be somewhat subdued, especially if a further strengthening of domestic currencies continues. On the other hand, exchange rate appreciation, together with continued domestic demand growth – though probably slower than in 1996 – in eastern Europe and the Baltic states, is likely to keep imports growing at a fast pace. While some observers argue that the expected import boom will concentrate on the import of machinery and equipment that will benefit the economies' production and eventually result in an increase in exports, others see a tendency of these investment good imports to consist of components for large infrastructure investment projects, which would not directly affect exports in the short to medium run.

If real wage growth remains strong throughout the region, the rapid growth of imports will become of increasing concern, the investment goods bias notwithstanding. Many governments in the east European and Baltic states, however, in pursuing further tight monetary policies are mainly depressing output but make little impact on demand – a combination which is bound to boost import growth and trade deficits.

On trade liberalization there are some mixed signals for 1997. In the area of international agreements, more liberal access to the mutual markets can be expected in the Baltic region where a free trade agreement for agricultural products entered into force on 1 January 1997. For Romania, expectations are high with its acceptance into CEFTA as from 1 July 1997. All three Baltic countries negotiated and signed bilateral free trade agreements with CEFTA member countries in the course of 1996 which take effect in 1997. Some relief also comes with the EU promise to ease anti-dumping rules for economies in transition. However, as regards trade policy changes at the level of individual transition countries, moves from liberalizing measures towards rising restrictions will probably remain a common feature in 1997. Bulgaria had already introduced a 5 per cent import surcharge in June 1996. Poland's restoration, with effect from 15 February 1997, of import duties on certain foodstuffs (including grain) which had been lifted in 1996 in an attempt to curb inflation is quite indicative.[284] The Baltic countries also seem to be resistant to further trade liberalization measures. Lithuania's government resisted IMF pressure to lower tariffs in March 1997, postponing the step until September, while Estonia is considering the introduction of certain protective tariffs as from 1 July 1997.[285] Hungary, on the other hand, after postponing

---

[284] The duties are restored at a rate of 10 per cent, half the rate suspended in 1996. A further problem in Poland is a remaining set of measures to protect various industries (including energy, steel, shipbuilding, etc.), for which the government has been repeatedly criticized by the EU.

[285] *BBC Summary of World Broadcasts*, SU/2841 E/1, 12 February 1997.

the 2 percentage point reduction of its temporary import surcharge scheduled for 1 January 1997, has announced the reduction of the 6 per cent surcharge to 4 per cent as of 10 March, with final abolition due on 1 July.[286] In Slovakia, an import surcharge of 7.5 per cent on non-capital goods was phased out on 1 January 1997, and Slovenia may start lowering its import duties for agricultural products under CEFTA requirements, while the Czech government confirmed its commitment not to introduce additional restrictions for imports. The Russian Federation annulled a September 1996 resolution that had called for the imposition of import quotas on alcoholic beverages and ethyl alcohol that was to come into effect on 1 January 1997. Nonetheless, all in all, trade liberalization most probably will proceed further under the transition economies' international commitments – the Uruguay Round requirements for those which are WTO members, the Europe Agreements or interim trade agreements with the EU, and under the CEFTA and the Baltic free trade agreements.

(ii) East European and Baltic countries

(a) Trade performance

The expansion of the east European and Baltic countries' foreign trade decelerated significantly in 1996. In *current dollar value*, exports stagnated in eastern Europe, after growth of some 25 per cent in 1995, and while in the Baltic states growth was still substantial at 16 per cent, this was less than half the rate of growth of the year before. The growth of *import* value also slowed significantly in both country groups, but remained high, at 11 and 24 per cent respectively, outpacing that of exports by large margins (table 3.5.1). In consequence, *trade balances* deteriorated further, the aggregate deficit widening from $20 billion in 1995 to $32 billion in eastern Europe, and from $2 billion to $3 billion in the Baltic countries (table 3.5.2).

Although the high dollar *value* growth rates in 1995 had to a substantial degree reflected the sharp depreciation of the dollar against other major currencies in that year, trade *volume* growth had also been quite high in 1995 at some 8-10 per cent in eastern Europe and growth probably in the same range in the Baltic group. Very approximative estimates for 1996 suggest that *exports* of eastern Europe contracted somewhat in volume, whereas in the Baltic states volume growth of exports may have remained unchanged at some 9 per cent. *Import* volume probably rose by about the same proportion as value in eastern Europe, at over 10 per cent, but is hard to estimate for the Baltic countries owing to the important share of the CIS market (with special price arrangements and a high share of fuels) in their trade.

Trade balances

A further widening of *trade deficits*, measured on the basis of customs data, was registered in the second half of 1996 in almost all east European and Baltic countries. In most, the deterioration of trade balances accelerated in the course of the year.

The most substantial deterioration was noted in the CEFTA countries (table 3.5.3). In *Poland*, the trade deficit reached $12.6 billion, more than doubling from 1995.[287] In the *Czech Republic* and *Slovakia*, the deficits widened by some $2 billion each as compared with 1995, and rose to over 11 per cent of GDP in both countries (table 3.1.1); in relative terms, the change in Slovakia's deficit was the more important, as it had been 2 per cent of GDP in 1995. *Hungary's* trade deficit also widened by some $0.5 billion, probably as a result of a softening of the austerity measures in place since March 1995.[288] Only *Slovenia's* trade deficit narrowed slightly in 1996. In the Czech Republic and Poland most of the deterioration was in trade with the western market economies, while in Hungary and Slovakia it was more substantial in trade with the other economies in transition, especially the CIS countries; in Hungary's trade with western partners it remained almost unchanged as compared with 1995. The trade balances of the Czech Republic and Poland with Germany, their main trading partner, swung into deficit ($0.4-$0.7 billion) in 1996 after surpluses of $1.5 billion and $1.0 billion in 1995. Hungary's and Slovakia's surpluses in trade with Germany shrank, amounting only to some $30 million in the former but to $270 million in the latter. Slovenia's surplus in trade with Germany grew by $0.2 billion as compared with 1995, but the deficit with Italy, the second major partner for Slovenia, widened by $0.1 billion.

---

[286] As of January 1997, Hungary, however, lowered its average import tariff on industrial products from 8 per cent to 7.5 per cent and on agricultural products from 38 per cent to 36 per cent, in line with its international commitments. In 1997, more than a half of industrial goods from the EU will enter Hungary free of customs duty, while tariffs for the remaining items are cut by 40 per cent. In addition, the Hungarian government scrapped a 1 per cent statistical fee and a 1 per cent customs clearance charge on imports from WTO members at the beginning of 1997. *BNA's Eastern Europe Reporter*, Vol. 7, No. 4 (Washington, D.C.), 24 February 1997.

[287] According to preliminary foreign trade data released by the Informatics Centre of the Foreign Trade Ministry of Poland, exports in 1996 amounted to $24.4 billion and imports to $36.9 billion, with growth of 6.3 and 27.2 per cent respectively (*Rzeczpospolita* (Warsaw), 25 February 1997). Data by partner region in table 3.5.3 refer to January-September 1996, as the geographical breakdown was not yet available for the full year at the time of writing.

[288] The March 1995 austerity measures included an import surcharge of 8 per cent, which was lowered by 1 percentage point from mid-1996 and by another point in October 1996, and a forint devaluation of an initial 9 per cent followed by a crawling peg adjustment at 1.9 per cent per month (March-June 1995), then 1.3 per cent (July-December 1995) and 1.2 per cent per month from the beginning of 1996.

TABLE 3.5.3

Foreign trade of CEFTA countries by direction, 1994-1996
(Growth rates in per cent, trade balances in billion dollars)

	Growth rates						Trade balances		
	Exports			Imports					
Country and trade partner groups [a]	1994	1995	1996	1994	1995	1996	1994	1995	1996
Czech Republic									
World	8.4	19.6	1.2	16.4	39.5	10.1	-0.7	-3.6	-5.9
Transition economies	-4.6	23.1	9.0	1.8	38.9	2.8	0.2	-0.5	-0.1
Slovakia	-17.4	18.0	4.1	-5.2	28.7	-10.9	0.2	–	0.5
Developed market economies	19.6	19.7	-2.5	23.9	40.6	11.9	-1.1	-3.2	-5.6
Developing countries	-6.5	3.2	5.5	26.0	29.8	26.0	0.2	0.1	-0.2
Hungary									
World	20.0	20.2	2.2	15.1	6.3	4.8	-3.9	-2.6	-3.1
Transition economies	5.2	26.7	1.8	-8.0	6.5	10.8	-1.0	-0.6	-0.9
Developed market economies	29.8	15.9	2.8	25.5	6.0	2.5	-2.6	-2.0	-2.0
Developing countries	-17.5	52.4	-3.6	17.3	8.5	8.1	-0.3	-0.1	-0.2
Poland									
World	21.4	32.8	6.3	15.0	34.7	27.2	-4.3	-6.1	-12.6
Transition economies	25.4	54.4	28.5[b]	22.1	44.6	24.6[b]	-0.9	-1.0	-0.9[b]
Developed market economies	21.8	32.2	2.9[b]	13.4	33.3	26.7[b]	-3.2	-4.4	-6.3[b]
Developing countries	12.5	0.7	17.8[b]	16.7	27.7	41.9[b]	-0.2	-0.7	-1.0[b]
Slovakia									
World	22.7	28.2	2.9	5.3	33.3	24.2	0.1	-0.3	-2.2
Transition economies	8.5	26.2	-2.9	-8.2	27.7	17.4	–	-0.1	-1.1
Czech Republic	8.2	20.9	-9.5	-13.2	24.9	9.8	0.5	0.6	–
Developed market economies	49.7	34.7	12.1	26.4	38.7	30.0	–	-0.2	-0.9
Developing countries	24.6	4.8	-7.0	48.9	61.6	52.2	0.1	–	-0.2
Slovenia									
World	12.2	21.4	-0.1	12.4	29.4	-1.0	-0.5	-1.2	-1.1
Transition economies	6.4	18.2	13.2	-2.3	30.0	-4.2	0.4	0.3	0.7
Developed market economies	17.7	22.5	-4.6	20.2	29.2	-1.2	-0.8	-1.3	-1.5
Developing countries	-34.9	21.1	1.0	-26.4	30.3	13.7	-0.2	-0.2	-0.3

Source: UN/ECE secretariat, based on national foreign trade statistics.

Note: Growth rates and trade balances are based on trade values in terms of dollars.

[a] Country groups as in table 3.5.1.

[b] 1996 growth rates and trade balance figures refer to January-September.

Widening trade deficits were registered also in most south European transition countries: *Albania, Croatia, The FYR of Macedonia* and *Yugoslavia*, where they generally are also very large in relation to GDP, but *Bulgaria's* trade balance swung into a small surplus on full-year data,[289] and the *Romanian* deficit narrowed slightly in 1996 (table 3.5.4).[290] Trade with western countries accounted for the bulk of these deficits except for Romania where trade deficits with CIS and other countries in transition were also important.

---

[289] Preliminary data announced by Minister of Trade D. Bobeva (*24 chasa* (Sofia), 19 March 1997). Table 3.5.4 refers to the first nine months for the regional components.

[290] Except for Croatia and Romania, foreign trade data for the south European transition countries are scarce, and often unstable. Data for Albania and The FYR of Macedonia are published only at irregular intervals, and Yugoslavia started releasing foreign trade figures only from the first quarter of 1996 (growth rates for 1996 in table 3.5.4 are from European Community Monitor Mission, Economics Section, *Weekly Economic Report* (Zagreb), 20-27 January 1997). Problems with Bulgarian foreign trade data were discussed in UN/ECE, *Economic Bulletin for Europe*, Vol. 48 (1996), box 3.1.1.

Estonia, Latvia and *Lithuania* followed the pattern of the CEFTA group: deterioration of trade balances accelerated in the second half of 1996 with substantial deficits for the year as a whole (table 3.5.5). Estonia's and Latvia's deficits widened by some 60 per cent from 1995 and accounted for almost two thirds of total export earnings. Relative to GDP, Estonia's deficit in 1996 (based on customs data) came to more than 26 per cent and was among the highest in the transition countries. In Lithuania, the widening of the deficit (by 20 per cent) in 1996 was less pronounced than in 1995, when it almost tripled. Again, the most rapidly growing deficits of the Baltic countries were registered in trade with western economies, while trade balances with other countries in transition improved in Estonia and Lithuania but deteriorated substantially in Latvia.

Balance of payments data, reflecting flows of actual payments for merchandise trade, generally indicate similar tendencies in trade balances in most countries, but in some instances the gap between the trade balances based on customs statistics and those

TABLE 3.5.4

Foreign trade of south-east European countries by direction, 1994-1996
(Growth rates in per cent, trade balances in billion dollars)

Country and trade partner groups [a]	Growth rates						Trade balances		
	Exports			Imports					
	1994	1995	1996	1994	1995	1996	1994	1995	1996
Albania									
World	15.4	45.5	4.4	30.1	18.4	40.0	-0.4	-0.4	-0.7
Transition countries	-46.6	71.3	-34.1	20.1	2.9	17.6	-0.1	-0.1	-0.1
Developed market economies	29.1	53.0	15.3	34.4	23.4	48.3	-0.3	-0.3	-0.6
Developing countries	-39.5	-16.4	-74.8	14.6	2.9	-14.1	–	–	–
Bosnia and Herzegovina									
World	..	..	..	..	..	..	..	..	..
Bulgaria									
World	5.8	28.3	-15.7	-21.7	20.7	-23.8	-0.2	-0.3	0.2
Transition economies	12.4	3.6	-12.5[b]	-41.7	17.9	-19.3[b]	-0.1	-0.6	-0.2[b]
Developed market economies	33.1	22.9	-14.4[b]	-15.3	20.9	-10.6[b]	–	-0.1	-0.1[b]
Developing countries	-42.6	117.4	-21.6[b]	14.0	28.8	-3.6[b]	–	0.4	0.1[b]
Croatia									
World	9.1	8.7	-2.6	12.1	43.6	3.7	-1.0	-2.9	-3.3
Transition economies	2.5	10.4	10.0	-9.5	36.9	9.6	0.2	-0.1	-0.1
Developed market economies	13.4	4.9	-13.6	19.2	48.3	-0.5	-0.8	-2.4	-2.7
Developing countries	1.9	51.0	36.9	27.4	28.1	21.5	-0.3	-0.4	-0.4
The FYR of Macedonia									
World	2.9	10.8	-29.0	23.7	15.8	-4.0	-0.4	-0.5	-0.8
Romania									
World	25.7	28.6	-3.2	9.0	44.6	-3.0	-1.0	-2.4	-2.3
Transition economies	-3.5	-2.7	-5.9	15.1	36.1	1.0	-0.6	-1.3	-1.3
Developed market economies	44.6	42.9	-3.7	15.2	44.2	-1.2	-0.9	-1.3	-1.4
Developing countries	18.3	20.3	-0.2	-15.8	59.6	-15.0	0.5	0.2	0.4
Yugoslavia									
World	..	..	20.3	..	..	53.9	..	..	-2.3

Source: UN/ECE secretariat, based on national foreign trade statistics.

Note: Growth rates and trade balances are based on trade values in terms of dollars.

[a] Country groups as in table 3.5.1.

[b] 1996 growth rates and trade balance figures refer to January-September.

derived from balance of payments data are very substantial (for comparison see tables 3.6.3 and 3.6.4 below).[291]

Exports and imports

The deceleration of *export* growth in 1996 was common to all countries of the region, but beyond that export performance varied quite widely. Among the CEFTA countries, only Hungary and Poland registered significant volume growth – of 6 and 9 per cent, respectively,[292] in spite of the slowing of value growth (table 3.5.3). Apart from Yugoslavia, where value growth also appears to have been substantial (but reflected a pick up from very low levels after the lifting of international sanctions), exports either stagnated or contracted in value – almost generally the latter in the south European transition countries (table 3.5.4). Among the Baltic countries, despite slowing down somewhat in the second half of the year, the fastest growth of exports was registered in Lithuania, by 21 per cent, but the 11-13 per cent rise in Estonia and Latvia was also substantially above the average growth in the CEFTA group (table 3.5.5).

---

[291] Among the causes of the gap are unregistered exports, largely the "cross-border" trade flows stimulated by still quite substantial price differentials in favour of the east European countries, but also time lags between the border crossing of trade flows and the arrival of payments. On this issue, see UN/ECE, *Economic Bulletin for Europe*, Vol. 48 (1996), p. 45.

[292] There is conflicting information on the volume growth of Hungarian foreign trade in 1996. For January-November, the Hungarian Central Statistical Office (HCSO) reported export growth of 5.5 per cent and import growth of 7 per cent in volume, while according to the National Bank of Hungary, which includes trade flows of Hungary's industrial duty free zones (these are not covered in the HCSO data), exports increased by 13 per cent and imports by 12 per cent in volume. It is difficult, however, to reconcile the volume estimates on the basis of the coverage difference alone, as trade from and to these zones amounted to only some $2 billion, or 16 and 11 per cent, respectively, of Hungary's total exports and imports. The analysis below is based on data of the Hungarian Central Statistical Office. HCSO, *Statistical Report*, November 1996, p. 14 and National Bank of Hungary, *Recent Economic Developments in Hungary*, February 1997, pp. 16-17.

TABLE 3.5.5

Foreign trade of the Baltic states by direction, 1994-1996

(Growth rates in per cent, trade balances in million dollars)

	Growth rates						Trade balances		
	Exports			Imports					
Country or partner groups [a]	1994	1995	1996	1994	1995	1996	1994	1995	1996
Estonia									
World	62.4	40.6	13.1	85.5	53.1	26.0	-354	-703	-1 122
Transition economies	64.1	19.4	17.7	71.5	41.7	19.4	155	86	90
Baltic states	80.2	25.8	29.8	34.1	36.7	22.8	111	133	178
CIS	61.9	16.4	13.0	76.1	40.7	12.8	56	-18	-19
Eastern Europe	1.0	49.3	2.9	134.4	67.3	70.3	-9	-19	-53
Developed market economies	64.9	59.1	8.4	90.2	57.3	25.4	-482	-745	-1 119
Developing countries	-13.2	45.0	78.7	121.2	56.5	99.2	-27	-44	-94
Latvia									
World	-2.0	31.9	10.6	30.1	46.5	27.6	-252	-513	-877
Transition economies	-10.0	23.1	9.0	9.3	39.2	28.6	-16	-117	-281
Baltic states	29.3	40.3	41.9	-8.6	63.9	44.9	-37	-79	-118
CIS	-11.4	18.3	3.6	4.3	35.4	15.4	49	-13	-75
Eastern Europe	-38.2	62.1	-20.6	139.8	53.5	61.8	-15	-24	-77
Developed market economies	19.6	47.0	8.1	98.1	79.2	26.0	-138	-383	-595
Developing countries	-35.3	-34.4	173.5	-22.6	-75.3	59.3	-97	-13	-2
Lithuania [b]									
World	1.9	33.3	21.2	3.0	31.0	20.7	-325	-944	-1 125
Transition economies	-14.6	16.2	30.2	-13.0	11.0	4.6	-136	-470	-97
Baltic states	14.1	12.8	53.5	97.9	64.9	33.2	121	71	146
CIS	-16.6	20.6	29.1	-23.3	2.7	-5.6	-235	-390	29
Eastern Europe	-21.3	-0.5	-1.5	102.8	34.8	38.9	-29	-152	-270
Developed market economies	56.2	67.8	9.2	50.0	63.8	39.9	-210	-450	-966
Developing countries	144.0	36.8	11.9	-0.3	74.8	53.8	22	-23	-63

Source: National statistics and direct communications from national statistical offices to UN/ECE secretariat.

Note: Growth rates and trade balances are based on trade values in terms of dollars.

[a] Country groups as in table 3.5.1.

[b] 1995 values according to the revised Lithuanian data under a new methodology, while 1995 growth rates are based on the previous data.

The actual export performance of most east European countries in 1996 turned out below the most bearish expectations of late 1995, but was slightly better than predicted in the case of the Baltic countries. The growth of export values had already lost momentum or turned into decline during the second quarter of 1996 in many of these countries, with a further growth deceleration during the second half of the year in Albania, the Czech Republic, Hungary, Lithuania and Poland. In the remaining countries of the region exports gained some momentum during the last two quarters of 1996, but not enough as to offset the earlier decline for year-on-year results in Slovenia or the south European transition economies. Improvement was more noticeable in Estonia and Latvia, where exports in dollar value terms rose by some 15-17 per cent in July-December 1996 as compared with the previous year, which assured double-digit growth rates for the year as a whole in these countries.

The growth of *import values* in east European and Baltic countries gained some impetus too in the second half of 1996 as compared with the relatively sluggish performance in March-June; however, for the year as a whole it decelerated significantly relative to the growth in 1995.

Among CEFTA countries, import growth was most rapid (at 24-27 per cent) in Poland and Slovakia (table 3.5.3). Imports into the Czech Republic increased by 10 per cent in 1996, while the import rise by 5 per cent in Hungary was all due to the resumed growth in July-December 1996. Import growth at above 20 per cent was registered in all three Baltic countries, but the highest growth rates in the region – 40-50 per cent – were in Albania and Yugoslavia, though from quite low levels. In some other east European countries imports fell in dollar value terms, and most considerably in Bulgaria (tables 3.5.4 and 3.5.5).

Estimates of trade *volume* (and trade prices) are still reported by only very few countries in the region. The available export volume growth rates are shown in table 3.5.6 (the two Baltic countries represented there estimate only export volume, while for the Czech Republic, Hungary and Poland import volume estimates are also available). Aggregate world market price changes, in dollar terms, were on the whole small for manufactured goods, but mostly negative for commodities other than fuels.[293] On this basis, and that

---

[293] For details of world commodity price developments, see chap. 2.4 above.

TABLE 3.5.6

Indicators of foreign trade performance for selected east European and Baltic countries, 1995-1996
(Percentage change over same period of the previous year, ratios in per cent)

	Demand conditions				Price competitiveness						Outcome			
	Domestic demand[a]		Partners' demand[b]		Export unit value (dollars)[c]		Real exchange rate (NC/DM)[d]				Export volume growth		Trade balance to GDP[e]	
							CPI		ULC					
	1995	1996	1995	1996	1995	1996	1995	1996	1995	1996	1995	1996	1995	1996
Czech Republic	4.9	7.8[f]	6.9	3.0	13.4	1.2	-3.1	-9.2	-5.9	-11.2	5.7	–	-7.7	-11.5
Hungary	-5.6	-3.9[f]	7.2	2.9	10.9	-3.4	7.1	-5.1	20.7	-1.4	8.4	5.8	-5.9	-6.9
Poland	5.3	10.7	7.0	2.9	13.8	0.9[f]	-4.1	-10.4	-4.8	-8.0	16.7	9.3[f]	-5.2	-9.6
Slovakia	3.7	18.4	7.3	3.0	..	..	-2.5	-5.9	-0.9	-12.3	..	..	-1.9	-11.1
Slovenia	7.1	..	7.2	2.2	..	..	-6.0	0.7	-8.5	4.0	..	..	-6.3	-5.9
Estonia	4.2	..	7.7	4.0	30.4	8.0	-21.2	-17.9	..	..	10.0	6.0	-19.4	-26.4
Latvia	–	2.3*	7.1	3.8	23.2	3.8	-13.2	-14.4	-15.0	-15.6	7.1	8.8	-11.5	-17.2
Lithuania	..	..	7.7	4.0	..	..	-17.0	-22.8	-8.0	-20.5	..	..	-15.8	-15.2
Russian Federation[g]	-5.5	..	8.9	6.2	16.1	2.0	-17.2	-29.3	6.4	-29.9	6.4	6.6	8.5	8.5

Source: UN/ECE secretariat computations, based on national statistics.

a Growth of final domestic demand in exporting country.
b Import demand in western partner countries, as for chart 3.5.1.
c Calculated from reported changes in export unit values in national currencies.
d Changes in nominal exchange rates of national currencies against the deutsche mark (NC/DM) deflated by changes in domestic and German consumer price indices (CPI) and unit labour costs in industry (ULC).
e Computed from trade levels and GDP expressed in dollars at current exchange rates.
f January-September 1996 over same period of 1995.
g Figures in outcome block above refer to trade excluding CIS partners.

of the available estimates, it seems that the changes in the volume of foreign trade of the *east European countries* in 1996 must have been rather similar to the change of dollar values, as already noted, i.e. stagnation of exports and growth of imports at over 10 per cent in the aggregate. In the *Baltic countries*, export volume growth appears to have been somewhat weaker than that of value, i.e. export unit values (in dollar terms) increased more substantially than in the east European countries for which data are available, as shown by the figures reported from Estonia and Latvia.[294] Assuming a somewhat similar relationship between value and volume in Lithuania, a very rough estimate of some 9 per cent for Baltic export volume growth can be obtained.

In the countries reporting import volume, it rose more rapidly in the second half of the year in the Czech Republic, but the pace was still far below that registered in 1995 (at 11 and 24 per cent, respectively). Hungary's imports resumed volume growth in the second half of the year after a decline posted in the first half. As a result, imports in volume terms rose by 7 per cent in 1996, following a fall of 4 per cent in the previous year. Polish import growth accelerated in 1996 with a 29 per cent rise in volume (20 per cent in 1995).

Directions of trade

As concerns developments in 1996 in the *geographical composition* of trade, they remained similar to those observed already in the first half of 1996.[295] Three developments in exports were among the more important in 1996. *First*, the fall of exports to the *developed market economies*: the share of developed market economies declined by 2-3 percentage points in the Czech Republic, Poland and Slovenia, by 3-4 percentage points in the Baltic countries, while among the south European transition countries the most affected exports to this partner group were in Croatia, down by 8 percentage points from 1995. In Romania's exports, the share of developed market economies remained roughly unchanged (60 per cent), while in total exports of Albania, Hungary and Slovakia, exports towards the west gained in share. *Second*, the faster than overall rate of growth of east European and Baltic exports directed to *CIS countries*. In aggregate, CEFTA countries' exports to these markets grew by 7 per cent (total exports, by 3 per cent), with the most rapid expansion in Poland (38 per cent in January-September 1996) but a substantial decline in Hungary (-13 per cent). The Baltic countries' exports to the CIS grew in aggregate by some 19 per cent, with growth at twice that rate in Lithuania, but much below in Latvia. South

---

[294] For some explanations on this issue, see A. Richards and G. Tersman, "Growth, non-tradeables, and price convergence in the Baltics", *Journal of Comparative Economics*, Vol. 23, No. 2 (Orlando, FL), October 1996.

[295] Analysed in some detail in UN/ECE, *Economic Bulletin for Europe*, Vol. 48 (1996).

European countries' penetration of the CIS markets was much less pronounced except probably for Bulgaria. *Third*, the continuing recovery of *intraregional trade*. In 1996, the value of intraregional trade grew by some 6-8 per cent in the CEFTA (and by some 15 per cent if the impact of contracting trade between the Czech Republic and Slovakia is excluded), and much faster, by 35-40 per cent, in the Baltic countries. Trade among the former members of the SFR of Yugoslavia also increased in 1996, with Croatia's exports up 15 per cent and Slovenia's up 19 per cent in face of an overall decline in exports.

On the import side, developments were quite different. Imports from *developed market economies* kept rising faster than overall imports in Albania, the Czech Republic, Slovakia and Lithuania, or declined less than overall imports in Bulgaria and Romania; in the remaining countries the change neutral. Imports from *CIS countries* rose only slowly or at below average rates; those from CIS countries other than the Russian Federation contracted by some 3 per cent, according to the preliminary data for 1996, in the case of the CEFTA and grew only slightly in the Baltic countries.

Commodity composition

The analysis of patterns of change of *exports by commodity* during 1996 shows that in spite of the overall slowing of value growth, exports of some commodity categories continued to expand at a respectable pace.[296] In particular, notable increases were recorded in exports of beverages and tobacco products (SITC 1) – in January-September 1996 they were up by 75 per cent in Romania, 40 per cent in aggregate in the Baltic countries, 15 per cent in the CEFTA group. Fast value growth of machinery and transport equipment (SITC 7) was also observed – up some 20 per cent on average in the Baltics and 15 per cent in the CEFTA; these were mostly fuelled by sharply rising exports of assembled passenger cars. Also fast growing but not as much as the above were miscellaneous consumer manufactures (SITC 8) – up 11 per cent in the CEFTA countries, and 15-20 per cent in Estonia and Lithuania in the first nine months of 1996. Among items in this category, the fastest growing remained textiles and apparel and footwear (mainly under outward processing trade (OPT) transactions).

The relatively poor overall export performance in the east European countries was caused primarily by a drop in exports of intermediate products classified by material (SITC 6) – down by 4 per cent in nominal terms in the CEFTA countries and by 11 and 13 per cent in Croatia and Romania, respectively. This is considered to be the most important commodity category in many east European countries (but not in the Baltic states), accounting for almost 30 per cent of total CEFTA export earnings in 1995. Exports of food and live animals (SITC 0) also performed very poorly in the Czech Republic, Hungary and Slovakia (down by 9-14 per cent), but their growth was sustained in Poland and better than expected in Croatia and Romania. There was a drop in exports of crude materials (SITC 2) – down by 2 per cent in dollar value in the CEFTA countries and 3-7 per cent in Croatia, Estonia and Romania, and stagnation in exports of chemicals and related products, except for the Baltic region.

In contrast to rather diverse commodity developments in exports, growth in imports was quite general across commodity groups except for crude materials (SITC 2), which broadly stagnated in dollar value throughout eastern Europe and grew notably slower than overall imports in the Baltics. The most substantial increase in dollar value of imports occurred in the category of machinery and transport equipment (SITC 7) – up some 27 per cent on average in the CEFTA countries, 25 per cent in the Baltic countries, and 6 per cent in Croatia, despite the overall import decline. This was mostly due to large increases in imports of telecommunication equipment, electric engineering and electronic products and vehicles. Imports of beverages and tobacco were up by 25-35 per cent in the countries which reported overall import growth, but went down at a faster pace than overall in the countries where imports were in decline. Imports of fuels also went up quite steeply in dollar value in the CEFTA group, at an average rate of 20 per cent, reflecting partly a substantial jump in fuels prices on world commodity markets (dollar prices were up 15.5 per cent for the year).

Among east European countries, only Polish and Hungarian involvement in OPT transactions, which accounted for nearly a quarter of their total exports in 1995, remained fast growing, by some 15 and 27 per cent, respectively, in January-September 1996, while in Croatia, the Czech Republic and Slovenia exports after inward processing went down by 3-8 per cent in 1996. Very rapid growth of such exports was registered in Latvia and Lithuania – up by 80-90 per cent in January-September 1996, but by only 7 per cent in Estonia. The Baltic countries became more deeply involved in OPT transactions later than eastern Europe and in some cases took over some orders from east European, and in particular, from the CEFTA countries. At present, the share of such exports in total export value accounts for 20-22 per cent in Estonia and Lithuania and for 15 per cent in Latvia.

---

[296] For a discussion of data problems confronting analysis of commodity patterns in east European and Baltic trade, see UN/ECE, *Economic Bulletin for Europe*, Vol. 48 (1996), chap. 3.

CHART 3.5.1
Specific western demand for selected transition countries' exports, 1992-1997
(Annual percentage change in volume terms)

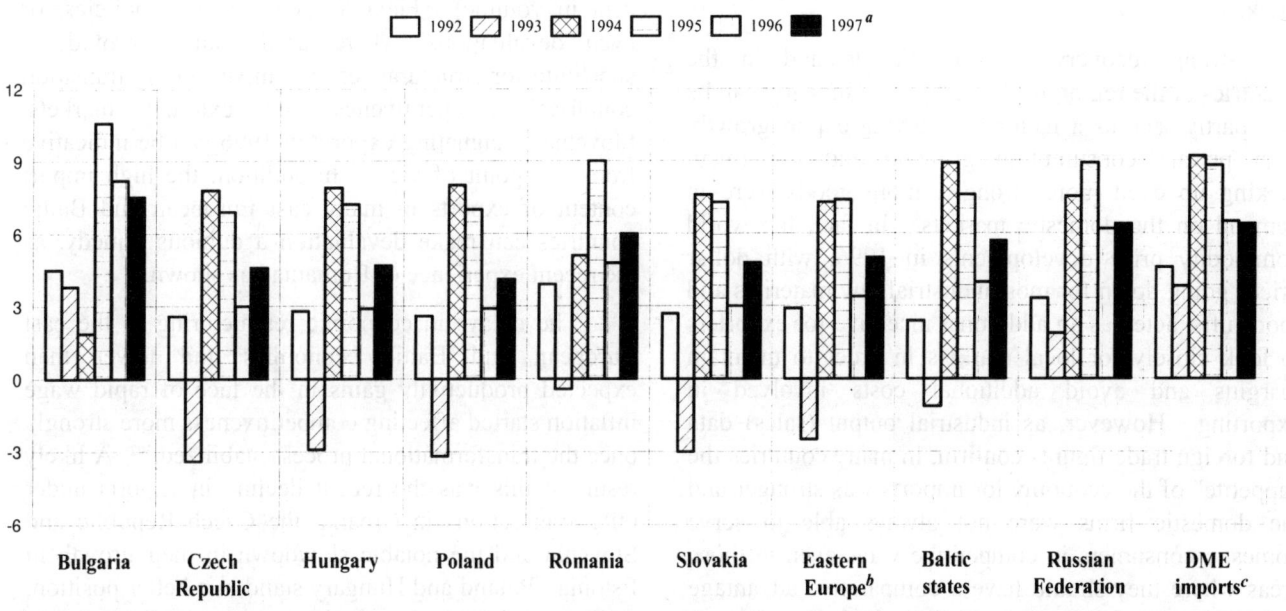

Source: UN/ECE secretariat computations: aggregation of observed import volume growth of individual western countries weighted with their share in the exports of each eastern country. Import data cover goods and services.

a Forecast.
b Six countries shown.
c OECD, excluding Australia, New Zealand, Mexico and South Africa.

(b) Factors behind the recent trade developments

The deterioration of the merchandise trade balance of most countries in eastern Europe and the Baltic region in 1996, the relative importance of which is shown in the trade deficit-to-GDP ratios (see the outcome block in table 3.5.6), reflected disappointing export results on the one hand and soaring imports, at least in some of them, on the other. Developments in both external and domestic demand were among the factors contributing to this situation (see the demand block in table 3.5.6 and chart 3.5.1).

As the review above shows, none of these countries was able to maintain its past *export* dynamics in 1996. The recession in western European resulted in weakening demand for imports of east European and Baltic goods and services; its growth can be estimated to have slowed from 7-8 per cent in 1995 to 3-4 per cent in 1996 (chart 3.5.1). This certainly became an export depressing factor for the east European and Baltic region, and more so for those countries which had most succeeded to reorient their trade towards the west in the period 1990-1995. CEFTA countries' exports were strongly oriented towards the EU market by 1996 (60-70 per cent, except for Slovakia), and towards Germany in particular (30-45 per cent), a concentration which made them vulnerable to cyclical developments on these markets. In 1996, aggregate exports of the CEFTA countries to the EU stagnated and exports to Germany went down by some 6 per cent in dollar value terms. Only Slovakia registered strong export growth to these markets, by 13 and 16 per cent, respectively. The Czech Republic, with nearly 44 per cent of its total exports directed towards Germany in 1995, was most exposed to the weak demand growth in the latter. The strong decline in Czech exports to Germany in 1996 (17 per cent) therefore had a large impact on the country's overall export performance.[297] For the Baltic countries, with their exports relatively less concentrated on the EU market (40-50 per cent in 1995) and especially on Germany (7-15 per cent in 1995), the impact of slowing western demand has been partly offset in 1996 by the continuing recovery of import demand in the region itself and in the European CIS countries.

The changes in the commodity composition of east European and Baltic total exports in 1996 also clearly reflect demand developments in the main export

---

[297] The notable contrast between Czech and Slovak outcomes in exports to the EU market (a 3.4 per cent decline in the former versus 13 per cent growth in the latter) stems primarily from developments in exports of "machinery and transport equipment" (SITC 6): in the Czech Republic, their value rose by 10 per cent (and the share in exports to the EU from 30 to 34 per cent), and in Slovakia by 62 per cent (with the share rising from 20 per cent in 1995 to 29 per cent in 1996, January-November data). The large jump in Slovakia suggests the impact of the coming on-stream of a new facility (probably a joint venture in the automotive sector), but also that an export gain of this size may not be easily repeatable.

markets – the falls were concentrated on exports of relatively energy-intensive and natural resource based products, the main product categories where these countries possess a comparative advantage on western markets.

Strong recovery of domestic demand in the countries of the region itself, except for Hungary, can be also partly seen as a factor for slowing export growth: with private consumption growing and investment picking up even more strongly, more goods were in demand on the domestic markets. In fact, the world commodity price developments in 1996, with dollar prices going down for most industrial raw materials and foodstuffs, acted as an additional incentive for exporters to look closely for local markets in order to maintain margins and avoid additional costs involved in exporting. However, as industrial output (sales) data and foreign trade figures confirm, in many countries the "appetite" of the economy for imports was stronger and the domestic firms were not always able to serve domestic consumers in competitive ways even in those areas where they should have a comparative advantage as in the case of many intermediate products and most final consumer goods.[298]

The deteriorating or not sufficiently improving *competitiveness* of east European and Baltic producers was another and, certainly, not less important factor for poor export performance in 1996. With the exception of Bulgaria, Romania and Slovenia,[299] east European and Baltic currencies continued to appreciate in real terms throughout 1996, although less strongly than in 1995, and less against the dollar than against the deutsche mark,[300] a development which probably affected price competitiveness of east European and Baltic goods on western markets and certainly squeezed exporters' margins. This resulted in pressure from exporters on the authorities in some countries for relief through the exchange rate, generally without success. Only Hungary with its crawling peg of the forint after the currency devaluation of 1995, and Slovenia with its free float of the tolar, could be said to be pursuing an export favouring exchange rate policy. However, in both countries export growth slowed significantly in 1996 (although in Hungary it remained around 6 per cent in volume). Flexible exchange rate policies, or even devaluations, where used, cannot provide a substitute for structural reforms in sustaining transition countries' competitiveness on external markets. Slovenia's stagnating exports in 1996 can be indicative from this point of view. In addition, the high import content of exports in many east European and Baltic countries can make devaluation a dubious remedy, as the recent experience of Romania has shown.[301]

The delays in economic restructuring in the east European and Baltic economies and lower than expected productivity gains in the face of rapid wage inflation started affecting competitiveness more strongly once the transformational process stabilized.[302] A likely result of this was the recent decline in exports under OPT transactions in Croatia, the Czech Republic and Slovenia and the notable slowdown in their growth in Estonia. Poland and Hungary stand in a better position, and many observers agree that structural adjustment in these two countries has progressed faster (the behaviour of FDI inflows in recent years and rapidly growing domestic investment can also be considered to attest to this).

The marked differences between real appreciation trends based on relative unit labour cost and consumer price deflators can be indicative on this issue (see the price competitiveness block in table 3.5.6).[303] For instance, the real appreciation of the Czech and Slovak currencies against the deutsche mark based on unit labour cost deflators was the strongest among CEFTA countries in the period 1994-1996, at 10 and 8 per cent, respectively, as compared with 3 per cent in Slovenia and 5 per cent in Poland, whereas the Hungarian forint depreciated in real terms. At the same time, real appreciation of the Czech and Slovak currencies based on consumer price deflators (6 and 4 per cent per year) was not much different from that of the Polish zloty or the Slovenian tolar (5 and 2 per cent annually). This observation leads to comparison of the pace of structural reform and modernization of the national technology

---

[298] For instance, four CEFTA countries traditionally were net exporters of commodities in section SITC 6, with surpluses in 1993-1995 of $2.4-$2.7 billion. In January-September 1996 this had fallen to $0.3 billion, and Poland for the first time was a net importer of these commodities.

[299] For currency developments in this region and in particular in Bulgaria and Romania, see sections 3.1(ii)-(iii) above.

[300] Most east European and Baltic currencies strengthened in real terms against the dollar over the past year (based on relative consumer price deflators): in the Czech Republic by 3.5 per cent, in Estonia by 15, in Latvia by 9.6, in Lithuania by 19.6, in Poland by 5.9, and in Slovakia by 1 per cent. The real appreciation against the deutsche mark in which a large portion of east European foreign trade is conducted (table 3.5.6) was stronger, as a result of changes in the dollar-deutsche mark relationship but also in reflection of individual currencies' exchange rate regimes (most often, managed peg against the dollar or a currency basket).

[301] In Romania, a rapid devaluation of the lei in the first half of 1996 sharply raised the cost of imported fuels, raw materials and intermediate goods, reducing the capacity of exporting industries to comply with contractual orders. Central bank administrative interventions in the forex markets caused hard currency shortages which further aggravated the problems.

[302] Section 3.4 above.

[303] On pros and cons of real exchange rate indicators based on CPI, export unit values and unit labour costs in the assessment of competitiveness, see I. Marsh and S. Tokarick, "An assessment of three measures of competitiveness", *Weltwirtschaftliches Archiv*, Vol. 132, No. 4 (Kiel), 1996.

base in these countries and suggests some lags in this respect in the Czech Republic and Slovakia, which can be partly attributed, according to many observers, to weaknesses of their mass privatization programmes.[304] A similar observation is, most probably, also valid for Lithuania. But inadequate progress in structural changes and in particular in privatization were recently characteristic throughout the region, including Hungary, Poland and Slovenia. In fact, the lag in structural adjustments is perhaps the key factor causing problems for the competitiveness of east European and Baltic products, and particularly for those of the CEFTA countries.[305]

On the other hand, as some observers suggest, the weakness of east European and (though less pronouncedly) Baltic export growth in 1996, in particular in relations with the west, may be not so much a temporary phenomenon caused by the above factors but a warning sign of the exhaustion of their trade potential for gaining additional market share in these markets, in particular in the EU.[306]

To an even more important extent than the weakness of export growth, the deterioration of the external position of the more advanced transition economies stems from the upswing of *imports*, boosted by exceptionally rapid growth in domestic demand (see section 3.2(ii) above). In 1996, economic policies of many of these countries focused on the reduction of inflation. A relatively slow pace of nominal currency depreciation under crawling peg or fixed rate regimes and the gradual reduction in import duties under the Europe Agreements and the CEFTA (or under the Memorandum agreements with the IMF as in Latvia and Lithuania) supported this aim. To supplement the impact of these policies, some governments (for instance, in Lithuania, Poland and Romania) stepped in with ad hoc measures, raising duty free import quotas for various commodities, most often food items. Others, like Hungary and Slovakia, loosened some temporary import restrictions: in Hungary the import surcharge of 8 per cent introduced in March 1995 was lowered by 2 percentage points by the fourth quarter of 1996, and in Slovakia a 10 per cent import surcharge on consumption related items was reduced to 7.5 per cent in July 1996. In Poland, a temporary import surcharge of 5 per cent was reduced by 2 percentage points from 1 January 1996. In combination, these measures reinforced the impact of rising domestic demand on import growth.

(iii) Russian Federation: trade with non-CIS countries

(a) Trade performance

In Russian foreign trade with the non-CIS countries,[307] export growth slowed sharply in value terms in 1996, but not in volume, while imports contracted on both measures (table 3.5.1). As a result, there was again a substantial rise in the trade surplus of the Russian Federation, one of the few transition countries in surplus in 1996. On customs data, it was well over $37 billion, $7 billion more than in 1995 (table 3.5.2), or some 8.5 per cent in relation to GDP (table 3.1.1). If adjustments for non-registered trade are taken into account, the surplus is smaller but nonetheless remains substantial at some $29 billion.[308]

In the course of 1996, Russia's trade surplus with non-CIS countries widened from quarter to quarter ($7.8 billion, $8.5 billion, $8.9 billion and $12.2 billion). About half of the surplus arose from trade with the developed market economies, that in trade with western Europe increasing most (up by $3.5 billion as compared with 1995), while that with non-European developed countries fell slightly (by $0.4 billion). The surplus in trade with transition economies also grew very rapidly, by $4 billion as compared with 1995, and exceeded import value from these countries by a factor of 2.3 in 1996. In trade with the developing countries, on the other hand, Russia's surplus fell by some $0.2 billion.

---

[304] "Czech economic monitor", *PlanEcon Report*, Vol. XII, Nos. 45-46, 16 January 1997; "Slovakia – external deficit in danger zone", CS First Boston, *Economic Comment, Economics Research – Europe*, 9 December 1996.

[305] UN/ECE, *Economic Bulletin for Europe*, Vol. 48 (1996), pp. 47-49.

[306] P. Havlik, "Growth in the CEECs slows down in 1996 as exports to the EU lose momentum", *The Vienna Institute Monthly Report*, No. 11 (Vienna), 1996; D. Gros and A. Gonciarz, "A note on the trade potential of central and eastern Europe", *European Journal of Political Economy*, Vol. 12, No. 4 (Amsterdam), 1996; C. van Beers and G. Biessen, "Trade possibilities and structure of foreign trade: the case of Hungary and Poland", *Comparative Economic Studies*, Vol. XXXVIII, No. 2/3 (Tempe, AZ), Summer-Fall 1996.

[307] In 1995, most CIS countries began to include intra-CIS trade in the foreign trade aggregates reported in their statistical publications, but this practice is not followed here. Although prices in intra-CIS trade are now converging rapidly towards world market prices, they are still generally (not always) below those reflected in the data on trade with the non-CIS world, and, owing to the price convergence, year-to-year price movements differ substantially from those on world markets. For this reason, and also to facilitate comparison with developments in previous years, the ECE secretariat finds it more useful to focus separately on the foreign trade with the non-CIS world, discussed in the following two subsections, and intra-CIS trade discussed in section 3.5(v) below.

[308] Russian statistical authorities report foreign trade data in three versions: data based on customs documents published by the State Customs Committee in the quarterly *Tamozhennaya statistika vneshnei torgovli Rossiiskoi Federatsii* (used in table 3.5.1); revised data based on the above but including adjustments of exports and imports for "not cross-border" operations published by the Russian Goskomstat (see table 3.5.7), and data with a broader revision, also by Goskomstat, including estimates of inflows of goods which entered the country without being registered by the customs authorities (memorandum line in table 3.5.7). These latter adjustments are large, accounting for 25 per cent of total imports in 1995 and 26 per cent in 1996. As no breakdown by countries or by commodities is available for the adjusted data, the analysis below is based on the unadjusted customs data.

The rapid rise of the surplus in the second half of 1996 was due to a further fall of imports from non-CIS countries as well as stronger export growth than in the first half of the year. If in January-June 1996 the value of Russian *exports* increased by only 1 per cent, in the second half of the year it gained momentum, rising by 8 per cent for 1996 as a whole. This compares with 29 per cent export value growth in 1995. At the same time, *imports* into the Russian Federation fell by 5 per cent in value, against an increase of 17 per cent in 1995 (table 3.5.1), the decline being more pronounced in the second and fourth quarters of the year.

Trade by direction

The upswing in Russian exports in the second half of the year was strongest in trade with the *transition economies*, notably with the Asian countries among them, in line with earlier expectations of a general shift in trade towards Asian markets.[309] The recovery of trade with *eastern Europe* was not as pronounced as in 1995: exports to these countries grew at a fairly rapid pace, again primarily in the second half after stagnating in the early part of the year, but imports went down steeply.[310] The value of Russian exports to the *developed market economies* rose by a mere 3 per cent in 1996, the lowest rate since 1992. Moreover, the positive rate for the year as a whole was fully due to the resumed export growth in dollar value terms in the second half, mainly in the last quarter of 1996. Imports from the developed market economies have lost some market share since the 1994 peak (appendix table B.13), falling by nearly 8 per cent in 1996. Imports from western Europe declined by 12 per cent, those from Germany contracting by 21 per cent.

Prices and commodities

The *volume* of Russian exports increased by some 6.5 per cent, at about the same rate as in 1995 (table 3.5.6), while imports most probably declined at a similar rate in volume as in value, following last year's decline of 3 per cent.

Rising world market prices for oil, natural gas and other energy products in the second half of the year increased Russian export earnings, and this, in contrast to developments in the first half of 1996, was not offset by declines or stagnation in the prices of other industrial raw materials. Average dollar unit values for Russian fuel exports to non-CIS markets increased by 18 per cent between 1995 and 1996, with increases in the unit revenues from crude oil and oil products of 21 and 26 per cent, respectively, and from natural gas of 9 per cent. Dollar export unit values for metals declined at the same time by 5 per cent on average.[311]

Recent changes in the *commodity structure* of non-CIS trade mirror clearly these price movements and indicate an increasing dependence of Russia on exports of energy products: in 1995, the share of fuels in total Russian exports was 40 per cent;[312] in 1996 it was 47 per cent. Customs data indicate that crude oil, natural gas and oil products remain the most important commodities. Exports of oil products doubled in value in the fourth quarter, and posted 53 per cent growth in 1996 as a whole. The value of base metal exports increased at a slower rate (5 per cent) under pressure of the decline in world market prices, but their share (22 per cent) remained almost unchanged. In *volume* terms, fuel exports increased some 8 per cent, mainly owing to the rise in the volume of oil products (21 per cent in 1996, with a 55 per cent jump in the last quarter alone), and metals by 10 per cent, owing to a rise in shipments of aluminium and copper (16 and 12 per cent, respectively), while ferrous metals fell sharply (-43 per cent). Declines were posted also in exports of some other traditional commodities – chemicals, timber, cellulose and paper, etc. The share of machinery and equipment in Russia's exports remained under 3 per cent in 1996, slightly lower than in 1994-1995, as exports of these goods went down by 1 per cent, despite attempts to boost sales to the developing and transition countries. Exports of foodstuffs, beverages and tobacco were among the fastest growing (30 per cent), albeit from a very low level. Fast growth, 29 per cent in value, was recorded also in the arms and ammunition group.[313]

With imports shrinking overall, the most rapid fall was registered in imports of some industrial consumer goods and of foods, beverages and tobacco (down by 20-25 per cent), followed by agricultural products (-17 per cent) – that fall seems to have been partially offset by increased imports from the CIS region. Imports of

---

[309] As the Asian socialist economies (China, Democratic People's Republic of Korea, Lao People's Democratic Republic and Viet Nam) are included in the group of transition economies in tables 3.5.1 and 3.5.2, changes in trade with developing countries only partially reflect developments in Russian trade with the Asian region. Exports to the Asian economies in transition grew by some 35 per cent in 1996, and imports by 21 per cent. Exports to developing Asian countries generally grew more slowly, but growth was very rapid in exports to the Republic of Korea (around 60 per cent), Iran (about 50 per cent), and Cyprus (80 per cent).

[310] A large discrepancy in mirror statistics should be noted: according to east European data, exports to Russia, rather than falling, increased by some 17 per cent in 1996. It is difficult to find a reasonable explanation for these differences. Barter transactions with undervalued imports, or the large import flows that escape registration by Russian customs can be suggested, but as Russian data do not provide the geographic breakdown of the figures for barter transactions or adjustments for non-registered flows, these are impossible to verify.

[311] Prices for Russia's copper and nickel declined by 10-16 per cent while those for aluminium went down only slightly, by 1 per cent. Computations based on data from *Tamozhennaya statistika vneshnei torgovli Rossiiskoi Federatsii*, Nos. 1-4 (Moscow), 1996.

[312] Harmonized Commodity Description and Coding System (HS), Chapter 27.

[313] HS, Chapter 93.

machinery and equipment fell by 12 per cent in value, reducing their share in total imports by 2 percentage points compared with 1995. Imports of cellulose and paper, chemicals and allied products, base metals and precious ores, stones and metals all grew in value quite steeply (10-30 per cent). These imports (except for chemicals), however, accounted for a relatively small share of total imports from non-CIS countries in 1996.

(b) Factors behind recent trade performance

The widening surplus in Russia's trade with non-CIS partners in 1996 was mainly the result of the steeper than expected decline in imports, but also somewhat improved performance of exports in the second half of the year.

The *decline of imports* from non-CIS countries came under the impact of developments on both the supply and the demand sides of the Russian economy, but their influence was strengthened by Russian policy to boost imports from CIS countries, in particular early in the year. However, the continuing decline of demand for non-consumer goods in general and, in particular, for machinery and equipment, from the export sectors as a result of the lower profitability of exporting in late 1995 and in the first half of 1996 was an important factor.[314] This was reinforced by the deepening payments crisis in the enterprise sector and rising difficulties in obtaining foreign credit and credit guarantees for imports from the west as a result of high indebtedness of the country and of its economic participants in general.[315] The continuing decline in private consumption (retail sales, for instance, declined by another 4 per cent in 1996; see table 3.2.12), in part the result of delays in salary and transfer payments from the budget, was reflected in shrinking import demand for consumer goods. Another factor working in the same direction was increasing protectionism and policies to boost imports of these goods from within the CIS area, even if not always at better terms, as domestic supplies of many of these products remained short because of further substantial declines of output in industries producing goods for final consumption and in agriculture (see tables 3.2.3 and 3.2.7 above).

The real appreciation of the rouble (by 25 per cent against the dollar and 30 per cent against the deutsche mark in 1996 as compared with 1995, based on CPI deflators) and generally falling world market prices for most of the commodities prevailing in Russia's non-CIS imports do not seem to have provided much of a boost for imports, or the impact may have been offset by that of increased import taxes.[316]

Among the key factors behind the rather uneven performance of Russian exports to non-CIS countries in 1996 was, first of all, world market price behaviour – a steep increase in the prices for fuels and a deep fall in the prices for most industrial raw materials (see chapter 2.4 above). Weak west European demand (chart 3.5.1) also was behind steep declines in exports of non-energy and non-metal sectors, but its impact was somewhat counterbalanced by stronger demand in the other transition economies and especially in Asia, particularly China, which is Russia's second largest export market.

Another factor for resumed growth was the response of exports to changes in the real exchange rate. The real appreciation of the rouble in 1995 and early 1996 has had a significant impact on the profitability of Russian exporters, and this affected export performance throughout the first half of 1996. But there has been a turn in the rouble exchange rate thereafter: it gradually depreciated in both nominal and real terms against the dollar from the end of April 1996 to the end of the year as the Central Bank of the Russian Federation has been operating a crawling peg since mid-1996.[317] Real appreciation of the rouble against the dollar in the second half of 1996 (compared with the same period of 1995) was only 6 per cent, and zero in the fourth quarter, as against 38 per cent in the first half of the year.[318]

Both above-mentioned developments in the export operating environment were in fact favourable for improving export performance noticeably in the second half of the year. However, results could have been even better but for constraints on the supply side. Russia's export output had weakened notably already by the end of 1995 and continued to decline or stagnate in 1996 (see table 3.2.7 for the decline of industrial output by main branches). Thus, despite the drastic rise in fuel prices, the

---

[314] According to V. Koneev, the aggregate "export effectiveness" coefficient (a ratio of foreign earnings to domestic costs) for major commodities dropped from 131 per cent in the first quarter of 1995 to 77 per cent in the last quarter, and remained at 85-87 per cent in the first half of 1996 after export duties were abolished and world market prices for energy products went up. V. Koneev, "Naskol'ko effektiven eksport", *Ekonomika i zhizn'*, No. 49 (Moscow), December 1996.

[315] For instance, until July 1994 the bulk of German non-consumer goods exports to Russia was covered by special credits usually guaranteed by Hermes under preferential conditions (10 years, 100 per cent of contract value was credited, first three years without repayment); deliveries under these credits lasted in general through 1995. Since mid-1994, the official credit lines have not been preferential any more and Hermes tightened its requirements for guarantees as well. Hence only a part of the credit line of DM 1.5 billion which was issued by the German government for exports by small and medium enterprises in eastern Germany (DM 0.6 billion) and all other companies in Germany (DM 0.9 billion) was actually used. *BIKI*, No. 27, 6 March 1997.

[316] For changes in Russia's foreign trade regime in 1995-1996, see UN/ECE, *Economic Bulletin for Europe*, Vol. 48 (1996), pp. 66-68.

[317] Section 3.1(ii) above.

[318] Owing to the pronounced strengthening of the dollar against the deutsche mark in the second half of 1996, and in particular in the last three months of the year, real appreciation of the rouble against the deutsche mark was higher: 13 per cent in the second half of 1996 and 9 per cent in the last quarter of 1996, on a year-to-year basis.

TABLE 3.5.7

CIS countries' trade with non-CIS countries, 1994-1996

(Value in million dollars, growth rates in per cent)

	Exports			Imports			Trade balances		
	Value	Growth rates		Value	Growth rates				
	1995	1995	1996	1995	1995	1996	1994	1995	1996
Armenia	104	80.6	53.8	340	80.5	61.2	-131	-236	-388
Azerbaijan	329	-9.2	-0.1	439	50.4	43.9	71	-110	-303
Belarus	1 776	72.2	1.4	1 887	93.6	23.5	57	-111	-530
Georgia	58	49.6	8.0	225	281.5	57.7	-20	-167	-292
Kazakstan	2 343	72.7	19.0	1 172	-15.3	5.9	-27	1 171	1 546
Kyrgyzstan	140	19.2	-27.7	169	56.9	134.1	10	-29	-294
Republic of Moldova	279	74.2	-9.5	271	48.0	51.9	-23	7	-160
Tajikistan	497	55.2	-13.1	321	0.9	-16.1	2	176	163
Turkmenistan	951	92.6	-44.1	619	-20.8	48.3	-288	332	-386
Ukraine	5 554	19.4	22.7	4 203	44.6	93.3	1 746	1 351	-1 308
Uzbekistan	1 712	77.2	65.0	1 630	35.6	91.1	-236	82	-291
Total above	13 742	43.8	17.1	11 276	34.3	62.6	1 159	2 466	-2 243
Russian Federation	65 666	23.9	8.1	33 155	17.0	-5.9	24 656	32 512	39 800
CIS total	79 409	26.9	9.7	44 431	20.9	11.5	25 815	34 978	37 557
Memorandum item:									
Russian Federation[a]	65 700	24.0	8.7	44 120	19.4	-4.1	16 200	21 580	29 100

Source: CIS Interstate Statistical Committee; for the Russian Federation: Russian Federation Goskomstat data based on customs statistics.

[a] Adjusted for non-registered trade. Data from *BIKI*, No. 21, 20 February 1997.

volume of exports increased by 8 per cent (partly at the expense of exports to CIS partners, which declined by 30 per cent in volume) but could probably have risen more if output had been higher. The insufficiency of infrastructure – pipeline and port facilities, roads and railways, etc. – as well as rising transport tariffs further impeded growth.

(iv) Other CIS countries: trade with non-CIS countries

(a) Trade performance

The trade of other CIS member countries with the rest of the world rose quite rapidly in dollar value in 1996: exports grew 17 per cent and imports 63 per cent, following increases of some 44 and 34 per cent, respectively, in 1995 (table 3.5.1). According to official statistics, the dollar value of trade with non-CIS partners now accounts for about half of total (non-CIS plus intra-CIS) exports and imports of these countries, but owing to the substantial price differences for the same commodities on world markets and in intra-CIS trade these proportions are probably not very meaningful.[319]

Trade balances with partners outside the CIS region deteriorated considerably across the countries except for *Kazakstan*, where a surplus widened, and *Tajikistan*, where it narrowed slightly. The aggregate trade surplus of $2.5 billion in 1995 swung into a deficit of $2.2 billion in 1996, the first deficit since 1992 (table 3.5.7). In absolute terms, the deterioration was most substantial in *Ukraine*, from a surplus of $1.4 billion in 1995 to a deficit of $1.3 billion a year later. But in relative terms, the deterioration was most severe in *Kyrgyzstan*, the *Republic of Moldova* and *Turkmenistan*, as evidenced by the changes in trade balance-to-GDP ratios which stood at -10 to -17 per cent in 1996, as against -2 to +11 per cent in 1995 (table 3.1.1). Relative to export earnings, deficits exceeded them by a factor of three in Kyrgyzstan and accounted for 64 and 73 per cent, respectively, in the latter two countries. In the European CIS countries, the rapid growth of deficits originated largely in trade with the developed market economies and in particular with western Europe. In January-September 1996, *Belarus'* and *Ukraine's* deficits with the west European partners amounted to $0.6 billion and $0.8 billion respectively (table 3.5.8). For Ukraine, most of this deterioration arose from its trade with Germany, where the deficit widened by some $0.4 billion between the first nine months of 1995 and 1996. In the case of the *Republic of Moldova*, deficits in trade with east European countries, mainly Bulgaria and Poland, also grew fast, while trade with its main partner, Romania, was nearly balanced (according to data for the first half of 1996). The only large surplus, in absolute terms and in relation to GDP, was reported in *Kazakstan*, but may in part reflect underrecorded or undervalued imports, as in the case of the Russian experience discussed above. Around 55 per cent of Kazakstan's surplus arose from trade with the developed market economies, the developing countries

---

[319] For developments in intra-CIS trade see section 3.5(v) below.

TABLE 3.5.8

Foreign trade of selected CIS countries by direction, 1995-1996
(Value in million dollars, growth rates in per cent)

	Exports			Imports			Trade balances		
	Value	Growth rates		Value	Growth rates		Value	Jan.-Sept.	
	1995	1995	1996[a]	1995	1995	1996[a]	1995	1995	1996
Belarus, *to and from*									
Non-CIS countries	1 776	72.2	-5.6	1 887	93.6	30.4	-111	1	-478
Transition economies	823	141.5	15.1	647	118.1	21.3	176	199	202
Baltic states	361	456.4	19.1	192	591.3	36.8	170	163	176
Eastern Europe	418	93.2	14.7	432	66.1	8.1	-14	19	42
Developed market economies	661	50.9	-21.8	1 097	94.2	39.6	-436	-287	-700
Western Europe	591	60.6	-25.2	958	90.4	37.0	-367	-223	-587
Germany	268	73.0	-26.6	424	42.1	30.2	-156	-112	-257
Developing countries	289	44.4	-25.5	111	347.2	11.5	178	111	41
CIS	2 931	98.2	29.1	3 676	75.8	27.8	-745	-533	-657
Russian Federation	2 089	80.5	44.9	2 965	58.2	26.1	-876	-664	-570
Republic of Moldova, *to and from*									
Non-CIS countries	279	74.2	-6.7	271	48.0	91.1	7	25	-71
Transition economies	158	43.3	-21.1	119	48.3	137.3	39	39	-30
Baltic states	17	103.7	110.2	12	61.3	8.5	5	1	7
Eastern Europe	139	37.2	-31.7	107	58.7	135.9	33	37	-31
Developed market economies	104	125.6	13.7	143	55.9	56.8	-39	-15	-40
Western Europe	91	118.5	5.0	124	77.5	51.2	-33	-14	-36
Germany	45	130.5	-4.5	46	28.1	9.2	–	-4	-7
Developing countries	17	338.5	59.6	10	-11.3	113.9	7	1	–
CIS	467	15.0	55.7	569	19.6	24.8	-102	-92	-62
Russian Federation	360	24.5	67.1	278	-10.1	13.1	82	-6	63
Ukraine, *to and from*									
Non-CIS countries	5 554	19.4	18.3	4 203	44.6	53.1	1 351	1 091	297
Transition economies	1 506	-10.2	58.0	881	15.5	139.6	625	469	275
Baltic states	198	-35.3	42.1	193	16.3	85.8	5	15	-28
Eastern Europe	835	5.1	65.7	646	17.2	124.4	189	124	-48
Developed market economies	2 547	35.9	-17.5	2 796	46.1	26.2	-249	-64	-897
Western Europe	1 807	28.1	-14.5	2 302	43.9	19.1	-495	-307	-812
Germany	474	67.6	-17.7	608	-7.1	83.8	-135	-54	-444
Developing countries	1 501	36.4	43.1	525	128.2	40.7	976	686	919
CIS[b]	6 012	7.0	19.0	7 133	-9.0	..	-1 120	-989	-3 107
Russian Federation	5 025	23.6	15.0	5 821	-8.3	..	-796	-751	-3 165

Source: CIS Interstate Statistical Committee; direct communications to UN/ECE secretariat.

Note: Values for non-CIS countries may not add to totals because of undistributed flows.

[a] January-September over same period of 1995, except Republic of Moldova: January-June over same period of 1995.

[b] Trade balances and growth rates for January-September 1996 are based on data including adjustments for crude oil and gas. Hence, they are not fully comparable with figures for 1995 and full-year data for 1996 presented in table 3.5.9 below.

---

accounting for another 12 per cent and the surplus in trade with non-European countries in transition, including China, other Asian transition countries and Cuba, for more than a quarter, a rising share. Kazak trade with east European countries, in contrast, swung into deficit (some $50 million).

The changes in trade values varied quite widely across the CIS countries, in particular on the export side. The *value of exports* fell by 44 per cent in *Turkmenistan*, by 28 per cent in *Kyrgyzstan*, and by 10-13 per cent in the *Republic of Moldova* and *Tajikistan*, while in *Azerbaijan* exports stagnated. On the other hand, some CIS countries reported very large increases in export value: by 55 to 65 per cent in *Armenia* and *Uzbekistan*, and by some 20 per cent in the two biggest CIS countries, *Kazakstan* and *Ukraine*. Within 1996, export growth accelerated in the second half of the year in *Belarus* and *Uzbekistan*, but lost impetus in the remaining CIS countries, most pronouncedly in *Kyrgyzstan* and the Republic of Moldova (tables 3.5.7 and 3.5.8).[320]

On the import side, developments were more uniform: except for *Tajikistan* and *Kazakstan*, the dollar *value of imports* grew quite rapidly in all the CIS countries in 1996. Imports more than doubled in value in *Kyrgyzstan*, almost doubled in *Ukraine* and *Uzbekistan*, and grew considerably in the other countries (table 3.5.7). These dollar value growth rates

---

[320] For comparison with the first half of 1996 see table 3.3.1 in UN/ECE, *Economic Bulletin for Europe*, Vol. 48 (1996), p. 56.

imply considerable increases in volume, given the fall or very moderate increase in world market prices for most of their imported commodities – foodstuffs, consumer goods, some intermediate goods and machinery – in 1996.

(b) Composition of trade

Comprehensive data for 1996 on the geographical and commodity breakdown of trade in the region are not yet available, but some preliminary observations can be derived from information on trade flows by direction in the European CIS countries (table 3.5.8) and in Kazakstan.

First, exports directed towards *developed market economies* slowed noticeably (Republic of Moldova, Kazakstan) or contracted (Belarus, Ukraine). Within this partner group, the value of exports to Germany was most affected: falls in all the European CIS countries and growth at a lower than overall rate in Kazakstan (7 per cent). On the other hand, imports from developed market economies grew steadily in the European CIS countries, accounting for 46-56 per cent of total non-CIS imports in January-September 1996, but import growth from this partner group into Kazakstan was only half the overall rate (3 per cent) in 1996. *Second*, exports to *developing countries* was the fastest growing segment of exports from the region (except in Belarus). In 1996, these exports increased in a range of 45-60 per cent in dollar value in Kazakstan, the Republic of Moldova (January-June) and Ukraine (January-September) and accounted for a substantial part of their total exports. Imports from this partner group grew at fairly high rates too, but their relative importance remained quite low in the European CIS countries (4-11 per cent). In Kazakstan, however, imports from developing countries accounted for a quarter of non-CIS imports in 1996. *Third*, exports to the *east European* and *Baltic* countries grew very rapidly in Belarus and Ukraine in January-September 1996, but contracted (by 32 per cent) in the Republic of Moldova. Trade with eastern Europe accounted for a substantial share of the total in the European CIS countries, from 20 to 43 per cent for exports and imports (with the lowest share in Ukraine, the highest in the Republic of Moldova). In Kazakstan, exports to the east European markets have rapidly lost share after peaking in 1994, while exports directed towards the Baltic states and the Asian countries in transition have grown steadily since then: by 17 and 56 per cent in 1996, following quadrupled and doubled exports, respectively, in 1995.

In the *commodity composition* of trade there have been only few changes since CIS economies opened up to non-CIS markets: in general, exports remain concentrated on a few categories of primary goods and fuels, while imports are dominated by agricultural products, prepared foodstuffs, and machinery and equipment. This concentration on a few commodities even strengthened in 1996 as under the impact of rising world market prices the share of fuels (including oil products) in total exports increased in the CIS countries engaged in fuels production or processing: in Belarus from 8 to 17 per cent, in Kazakstan from 22 to 28 per cent.[321] In volume terms exports of crude oil from the two countries increased by 25-35 per cent,[322] while sales of oil products grew more than threefold in Kazakstan and by 77 per cent in Belarus in January-September 1996. Similarly for exports of metals and metal products: exports of this commodity category grew fast in Kazakstan, Kyrgyzstan, Tajikistan and Ukraine in volume terms, but much less in dollar value terms as prices for metals fell considerably on world markets in 1996. Nevertheless, exports of metals accounted for about one quarter of total non-CIS exports in Kyrgyzstan, for about one third in Ukraine, for 55-65 per cent in Kazakstan and Tajikistan. In the other two central Asian republics, Turkmenistan and Uzbekistan, raw cotton remained the primary export commodity (70-80 per cent); in Kyrgyzstan its share in total exports declined (from 38 to 30 per cent), while in Tajikistan it increased to above one third in 1996.[323] Commodity concentration was also high in the Republic of Moldova, which is not endowed with natural resources: its exports from the agro-processing sector accounted for 63 per cent (foods, beverages and tobacco for 47 per cent), followed by textiles exports with a share of 16 per cent, mainly under OPT transactions.[324]

(c) Factors behind trade expansion

The review above shows that aggregate trade flows with non-CIS markets were growing at a rapid pace despite the fact that external markets, especially in western Europe, were not favourable in general and domestic demand conditions also remained weak. As noted in section 3.5(ii), demand for most raw materials and primary processed goods (except fuels) was weak in 1996, and most commodity prices fell, particularly for industrial and agricultural raw materials which account for the bulk of CIS exports. In fact, the export data for the smaller CIS economies, in particular among the European CIS countries, appear to reflect these developments. The larger countries, however, including

---

[321] Data for Turkmenistan are not yet available.

[322] The reported crude oil exports from Belarus probably are re-exports of raw materials, obtained in payment for the activities of the country's well-developed oil-refining facilities.

[323] These scattered figures on recent changes in commodity structure in CIS countries are based on data published by CIS Interstate Statistical Committee, *Sodruzhestvo Nezavisimykh Gosudarstv v 1996 g., Statisticheskii Spravochnik* (Moscow), 1997.

[324] Government of Moldova/European Expertise Service, *Moldovan Economic Trends*, *Quarterly Issue* (Chisinau), February 1997, pp. 50-51.

Kazakstan, Ukraine and Uzbekistan, seem to have been affected less.[325] On the one hand, the impact of weakened western demand on their exports was probably offset to some extent by a recently pronounced turn (in Ukraine, in particular) towards developing economies' markets and markets of other countries in transition where demand conditions were much better. CIS countries less constrained by supply-side factors (in particular Kazakstan) expanded their volume exports of fuels and industrial raw materials including metals as soon as there was any demand for them. They also benefited from the long-term agreements with foreign investors in oil extraction and metal industries, which generally involve the repatriation of a part of the revenues in kind.[326] On the other hand, it is likely that the expansion of CIS trade reflects a "late-comer" effect, combined with the continuing liberalization of foreign trade and domestic prices, and recent improvements in access to western markets (owing to the interim trade agreements with the EU which came into effect in 1996).[327]

However, trade liberalization and recent trade agreements with the EU were more important for the surge in imports.[328] Among other important factors behind the growth of imports which might be mentioned were increased hard currency inflows from exports and, in some cases, increased foreign direct investment, together with improvements in the availability of western credit for specific imports. Changes in exchange rate policies, resulting most often in real appreciation of local currencies, also had a positive impact.[329] On the other hand, as some observers note, the surge in registered imports could also be a result of better recording and declining use of barter, at least in the European CIS countries. Finally, it should be noted that, with import levels still relatively low in value, any shipment of more sophisticated machinery in the course of a foreign financed project tends to show up as a discrete jump in the value of imports into some of these countries (mainly in central Asia).

(v) Intra-CIS trade

The fast pace of growth in the *mutual trade* of the CIS countries reported in the first half of 1996 (some 25-30 per cent) was not sustained through the second half of the year;[330] the preliminary full-year figures show growth of only 3 to 8 per cent (table 3.5.9).[331] The slowing was mainly due to a sharp downturn in Ukraine's intra-CIS trade flows (exports and imports declined by more than 12 per cent),[332] as well as a noticeable slowdown in Russia's trade flows. As a result, the dollar value of intra-CIS trade of the CIS members other than Russia grew at a lower rate than that of their trade with third countries, in particular on the import side. In the case of Russia, the growth of exports to the CIS and third countries was about the same, whereas CIS imports rose while those from the rest of the word contracted (tables 3.5.1 and 3.5.9).

The *volume* growth was, probably, very moderate or even negative, as the increase in trade value was largely due to the continued rise of prices in intra-CIS trade. A precise assessment is not possible since aggregate price indicators are not available for any of these countries. For

---

[325] The very high non-CIS export growth reported for Uzbekistan poses some questions (65 per cent, table 3.5.7). The country depends mainly on raw cotton (about 80 per cent), but also exports gold (about 15 per cent). A bad harvest and declining world market prices for cotton in 1996 should have affected Uzbekistan's exports in both real and value terms, even considering the fact that Uzbekistan reoriented its exports away from the CIS (a fall of 24 per cent, table 3.5.9).

[326] On this issue see, for instance, *Financial Times*, "A little liquidity transforms Kazakhstan's metals industry", 20 December 1996, and "Kazakstan-Iran oil swap begins", 6 January 1997.

[327] There are contradictory signals about the impact of the EU agreements. For instance, most recently, protectionist policies of the EU regarding the "sensitive" sectors (textiles and steel) were blamed by Ukrainian authorities for the deficit of over $1 billion recorded in 1996 (*BBC Summary of World Broadcasts*, SU/2861 D/4, 7 March 1997). The European Commission began talks with Ukraine at the end of October 1996 on a so-called "steel agreement" for a five-year period, which is foreseen to enter into force as from 1 July 1997 (Agence Internationale d'Information pour la Presse, *Europe*, No. 6853, 15 November 1996).

[328] Faced with a rapidly widening trade deficit in late 1996 as a result of surging imports, the Ukrainian authorities made some moves to tighten export and import restrictions, raising import duties on some agricultural products and introducing a temporary ban on raw material exports for outward processing and later re-import (*BBC Summary of World Broadcasts*, SUW/0465 WA/1, 13 December 1996). For more detail on changes in trade policies of CIS countries in 1996, see UN/ECE, *Economic Bulletin for Europe*, Vol. 48 (1996), chap. 3.3(v).

[329] The experiences of Kazakstan and Ukraine are indicative. In Uzbekistan, however, attempts to keep the exchange rate artificially low went awry in the summer of 1996, leading to a shortage of hard currency which the government tried to resolve through import-impeding administrative measures such as currency rationing and import licensing. *Financial Times*, 6 September 1996; CS First Boston, *Emerging Economics Research – Europe*, 28 February 1997.

[330] For the first half data, see UN/ECE, *Economic Bulletin for Europe*, Vol. 48 (1996), p. 64.

[331] The range is that of reported growth of exports (8 per cent) and imports (3 per cent). The data on intra-CIS trade flows reported by the CIS Interstate Statistical Committee continue to show internal inconsistencies, as in this closed trading circle the two values should of course be identical. In addition, the absolute growth of Russian exports in 1996 ($1.2 billion) is much higher than the absolute increase in aggregate imports of the other CIS countries ($0.5 billion). This inconsistency is reflected in the reported aggregate imbalance in the mutual trade flows.

[332] Data on Ukraine's trade with CIS partners reported by the CIS Interstate Statistical Committee in the course of 1996 tended to be conflicting. Thus, the reported dollar levels of Ukraine's exports and imports in January-September 1996 ("adjusted for gas and crude oil"; CIS, *Statisticheskii Bulleten*, No. 3 (163), February 1997) are higher than the full-year values (by $86 million and $2.2 billion, respectively) reported in *Sodruzhestvo Nezavisimykh Gosudarstv v 1996 g., Statisticheskii Spravochnik*, 1997. Hence generalizations on intra-CIS trade developments below can only be very tentative.

TABLE 3.5.9

CIS countries' trade within the CIS region, 1994-1996
(Value in million dollars, growth rates in per cent)

	Exports			Imports			Trade balances		
	Value	Growth rates		Value	Growth rates				
	1995	1995	1996	1995	1995	1996	1994	1995	1996
Armenia	167	6.3	-32.4	334	62.3	-15.5	-48	-166	-169
Azerbaijan	218	-20.6	77.4	228	-53.0	44.5	-212	-11	56
Belarus	2 931	98.2	18.4	3 676	75.8	21.2	-613	-745	-983
Georgia	97	-16.9	23.5	154	-42.4	44.5	-151	-57	-103
Kazakstan	2 631	40.5	35.5	2 609	23.7	16.1	-237	22	537
Kyrgyzstan	269	20.6	50.7	353	68.8	40.1	14	-84	-90
Republic of Moldova	467	15.0	12.1	569	19.6	11.7	-70	-103	-113
Tajikistan	252	170.7	32.7	478	105.1	-19.7	-140	-226	-50
Turkmenistan	930	-43.7	26.0	745	8.6	-53.0	965	185	822
Ukraine	6 012	7.0	-12.6	7 133	-9.0	-12.4	-2 218	-1 121	-989
Uzbekistan	1 109	-30.0	-23.6	1 118	-20.2	30.2	182	-9	-608
Total above	15 083	11.7	7.4	17 398	8.7	2.8	-2 528	-2 315	-1 690
Russian Federation	14 244	2.8	8.3	13 526	31.1	4.2	3 544	718	1 332
CIS total	29 327	7.3	7.8	30 924	17.5	3.4	1 016	-1 597	-358
Memorandum item:									
Russian Federation[a]	15 410	4.0	9.7	16 760	23.2	4.4	1 400	-1 350	-600

Source: CIS Interstate Statistical Committee; for the Russian Federation: Russian Federation Goskomstat data based on customs statistics.

[a] Adjusted for non-registered trade. Data from *BIKI*, No. 21, 20 February 1997.

---

intra-CIS trade in bulk commodities, the developments in Russian export prices for fuels and most metals sold on intra-CIS markets should be fairly indicative: in 1996 these rose by 42 and 24 per cent, respectively.[333] According to one estimate,[334] export prices in Russia's trade with its CIS partners increased in dollar terms on average by 23 per cent in 1996, which implies that Russia's exports to them declined in volume quite considerably, in contrast to the 10 per cent volume growth reported for the first half of the year.[335] Fast growing unit dollar values (in a range of 25-70 per cent) were reported also for the Republic of Moldova's exports of foods, beverages and tobacco, which comprise almost two thirds of its intra-CIS exports.[336]

Trade with CIS partners declined in dollar value terms not only in Ukraine, but also in Armenia, on both the export and import sides, while Uzbekistan reported rapidly decreasing exports but strongly growing imports. Tajikistan and Turkmenistan, in contrast, posted fast growth on the export side (26-33 per cent) and a steep decline in imports from within the region.

All other CIS countries reported growth in both exports and imports in 1996, and at relatively high rates (table 3.5.9). Russia's exports and imports within CIS lost pace in the second half of 1996 resulting in a mere 4 per cent import value growth for the year as a whole, in contrast to the 50 per cent rise reported for January-June (year-on-year basis). Nevertheless, from data presented in tables 3.5.8 and 3.5.9 it is evident that intra-CIS trade in most cases remains centred on trade with the Russian Federation, which accounts for almost one half of the mutual trade flows. Russia acquires three quarters of the exports from Belarus, Kazakstan and Ukraine. These three countries account for more than a third of mutual trade and are the main partners for imports from the Russian Federation.

The trade imbalances, which have gradually narrowed in 1995 and the first half of 1996, seem to be increasing again (table 3.5.9). As a result of the slowdown in import growth in the second half of 1996, the Russian surplus of $0.3 billion in January-June (reported by Goskomstat) had quadrupled and reached $1.3 billion for the year as a whole, almost twice the level of 1995. Trade surpluses increased substantially in three of the CIS energy exporters, Azerbaijan, Kazakstan and Turkmenistan, while the fourth, Uzbekistan was an exceptional case: it reduced its exports to CIS partners by a quarter but noticeably increased its imports from them and, as a result, posted a considerable deficit in its intra-CIS trade. All other CIS countries reported widening deficits in 1996 as compared with 1995.

---

[333] Russian export dollar unit values for aluminium sold on the CIS markets in 1996 were 1.5 times higher than those for sales on world markets, and the difference for pig iron was 1.8 times. The export unit values for all energy products (except crude oil) sold on CIS markets were higher as well. UN/ECE secretariat computations on data in *Tamozhennaya statistika vneshnei torgovli Rossiiskoi Federatsii*, Nos. 1-4 (Moscow), 1996.

[334] *BIKI*, No. 21, 20 February 1997.

[335] *BIKI*, Nos. 116-117, 5 October 1996.

[336] Government of Moldova/European Expertise Service, op. cit.

Intra-CIS trade deficits have a tendency to lead to large payments arrears, which imposes a heavy burden on the energy suppliers in the group since the main commodities traded within the region are oil and gas.[337] However, starting in 1994, net exporters, and especially the Russian Federation, became more strict on repayment, and also in their credit policies. Although debt repayments improved in 1996, the Russian Federation has repeatedly proposed debt-for-equity swaps, mainly the acquisition of pipelines and storage facilities, as well as oilfields or mines, in the debtor countries, and, in some cases, has cut off supplies of gas, electricity and heating oil to debtors.

With credit restructuring and the extension of new trade credit facilities, as well as the proposed debt-for-equity swaps, the Russian Federation is pursuing its initiative to reintegrate the CIS countries into one economic space.[338] Two countries, Belarus and Kazakstan, had already entered into a customs union with Russia in January 1996 which, according to some observers, brought some positive changes in their foreign trade developments by the end of 1996, at least on the administrative side.[339] More recently, Kyrgyzstan, Tajikistan and Uzbekistan expressed their willingness to join the customs union, while others prefer to intensify their trade links through bilateral free trade agreements across the region.

3.6 External financial developments in the transition economies

(i) Overview

In 1996, the balance of payments[340] of most European transition economies weakened, but their debt burdens lightened and they made further substantial progress in integrating into the international financial system. There were, however, still considerable differences in the financial situations of individual economies. The trade in goods and services of most countries continued to expand in the first three quarters of 1996, but generally at a slower pace than in 1995. The growth in the value of imports of merchandise and services outpaced that of exports and current account balances generally deteriorated in 1996, with most countries moving more deeply into deficit (although the Russian Federation registered another large surplus). Further improvement in the balances on services partially offset larger merchandise trade deficits.

Current account deficits in the region were generally easily financed through capital inflows and the use of official reserves. However, some countries have failed to attract capital and this has constrained their economic growth and more generally limited their room for manoeuvre.

Debt burdens generally continued to decline, although they rose among some of the least indebted countries. Eastern Europe is considered to be generally a low debt area, a major change from the beginning of the decade. In contrast to the strong build-up in 1995, the accumulation of international reserves ceased in 1996, due to higher current account deficits and smaller capital inflows. In eastern Europe, those countries whose reserves in 1995 had fallen short of recommended levels generally managed to strengthen them in 1996, but this was not the case for the Baltic states and the European CIS countries, although the Republic of Moldova was an exception.

Overall, the macroeconomic and external financial developments in the countries covered here have been favourably assessed by the international financial community. In consequence, they were able to gain further access to the international financial markets in 1996 and early 1997. The number of countries ranked by the international credit agencies tripled, increasing to 13, more than half of which are ranked investment grade. A noteworthy development is the extension of the group of rated countries beyond the initial group of rapid reformers (Czech Republic, Hungary, and Poland) to certain republics of the former Soviet Union and SFR of Yugoslavia. The transition economies were able to raise a record $9.8 billion in medium- and long-term debt in 1996, despite the fact that the most creditworthy governments had little need of new funds. Borrowing continued at a record pace in early 1997. Initial private

---

[337] On 1 October 1996 the total debt of CIS countries for energy deliveries from the Russian Federation amounted to R8,962 billion (approximately $1.7 billion), with Ukraine's debt at R3,548 billion ($0.7 billion), Kazakstan's and the Republic of Moldova's at R2,356 billion and R2,055 billion, respectively (about $0.4 billion for each). *BIKI*, No. 11, 28 January 1997.

[338] See UN/ECE, *Economic Bulletin for Europe*, Vol. 48 (1996), pp. 65-66 and A. Tuleev, "Net al'ternativi integratsii", *Ekonomika i zhizn'*, No. 51 (Moscow), December 1996, on Russia's credit policies for the CIS reintegration process.

[339] *BIKI*, No. 3, 9 January 1997.

[340] The data in this section derive from national balance of payments statistics, which in most cases are reported on a cash basis, i.e. cash settlements made through the banking system. Thus, the merchandise trade data used here as the point of entry into the discussion of the current account differ somewhat from the customs reported statistics used in section 3.5. In general, the two sets of data give broadly consistent results. However, the differences for Hungary in the first three quarters of 1996 are significant, the balance of payments data showing a much faster growth of merchandise trade.

The general decline in growth rates of trade in the first three quarters of 1996 is due partly to the valuation effect of the appreciation of the dollar in that period. The movement of exchange rates also had an impact on the debt and reserves of the transition economies because their European currency and, in some cases, yen components are relatively high. The appreciation of the dollar lowered the dollar value of the non-dollar component of these stocks thereby lowering the dollar value of the total debt. In contrast, in 1995 the depreciation of the dollar had the opposite effect on the growth rates of trade and financial stocks.

offerings, although still generally modest ($1.4 billion), have grown quickly as companies (generally "blue chips") have recognized the benefits of foreign placements. In general, borrowing terms have improved, and the sources of finance have broadened. Foreign investors contributed to the strong performances of the Russian, Budapest, and Warsaw stock exchanges in 1996 and early 1997. However, major differences persist among the transition economies in their ability to attract foreign private capital.

Despite the improvement in market access, however, the net capital imports of eastern Europe declined from $24 billion in 1995 to $15 billion in 1996: FDI fell because the exceptionally large privatizations of 1995 were not repeated; Hungary reduced sovereign borrowing and made net repayments on sovereign debt; and the Czech Republic took policy measures to deter additional inflows of short-term funds. In general, short-term and other potentially volatile funds have played a large role in the financing of the balance of payments of the east European and Baltic countries in the past two years. Private capital has accounted for the bulk of the new funds attracted by both groups of countries.

The *Baltic states* posted the best performance in the first three quarters of 1996 of the economies under consideration, their trade in goods and services growing by over 20 per cent. Although smaller than in 1995, the expansion has been in progress for four consecutive years. Current account deficits worsened – all increased to some 7 per cent of GDP – but they were fully covered by larger capital inflows, which also allowed foreign exchange reserves to increase. Estonia and Latvia remain net creditors but Lithuania has become a (small) net debtor vis-à-vis the rest of the world.

The combined exports of goods and services of *Belarus*, the *Republic of Moldova* and *Ukraine* also expanded by around 20 per cent in the first three quarters of 1996, and imports somewhat less. Except for Ukraine, current account deficits have been significant and have deteriorated. Although financial inflows exceeded the current account imbalances, only the Republic of Moldova significantly strengthened its foreign currency reserves. Foreign debt remains low to moderate in all three countries. The Republic of Moldova has made the most progress in gaining access to the international capital markets.

After some improvement in the external position of the *Russian Federation* in 1995, there was little further change in the first three quarters of 1996. The expansion of exports of goods and services slowed to 6 per cent and imports to 2 per cent and the current account surplus grew to over $10 billion. The opening of the government debt market to foreigners, together with FDI inflows, appear to have resulted in larger imports of private capital during the course of the year. Nonetheless, foreign exchange reserves declined during the year, although, according to government statements, they remain high enough to meet the country's needs. In the four months since the Russian Federation received an international credit rating (November 1996), it has taken the lead among the transition economies in raising funds in foreign markets (over $5 billion), including bond and equity issues and syndicated loans.

In *eastern Europe* the growth of exports of goods and services slowed sharply in the first three quarters of 1996, to some 3 per cent. The expansion of imports slowed as well, but at 11 per cent continued to exceed that of exports. The aggregate current account deficit is estimated at over $13 billion for 1996, compared with $1 billion in 1995. Although the balances of most countries deteriorated, the deficits of Croatia and Hungary declined substantially. Eastern Europe has run a current account deficit since the beginning of the decade, chiefly with western Europe, its major trading partner. A major feature of east European developments in 1996 has been the divergence in trade performance. The trade in goods and services of Albania, Croatia, the Czech Republic, Hungary and Poland continued to expand at varying rates,[341] but those of the other countries stagnated or declined, sometimes by substantial amounts.

A general deterioration of the current account balances of the east European and Baltic countries had been anticipated for 1996, due to expectations of continuing (or an upturn in) economic growth, real exchange rate appreciation (in certain countries), selected tariff reductions, and some slowing of import demand in western Europe. In several cases, however, the deterioration was greater than anticipated at the start of the year. Surges in capital inflows, such as those widely experienced by the transition economies in 1993-1995 (see section 3.6(ii)), tend to raise real exchange rates and to increase domestic demand, thereby causing current accounts to worsen. The lagged effects of these capital surges may have continued in 1996 even though the inflows weakened.

(ii) Balance of payments, reserves and debts

(a) Eastern Europe and the Baltic states

Current account

In 1996, the aggregate current account of the countries of *eastern Europe* deteriorated by $12 billion to a deficit of $13 billion (table 3.6.1).[342] The current account has been in deficit since the beginning of the

---

[341] As noted above, customs statistics show Hungarian merchandise trade growing more slowly than do the balance of payments data.

[342] Detailed balance of payments figures for all European transition economies are presented in table 3.6.18 at the end of this section.

TABLE 3.6.1
Current account balances of eastern Europe, the Baltic countries and European members of the CIS, 1994-1997
(Million dollars and per cent)

	1994	1995	Jan.-Sept. 1996	1996	1997[a]	Per cent of GDP 1994	1995	QI-QIII 1996	1996	1997[a]
Eastern Europe[b]	..	-1 389	-7 387	-13 466	..	..	-0.4	-3.1	-4.0	..
Former presentation[c]	-4 203	-9 143	-8 521	-20 619	..	-1.6	-2.8	-5.4	..	..
Albania	-43	-15	-85	-100*	61	-2.4	-0.6	-4.3	-3.8	2.3
Bosnia and Herzegovina	-189	-134	..	..	..	..	..	..	..	..
Bulgaria[d]	-25	-26	-34	-50*	..	-0.3	-0.2	-0.5	-0.5	..
Croatia	103	-1 711	-745	-1 129[a]	-1 239	0.7	-9.5	-5.3	-6.1	-6.6
Czech Republic	-50	-1 362	-3 083	-4 476[e]	-6 500 - -7 500[f]	-0.1	-2.9	-8.0	-8.7	-11.1 - -12.8[f]
Hungary[d]	-3 911	-2 480	-1 104	-1 678	-1 500 - -2 500[f]	-9.4	-5.7	-3.3	-3.8	-1.8 - -3.0[f]
Poland[c d]	..	5 455	-311	-1 352	-3 900 - -7 800[f]	..	4.6	-0.3	-1.0	-2.6 - -5.5[f]
Former presentation	-944	-2 299	-5 828	-8 505	..	-1.0	-1.9	-6.1	..	..
Romania	-428	-1 639	-713	-2 336[e]	-1 375	-1.4	-4.6	-2.7	-6.6	-4.5
Slovakia	712	646	-1 032	-1 900[e]	-2 000[f]	5.2	3.7	-7.2	-10.0	-10.5[f]
Slovenia	540	-36	68	46	-35	3.8	-0.2	0.5	0.3	-0.2
The FYR of Macedonia	-158	-222	-348	-491	..	-5.7	-6.2	-13.2	-14.0	..
Baltic states	-67	-827	-908	..	..	-0.7	-5.9	-7.2	..	..
Estonia	-178	-185	-223	..	..	-7.7	-5.1	-7.0	..	..
Latvia	201	-27	-273	..	..	5.5	-0.6	-7.1	..	..
Lithuania	-90	-614	-412	..	-481	-2.1	-10.3	-7.4	..	-5.5
CIS										
Belarus	-506	-567	-752	..	..	-10.3	-5.5	-7.6	..	..
Republic of Moldova	-82	-115	-134	..	-180	-5.8	-6.8	-11.2	..	-11.3
Russian Federation	11 378	9 306	10 243	..	..	4.1	2.6	3.1	..	..
Non-CIS	8 205	9 410	11 090	..	..	..	..	..	..	..
Ukraine	-1 161	-1 152	-553	..	..	-3.1	-3.2	-1.7	..	..

Source: National balance of payments statistics; press reports; UN/ECE secretariat estimates.

[a] Official forecasts.
[b] Includes revised data for Poland. Eastern Europe aggregate excludes Bosnia and Herzegovina.
[c] Revised data for Poland. See table 3.6.2.
[d] Convertible currencies; Hungary until 1995.
[e] Preliminary.
[f] Independent forecasts.

decade, but the imbalance in 1996 is the largest yet.[343] In the first three quarters of 1996, the growth of *exports of goods and services* of the region slowed sharply from 36 per cent in 1995 to 3 per cent (tables 3.6.3 and 3.6.4). Receipts from non-factor services outpaced those from merchandise. The growth of *imports of goods and services* also declined, from 40 per cent in 1995 to 11 per cent in the first three quarters of 1996, merchandise imports continuing to expand faster than services. Virtually all east European countries had deficits on reported merchandise trade, totalling nearly $17 billion in the first three quarters of 1996. The growing merchandise imbalance has been partially offset by surpluses on transfers and services, that for services increasing again, to nearly $5 billion in all of 1996.

Most of the increase in the services balance was accounted for by the growth of net receipts from *tourism* in Hungary and Croatia, which raised the combined net earnings of the region to $4.8 billion. Eastern Europe also earned over $1 billion[344] in net revenues from *transport services*. In most east European countries, service exports have been in surplus, but Albania, Romania and The FYR of Macedonia have been in deficit for at least two years while in Poland the deficit is comparatively recent. Services account for some 21 and 17 per cent, respectively, of the aggregate exports and imports of goods and services of eastern Europe. These shares are higher than in the other transition economies discussed below.

---

[343] Attention is drawn to the fact that the current account of Poland in 1995 has been significantly revised due to the change in Poland's balance of payments methodology. Poland now reports a large current account surplus for 1995 instead of a sizeable deficit (see table 3.6.2), and, as a result, the aggregate current account deficit of eastern Europe in 1995 now amounts to $1.4 billion instead of the nearly $9 billion originally reported. The methodological change results in a shift of similar magnitude in the current account in 1996, although Poland was in deficit according to both measures. See box 3.6.1 in UN/ECE, *Economic Bulletin for Europe*, Vol. 48 (1996).

[344] This figure excludes Poland which in 1995 realized net earnings on transport of some $600 million.

The decline in net factor *income* paid to foreigners appears have slowed or bottomed out in 1996 at somewhat over $3 billion.[345] Net income payments have fallen in line with the reduction in the net foreign debt of the region (see below) and lower international interest rates. However, several countries are now reporting increasing payments abroad of profits generated by foreign investments: such payments can be expected to increase significantly in the near to medium term.

Transfers to eastern Europe remained unchanged at some $3.7 billion, all countries reporting positive transfers, and about half of them posting increases in 1996. In general these mainly reflect remittances from the sizeable number of citizens of eastern countries employed abroad and aid from the international community to support the transition process. In Albania, Croatia, the Czech Republic, Hungary, Romania and The FYR of Macedonia transfers have become an important source of balance of payments financing.

Country notes

In 1996, the deterioration in the aggregate current account of eastern Europe was due chiefly to a worsening in the accounts of Poland ($7 billion), the Czech Republic ($3 billion), Slovakia ($2.5 billion), Romania ($0.7 billion) and, to a lesser extent, The FYR of Macedonia. The balances of the other countries improved, especially those of Hungary and Croatia. With the exception of Bulgaria, all the east European countries had current account deficits in the January-September period of 1996, that of the Czech Republic being the largest in dollar terms (table 3.6.1). Persistent current account deficits exceeding 5 per cent of GDP are often viewed as a warning of the need for policy action (see section 3.6(ii)(c) below and chapter 1). In 1996, the number of east European (and Baltic) countries posting deficits in excess of this threshold increased markedly (table 3.6.1), although only a few have been above it for several consecutive years. Hungary has managed to reduce a deficit of some 9 per cent of GDP to less than 4 per cent of GDP in a relatively short period.

In contrast to 1995, fewer than half the countries contributed to the expansion of trade in 1996, the trade in both goods and services of the others stagnating or declining, in certain cases by substantial amounts. During the first three quarters of 1996, the growth of east European exports of goods and services slowed, reflecting similar tendencies in the exports of Bulgaria, Romania and Hungary. However, there may have been some pick-up in the exports of Slovakia in the third quarter and there was a sharp increase in the service receipts of Croatia. On the import side, there was little change in the aggregate, but the decline in the imports of Bulgaria and probably, Romania, quickened during the year.

Albania recorded the highest rate of growth of trade in goods and services in the first three quarters of 1996. Merchandise imports rose significantly faster than exports, increasing the trade and current account imbalances further. Exports of services increased by nearly one third, due chiefly to the doubling of revenues from transport. There was little increase in receipts from tourism, the surplus on which remained modest. Current transfers, which have been a major source of external finance, remained roughly unchanged since 1995, although their composition has changed sharply. Private transfers, mainly workers' remittances, fell from $264 million in the first three quarters of 1995 to only $47 million in the comparable period in 1996, while official transfers rose to over $300 million.[346] Balance of payments statistics are believed to record only part of the foreign currency brought into the country (in cash). These funds, which are held by the population and generally not transferred to the banking system, are likely to have supplied part of the inflow into the recently collapsed pyramid schemes. It is therefore possible that the decline in recorded private transfers reflects the rechannelling of funds from abroad directly into these schemes. The pyramid funds, in turn, are likely to have moved the cash out of the country (also without using the banking system and therefore avoiding recording).[347]

In the *Czech Republic*, the growth of the current account deficit exceeded expectations in 1996, the imbalance reaching $4.5 billion (at the beginning of the year a current account deficit of some $2 billion was anticipated).[348] The growth of exports of goods and services slowed to 5 per cent, while that of imports fell to 14 per cent in the first three quarters of the year. A much larger merchandise trade deficit was partially offset by a surplus on services, but to a much lesser degree than in 1995. Although earnings from tourism declined in 1996, the surplus still exceeds that on transport, also a major earner of foreign currency. A sizeable and increasing deficit on net income paid abroad – due to rising net debt

---

[345] Net income payments due fell from some $6 billion at the beginning of the decade to $3.2 billion in 1995. The decline reflects the interest obligations of Bulgaria and Poland which were subsequently restructured or written off.

[346] Official transfers reflect chiefly international financial assistance, especially from the European Union.

[347] The functioning of the pyramid funds is known to have resulted in changes in the channels through which foreign earnings were brought into Albania. For example, Albanians used to repatriate part of their foreign earnings in kind (e.g. electronic goods). With the rise of the pyramid schemes, there was a switch to cash since the returns promised on the investments were so high.

[348] Attention is drawn to the revision to the 1995 current account deficit from $1.9 billion to $1.4 billion. Czech National Bank, *Report on the Monetary Development in the Czech Republic for the First Half 1996* (Prague), 1996.

— and smaller net transfers also contributed to the worsening current account.

In *Hungary*, the current account balance continued to improve during 1996, the deficit falling by more than one half since 1994.[349] This was largely the result of the government's austerity programme of March 1995,[350] a key objective of which was to cut the record external deficit. Hungary posted one of the fastest growth rates of exports of goods and services in eastern Europe in 1996, the doubling of net service revenues more than offsetting a larger merchandise trade deficit. The improvement was broadly based, led by a surge in net tourist receipts, which rose to a record $1.3 billion, some increase in net earnings on transport (in contrast to historical deficits prior to 1995), and a surplus on other services. There was also a marked decline in net income payments abroad as a result of declining net external indebtedness. However, there has been a growing repatriation of profits by foreign investors ($252 million in 1996), this flow representing the return on Hungary's stock of foreign investment, the largest in the region (see below).

The current account of *Poland* has deteriorated sharply, moving from a $5.5 billion surplus in 1995 to a deficit of over $1 billion in 1996. This markedly exceeded expectations early in 1996 when a change of only about $2 billion was foreseen. It was due almost entirely to a widening of the deficit on *recorded merchandise trade*, net service flows remaining small. Also the net outflow on the income account diminished, mainly because of higher earnings on reserves, and net receipts from "non-classified current transactions"[351] declined moderately to $7.2 billion (table 3.6.2). As noted above, the change in methodology has transformed the balance of payments, changing the current account deficit originally reported in 1995 to a large surplus.

During most of 1995, *Romania* was unable to reach agreement with the IMF on a stand-by arrangement, in part because of the slow pace of market reforms and structural change. Private capital inflows were small and official reserves were inadequate and declining. However, an arrangement with the IMF was concluded in December 1995.[352] The agreed stabilization programme

TABLE 3.6.2

Revisions to the Polish balance of payments, 1995-1996
(Million dollars)

	1995		1996	
	Former	New	Former	New
Current account	-2 299	5 455	-8 505	-1 352
of which:				
Non-classified items (net)	–	7 754	–	7 153
Capital and financial account	10 513	2 759	11 936	4 783
of which:				
Other short-term capital	9 381	1 627	9 077	1 924

Source: National Bank of Poland.

and the subsequent large real depreciation of the leu improved the prospects for a smaller current account deficit in 1996 (officially projected at $1.1 billion) and for rebuilding foreign exchange reserves.[353] In March, Romania received an international credit rating. However, in the first three quarters of 1996, the country's trade in goods and services faltered, exports perhaps because the real exchange rate still remained too high. Imports fell more than exports, initially resulting in a smaller current account deficit. However, the deficit widened quickly in the last quarter of 1996, resulting in a deficit of $2.3 billion (preliminary) for the full year. There was a modest deficit on services due to the (growing) deficit on tourism (Romania used to enjoy significant earnings from tourism) and, especially, a large deficit on other services. Only transport generated (higher) net revenues. The modest level of net transfers has continued to rise, but they are increasingly offset by higher net interest payments due to growing indebtedness.

In the closing months of 1995, the external financial position of *Bulgaria* appeared increasingly precarious. Also because of the slow rate of structural change and market reform (see section 3.1(iii)), the country had been unable to reach agreement with the IMF. There were virtually no private capital inflows. Although foreign currency reserves had been increased somewhat, they still felt short of recommended levels, and debt service payments of $1.3 billion were to fall due in 1996. Although Bulgaria reached an agreement with the IMF on a stabilization package in July 1996, only the first tranche of the stand-by credit was released. In the first months of 1996, debt repayments, official intervention in defence of the lev and residents' withdrawal of cash from foreign exchange deposits[354] lowered foreign exchange reserves

---

[349] Although the current account deficit was less than the official forecast of $2 billion for 1996, the outcome was less favourable than had come to be expected in the latter half of the year, due to an exceptionally large deficit in December (some $500 million). The latter resulted chiefly from a large trade deficit and a greater outflow of factor income.

[350] The government's programme included a devaluation of the forint (9 per cent in March 1995 and monthly devaluations thereafter), imposition of an 8 per cent import surcharge, and tighter fiscal policy.

[351] Net receipts from unregistered cross-border trade in goods and services are included in the current account for the first time in 1996 with comparable figures being made available by the National Bank of Poland for 1995 (see table 3.6.2).

[352] This enabled Romania to make an immediate (and only) drawing on the stand-by credit, which was subsequently suspended.

[353] The IMF agreement paved the way for approval by the World Bank of a $280 million Financial and Enterprise Sector Adjustment Loan (FESAL) in January 1996, half of which was disbursed in the first six months of the year.

[354] The Bulgarian National Bank reported significant internal capital flight during the first three quarters of 1996, which substantially reduced its own foreign exchange reserves and those of commercial banks

TABLE 3.6.3
Foreign trade in goods and non-factor services of eastern Europe, the Baltic countries and European members of the CIS, 1994-1996
(Per cent and million dollars)

	Growth rates						Balances			
	Exports			Imports					January-September	
	1994	1995	Jan.-Sept. 1996[a]	1994	1995	Jan.-Sept. 1996[a]	1994	1995	1995	1996
Eastern Europe										
Goods and services[b][c]	13	36	3	8	40	11	-4 870	-9 670	-5 875	-13 118
Goods[c]	13	38	2	8	42	12	-7 634	-13 905	-9 221	-16 785
Services	12	30	9	13	27	9	2 765	4 235	3 345	3 667
Baltic states										
Goods and services	21	35	24	32	43	28	-464	-1 138	-671	-1 099
Goods	15	34	16	28	49	21	-880	-1 970	-1 293	-1 778
Services	42	37	47	54	13	74	416	832	622	679
Russian Federation										
Goods and services	..	19	6	..	25	2	13 218	12 470	10 629	13 777
Goods	13	20	7	20	24	11	19 712	22 173	17 726	17 474
Services	..	12	-1	..	28	-24	-6 494	-9 703	-7 097	-3 697
Belarus, Republic of Moldova, Ukraine										
Goods and services	..	15	22	..	14	20	-1 970	-1 922	-1 406	-1 438
Goods	..	16	14	..	15	21	-3 185	-3 534	-2 549	-4 150
Services	..	13	67	..	1	7	1 215	1 621	1 143	2 713

Source: UN/ECE secretariat, based on national balance of payments statistics.

[a] Over same period of 1995.

[b] Excludes "non-classified current transactions" reported by Poland. See table 3.6.2.

[c] Reflects a break in the merchandise trade series of the Czech Republic in 1995.

even further. Thus, the prospects were for a continuing *ex ante* balance of payments constraint, which had already forced Bulgaria's current account to move roughly into balance in 1994.

This weak financial position is likely to have contributed to the overall contraction of the economy and to the decline in trade in goods and services in the first three quarters of 1996 (exports and imports fell by some 12-14 per cent). The surpluses on both merchandise trade and services declined, the latter reflecting a slowing of cross-border flows of transport and travel services (in both directions). Transfers remained modest, suggesting only small inflows of aid and remittances. The large net outflow of funds was mirrored in the depletion of foreign exchange reserves (see below). However, Bulgaria did manage to fulfil virtually all of its external debt servicing obligations.[355]

After registering one of the few surpluses in the region in 1995, the current account of *Slovakia* swung into substantial ($1.9 billion)[356] deficit in 1996, the $2.5 billion turn around being equivalent to around 14 per cent of GDP. The deterioration of the current account appears to have gathered pace in the latter part of the year, a deficit of nearly $500 million being recorded in December. The combination of stagnating merchandise exports and booming imports resulted in a large trade deficit, while the sharp drop in the exports of services nearly eliminated what had been a substantial surplus.[357] Although the transport and travel sectors continued to earn modest surpluses, "other services" swung into deficit. Slovakia's earnings from transfers increased.

In 1995, in Croatia, Slovenia and The FYR of Macedonia current account balances deteriorated, and all three countries posted deficits.[358] In the first few months of 1996 their trade faltered, but their current accounts developed differently: Croatia's deficit declined substantially, that of The FYR of Macedonia expanded further, while Slovenia moved into surplus. In Croatia and The FYR of Macedonia, the deficits remain large in relation to GDP.

(deposits of individuals and private companies declining by $622 million, mostly in the first half of the year). This was prompted by expectations of a large nominal devaluation of the lev (which actually occurred in the second half of the year) and concerns about the ability of the country to meet its debt servicing obligations, both factors being exacerbated by the erosion of confidence in the banking system. A large negative interest differential in several months also contributed to the outflow. Bulgarian National Bank, *Monthly Bulletin*, No. 11 (Sofia), 1996, pp. 20-21. Also see section 3.1(iii).

[355] The exception appears to have been a default on a small private placement of bonds in yen.

[356] This figure is preliminary.

[357] Initially, the setback in export performance appears to be due to the termination on 1 October 1995 of the bilateral agreement governing external transactions with the Czech Republic.

[358] Of the successor states of the former SFR of Yugoslavia, current balance of payments data are available only for Croatia, Slovenia and The FYR of Macedonia.

TABLE 3.6.4

Foreign trade in goods and non-factor services of eastern Europe, the Baltic countries and European members of the CIS, 1994-1996
(Per cent and million dollars)

| | Growth rates | | | | | | Balances | | | |
| | Exports | | | Imports | | | | | January-September | |
	1994	1995	Jan.-Sept. 1996[a]	1994	1995	Jan.-Sept. 1996[a]	1994	1995	1995	1996
Eastern Europe										
Goods and services[b]	13	36	4	8	40	11	-4 870	-9 670	-5 875	-13 118
Goods	13	38	2	8	42	12	-7 634	-13 905	-9 221	-16 785
Services	12	30	9	13	27	9	2 765	4 235	3 345	3 667
Albania										
Goods and services	19	34	21	1	8	31	-548	-534	-367	-425
Goods	26	45	17	–	13	31	-460	-475	-341	-468
Services	7	17	31	6	-9	34	-88	-59	-26	-37
Bulgaria										
Goods and services	6	31	-14	-11	25	-12	-6	274	417	254
Goods	6	36	-16	-14	32	-14	-17	121	243	132
Services	7	14	-9	1	3	-5	11	153	174	122
Croatia										
Goods and services	15	10	6	16	40	1	-231	-2 264	-1 551	-1 304
Goods	9	9	-9	12	44	-3	-969	-2 877	-2 077	-2 241
Services	27	12	34	32	26	18	738	614	526	937
Czech Republic										
Goods and services[c]	7	49	5	12	57	14	-156	-1 836	-1 020	-2 989
Goods[c]	8	53	3	12	69	13	-889	-3 678	-2 400	-4 118
Services	4	37	10	12	17	21	733	1 842	1 380	1 129
Hungary										
Goods and services	-6	66	15	-2	38	6	-3 375	-1 788	-1 675	-722
Goods	-6	68	12	-1	36	9	-3 635	-2 442	-2 048	-1 928
Services	-5	58	23	-7	48	-5	260	654	373	1 206
Poland										
Goods and services[b]	23	37	7	14	40	29	-779	-1 677	-944	-5 446
Goods	25	35	6	12	39	30	-836	-1 827	-930	-5 260
Services	14	52	12	38	49	19	57	150	-14	-186
Romania										
Goods and services	26	31	-6	12	44	-9	-582	-1 767	-1 036	-762
Goods	26	29	-6	9	45	-9	-411	-1 577	-1 021	-695
Services	31	43	-10	33	39	-5	-171	-190	-15	-67
Slovakia										
Goods and services	24	22	-3	4	26	20	762	570	605	-1 143
Goods	27	27	3	4	29	24	105	24	68	-1 222
Services	16	5	-21	4	14	4	657	546	537	79
Slovenia										
Goods and services	16	19	1	14	28	1	385	-228	-75	-23
Goods	12	22	–	15	30	–	-338	-953	-628	-617
Services	33	9	6	11	15	5	723	725	553	595
The FYR of Macedonia										
Goods and services	10	10	-27	28	13	-2	-340	-420	-229	-479
Goods	3	11	-29	26	12	2	-185	-220	-87	-368
Services	105	8	-10	37	18	-16	-155	-200	-142	-112
Estonia										
Goods and services	68	42	16	83	39	20	-268	-314	-199	-309
Goods	71	34	8	88	45	18	-378	-692	-464	-678
Services	61	63	33	66	16	28	110	378	265	369
Latvia										
Goods and services	6	24	26	29	35	44	60	-113	-34	-329
Goods	-3	34	12	26	47	26	-301	-580	-380	-614
Services	23	8	51	45	-17	179	360	466	347	284
Lithuania										
Goods and services[b]	7	36	30	8	50	26	-256	-711	-439	-461
Goods	1	33	23	3	53	20	-202	-698	-448	-468
Services	63	51	66	49	32	63	-54	-13	10	25
Belarus										
Goods and services	..	87	19	..	76	27	-504	-595	-371	-750
Goods	..	87	15	..	78	28	-556	-777	-476	-1 046
Services	..	85	61	..	42	1	52	182	104	296
Republic of Moldova										
Goods and services	..	33	12	..	33	7	-100	-137	-149	-136
Goods	57	19	10	27	18	17	-54	-55	-64	-106
Services	..	285	23	..	164	-25	-46	-82	-85	-30
Ukraine										
Goods and services	24	3	23	26	2	19	-1 366	-1 190	-885	-552
Goods	28	3	14	31	3	20	-2 575	-2 702	-2 009	-2 998
Services	6	4	69	-8	-13	14	1 209	1 512	1 124	2 446

Source: UN/ECE secretariat, based on national balance of payments statistics.

[a] Over same period of 1995.
[b] Excludes "non-classified current transactions" reported by Poland, (see table 3.6.2).
[c] Reflects a break in the merchandise trade series of the Czech Republic in 1995.

Croatia's exports of goods and services turned up in the third quarter of 1996 resulting in a growth rate of 6 per cent for the nine month period as a whole. This upturn was due to a surge in tourist receipts, net receipts doubling to $1.3 billion for the full year. A recovery in revenues from tourism was expected after the Dayton Accord, although the continuing overvaluation of the exchange rate was seen as a possible drawback. The overall improvement of the service balance and higher private transfers more than offset the further growth in the merchandise trade deficit. The resulting $1.1 billion current account deficit was some one third less than in 1995.

Slovenia ended 1996 with a small current account surplus, although for much of the year it looked as if it would be in deficit. The outcome was due partly to recent revisions in the data and methodology.[359] Unrevised data for the first three quarters of the year show exports of goods and services falling by 3 per cent (although revisions are likely to show a better result). Imports of goods and services stagnated, although services showed some growth. The increased surplus on services was due to a 13 per cent rise in receipts from tourism and higher net revenues from transport.

There was a sharp deterioration in the current account of *The FYR of Macedonia* in the second half of 1995 which carried over into 1996. The deficit for all of 1996 was $491 million (14 per cent of GDP). Exports of goods and services fell by 24 per cent in the year as a whole, chiefly because of a similar fall in merchandise exports (services fell by 11 per cent), while the decline in imports of goods and services was 4 per cent (due entirely to a 17 per cent fall in service imports). Thus a substantial widening of the trade deficit was partially offset by a somewhat smaller deficit on services. Expenditures on transport declined reducing the country's comparatively large deficit on that item, presumably because of the reopening of transit routes through Greece.[360] Private transfers remained relatively high.

Among the *Baltic states,* Estonia and Latvia moved further into current account deficit in the first three quarters of 1996 (table 3.6.1).[361] Although remaining significant, their export growth declined during the course of the year, but the pace of imports of goods and services seemed to pick up in the third quarter. All three countries earned surpluses on *services* in 1996, but only those of Estonia and Latvia were relatively substantial. Relative to GDP, all three current account deficits are comparatively high, and in the case of Estonia, have persisted for several years.

The current account deficit of *Estonia* more than doubled in the first three quarters of 1996, exports of goods and services decelerating much more sharply than imports. The much larger merchandise trade deficit was partly offset by larger net receipts from services, which now rank among the highest of the transition economies. Net earnings on tourism have grown from virtually zero a few years ago to a surplus estimated at over $300 million in 1996.[362] There was also a growing surplus on transport, and transfers remained substantial.

Services also play an important role in the balance of payments of *Latvia*. There was a sizeable surplus in the first three quarters of 1996, although noticeably smaller than in 1995. The comparatively large transport surplus continued to grow and the balance of other services moved into surplus. However, the deficit on tourism has grown quickly. Overall, the expansion of Latvia's trade in goods and services was among the fastest in the region and seemed to quicken as the year progressed.

In *Lithuania*, the growth of exports slowed only modestly in the first three quarters of 1996, to 30 per cent. However, imports of goods and services accelerated in the third quarter of the year, reversing the improvement in the current account earlier in the year. The large merchandise trade deficit has been partially offset by sizeable transfers. Services have been roughly in balance, although a rapidly growing travel surplus emerged in 1996. Outflows of net property income have increased.

Reserves and external debt

In 1996, a foreign capital inflow of some $15 billion (section 3.6(iii)) roughly covered the combined current account deficits of the countries of *eastern Europe*. In consequence, there was no further rise in aggregate *official foreign exchange reserves* (table 3.6.5). In 1995, official reserves had risen by $24 billion, the largest increment since the rebuilding of reserves began early in the decade. This was made possible by a record inflow of funds and, compared with 1996, a much smaller current account financing requirement. In 1996 reserves increased in about half the east European countries, led by those of Poland ($2.9 billion). However, these

---

[359] Until the recent revisions, exports, including those involving subcontracting, had been undervalued (also see section 3.5). Institute of Macroeconomic Analysis and Development, *Slovenian Economic Mirror* (Ljubljana), January 1997, p. 6 and *Autumn Report, Republic of Slovenia* (Ljubljana), December 1996, pp. 7-8. Data for the first 11 months of 1996, prior to the revision, show a $271 million current account deficit. Bank of Slovenia, *Monthly Bulletin*, December 1996.

[360] In September 1995, Greece lifted the prohibition on the transport across its territory of goods to and from The FYR of Macedonia.

[361] Attention is drawn to the substantial revision of the current account balances of Latvia and Lithuania for 1995, the originally reported surpluses changing to deficits.

[362] A large part of this may reflect "shopping tourism" from neighbouring countries, particularly Finland.

TABLE 3.6.5

Foreign exchange reserves of eastern Europe, the Baltic countries and European members of the CIS, 1990, 1993-1997

(Million dollars and months of imports, end of period)

	Reserves						Months of imports[a]				
	1990	1993	1994	1995	1996	1997	1990	1993	1994	1995	1996
Eastern Europe	6 802	18 202	26 493	50 797	50 823	..	1.5	2.4	3.2	4.5	4.1
Albania	199	147	205	241	280	..	5.0	2.2	3.0	3.3	3.0
Bosnia and Herzegovina	..	..	17	66	111[b]	..	..	..	..	..	..
Bulgaria[c]	–	701	1 002	1 236	490	446[d]	–	1.4	2.2	2.1	0.9
Croatia	–	612	1 405	1 898	2 314	2 174[e]	–	1.2	2.4	2.3	2.8
Czech Republic	400	3 781	6 145	13 843	12 352	11 846[e]	0.7	2.6	3.7	5.3	4.1
Hungary	1 070	6 691	6 727	11 968	9 714	..	1.4	5.2	5.1	6.7	5.1
Poland	4 491	3 985	5 728	14 657	17 541	..	3.8	2.2	3.1	6.0	5.5
Romania	524	994	2 031	1 523	2 099	1 799[e]	1.1	1.7	3.0	1.6	2.4
Slovakia	118	415	1 605	3 306	3 403	3 400[d]	0.5	0.6	2.3	3.7	3.2
Slovenia	–	770	1 480	1 802	2 279	2 239[e]	–	1.2	2.1	2.0	2.5
The FYR of Macedonia	–	105	149	257	240	..	–	1.1	1.0	1.7	1.6
Yugoslavia	..	..	..	..	..	..	..	..	..	..	..
Baltic states	..	938	1 497	1 822	2 050	..	..	2.3	2.7	2.3	2.0
Estonia	..	329	442	580	637	..	..	3.2	2.3	2.2	2.0
Latvia	..	334	545	503	652	..	..	3.2	3.9	2.7	2.4
Lithuania	..	275	510	739	762	731[e]	..	1.4	2.3	2.2	1.8
CIS	1 600	6 107	4 720	15 798	13 132	..	..	..	0.6	1.7	1.3
Belarus	..	75	115	400	469	..	..	..	0.4	0.8	0.7
Republic of Moldova	..	42	159	227	304	279[d]	..	1.0	2.4	2.6	3.4
Russian Federation	1 600	5 829	3 976	14 265	11 272	..	..	..	0.7	2.0	1.5
Ukraine	..	162	470	906	1 087	..	..	0.1	0.3	0.6	0.6

Source: National statistics; IMF, *International Financial Statistics* (Washington, D.C.), February 1997; press reports; UN/ECE secretariat estimates.

[a] Imports of merchandise and services and income payments.
[b] April.
[c] Includes holdings of SDRs and reserve position with the IMF.
[d] February.
[e] January.

increases were fully offset by declines elsewhere, particularly in Hungary ($2.2 billion) and the Czech Republic ($1.5 billion). Due to the continuing expansion of imports, the aggregate *liquidity* or *import-coverage ratio* fell to around four months of imports of goods and services and income payments in 1996 (table 3.6.5). In general, liquidity ratios fell, but in most cases they remained comfortably above the minimum recommendation of three months of import coverage.

Since the authorities in several countries had considered the build-up of reserves in 1995 as excessive, some scaling back was expected in 1996. In Hungary the reserves were used to repay external debt (see below); in the Czech Republic they helped to finance the rapidly increasing current account deficit which could no longer be fully covered by the diminished inflow of capital.[363] Both countries, nevertheless, have retained more than adequate levels of reserves. In Bulgaria, however, official foreign currency holdings fell below $500 million in December 1996 and further in the new year, for the reasons noted above. The growth in Poland's reserves occurred largely in the first half of the year. Thereafter, smaller capital inflows were largely offset by the growing current account deficit. The Romanian authorities managed to reverse the depletion of reserves in 1996 – foreign exchange reserves rose to $2 billion by year's end – thanks to record borrowing in the international financial markets. In Croatia and Slovenia, reserves and the liquidity ratio continued to increase,[364] although import coverage remained below three months (as was also the case in The FYR of Macedonia). It is worth recalling that the official reserves of these three countries reflect accumulation only since the dissolution of the former SFR of Yugoslavia, and thus they exclude claims on the foreign reserves of that entity.[365]

The combined *gross external debt* of the east Europe countries decreased somewhat in 1996 to $122 billion (table 3.6.6), despite the region's large current

---

[363] The Czech National Bank spent $660 million defending the koruna in the three days following the introduction of the ±7.5 per cent exchange rate band in late February 1996. Czech National Bank, *Report on the Monetary Development of the Czech Republic for the First Quarter 1996* (Prague), 1996.

[364] Slovenia's official reserves increased due to the transfer of assets from the country's fiduciary account and a Eurobond issue (see below). *Slovenian Economic Mirror*, op. cit., p. 7.

[365] The possible division of the holdings of the former SFR of Yugoslavia at the BIS is to be decided at the June 1997 annual meeting.

TABLE 3.6.6

Total external debt of eastern Europe, the Baltic countries and European members of the CIS, 1990, 1993-1996
(Dollars, end of period)

	Gross debt (millions)					Net debt (billions)				
	1990	1993	1994	1995	1996	1990	1993	1994	1995	1996[a]
Eastern Europe[b]	93 376	109 389	111 095	124 008	121 780	86.6	91.2	84.6	73.2	71.1
Albania	297	790	820	600*	600*	0.1	0.6	0.6	0.4	0.3
Bosnia and Herzegovina	..	..	..	3 361	..	..	..	..	..	..
Bulgaria	10 890	13 889	11 411	10 513	9 861[c]	10.9	13.2	10.4	9.3	9.4
Croatia[d]	2 500	2 638	3 007	3 661	4 881[e]	2.5	2.0	1.6	1.8	2.6
Czech Republic	4 400	9 605	12 210	17 190	18 458[c]	4.0	5.8	6.1	3.3	6.1
Hungary[f]	21 270	24 560	28 521	31 660	27 646	20.2	17.9	21.8	19.7	17.9
Poland[f]	48 475	47 246	42 174	43 886	40 661[e]	44.0	43.3	36.4	29.2	23.1
Romania	1 140	4 249	5 509	6 492	8 200[g]	0.6	3.3	3.5	5.0	6.1
Slovakia	2 004	3 682	4 285	5 800[h]	6 300[c]	1.9	3.3	2.7	2.5	2.9
Slovenia[d]	1 900	1 873	2 258	2 970	4 001	1.9	1.1	0.8	1.2	1.7
The FYR of Macedonia[d i]	500	857	900	1 236	1 172	0.5	0.8	0.8	1.0	0.9
Yugoslavia	..	..	..	8 500	9 000[j]	..	..	..	..	..
Baltic states	..	475	1 166	1 498	1 923	..	-0.5	-0.3	-0.3	-0.1
Estonia	..	120	276	269	297	..	-0.2	-0.2	-0.3	-0.3
Latvia	..	237	347	403	409	..	-0.1	-0.2	-0.1	-0.2
Lithuania	..	118	543	826	1 217	..	-0.2	–	0.1	0.5
CIS										
Belarus[k]	..	463	1 159	2 022	1 357	..	0.4	1.1	1.6	0.9
Republic of Moldova	..	251	484	624	651[c]	..	0.2	0.3	0.4	0.3
Russian Federation	..	112 700	121 600	120 400	127 000[j]	..	106.9	117.6	106.1	115.7
Debt to west	59 800	83 700	93 600	102 400	..	58.2	78.7	89.6	88.1	..
Ukraine	..	..	7 500	8 004	8 840	..	..	7.0	7.1	7.8

Source: National statistics; IMF and World Bank data for Albania; UN/ECE secretariat estimates; table 3.6.5 for official foreign exchange reserves.
Note: Net debt equals gross debt less foreign exchange reserves.

[a] Net debt reflects the latest months for which gross debt and reserves are available.
[b] Excludes Bosnia and Herzegovina and Yugoslavia.
[c] September.
[d] Allocated debt only (see text). In Croatia and The FYR of Macedonia unallocated debt is included in 1996.
[e] November.
[f] Convertible currencies (Hungary through 1994 only).
[g] Medium- and long-term debt plus estimated short-term debt.
[h] Reflects change of methodology.
[i] Medium and long-term loans only.
[j] Projection for end-1996.
[k] Excludes arrears of enterprises on payments for imported natural gas.

account deficit. The valuation effect of dollar appreciation tended to reduce the dollar value of the total debt as did some repayment of principal.[366] Hungary and Poland report the largest reductions. In most countries, however, and especially for Romania and the Czech Republic, debt increased, due to borrowing in the international financial markets (see below). The aggregate *net debt* of the countries of eastern Europe continued to fall in 1996 (table 3.6.6), but more slowly than in 1995 when there was a surge in foreign exchange reserves.

The expansion of exports of goods, services and investment income contributed to a further reduction in the debt burden, measured here by the *ratio of gross* and *net debt to exports* (table 3.6.7). Eastern Europe has made considerable progress in this respect, since it was regarded as a "highly indebted region" at the beginning of the decade. The recent fall in the average gross debt-export ratio to around 100 means that eastern Europe now rates as a comparatively low debt area. Initially, highly indebted countries such as Albania, Bulgaria, Hungary and Poland[367] have seen their debt indicators improve dramatically in recent years,[368] although those

---

[366] It is estimated that the valuation effect of dollar appreciation on eastern Europe's $121 billion stock of debt during 1996 amounted to a reduction of some $4 billion.

[367] The debt-export ratios for Poland do not reflect net receipts from "non-classified current transactions". Incorporation of that $7.8 billion item in exports of goods, services and income would lower the gross debt-export ratio in 1995 in table 3.6.7 from 162 to 126.

[368] In 1994, the external debts of Bulgaria and Poland were substantially reduced as a result of debt reduction agreements with their commercial creditors. Previously, the Paris Club reduced and

TABLE 3.6.7

Measures of the debt burden of eastern Europe, the Baltic countries and European members of the CIS, 1990, 1994-1996

(Per cent)

	Gross debt/exports[a]				Net debt/exports[a]				Gross debt/GDP			
	1990	1994	1995	1996[b]	1990	1994	1995	1996[b]	1990	1994	1995	1996[b]
Eastern Europe	194	125	102	96	180	95	60	56	38	43	39	36
Albania	87	293	161	135	28	220	96	72	14	47	25	23
Bulgaria	154	216	152	165	154	197	134	157	55	118	81	103
Croatia[c]	56	45	50	62	56	24	24	32	10	21	20	26
Czech Republic	70	62	59	60	63	31	11	20	14	34	36	36
Hungary	243	261	178	133	231	200	111	86	60	69	72	62
Poland	380	216	162	139	344	186	108	79	82	46	37	31
Romania	28	75	68	92	15	48	52	69	3	18	18	23
Slovakia	110	47	52	58	103	29	22	27	13	31	33	33
Slovenia[c]	32	25	28	37	32	9	11	16	11	16	16	22
The FYR of Macedonia[c]	42	71	89	115	42	59	70	91	12	33	35	33
Baltic states	..	19	18	19	..	-5	-4	-1	..	11	11	11
Estonia	..	14	10	9	..	-8	-11	-10	..	12	7	7
Latvia	..	20	19	15	..	-11	-5	-9	..	10	9	8
Lithuania	..	23	25	29	..	1	3	11	..	13	14	16
CIS	..	130	110	105	..	126	97	95	..	41	32	28
Belarus	..	42	39	20	..	38	31	13	..	24	20	10
Republic of Moldova	..	73	70	59	..	49	45	32	..	34	37	41
Russian Federation	..	152	126	125	..	147	111	114	..	44	34	29
Russian Federation[d]	78	189	159	..	76	181	136	..	..	..	..	..
Ukraine	..	45	46	42	..	42	41	37	..	20	22	20

Source: UN/ECE secretariat estimates, based on national statistics.

Note: Net debt equals gross debt less foreign exchange reserves.

[a] Exports of merchandise and services and income receipts. For Poland, excludes net receipts from non-classified current account items.

[b] Reflects latest available data in table 3.6.6 (see notes).

[c] See comments in table 3.6.6.

[d] Debt to the west and exports of merchandise to non-CIS only.

of Bulgaria deteriorated in 1996.[369] On the other hand those of relatively low-medium debt countries such as Croatia, Romania, Slovakia and Slovenia rose in 1996 due to the combination of increasing debt and weaker exports. This was also the case in The FYR of Macedonia, although its debt burden is higher.

In Hungary and the Czech Republic there has been a marked shift from public towards private debt, the absolute amount of public obligations actually falling (table 3.6.8). Enterprises in both countries have borrowed heavily, taking advantage of lower interest rates abroad. In the Czech Republic, private borrowing has exceeded the reduction in government debt and the total debt has risen. In contrast, in Hungary, repayments of public debt (mostly of the National Bank of Hungary) have exceeded net borrowing by business and the total gross debt fell by $4 billion in 1996. Since Hungary's current account deficit was fully covered by foreign investment, there was no net flow of debt creating funds. In consequence the strong upward trend in its indebtedness was reversed.

The debt statistics of Croatia and The FYR of Macedonia (table 3.6.6) have only recently incorporated their respective shares of the so-called "unallocated" debt of the former SFR of Yugoslavia to commercial banks (some $5.6 billion), along with the "allocated" debt.[370] Croatia took this step in July 1996 raising the (total) debt then to $4.5 billion.[371] The 1996 debt statistic for The FYR of Macedonia includes its share of the unallocated debt, but data for the previous years do not (as is also the case for Croatia).[372] However, the debt figure published by Slovenia ($4.0 billion) at the end of 1996 does not yet include the $1.1 billion it has assumed under this procedure.[373] The unallocated debts of Bosnia and Herzegovina and Yugoslavia will be

restructured Poland's official obligations. In 1995, some $480 million of Albania's debt was written off as part of the debt restructuring accord concluded with commercial bank creditors.

[369] Attention is drawn to the revision in the external debt statistics of Bulgaria, which resulted in higher estimates of debt since 1991. Bulgarian National Bank, *Monthly Bulletin* (Sofia), December 1996.

[370] This is the debt of the former SFR of Yugoslavia which was directly attributable to legal entities based on the territories of the individual republics and autonomous provinces.

[371] National Bank of Croatia, *Bulletin*, No. 8 (Zagreb), October 1996.

[372] Bank of The FYR of Macedonia, direct communication.

[373] The central bank notes that the country is currently in the process of ratifying the agreement under which it assumes this amount of the "unallocated" debt. Bank of Slovenia, *Monthly Bulletin* (Ljubljana), January 1997, p. 50.

TABLE 3.6.8

Czech Republic and Hungary: foreign liabilities by type of debtor, 1992-1996
(Million dollars)

	1992	1993	1994	1995	1996[a]
Czech Republic					
Public	3.3	3.9	3.0	2.8	2.1
Government	2.1	2.0	2.2	2.0	1.7
National Bank	1.2	2.0	0.8	0.8	0.4
Private	1.9	2.5	4.8	8.7	10.9
Banks	0.5	0.5	1.0	3.6	5.1
Enterprises	1.4	2.0	3.8	5.1	5.7
Total	5.3	6.5	7.8	11.5	12.9
Hungary					
Public	..	20.2	22.2	22.7	18.3
Government	..	2.0	2.2	1.9	2.1
National Bank	..	18.2	20.0	20.8	16.2
Private	..	2.4	3.9	5.7	6.0
Banks	..	0.9	1.2	1.5	1.2
Enterprises	..	1.5	2.7	4.3	4.8
Total	..	22.6	26.1	28.5	24.3

Source: Czech National Bank and National Bank of Hungary.
Note: Medium- and long-term debt only.
[a] For Czech Republic end-September.

determined in future negotiations with commercial bank creditors, after they reach agreement with the Paris Club on bilateral obligations.

Capital inflows into the *Baltic states* in the first three quarters (see below) were sufficient to fully finance their current account deficits and boost *foreign exchange reserves* (table 3.6.5). Except for Latvia the increases were not large, and due to the relatively rapid growth of imports of goods and services, the reserve-import ratios weakened. Although import coverage remains short of the recommended three months, all three countries are very lightly indebted. On a net basis Estonia and Latvia remain creditors vis-à-vis the rest of the world, in both cases official reserves exceeding *gross debt* (table 3.6.6).

(b) European CIS

Current account developments

The trade in *goods and services* of the European CIS countries continued to expand in 1996 but, with the exception of Ukraine, at a slower rate. There was a tendency for the pace to slacken as the year progressed, chiefly in goods trade (tables 3.6.3 and 3.6.4). The goods and service trade of *Russia* expanded considerably less, exports and imports rising by 6 and 2 per cent respectively.

Current account balances, however, moved in different directions (table 3.6.1), the surplus in Russia increasing, as did the deficits of Belarus and the Republic of Moldova. The deficit of Ukraine declined. In Belarus and the Republic of Moldova current account deficits have been persistently high in relation to GDP for several years.

The current account surplus of the *Russian Federation,* which had fallen in the first half of 1996, rebounded in the third quarter of 1996 resulting in a $10 billion surplus for the first nine months of the year (compared with $8.6 billion in the same period in 1995). Since the merchandise trade surplus remained roughly unchanged, the improvement was due primarily to much lower expenditures on foreign travel. In 1995 this item had swelled to nearly $12 billion (implying a net outflow of $7.2 billion) becoming a significant influence on the Russian balance of payments.[374] The reasons for the recent decline in such expenditures are uncertain, but they may be related to falls in real incomes, which may have also affected the profitability of shuttle trade. The limitations placed on the value of goods which can be imported in conjunction with foreign tourism, which were to go into effect in August 1996,[375] are unlikely to have influenced these figures. Outflows of net property income (chiefly net interest due) have increased, presumably because of rising net external debt and rapidly growing interest payments on local currency instruments.[376]

In *Belarus,* the current account deficit widened considerably because of the strong growth of merchandise imports. However, the larger deficit on goods was partially offset by a 61 per cent increase in service exports. Belarus now earns a sizeable surplus on services, but transfers are modest. In the *Republic of Moldova* the growth of exports of goods and services slowed to 12 per cent, but this still exceeded that of imports. However, the merchandise trade deficit widened and more than offset a smaller services deficit. The latter was due primarily to sharply lower transport expenditures and a small increase in receipts.

The current account balance of *Ukraine* improved considerably in the first nine months of 1996, the deficit stemming entirely from developments in the first quarter of the year. The growth of the merchandise trade deficit and net income payments abroad was less than the doubling of net service receipts. Export growth recovered strongly from the stagnation of 1995. Of all the transition economies, Ukraine has the largest surplus on services,

---

[374] This spending also includes some expenses incurred in the so-called "shuttle trade".

[375] A limitation of $2,000 per person per trip abroad was to be implemented on 1 August 1996. Enforcement of this regulation would have cut tourism related imports, but tourism expenditures had already fallen in the second quarter of the year.

[376] In accordance with IMF recommendations, Russia records interest due in its current account statistics, but the actual flow of cash is uncertain. Russia reports foreign assets of some $140 billion, chiefly claims on highly indebted countries, which implies considerable interest obligations on their part. However, only a small part of these obligations is being met by the debtors.

amounting to $2.4 billion in the first three quarters of 1996. Transit fees from natural gas pipelines appear to provide the bulk of these earnings. Ukraine receives various types of international assistance which are reflected in relatively high transfers (these are likely to exceed over $400 million for all of 1996).

External debt and reserves

The *total external debt* of the Russian Federation is reported to have fallen from $122 billion in 1994 to $120 billion in 1995 (table 3.6.6) and was projected to increase to $127 billion at the end of 1996.[377] The decline in the debt in 1995 is attributed to reductions in obligations to CMEA countries from $28 billion in 1994 to $17.4 billion in 1995 (this was part of the debt of the former USSR). Meanwhile, Russia's (new) debt has increased (to $25 billion in 1996), mainly because of borrowing from the multilateral institutions, but its debt burden (as measured by the *debt-exports ratio*) is moderate and has declined (table 3.6.7). The debt burdens of Belarus, the Republic of Moldova[378] and Ukraine are comparatively low and have fallen; the debt of Belarus has also fallen in absolute terms.

After reaching over $14 billion at the end of 1995, the *foreign exchange reserves* of the Russian Federation fell to about to $11.2 billion at end-1996 (table 3.6.5). Bouts of political uncertainty during the course of the year put the rouble under pressure, necessitating central bank intervention.[379] However, Russian officials consider the recent level of reserves to be adequate. The overall loss of reserves despite a current account surplus, $5.7 billion in IMF loans, various bilateral credits, and foreign investment in domestic securities (see below) suggests sizeable currency substitution (capital flight) by residents.[380] The reserves of Belarus and Ukraine increased in 1996, but modestly, and import coverage ratios still fall short of the recommended three months. However, the official reserves of the Republic of Moldova have exceeded this level, thanks to loans from multilateral institutions and the placement of a $30 million floating rate note.

(c) The outlook for current accounts in 1997

Although prospects vary considerably within the region, in general, the large current account deficits are expected to persist and even worsen. Various forecasts suggest that the combined current account deficit of *eastern Europe* could rise to some $18-$20 billion in 1997, chiefly because of current expectations regarding Poland and the Czech Republic (table 3.6.1).[381] Official forecasts of GDP for 1997 point to a continuation of recent growth rates of output and domestic demand. Under current policy stances, further real exchange rate appreciation and trade liberalization can be expected in a number of countries (see section 3.5). Although a pick-up in western import demand (section 3.5) would support eastern export growth in 1997, it may not be as strong as some countries have assumed in their forecasts. It should be recalled that the deterioration of the current accounts in a number of countries quickened as the year progressed and in some there was a sharp and unexpected deterioration in December 1996 (e.g. Slovakia, Romania, and Hungary). The possible implications of these recent developments have not yet been assessed, but they underline the potential difficulties of bringing the current account imbalances under control in some countries.

The rapid expansion of the current account deficit of the *Czech Republic* has caused increasing official concern.[382] According to official forecasts, the economy is expected to grow by some 4 per cent in 1997. Independent forecasts show the trade deficit rising to CKR 195-220 (some $8-$9 billion),[383] implying a current account deficit of some $6.5-$7.5 billion or over 10 per cent of GDP. It is not clear to what extent these estimates take into account the impact of the tightening of monetary policy during 1996, but they do exclude the effect of the depreciation of the koruna (within the ± 7.5 per cent fluctuation band) in March 1997. Official pronouncements have been strongly opposed to any devaluation of the exchange rate (i.e. the central reference rate). An external imbalance of recent proportions can probably be financed for some time given the expected buoyancy of foreign investment, the large scope for further foreign borrowing, and large international reserves.

---

[377] Ministry of Finance estimates. It might be noted that some higher figures for the Russian debt have been mentioned. In early 1995 the debt was projected to rise to $130 billion by the end of the year. Subsequently, an estimate of $129.4 billion for end-1995 was published in April 1996, of which $27 billion were obligations to the former CMEA countries. (According to A. Illarionov, Director of the Institute of Economics. *BIKI*, 9 April 1996). In the second half of 1996, total debt was projected to increase to $136.2 billion by the end of the year, according to the Budget Committee of the State Duma.

[378] The debt statistics for the Republic of Moldova have been revised upwards.

[379] According to Y. Yasin, Minister of Economics. *AP-Dow Jones News Service*, 10 September 1996.

[380] In the first three quarters of 1996, the errors and omissions item of the balance of payments amounted to some $7 billion.

[381] Official current account forecasts have not been published by many transition economies. Even where they have, some were made in 1996 and thus may not adequately reflect recent developments. Independent forecasts are also noted in table 3.6.1.

[382] At the recent Zofin Forum in Prague, the Czech Prime Minister, V. Klaus, said that foreign trade may well be the largest problem facing the Czech economy. *Hospodarske noviny* (Prague), 12 March 1997.

[383] The range of forecasts of the trade deficit is from *Svet hospodarstvi* (Prague), 20 February 1997. The conversion into dollars assumes an exchange rate of CZK 25/$.

A current account deficit of around $4 billion has been officially forecast for *Poland* in 1997. However, more recent, independent forecasts suggest a deficit of $5.5-$8.0 billion (some 4-6 per cent of GDP). The offset to the widening merchandise trade deficit provided by the surplus on "non-classified current account transactions" (some $7 billion) is expected to continue to diminish as the large price differential between Polish goods and services and those of its neighbours declines. Increased financial inflows are likely in the form of FDI and the large foreign exchange reserves can be tapped. Official forecasts for *Hungary* show some pick-up in economic growth and a current account deficit of some $1.5 billion.[384] However, recent policy changes and the growth of the trade deficit in January 1996 suggest a current account deficit larger than the $1.7 billion posted in 1996. It should be easily financed, largely by inflows of foreign investment, and foreign exchange reserves should remain high.

In *Slovakia* there was a sharp deterioration of the current account at the end of 1996. This, and the economic policies currently in place, suggest that the current account deficit in 1997 will exceed that in 1996. Cutting the trade deficit is one of the government's main priorities for 1997. However, aside from plans to step up export promotion, including financing through the new Eximbank, no coherent strategy to achieve this has been announced. The tight monetary policy is expected to continue and domestic growth is likely to slow, but no devaluation seems be under consideration.[385] Moreover, the remaining 7.5 per cent import surcharge was lifted at the beginning of 1997. Nearly half of the current account deficit in 1996 was financed by short-term credits while foreign direct portfolio investment has remained low (see below). However, Slovakia's external debt is modest and the country's access to financial markets has improved with the upgrading of its credit rating. Foreign currency reserves remain more than adequate.

According to official forecasts, the current account deficit of *Croatia* will increase marginally in 1997. Some widening of the merchandise trade deficit is expected to be largely offset by a further marked increase in revenues from tourism.[386] The imminent signing of an extended arrangement will make some $160 million available from the IMF and trigger funds from other multilateral institutions. The recent award of an international credit rating has improved the country's access to long-term capital and FDI has shown signs of rapid growth. Overall, these funds should further reduce the country's dependence on unrecorded capital inflows and short-term financing. Reserves are expected to rise by some $400 million.

The *Slovene* authorities expect exports of goods and services to pick up in 1997, in line with their expectations of stronger import demand in western Europe. Somewhat more buoyant domestic economic growth should boost imports, but the current account is expected to remain roughly in balance. Accordingly not much change is expected in the country's favourable external financial situation.

The FYR of Macedonia has run substantial current account deficits for several years now and has been heavily dependent on official assistance. In 1997 the balance of payments financing gap[387] is expected to be about $85 million which should be fully covered by the multinational institutions and bilateral assistance, including some $46 million from the European Union.[388] The financial package, which was approved by a donors' conference in February, also provides for some increase in foreign exchange reserves. As noted below, The FYR of Macedonia hopes to enter the euromarkets in 1997.

Although *Bulgaria* has elaborated a new economic stabilization programme, the current account target for 1997 has not been released. Given recent external constraints, a balanced or (small) surplus position would seem to be likely (as was the case in 1996). However, prospects of significant foreign capital inflows from privatization sales, a G-24 assistance programme (including up to $400 million from the EU), and implementation of the IMF arrangement agreed in mid-March 1997, would permit a modest current account deficit. The IMF package consists of a $510 million stand-by credit and a $148 million loan to finance grain imports.[389] Signing this agreement would also trigger the release of $290 million in World Bank funds, for emergency assistance[390] and structural reforms. For their part, the Bulgarian authorities have undertaken to speed up the privatization of state owned enterprises and banks and to establish a currency board to restore confidence in the currency (see section 3.1). This new commitment to structural change should boost the chances of the government being able to privatize selected major enterprises and meet the $1 billion target for 1997 (see

---

[384] The forecasts appear to incorporate a reduction in the rate of downward crawl of the forint and the removal of the remaining import surcharge in stages through the first half of 1997.

[385] According to Mr. Masar, Governor of the National Bank of Slovakia, citing the country's high level of reserves, 11 November 1996.

[386] Croatia's merchandise trade balance improved in early 1997: if this continues the outcome would be more favourable than current expectations.

[387] The sum of the current account (deficit), private capital inflows and target for increase in official reserves.

[388] OMRI, *Daily Digest*, 19 February 1997. $30 million of the EU loan will be used to pay off the country's debt to the European Investment Bank.

[389] *Financial Times*, 18 March 1997.

[390] The EU is also considering extending some $23 million to finance emergency food imports.

below).[391] Bulgaria has to meet nearly $1 billion in debt service obligations in 1997,[392] which has prompted the government to request Paris Club members to reschedule part of the repayments of official debt coming due.[393] Reserves need to be rebuilt from their current low levels. In late January 1997, the EU Commission offered to organize international assistance for Bulgaria within the framework of the G-24.[394] Unofficial estimates suggest that up to $500 million would be sought for balance payments support.[395]

The government of *Romania* has announced a target current account deficit of $1.4 billion for 1997, almost $1 billion less than in 1996. Export growth should recover given the large depreciation of the real exchange rate and the recent recovery in industrial production. Repayments of principal coming due are estimated at $1.5-$2.0 billion. Thus the country's financing requirement should be broadly covered by the $3.2 billion in medium- and long-term official borrowing (i.e. by the national bank and government) authorized by the parliament.[396] However, approval of the government's economic programme by the IMF and World Bank will be essential. If it is forthcoming, an IMF stand-by loan of some $400-$500 million and World Bank funds of $500 million would become available. A vote of confidence by these institutions is important for the country to be able to maintain access to the international financial markets and attract the larger amounts of FDI that the government is hoping for (see below). The programme also aims at a boost in foreign exchange reserves in 1997.

The developments in 1996 and the current outlook for 1997 raise the related issues of the sustainability of the current account deficits in the region and the possible constraints they could impose on economic growth. In 1996 more than half of the east European transition economies posted large current account deficits, of 5 per cent or more of GDP. In some cases, these have persisted for several years, and are likely to continue in 1997. According to the conventional wisdom, persistent deficits of this magnitude over a period of 3 to 4 years are usually seen as a warning signal of a potential balance of payments crisis, particularly if the deficits are financed by short-term or other volatile funds, or by depleting foreign currency reserves. In such circumstances, governments are generally recommended to make adjustments in order to avoid the "hard landing"[397] which could be precipitated by a loss of confidence by foreign investors.

In reality the relationship between current account deficits and crises is not so straightforward. Deficits are considered acceptable, indeed desirable, if they are accompanied by higher domestic investment, particularly in export industries. In this way, a country has a better chance of achieving higher growth and remaining solvent. A recent study suggests that in addition to the maintenance of solvency (i.e. the ability to pay), the actual occurrence of a crisis also depends on the country's willingness to service its debt[398] and on the willingness of foreign investors to continue to lend on current terms.[399] The current account imbalance also needs to be considered in conjunction with exchange rate policy, the external financial position and various structural factors, including the degree of openness of the economy, the levels of domestic savings and investment, and the health of the financial system (see chapter 1).

The second, related issue is to what extent the balance of payments is likely to be a constraint on growth in the transition economies. In the extreme case of a lack of external finance (e.g. Bulgaria), imports fall and the current account moves into balance (or even surplus), with adverse implications for output. Another situation is that of several countries with large (unsustainable) current account deficits but which, however, dispose of adequate financing for at least the next year or so (private capital has flowed in, foreign currency reserves are high and there is scope for increasing external debt). But even in these cases, the deficits have evoked official concern, leading to the implementation, or at least the consideration, of restrictive policy measures.

---

[391] The sale of a chemical plant for $160 million (see below) has been completed and the transfer of funds, which are to boost foreign currency reserves, is reported to be imminent.

[392] According to the Bulgarian National Bank. Bulgaria successfully met payments of $130 million on Brady bond debt in January, but at the expense of a further depletion of official reserves.

[393] The authorities are reported to have requested the rescheduling of $50 million of the total 1996 repayment of $126 million. *Financial Times*, 3 March 1997.

[394] *Europe* (Brussels), 31 January 1997. Among the conditions for granting the assistance is the requirement that local political groups reach a political consensus which would allow Bulgaria to achieve a certain degree of stability (preferably by organizing early elections). Furthermore, Bulgaria was expected to reach agreement with the international financial institutions on a stabilization programme and with its creditors on debt servicing. The European Commission has been the coordinator of G-24 assistance to a group of transition economies. The G-24 assistance programme has been discussed in past issues of this *Survey*, as well as in the *Economic Bulletin for Europe*.

[395] *Financial Times*, 31 January 1997.

[396] *BBC Summary of World Broadcasts*, EEW/0469 WB/6, 16 January 1997.

[397] That is, a sharp, perhaps excessive, depreciation of the exchange rate and a rapid and costly adjustment.

[398] Adoption of the necessary adjustment measures may not be politically feasible.

[399] G. Milesi-Ferretti and A. Razin, "Current account sustainability", *Princeton Studies in International Finance*, No. 81, October 1996. In a study of seven countries, five were able to sustain large current account deficits for several years, but two were unable to do so and suffered severe external crises.

TABLE 3.6.9

Net capital flows into eastern Europe, the Baltic countries and the European members of the CIS countries, 1993-1996
(Billion dollars and per cent)

	Capital account flows [a]			Jan.-Sept.			Share of GDP			Jan.-Sept.		
	1993	1994	1995	1995	1996	1996	1993	1994	1995	1995	1996	1996
Eastern Europe	13.7[b]	8.7[b]	24.0	14.8	9.4	15.4	5.9	3.4	7.5	6.2	3.8	4.6
Albania	0.1	0.1	–	–	0.1	0.1[c]	10.4	6.8	1.8	-0.5	4.9	4.9[c]
Bulgaria	0.9	0.4	0.3	0.3	-0.7	-0.7[c]	7.9	3.9	2.0	3.4	-10.2	-10.0[c]
Croatia	0.3	0.7	2.2	1.7	1.1	1.7	2.9	4.9	12.2	12.5	8.2	9.2
Czech Republic	2.9	2.4	8.8	5.9	2.3	3.7	9.3	6.7	18.7	16.7	5.9	7.1
Hungary	6.1	3.3	7.0	3.5	0.1	0.2	15.8	7.8	16.0	10.7	0.2	0.5
Poland	1.6[b]	-0.1[b]	2.7	2.1	3.5	5.3	2.0	-0.1	2.3	2.4	3.6	4.0
Romania	1.1	1.0	1.4	0.8	0.9	2.6	4.2	3.5	3.9	2.8	3.3	7.2
Slovakia	0.6	0.6	0.9	0.3	1.4	1.6	5.1	4.2	5.3	2.5	9.9	8.3
Slovenia	-0.1	0.1	0.3	–	0.5	0.5	-0.6	0.7	1.4	0.1	3.5	3.0
The FYR of Macedonia	–	0.2	0.3	0.2	0.3	0.5	1.8	7.2	9.1	5.8	12.2	13.7
Baltic states	0.5	0.4	1.1	0.6	1.1	..	7.2	3.8	8.0	6.1	8.6	..
Estonia	0.1	0.2	0.3	0.2	0.3	..	7.4	9.1	8.0	7.1	8.5	..
Latvia	–	-0.1	–	-0.1	0.4	..	-2.1	-2.7	-0.2	-2.3	10.1	..
Lithuania	0.4	0.3	0.8	0.5	0.4	..	15.1	6.4	14.2	11.7	7.7	..
CIS	1.6	2.4	2.4	1.9	1.6	..	4.6	5.5	5.0	5.3	3.8	..
Belarus	–	0.6	0.7	0.5	0.8	..	–	12.1	6.5	7.0	7.6	..
Republic of Moldova	0.1	0.1	0.1	0.1	0.1	..	6.6	8.1	7.6	8.5	9.4	..
Ukraine	1.5	1.7	1.6	1.3	0.8	..	4.9	4.5	4.5	4.6	2.4	..

Source: UN/ECE secretariat estimates, based on national balance of payments statistics.

[a] Including errors and omissions.
[b] Excluding for Poland non-classified current account items estimated at $1 billion and $1.7 billion in 1993 and 1994, respectively.
[c] January-September.

Curbing the growth of current account imbalances in 1997 is likely to require cuts in the growth of domestic demand, below the rates officially forecast. Indeed various independent forecasts show growth rates which are lower than the official forecasts or below those of 1996. It should be borne in mind that even recent rates of GDP growth in eastern Europe – 3.9 per cent on average in 1996 – are not particularly high, especially if compared with the performance of the more dynamic emerging market economies. Current account considerations might therefore limit the region to relatively modest growth rates which are below the expectations of the population and short of what is required to substantially narrow the income gap between them and their western trade partners.[400] An important aspect of the problem (in eastern Europe at least) is that exports of goods and services have become sluggish while the growth of imports has remained buoyant. In 1996, this was due partly to the slowdown of import demand in western Europe. However, more fundamental factors are also at work: a slow rate of restructuring, declining competitiveness, and high dependence on the markets of western Europe, which is currently a relatively slow growth area.

(iii) External financing

(a) Net financial flows to the transition economies

In 1996 the flow of net capital imports into *eastern Europe* slowed to $15 billion after reaching a record $24 billion in 1995 (tables 3.6.9 and 3.6.10).[401] The change reflects the virtual elimination of the flow into Hungary ($7 billion in 1995) and a large decline ($5 billion) in capital imports into the Czech Republic. In Croatia, there was also a decline, albeit small, but in Bulgaria, a small inflow in 1995 became a sizeable outflow in 1996 (amounting to nearly 10 per cent of GDP).[402] In the other countries capital imports increased, above all in Poland where they amounted to over $5 billion. Poland has now moved ahead of Hungary and the Czech Republic as the principal destination for new flows of foreign capital. Compared with GDP, however, inflows were particularly large in The FYR of Macedonia, Croatia and Slovakia.

The *composition of capital flows* changed in 1996. Foreign direct investment continued to lead the way, its share of net inflows into eastern Europe rising to almost

---

[400] This was already the case in Hungary in 1994-1996 and Bulgaria in 1996 (among others).

[401] These figures include estimates of unrecorded capital movements (measured by "errors and omissions" in the balance of payments).

[402] The reasons for this turn around are discussed above.

TABLE 3.6.10

Net capital flows into eastern Europe, the Baltic countries and the European members of the CIS, by type of capital, 1994-1996

(Billion dollars)

	Eastern Europe					Baltic states				European CIS[a]			
			Jan. -Sept.					Jan. -Sept.				Jan. -Sept.	
	1994	1995	1995	1996	1996[b]	1994	1995	1995	1996	1994	1995	1995	1996
Capital account	9.0[c]	21.5	13.7	7.5	13.1	0.9	1.5	1.0	0.7	2.1	2.2	1.2	1.1
Capital account[d]	8.7[c]	24.0	14.8	9.4	15.4	0.4	1.1	0.6	1.1	2.4	2.4	1.9	1.6
of which:													
FDI	3.3	9.1	4.1	4.8	7.1	0.5	0.5	0.4	0.3	0.2	0.3	0.2	0.4
Portfolio	3.0	4.7	3.5	0.7	0.6	–	–	–	-0.1	–	–	–	0.1
Medium-, long-term funds	2.0	5.4	3.4	0.9	3.4	0.4[e]	0.5	0.3	0.4	1.2	–	-0.2	0.2
Short-term funds	0.9[c]	4.5	4.2	1.3	1.8	..	0.6	0.4	0.1	0.3	0.4	0.2	0.1
Errors and omissions	-0.2	2.4	1.1	1.9	2.3	-0.5	-0.4	-0.3	0.4	0.3	0.2	0.7	0.6

Source: UN/ECE secretariat estimates, based on national statistics.

[a] Belarus, Republic of Moldova and Ukraine.
[b] January-September for Albania and Bulgaria; January-November for Slovakia.
[c] Excluding for Poland non-classified current account items estimated at $1.7 billion in 1994.
[d] Including errors and omissions and excluding valuation effects for Poland.
[e] Includes short-term funds.

one half (table 3.6.10) of total inflows. However, medium-term and, especially, portfolio flows diminished. Short-term and unrecorded flows continued to be an important source of funds in 1996.

In the first half of the 1990s *Hungary* generally attracted the most capital in eastern Europe.[403] External borrowing and inflows of FDI financed a large current account deficit and increases in official reserves. The elimination of a $7 billion surplus on capital account in 1996 partly reflects a drop in privatization revenues, which reduced FDI from $4.4 billion in 1995 to $2 billion, and a shift in portfolio investment from a significant inflow to an outflow of nearly $1 billion, the turn around being associated with a smaller current account deficit and repayment of sovereign debt (both external bonds and loans; table 3.6.8). As noted above, these net repayments of public debt conceal an increase in borrowing by enterprises. The inflow of short-term funds has slowed, but it remains significant.

In the *Czech Republic*, net capital imports fell in 1996 by over one half to $3.6 billion. Inflows were particularly small early in the year, but they gradually accelerated reaching $2 billion in the final quarter of 1996. A large part of the overall decline was due to a $2 billion reversal of the inflow of short-term funds (to a $0.8 billion outflow). The particular combination of Czech monetary and exchange rate policies had attracted large inflows of short-term funds in 1994 and 1995. However, foreign investors withdrew deposits following the introduction of a wider exchange rate band at the end of February 1996, a measure which was intended to increase exchange rate risk and discourage speculation on the koruna. (Later in the year, there was some recovery of short-term inflows.) FDI also declined substantially in 1996, inflows in 1995 having been boosted by exceptional privatization sales (see below). The comparatively large long-term borrowing by Czech banks and corporations remained roughly unchanged.

In *Bulgaria*, the reversal of flows was due chiefly to residents' withdrawals of over $600 million from deposit accounts as a result of mounting concern about the country's banking system and external finances. Net long-term borrowing was also negative due to the repayment of loans, including some to the IMF. The country's small FDI inflows declined further and residents' investment abroad led to a sizeable outflow of portfolio investment.[404] However, Bulgaria attracted some short-term capital in the second quarter of the year due to the emergence of positive interest rate differentials. The decline in *Croatia's* capital imports was due to smaller net medium- and long-term borrowing and, especially, a reduction in unrecorded flows.[405] Nevertheless, the latter remained the primary source of external financing. Short-term funds, also an important source of finance, flowed out in the first half of the year, but they rebounded in the second half. FDI boomed in the third quarter of 1996 and is estimated to have surpassed $300 million for the full year.

---

[403] See table 3.6.18 for changes in the composition of capital account flows of the individual transition economies.

[404] Currently only outflows of portfolio investment are recorded (because they require prior approval), and presumably inflows have been small (if there have been any at all).

[405] In Croatia, unrecorded capital flows (errors and omissions) reflect a variety of sources including trade credits under three months, earnings from transport, workers' remittances, unrecorded use of foreign accounts and drawing of foreign interest income, etc.

In contrast, the other transition economies received more foreign capital in 1996 than in 1995. In *Poland*, FDI led the way, more than doubling to $2.7 billion, and net inflows of short-term funds recovered, to over $2 billion. Also, net repayments of long-term debt fell by about $1 billion. However, portfolio investment, which had risen to over $1 billion in 1995, fell to virtually zero in 1996 (despite external bond issues of $500 million). This suggests a fall in net foreign sales of domestic securities, presumably of T-bills, in response to the fall in yields. Nearly $1 billion in short-term funds accounted for much of the surge in capital flows into *Slovakia* in 1996. Until 1995 this item had been negative. The country also attracted somewhat higher amounts of long-term funds, about equally split between new liabilities and the return of assets held abroad. However, foreign direct investment remained weak.

In *Romania*, larger inflows of long-term funds (despite repayments of IMF debt) reflect the country's re-entry into the international capital markets and completion of the government's borrowing programme (see below).[406] However, inflows of short-term funds and FDI fell, and outflows of unrecorded capital, already rising in 1995, increased further, presumably because of the unsettled domestic financial situation.[407] A surge in portfolio investment, including nearly $360 million in external bond issues, and marginally higher FDI flows underpinned the growth of capital imports into *Slovenia*. However, new long-term loans fell sharply and there was a further substantial outflow of short-term funds. In *The FYR of Macedonia*, new loans obtained by the government and, to a lesser extent, some increase in portfolio flows, accounted for most of the increase in capital imports.

Of the *Baltic states*, Latvia and Estonia attracted larger foreign capital flows in the first three quarters of 1996, and in all three countries the flows amounted to some 8-10 per cent of GDP. In *Latvia*, the surge was mainly due to unrecorded capital, which changed from a significant outflow in 1995 to a sizeable inflow in 1996. Latvia also raised more foreign loans, but these were fully offset by a surge in portfolio investment abroad. FDI remained buoyant, being only slightly lower than in 1995. In *Estonia*, short-term funds and, to a lesser extent, unrecorded capital flows, increased. Portfolio investment turned positive, possibly because of foreign investment in the newly inaugurated stock exchange, but FDI fell sharply. In *Lithuania*, the combined inflows of foreign direct and portfolio investment more than doubled. Long-term borrowing remained buoyant, but short-term flows reversed, from a sizeable inflow to an outflow. There was also a fall in the inflow of unrecorded capital.

In the first three quarters of 1996, foreign capital flows in Belarus and the Republic of Moldova strengthened, to some 8-9 per cent of GDP. Unrecorded and short-term capital continue to dominate the flows into *Belarus*, long-term capital flows remaining small and FDI negligible. FDI, however, remained the leading source of capital in the *Republic of Moldova*, but the inflow of medium- and long-term funds increased as well. In *Ukraine*, which has attracted comparatively less foreign capital, inflows declined in 1996. Net medium- and long-term inflows fell (as did unrecorded capital inflows), and short-term funds left the country. However, FDI rose further and Ukraine started to receive some portfolio investment.

(b) Funds raised on the international financial markets

Medium- and long-term debt

The volume (gross) of medium- and long-term funds raised by the transition economies in the international markets continued to rise, reaching $9.8 billion in 1996 (table 3.6.11). Several countries which had gained access to the market within the past year accounted for a sizeable part of this. In the first three months of 1997, preliminary data indicate that their borrowing amounted to $4.4 billion, considerably above the pace of 1996. Most of this was accounted for by the Russian Federation.

Due to relatively high yields, demand continued to be strong for emerging market debt. The transition economies have become more familiar to international investors, particularly since many of them have recently received favourable credit ratings. The transition economies have also benefited from the global diversification of portfolios, which has been reflected in the growth of regional mutual funds. However, the supply of new emerging market debt has been limited. In the case of the transition economies, those with the largest borrowing potential – the Czech Republic, Hungary, and Poland – have had little need of additional funds, all enjoying various other forms of capital inflow and able to draw on large official reserves (see above). This situation has resulted in the oversubscription of new issues and better terms for the most creditworthy issuers.

In 1996, the Czech Republic and Hungary each raised somewhat over $2 billion. This is Hungary's lowest level of borrowing in several years, but for the Czech Republic it represents a doubling of new funds from 1995. Borrowing by Romania and Slovakia increased sharply, that of the former reaching the $1.4 billion target set by the government. Funds raised by

---

[406] Note that the Romanian balance of payments data include bond issues ($1,025 million in 1996) in long-term capital rather than in portfolio investment.

[407] The Romanian government estimates that unauthorized holdings abroad of funds amount to some $2.5 billion, and it intends to encourage their repatriation.

TABLE 3.6.11

Medium- and long-term funds raised on the international financial markets by eastern Europe, the Baltic countries and the CIS, 1992-1997

(Million dollars)

	1992	1993	1994	1995	1996 Jan.-Dec.	1996 of which: Bonds	1996 of which: Loans	1997 Jan.-Mar.[a]	1997 of which: Bonds	1997 of which: Loans
Eastern Europe	1 494	6 314	3 587	6 483	8 266	3 033	5 233	565	375	190
Bulgaria	–	–	–	–	–	–	–	–	–	–
Croatia	..	–	–	60	317	–	317	300	300	–
Czech Republic	..	903	638	1 000	2 191	546	1 645	–	–	–
Hungary	1 446	5 071	2 541	4 178	2 108	326	1 782	–	–	–
Poland	9	–	3	324	526	500	26	190	–	190
Romania	–	–	–	268	1 400	1 025	375	75	75	–
Slovakia	..	240	331	427	1 130	280	850	–	–	–
Slovenia	–	100	75	226	594	356	238	–	–	–
Former Czechoslovakia	40	..	..	..	..	..	..	..	..	..
Baltic states	–	–	–	101	189	89	100	–	–	–
Estonia	–	–	–	–	64	39	25	–	–	–
Latvia	–	–	–	41	–	–	–	–	–	–
Lithuania	–	–	–	60	125	50	75	–	–	–
CIS	–	28	75	1 345	1 315	1 230	85	3 795	1 180	2 615[b]
Georgia	–	20	–	–	–	–	–	–	–	–
Kazakstan	–	–	–	–	200	200	–	–	–	–
Kyrgyzstan	–	–	–	140	–	–	–	–	–	–
Republic of Moldova	–	–	–	–	30	30[c]	–	–	–	–
Russian Federation	–	8	75	1 205[d]	1 085	1 000	85	3 795	1 180	2 615[b]
Total above	1 494	6 341	3 662	7 929	9 770	4 352	5 418	4 360	1 555	2 805
of which:										
Bonds	1 250	5 751	2 445	4 120	4 352	4 352	..	1 555	1 555	..
Bank loans[e]	244	590	1 217	3 809	5 418	..	5 418	2 805	..	2 805
Memorandum item:										
Eastern Europe, Baltic and CIS countries' share of funds raised globally *(per cent)*[f]	0.33	1.01	0.55	0.94	0.92	0.61	1.57	..	..	..

Source: UN/ECE secretariat based on press reports; OECD, *Financial Statistics Monthly*, Part I (Paris), January 1997 and previous issues.

Note: Funds are recorded as of the date on which the deal was signed.

[a] To 15 March, preliminary.
[b] Includes a $2,500 million loan to Gazprom and a $115 million loan to JSC Nizhnekamsbneftekhim of the Republic of Tatarstan.
[c] Floating rate note, private placement.
[d] Includes convertible bonds converted to equity in April 1996.
[e] International bank loans in Eurocurrencies and in domestic currency of lending countries, excluding guaranteed loans and rescheduled debt.
[f] As a share of bonds, syndicated loans and other debt facilities.

Croatia, Lithuania and Slovenia rose sharply although the absolute amounts were modest. Despite Russia's large debut bond issue (see below), the country's overall borrowing fell in 1996.[408] This was reversed in early 1997, the amount raised surpassing all of its previous borrowing. Overall, 12 transition economies obtained funds on the international capital markets in 1996 and early 1997, about the same as in 1995. However, the number of potential borrowers has increased. Estonia and Kazakstan entered the market for the first time, and debut bond issues were launched by Romania, Russia, Slovenia and Croatia. These initiatives by sovereign entities have cleared the way for other potential resident borrowers.

Syndicated loans raised by the transition economies rose to $5.4 billion in 1996, placing them ahead of bonds (the volume of which also increased) for the first time (table 3.6.11). The Czech Republic and Hungary account for most of the growth in new loans, in the latter case substituting for bond financing. In early 1997, new lending was dominated by a $2.5 billion syndicated loan to Russia. The $4.4 billion raised through *bonds* in 1996 includes issues by most countries, in several cases shortly after receiving credit ratings. Russia's $1 billion bond issue, more than twice the amount that the authorities had originally sought, was the largest ever for a new market entrant. The country's second, $1.2 billion deutsche

---

[408] In 1995, the total borrowing of Russia was heavily influenced by a single $885 million loan to Gazprom, the gas utility, to finance its share of certain new gas pipelines in Germany.

TABLE 3.6.12

International credit ratings of eastern Europe, the Baltic countries and the CIS

Standard and Poor's			Moody's		
Rating	Country	Date received	Rating	Country	Date received
Investment grades					
AAA			Aaa		
AA			Aa		
A	Czech Republic	November 1995	A2		
	Slovenia	May 1996			
A-			A3	Slovenia	May 1996
BBB+			Baa1	Czech Republic	September 1995
BBB	Latvia	January 1997	Baa2		
BBB-	Croatia	January 1997	Baa3	Croatia	January 1997
	Hungary	October 1996		Hungary	December 1996
	Poland	April 1996		Poland	June 1995
	Slovakia	April 1996		Slovakia	May 1995
Sub-investment (speculative) grades					
BB+			Ba1		
BB			Ba2	Republic of Moldova	January 1997
				Russian Federation	October 1996
				Lithuania	September 1996
BB-	Romania	March 1996	Ba3	Romania	March 1996
	Russian Federation	October 1996		Kazakstan	November 1996
	Kazakstan	November 1996			
B+			B1		
B			B2		
B-			B3	Bulgaria	September 1996
Memorandum item:					
Recent Municipal Ratings					
BB			Ba2	City of Moscow	February 1997
				Nizhniy Novgorod	February 1997

Source: Press reports.

Note: Foreign currency, long-term, sovereign debt ratings.

marks bond issue in early 1997 was also oversubscribed and accounts for the bulk of the bond issues by the transition economies.

In general the trend towards borrowing by banks and corporations (as opposed to sovereign entities) continued, prompted by the availability of cheaper and longer-term credits than are available on domestic markets. In several cases companies raised funds for the first time, further broadening the range of eastern borrowers. As in 1995, banks accounted for most of the borrowing of the Czech Republic although SPT Telecom obtained a $750 million loan. Banks were also the most active borrowers in Croatia. Enterprises in the energy-fuel, telecommunications and steel sectors dominated the borrowing of Slovakia, while the telecommunications, oil refining and banking sectors accounted for most of the funds raised by Hungary. The $2.5 billion loan obtained by Russia's Gazprom in early 1997[409] exceeded the recent bond issues of the government. The borrowing of Romania and Slovenia, however, was mainly sovereign.

International credit ratings and borrowing terms

The transition economies as a group continued to improve their *international credit standing* in 1996 (table 3.6.12).[410] The generally positive credit assessments reflect a better macroeconomic outlook, progress in economic restructuring and generally favourable external financial positions. The most creditworthy eastern countries have now achieved credit rankings on a par with some western market economies.[411] However, given their prevailing economic and political climate, several transition economies have little prospect of obtaining favourable credit ratings in the near future.

In the 12-month period ending January 1997, nine transition economies – the latest being Croatia, Latvia and the Republic of Moldova – received international

---

[409] The funds are intended to help finance the construction of a gas pipeline from the Yamal Peninsula to western Europe via Belarus and Poland.

[410] *Institutional Investor* (London), October 1996. A survey of selected sovereign borrowers published in October 1996 shows that the average credit rating of 19 east European, Baltic and CIS countries improved further during the preceding six months, and, once again, by more than that of any other regional group of countries.

[411] The Czech Republic, Hungary and Poland have become members of the OECD which has been a factor in obtaining better borrowing terms.

credit ratings for the first time, raising to 13 the number of rated countries.[412] Of the new recipients, Slovenia, Latvia and Croatia were rated investment grade, while Russia, the Republic of Moldova, Lithuania, Romania and Kazakstan received various sub-investment (speculative) ratings.[413] In 1996 Hungary was uprated to investment grade by both major agencies, and Poland and Slovakia were upgraded by one agency. Overall, seven transition economies have obtained the widely sought-after investment grade ratings from the leading agencies, which gives them access to the large resources of certain institutional investors. This is a significant development considering that only Hungary was rated several years ago.[414] Their experience compares very favourably with other groups of emerging market economies.

High liquidity and greater competition in the international markets has led to better *terms* for most borrowers. The transition economies have also benefited from their improved ratings and the relative scarcity of their issues. Those countries with investment grade ratings have obtained narrower margins and longer maturities. For example, within the past year, margins obtained by Hungary have fallen from 185 basis points to around 50 basis points. Czech and Slovene borrowers have paid only 22-25 basis points, terms comparable with those obtained by some large companies in the western market economies. However, yields on the debt of transition economies with sub-investment grade ratings have remained high, contributing to the large oversubscriptions of their offerings. In mid-1996, for example, Romania paid a margin of 280-305 basis points on new bond issues (compared with base interest rates on yen and dollars of some 2.7 and 6.2 per cent, respectively), while Russia paid 345 basis points on its first eurobond in November. Maturities obtained by the transition economies vary, but 2-5 years is common and Hungary has obtained as much as 15 years. Overall, the international markets continue to offer better conditions than domestic markets, where medium- and long-term funds are unlikely to be available at all.

International equity offerings

International private offerings have become more popular among the corporations of the transition

---

[412] Previously only the Czech Republic, Hungary, Poland and Slovakia had credit ratings.

[413] The B3 rating given to Bulgaria (by Moody's), reflecting the country's difficult financial situation, will not enable it to raise new funds.

[414] In the 1980s, Hungary was the first eastern country to receive a (sub-investment) grading. It has enjoyed an investment grade rating from the Japan Bond Research Institute which has facilitated its comparatively large borrowings in yen ($3.8 billion in 1995). It received an investment grade from IBCA early in 1996.

TABLE 3.6.13

International equity issues of eastern Europe, the Baltic countries and the CIS, 1993-1997

(Million dollars)

	1993	1994	1995	1996[a]	1997[b]
Eastern Europe	13	274	644	571	25
Croatia	–	–	–	175	–
Czech Republic	–	26	32	104	–
Hungary	13	247	334	243	–
Poland	–	–	218	17	–
Romania	–	1	–	10	–
Slovakia	–	–	60	–	–
Slovenia	–	–	–	–	25
Baltic states	–	7	7	21	–
Estonia	–	7	3	–	–
Lithuania	–	–	4	21	–
CIS					
Russian Federation	–	48	23	808	–
Total above	13	329	674	1 400	25

Source: Press reports; World Bank, *Financial Flows and the Developing Countries* (Washington, D.C.), February 1997 and previous issues; for data through 1995: World Bank, *World Debt Tables 1996*, Vol. 1 (Washington, D.C.), 1996.

[a] Preliminary.

[b] January-February.

---

economies.[415] As in other regions, the bulk of their equities has been issued in conjunction with the privatization of state owned enterprises (table 3.6.13). Access to the international equities markets is generally limited to large "blue chip" enterprises.[416] In 1996, corporations from seven transition economies issued stock worth $1.4 billion in the form of American depository or global depository receipts (ADRs or GDRs).[417] Most of this was accounted for by Russian enterprises which raised a total of $800 million, up from virtually nothing in 1995. Gazprom, the natural gas production and distribution company, led the way with the largest single issue amounting to $429 million.[418] This was the company's first issue, and as in some other recent flotations, was oversubscribed by foreign investors.

Prospects for new emissions and related issues

Borrowing by the transition economies is expected to remain buoyant in 1997 (table 3.6.14). Given the generally underdeveloped state of their domestic financial

---

[415] The global issuance of international equities has risen rapidly, from around $10 billion at the turn of the decade to $58 billion in 1996. OECD, *Financial Market Trends* (Paris), March 1997.

[416] Potential corporate borrowers must meet rigorous reporting standards to enter the international equity market.

[417] ADRs and GDRs are negotiable equity based certificates, internationally traded as proxies for local equity shares. ADRs are certificates of ownership of shares in a non-United States company which are issued by United States banks and are traded in the United States securities markets. A GDR involves the issuance of such certificates in several international (public or private) markets.

[418] The sale involved 1.15 per cent of the company's stock.

TABLE 3.6.14

Sample of proposed international issues of eastern Europe, the Baltic countries and the CIS in 1997
(Million dollars)

	Issuer	Size	Launch date
Bonds			
Czech Republic	City of Prague	n.a.	1 half
	Czech Export Bank	n.a.	1997
Hungary	National Bank of Hungary	500	1 half
Republic of Moldova	Republic of Moldova	70	1 half
Romania	City of Brasov	n.a.	1997
	City of Bucharest	150	1 half
	City of Constantza	n.a.	1997
Russian Federation	City of St. Petersburg	300	2 half
	Moscow City Telephones	n.a.	1997
	Nizhniy Novgorod	50	1 half
	Rostelekom	n.a.	1997
	City of Moscow	500	1 half
	Sverdlovsk Region	100	1997
	AO Tatneft (Tatarstan)	n.a.	1997
Slovakia	Vodohospodarska Vystavba	n.a.	1997
The FYR of Macedonia	The FYR of Macedonia	n.a.	1997
Ukraine	Ukraine	n.a.	1997
Equities			
Bulgaria	Expressbank	n.a.	1 half
Czech Republic	Investicni a Postovni Banka	100	1 half
Hungary	Matav	300-500	1 half
	MOL	150-200	IIIQ
Poland	KGHM	750	IQ
Russian Federation	Inkombank	100	2 half
	Lukoil	500-1 000	2 half
Slovenia	Slovenia Telecom	n.a.	IIIQ

Source: Emerging Markets Investor, January 1997 and press reports.

markets with their high interest rate differentials, creditworthy entities will continue to seek lower cost and longer-term finance abroad. The dominance of corporate and bank borrowers appears likely to continue, especially since those sovereign entities with the largest borrowing potential do not appear to need the funds for the time being (although there may be the occasional issue to establish market benchmarks). So far at least, governments of countries with large current account deficits have not announced large borrowing programmes, essentially corroborating the view that they can be financed by other means (see above). The Republic of Moldova, Ukraine (expected to receive an international credit rating soon) and a host of subnational entities (see below) are expected to enter the Eurobond market for the first time shortly (table 3.6.14).

The *Russian Federation* appears poised to retain its position as the most active transition economy borrower in the region. The award of a sovereign credit rating in November and the successful issuance of two eurobonds have paved the way for a host of companies and municipalities intending to tap the markets for foreign funds. The government has announced plans to offer a further $2 billion of Eurobonds in 1997. Some observers foresee around $10 billion in bond issues from the CIS (mainly from Russia) over the next two years.[419]

In the *equity markets*, the $429 million issue by Gazprom is seen as the first of several large international offerings by private sector entities. Gazprom itself has been authorized to sell up to an additional 9 per cent of its stock abroad, although the recent fall in the ADR price has led to a postponement of a second tranche. Lukoil, the large Russian petroleum company, also hopes to raise some $1 billion from the sale of 15 per cent of its shares on the New York Stock Exchange.[420] Overall, at least a dozen large Russian corporations are reported to be working on or considering ADR placements. During the next year or so, MATAV, the Hungarian telephone company, plans an international public offering of at least 16 per cent of the company's shares. If realized this would be eastern Europe's largest equity offering to date.

A growing number of *municipalities and regional governments* are preparing to enter the international debt markets (table 3.6.14). Prague and Ostrava have already done so, and in Russia, the region of Nizhniy Novgorod appears to be the next in line. There are several reasons why subnational entities may seek foreign funds. Federal governments facing fiscal constraints may shift the burden of financing infrastructure investment to local governments which, because of decentralization, are increasingly getting the right to engage in such activity. Municipalities can also benefit from cheaper foreign financing for the reasons already mentioned. However, the risks for the foreign investor involve the lack of sovereign guarantees on such issues. Local authorities generally do not have their own sources of foreign currency revenue to service the debt, leaving them dependent on the local currency market or on the discretion of the central bank. For the borrowing country the risk is that a local authority will default, or that uncontrolled and excessive borrowing by such authorities will threaten the (favourable) sovereign credit rating. To lessen these risks, federal governments tend to limit the volume of subnational issues. In Russia, where a host of local authorities hope to tap the international markets, a recent government draft resolution would make it mandatory for cities and regions to obtain an international credit rating before issuing bonds.

---

[419] *Financial Times*, 9 September 1996.

[420] *International Herald Tribune*, 16 November 1996.

TABLE 3.6.15

Flows of net foreign direct investment[a] into eastern Europe, the Baltic countries and the CIS, 1994-1996
(Million dollars and per cent)

Reporting country	1994	1995	January-September 1995	1996	1996[c]	FDI flow per capita[b] 1995	1996	FDI flow /GDP (per cent) 1995	1996
Eastern Europe	3 403	9 073	4 123	4 756	7 105	84	66	2.9	2.1
Albania	53	70	54	57	76	20	21	2.9	2.2
Bulgaria	105	98	90	82	109	12	13	0.8	0.9
Croatia	98	81	57	254	340	17	78	0.4	1.8
Czech Republic	842[d]	2 526	1 965	947	1 388	244	134	5.4	2.7
Hungary	1 097	4 410	797	1 340	1 986	431	194	10.1	4.5
Poland	542	1 134	577	1 731	2 741	29	71	1.0	2.1
Romania	341	417	292	160	210	18	9	1.2	0.6
Slovakia	170	157	140	79	70	29	13	0.9	0.3
Slovenia	131	170	144	102	180	85	90	0.9	1.0
The FYR of Macedonia	30	9	8	5	6	4	3	0.3	0.2
Baltic states	535	518	375	290	387	67	50	3.7	2.3
Estonia	225	202	165	44	58	136	39	5.6	1.4
Latvia	279	245	162	161	215	98	86	5.5	4.2
Lithuania	31	72	48	86	114	19	31	1.2	1.5
European CIS	808	2 046	1 087	1 553	2 071	9	10	0.5	0.4
Belarus	9	15	5	12	16	1	2	0.1	0.1
Republic of Moldova	12	64	59	68	91	15	21	3.8	5.7
Russian Federation	636	1 711	841	1 167	1 556	11	10	0.5	0.4
Ukraine	151	257	182	306	408	5	8	0.7	0.9
Total above	4 746	11 637	5 584	6 600	9 563	35	29	1.6	1.1
Asian CIS	783	1 596	..	662	..	6	..	4.2	..
Armenia	3	19	..	..	..	5	..	1.5	..
Azerbaijan	22	155	..	..	..	22	..	6.4	..
Georgia	8	6	..	..	..	1	..	0.3	..
Kazakstan	519	859	614	662	884	50	54	5.1	5.0
Kyrgyzstan	45	191	..	..	..	41	..	12.8	..
Tajikistan	12	13	..	..	..	2	..	2.2	..
Turkmenistan	103	233	..	..	..	57	..	7.8	..
Uzbekistan	72	120	..	..	..	5	..	1.2	..

Source: UN/ECE secretariat Foreign Investment Database, based on national balance of payments statistics. For the Asian CIS (except Kazakstan), The World Bank, *Statistical Handbook 1996, States of the Former USSR* (Washington, D.C.), 1996.

a Cash basis; FDI in reporting country less its FDI abroad.
b Dollars per capita.
c Annualized rate (January-September) except for Slovakia (January-November) and the Czech Republic, Hungary, Poland, Romania, Slovenia and The FYR of Macedonia which are full-year figures.
d Excluding flows between the Czech Republic and Slovakia.

Several borrowers have announced their intention to raise external funds to finance part of their budget deficits.[421] This includes not only the funding of infrastructure investment, but also various current expenses, including the clearing of wage arrears, etc. Since foreign funds are cheaper than local funds, the cost of debt servicing and limiting the budget deficit are reduced. However, such borrowing is likely to do little to strengthen the capacity to service external debt.

(c) Foreign investment

Recent developments

After reaching a record $12 billion in 1995, *net foreign direct investment*[422] flows into the east European

---

[421] In 1996, Russia already funded one half of its budget deficit with foreign funds.

[422] Attention is drawn to the fact that the FDI flows in tables 3.6.15 and 3.6.16 are reported on a cash basis. They can differ considerably from data on the value of foreign investment deals compiled by various national agencies. Poland also publishes FDI data on an accrual basis (table 3.6.16). They show considerably higher FDI inflows than cash data do because they incorporate reinvested profits and investment in kind, but they are available only on an annual basis and with a longer lag. The methodological issues involved are discussed in UN/ECE, *Economic Survey of Europe in 1995-1996*, p. 151, box 3.6.1.

TABLE 3.6.16

Foreign direct investment of eastern Europe, the Baltic countries and the CIS, 1992, 1994-1996
(Million dollars)

Reporting country	FDI inflows[a]			Cumulative FDI[b]			Cumulative FDI (per capita)[b,c]			
	1994	1995	1996	1994	1995	1996*	1992	1994	1995	1996
Eastern Europe	3 469	9 152	7 171	14 031	23 182	30 304	61	130	215	281
Albania[d]	53	70	76	131	201	258	6	37	56	72
Bulgaria	105	90	82	247	337	399	12	29	40	48
Croatia[d]	98	81	340	188	269	609	3	40	57	129
Czech Republic	862	2 562	1 428	3 381	5 943	7 371	189	327	574	712
Hungary	1 146	4 453	1 983	6 941	11 394	13 377	335	676	1 113	1 307
Poland[d]	542	1 134	2 741	1 617	2 751	5 492	13	42	71	142
Poland (accrual basis)	1 875	3 659	..	4 674	8 333	..	..	..	..	..
Romania	341	419	210	555	974	1 184	5	24	43	52
Slovakia[d]	170	157	119	523	680	789	40	98	127	147
Slovenia	128	176	186	424	600	786	92	212	300	393
The FYR of Macedonia[d]	24	9	6	24	33	39	–	11	16	19
Baltic states	536	457	443	888	1 344	1 787	14	114	175	233
Estonia	225	205	104	444	648	752	38	296	437	507
Latvia	279	180	224	371	551	775	16	146	221	310
Lithuania	31	73	115	73	145	261	3	20	39	70
European CIS	817	2 363	2 436	4 191	6 554	8 989	8	19	30	41
Belarus[d]	9	15	16	26	41	57	1	3	4	6
Republic of Moldova	12	64	91	12	76	167	–	3	17	38
Russian Federation	637	2 017	1 907	3 595	5 612	7 519	10	24	37	50
Ukraine[d]	159	267	421	557	824	1 245	4	11	16	24
Total above	4 822	11 972	10 050	19 109	31 080	41 080	25	57	93	123
Asian CIS[d]	783	1 596	..	1 310	2 905	..	2	19	41	..
Armenia	3	19	..	3	22	..	–	1	6	..
Azerbaijan	22	155	..	22	176	..	–	3	25	..
Georgia	8	6	..	12	18	..	–	2	3	..
Kazakstan	519	859	884	847	1 706	2 591	6	50	99	150
Kyrgyzstan	45	191	..	55	245	..	–	12	53	..
Tajikistan	12	13	..	29	42	..	1	5	7	..
Turkmenistan	103	233	..	182	415	..	–	45	101	..
Uzbekistan	72	120	..	160	280	..	2	7	13	..

Source: UN/ECE secretariat Foreign Investment Database, based on national balance of payments statistics; IMF, *Balance of Payments Statistics Yearbook* (Washington, D.C.), 1995, for Poland (accrual basis). For the Asian CIS (except Kazakstan), The World Bank, *Statistical Handbook 1996, States of the Former USSR* (Washington, D.C.), 1996.

[a] In reporting country.
[b] Cumulative foreign direct investment in reporting countries from 1988, balance of payments cash basis, end of period.
[c] Dollars per capita, end of period.
[d] FDI inflows are net of residents' FDI abroad.

countries, the Baltic states and the European CIS fell to an estimated $9.4 billion in 1996 (table 3.6.15). Some decline had been expected in 1996, if only because there had been several exceptional sales of large companies to strategic investors in 1995 (primarily in Hungary and the Czech Republic).[423] Indeed net FDI attracted by both countries fell by about one half in 1996.

Foreign direct investment *inflows* increased in only a handful of countries in 1996 (table 3.6.16). The emergence of Poland as the main recipient of FDI is notable, its $2.7 billion inflow surpassing that of Hungary and the Czech Republic, the previous leaders. The surge into Poland had already begun in the second half of 1995, most probably because of the improving macroeconomic environment and the normalization of relations with foreign creditors (London Club) towards the end of 1994. Although prospects for FDI in the republics of the former SFR of Yugoslavia appeared to have improved following the Dayton Accord, the results so far have been mixed. Inflows into Croatia increased significantly, placing it ahead of most European transition economies as a host for FDI. In Slovenia, FDI recovered sharply in the second half of the year, but for all of 1996 there was only a small increase compared with 1995. The small inflow into The FYR of Macedonia has declined. This was also the case in Bulgaria, Romania and Slovakia which have also received relatively little FDI.

---

[423] See UN/ECE, *Economic Survey of Europe in 1995-1996*, p. 154. In Hungary the surge in privatization occurred in the last quarter of 1995, while in the Czech Republic the large SPT Telecom sale occurred in the third quarter. These sales significantly affected the flows recorded in tables 3.6.15 and 3.6.16.

In the Baltic states, FDI flows into Lithuania have increased rapidly, exceeding those of Estonia in 1996. Estonia has been one of the most attractive destinations for FDI among the transition economies but the inflow has diminished in the last two years. Flows into Latvia have remained buoyant.

FDI inflows into Russia amounted to some $2 billion in 1995-1996, considerably higher than in preceding years but far below expectations. Flows into Ukraine and the Republic of Moldova have increased rapidly, but Belarus continues to receive little FDI.

Net FDI inflows have generally been small relative to the GDP of the host countries. Only in Hungary, the Czech Republic, Estonia, Latvia and, more recently, the Republic of Moldova have FDI flows exceeded 5 per cent of GDP (a high figure by international standards), and there was a tendency for the ratios to decline in 1996.

Cumulative inflows of FDI into eastern Europe, the Baltic states and the European CIS countries reached an estimated $41 billion by the end of 1996 (table 3.6.16). Of this, nearly three quarters has been invested in eastern Europe, somewhat over 4 per cent in the Baltic States, and around 22 per cent in the European CIS countries. These shares have been fairly constant over the past few years. Hungary has accumulated the largest stock of FDI, followed by Russia, which appears to have passed the Czech Republic in 1996. Hungary also leads the region on a per capita basis. The Czech Republic, Estonia Slovenia and Latvia also have comparatively high FDI intensities but they lag considerably behind Hungary in this regard.

FDI data (tables 3.6.15 and 3.6.16) reflect proceeds from privatization as well as greenfield investments and additions to the statutory capital of companies. Separate data, which cover receipts in domestic and foreign currencies (table 3.6.17), indicate that proceeds from privatization fell sharply in 1996 in Hungary and Poland. This implies that their FDI was driven chiefly by greenfield investments and increases in existing capital assets. In Hungary these two sources appear to have contributed some $1.5 billion to a total FDI inflow of $2 billion in 1996. Privatization revenues in foreign currencies amounted to only about $500 million[424] (out of total proceeds of $880 million). In contrast, in 1995 greenfield investments and increases in existing capital assets amounted to only $700 million (total FDI of $4.5 billion less privatization revenues of $3.8 billion,

TABLE 3.6.17

Foreign direct investment inflows and proceeds from privatization,[a] 1993-1997
(Million dollars)

	1993	1994	1995	1996[b]	1997[c]
Czech Republic					
FDI	568	862	2 562	1 428	..
Privatization proceeds	..	1 077	1 205	944	700
Hungary					
FDI	2 339	1 146	4 453	1 983	..
Privatization proceeds	1 842	1 017	3 813	880	800[d]
Poland					
FDI[e]	580	542	1 134	2 741	..
Privatization proceeds	734	642	1 516	495	3 500

Source: FDI: table 3.6.16; privatization revenues: OECD, "Privatization: recent trends", *Financial Market Trends* (Paris), March 1997, table 1.

[a] Gross proceeds in domestic and foreign currency.
[b] Privatization proceeds are preliminary.
[c] Estimates.
[d] Projection of APV, the State Privatization Agency.
[e] Net flows.

virtually all of the latter in foreign currency).[425] The figures for Poland are even more striking. In 1996 proceeds from privatization amounted to only about $500 million (including stakes worth $350 million in two tobacco companies), compared with FDI inflows of $2.7 billion.

This development suggests that FDI flows to Hungary and Poland have become increasingly determined by expectations of growth and profitability and the other usual arguments in the investment function rather than by the government's timetable for privatization. The development is also important from the viewpoint of the potential sustainability of FDI flows, since those generated by sales to strategic foreign investors cannot continue indefinitely. Depending on the country and its policies, the stock of saleable state enterprises may soon be exhausted, as is nearly the case in Hungary (see below).

So far, *outward* foreign direct investment from the transition economies has been small, increasing to an estimated $500 million in 1996. Russia has led the way with reported FDI of over $600 million in 1995-1996, although the actual figures are likely to be higher.[426] In eastern Europe the largest outflows have been from Hungary and the Czech Republic, and in 1996 from Slovakia and Latvia. FDI in the transition

---

[424] Mr. P. Szabo, managing director of APV, the privatization agency. The figures are for the first 11 months of 1996.

[425] The Privatization Research Institute has estimated that greenfield investments amounted to $2.5 billion in the period 1992-1995.

[426] Unauthorized transfers of money abroad (capital flight) have been considerable in Russia's case. It is likely that some of this has been used to purchase stakes in foreign companies, but has gone unreported to the Russian authorities.

economies is likely to increase as local enterprises develop their financial capacity and increasingly turn to foreign subsidiaries to boost their exports. For the new OECD member countries, the recent relaxation of capital controls may also foster direct investment abroad.

As regards *portfolio investment in local securities*,[427] Russia led the transition economies in 1996, some $6.7-$13 billion flowing into *government treasury bills* (GKOs).[428] Foreign interest in the GKO market has been keen on account of the prevailing high yields.[429] During 1996, rules governing non-residents' access to the market were liberalized, and as a result their holdings of outstanding GKOs rose from 1 per cent of the total in early 1996 to 15-30 per cent at the end of 1996. Russia has been encouraged, *inter alia* by the IMF, to provide greater access to foreign participants, since additional foreign funds would lower interest rates on treasury bills and reduce pressure on government spending. Additional foreign funds entered the stock market, driving it to record levels. Russian stock market performance was among the world leaders in 1996, and again in early 1997.[430]

Outlook for FDI

Prospects for increased foreign direct investment in the transition economies appear to have improved. As discussed above, FDI inflows unrelated to new privatization have developed a certain momentum in a number of countries. Also, many transition economies intend to carry out (in some cases, accelerate or resume) existing privatization programmes, including sales of strategic stakes to foreigners. In Bulgaria and Romania, for example, the new governments have reaffirmed previous commitments to privatize state assets and have pledged to pursue policies to stabilize and restructure their economies, and otherwise to improve the investment climate. Any acceleration of FDI inflows into these countries, which have so far attracted only small flows, would result in a greater geographical diversification of FDI than has heretofore been the case.

The new *Bulgarian* government (with the agreement of the political opposition) has taken steps to accelerate the sale of state owned companies valued at some $1-$1.3 billion.[431] It is also willing to offer larger stakes in companies to be privatized and otherwise make such deals more attractive to foreign investors. The government also hopes to generate more revenue than would have been forthcoming under the previous policy.[432] In early 1997, the sale of a stake in Sold Sodi Devnaya to Solvay of Belgium was reported to be near finalization: the proceeds, some $160 million, were to be used to boost foreign currency reserves. Negotiations for the sale of a 60 per cent stake in fertilizer producer Chimko-Vratsa for $100 million were also underway, and the Bulgarian authorities have asked foreign investment banks to bid for the mandate to privatize the Bulgarian telecommunications industry.

In *Romania* the new government intends to submit a series of bills to parliament to speed up the privatization of companies and banks (with sales to strategic investors) and amend those laws which have been seen as hindering the restructuring process.[433] Development of the capital market, including the listing of treasury bills, has been announced as one of the priorities.[434] The macroeconomic measures and other policies to be implemented should also improve the climate for investment. Since the State Ownership Fund, the main privatization body, has retained 51 per cent of some of the most attractive enterprises, it is in a position to offer controlling stakes to strategic investors. The Romanian Development Agency has

---

[427] Portfolio investment in the domestic securities markets of the transition economies is discussed more extensively in UN/ECE, *Economic Bulletin for Europe*, Vol. 48 (1996), chap. 5.

[428] GKOs are short-term treasury bills denominated in roubles and sold at a discount. Early in 1996, foreign access to this market was on highly restrictive terms, including the obligation to hold them until maturity. The $6.7 billion pertains to foreign investments through the so-called "s" accounts, according to the National Bank of Russia (14 February 1997). The (higher) $13 billion figure also includes an estimate of foreign investments made through "grey" accounts. Government of the Russian Federation/Working Centre for Economic Reform, *Russian Economic Trends* (Moscow), February 1997.

[429] Annualized yields on GKOs for residents were as high as 250 per cent in the spring of 1996, prior to the presidential elections, declining thereafter to around 75 per cent in mid-September. In principle, foreigners have been subject to a 19 percentage points cap on yields. Nonetheless, given the limited depreciation of the rouble, the yield in terms of foreign currencies was substantial.

[430] Current data for portfolio investments in *equity* markets are not available. Some estimates put foreign acquisitions of shares in Russian companies at some $1.5-$2.5 billion in 1994 and $1-$1.5 billion in 1995.

[431] Stakes in some 22 companies in telecommunications, mining, engineering, metallurgy, banking, chemicals and other sectors are to be sold to foreign strategic investors.

[432] For example, the former government intended to sell off a maximum of 25 per cent of the telecommunication sector and retain management control. However, the new government is reported to be willing to offer foreign investors larger stakes so as to maximize revenue per share. *Financial Times*, 3 March 1997.

[433] Address of Prime Minister V. Ciorbea, *Romanian Business Journal*, 20 February 1997.

[434] The proposed changes in economic policies and the sharp depreciation of the leu have vitalized the stock market. In 1996 trading was weak and prices fell, but in January 1997 the Bucharest Stock Exchange rose by nearly 30 per cent, boosted by the entry of foreign funds.

submitted legislation to amend the foreign investment law to simplify procedures and increase incentives.[435]

In 1996, the government of *Ukraine* proposed an economic reform package which included, among other things, tax cuts, trade liberalization, and various structural measures. A new foreign investment agency and the simplification of legislation pertaining to company registration are also intended to help attract FDI. The mass privatization programme is to be completed by July 1997. Meanwhile, the government is preparing the second phase of the privatization programme, involving the sale of assets to foreign investors.[436] The large AvtoZaz motor company is to be among the first to be sold off. So far Ukraine has received little FDI despite having a population of 52 million, the second largest of the transition economies.[437] As of mid-March 1997, the reform and budget proposals had not been acted on by parliament, also delaying disbursements from the $3.1 billion package of international financial assistance.

In *Hungary*, which has led the way in selling state assets to foreign investors, the privatization process is entering its final stage in 1997.[438] Privatization revenues, including the foreign currency component, are projected to decline to $800 million in 1997. This includes the sale of shares in Matav ($500 million, see above), and two terrestial TV channels. Additional privatizations in the utility sector are planned for 1998.[439]

In *Poland* administrative changes have resulted in the Treasury taking over the responsibility for the sale of state assets, through the medium of a new privatization agency.[440] Meanwhile, the introduction of the new privatization law has been delayed from January until April. Overall, receipts from privatization are to rise considerably in 1997, to an estimated $3.5 billion (table 3.6.17). Among other deals, strategic investors have been, or are to be, offered stakes in Orbis (travel agency), KGHM (copper refining), Bank Handlowy and Powszechny Bank Kreditowy, and two pharmaceutical companies.

There are several reasons why a number of governments have sought to give a new impetus to restructuring their domestic industries and attracting foreign capital to help them do so. First, the view is widespread (including in some of the slow-reforming countries), that to achieve growth there is no credible alternative to pushing ahead with privatization and involving foreign capital. It is becoming increasingly evident that those countries which have reformed the most, have, on the whole, outperformed the others in terms of economic growth, etc. Moreover, economic crises have been precipitated by policy immobility and partial reforms, and this has been noted by governments in other countries.[441] There also appears to be greater appreciation of the liabilities of state ownership, with disclosure of mismanagement or corruption becoming an embarrassment to governments.

Second, the elaboration and implementation of restructuring programmes has been a condition for access to multilateral and various bilateral credits, some of which are earmarked for that very purpose. Indeed the dependence of countries on official sources of finance underlines their inability to attract private funds. However, recent experience (see above)[442] has shown that countries committed to economic reform and the improvement of the investment environment have been able to attract large amounts of private capital in a relatively short period. Of course, the potential benefits of capital inflows may go beyond fostering structural change. In several cases, surges in capital inflows have helped to markedly improve the external financial situations of these countries. They have significantly boosted foreign currency reserves, relaxed constraints on the current account, allowed currency restrictions to be liberalized, and contributed to the reception (or upgrading) of favourable international credit ratings.

---

[435] *Financial Times*, 13 March 1997.

[436] Ibid.

[437] The low level of FDI has been explained by the country's generally poor investment climate: conflicting regulations, customs problems, high taxation and corruption; difficulties faced by foreigners in buying stakes in privatized enterprises; and the suspension by the parliament of the sale of over 200 firms specially targeted for privatization with the participation of strategic foreign investors. See, for example, *Business Eastern Europe*, 30 December 1996.

[438] APV, the privatization agency, retains assets worth some $5 billion. However, this includes some strategic assets which will remain in the public domain. APV holds minority stakes in firms already privatized, but their sale would require a modification of the privatization law.

[439] Presumably questions regarding the pricing of energy need to be resolved. In 1996, energy price increases were postponed, and foreign investors claimed that this prevented the achievement of a rate of return of 8 per cent stipulated in the privatization contracts. In consequence these investors are reported to have postponed their intended investments.

[440] EIU, *Country Report. Poland* (London), First Quarter 1997.

[441] For example, Prime Minister Larzarenko was recently quoted as saying "the terrible Bulgarian lesson will not be repeated in Ukraine. The government is very keen on structural reform", *Financial Times*, 13 March 1997.

[442] The capital surge experienced by the most reformed transition economies has been discussed in UN/ECE, *Economic Bulletin for Europe*, Vol. 48 (1996), chap. 5 and previous volumes of this *Survey*.

(d) Concluding comments

Several observations can be made on the basis of the discussion above. First, several countries have had strong and sudden inflows of foreign capital, generally since 1993 and peaking in 1995. In a number of cases they continued, although mitigated, in 1996. Unless these flows are offset by current account deficits, or channelled directly into some form of asset denominated in foreign currency, they are likely to pose serious problems for economic policy (see section 3.1). In particular, they can raise the real exchange rate and boost domestic demand, and thus cause current account balances to deteriorate.

Second, although medium- and long-term funds raised by the transition economies reached a record in 1996, *net* capital flows into eastern Europe declined. Only in a few cases – Romania, Slovakia, Slovenia, and Lithuania – did recent gains in access to the international markets result in higher borrowing (table 3.6.11). In general, however, such borrowing declined (or increased only modestly). In fact few of the larger creditworthy countries needed additional long-term funds, their requirements being covered by FDI, short-term capital, and drawings on foreign exchange reserves. In Hungary, the authorities sharply cut borrowing and repaid large amounts of public debt, which more than offset a higher volume of borrowing by the private sector. The Czech authorities aimed to discourage further inflows of short-term funds, by widening the permitted fluctuation margins of the exchange rate. In Poland, funds raised in the international markets accounted for only a very small share of the record inflow of capital.

The third observation concerns the maturity and the potential volatility of inflows during the past two years. On the one hand, FDI has emerged as the largest source of foreign capital in eastern Europe, and is considered to be the most stable type of inflow, the least subject to reversals. On the other hand, the role of short-term and unrecorded capital flows (and in some cases portfolio investment as well) has increased, their combined total rising to $7 billion in 1995. A further $4 billion inflow was received in 1996. These flows represent some 25-30 per cent of total net inflows, up from less than 10 per cent in 1994. Such potentially volatile capital has also been important in some of the Baltic states and the European CIS. As in other parts of the world, such funds seem to have responded quickly to differential rates of return, changes in foreign investor sentiment, and various domestic structural and policy changes. Residents in many transition economies appear to be increasingly willing to repatriate funds held abroad as domestic conditions improve. Presumably these capital inflows have also responded to perceptions of improved creditworthiness, although this is less sure than in the case of long-term private flows.[443] Aside from Bulgaria, which was already in a precarious financial situation, reversals and/or outflows[444] of funds did not appear to endanger the financial position of any country.[445] All were able to rely on other sources of capital, including unrecorded flows, or ample international reserves.

---

[443] It should be borne in mind that unrecorded flows may include investments in local securities markets (not properly registered as such by commercial banks), which also depend on the confidence of foreign investors in the overall creditworthiness of the receiving country.

[444] Strictly speaking the outflows from enterprise and domestically held foreign currency accounts, as in Bulgaria, involve demonetization rather than cross-border movements which are the focus of this section.

[445] The aggregate figures conceal large movements in volatile flows in individual countries. For example, the introduction of wider fluctuation bands for the koruna to ± 7.5 per cent in late February 1996, and the increase in exchange rate risk that this engendered, prompted non-residents to sell $490 million (CZK 13.4 billion) of short-term treasury and NPF bills and buy $1.1 million (CZK 28.5 billion) of local equities. (Czech National Bank, *Monthly Bulletin* (Prague), September 1996.) In Poland portfolio flows continued to flow into high-yielding government securities in the first half of 1996 (some $450 million). However, a subsequent decline in interest rates resulted in a net outflow in the second half of the year.

TABLE 3.6.18

Balance of payments of selected transition economies, 1994-1996

(Million dollars)

	Eastern Europe				Albania			Bulgaria		
		Jan.-Sept.				Jan.-Sept.			Jan.-Sept.	
	1995	1995	1996	1996[a]	1995	1995	1996	1995	1995	1996
Current account balance	-1 389	-179	-7 387	-12 935	-15	26	-85	-26	103	-34
Trade balance	-13 904	-9 223	-16 785	-25 357	-475	-343	-468	121	243	132
Exports	93 338	68 022	69 355	93 781	205	150	175	5 345	3 956	3 337
Imports	107 242	77 242	86 140	119 138	680	491	643	5 224	3 713	3 205
Services	4 235	3 345[b]	3 667[b]	4 736[b]	-59	-26	-37	153	174	122
Transport	1 577	694	771	1 094	-42	-30	-34	-37	-13	-14
Travel	3 546	2 872	3 579	4 779	58	46	49	278	215	172
Other services	-888	-206	-508	-927	-75	-42	-52	-87	-28	-36
Non-classified transactions[c]	7 754	5 549	5 537	7 153	–	–	–	–	–	–
Investment income	-3 194	-2 417	-2 602	-3 167	42	30	49	-432	-415	-383
Transfers	3 693	2 565	2 787	3 698	477	362	370	132	101	95
Capital and financial account	21 539	13 662	7 495	13 063	-12	-18	66	148	105	-705
Capital transfers	665	427	237	522	390	388	5	–	–	50
Foreign direct investment	9 073	4 123	4 756	7 080	70	54	57	98	90	82
Portfolio investment	4 725	3 512	702	606	–	–	–	-66	-36	-184
Long-term capital	3 521	1 926	568	2 992	97	78	64	-14	60	-837
Assets	517	657	-203	-377	–	–	-1	293	212	-640
Liabilities	3 005	1 270	771	3 368	97	78	66	-307	-153	-197
Short-term funds	4 497	4 169	1 256	1 822	-100	-70	-61	130	-8	185
Other funds	-941	-495	-24	42	-468	-468	–	–	–	–
Valuation effects[d]	748	703	-707	-846	–	–	–	–	–	–
Errors and omissions	2 429	1 101	1 935	2 312	57	8	30	111	222	-25
Overall balance	23 327	15 287	1 336	1 594	31	16	11	234	430	-764
Reserves (net)	-23 327	-15 287	-1 336	-1 594	-31	-16	-11	-234	-431	764

	Croatia			Czech Republic			Hungary			Poland		
	1994	1995	1996[e]	1994	1995	1996	1994	1995	1996	1994	1995[f]	1996
Current account balance	103	-1 711	-1 129	-50	-1 362	-4 476	-3 911	-2 480	-1 678	-944	5 455	-1 352
Trade balance	-969	-2 877	-3 148	-889	-3 678	-5 972	-3 635	-2 442	-2 645	-836	-1 827	-8 154
Exports	4 260	4 632	4 364	14 016	21 463	21 703	7 613	12 810	14 183	16 950	22 878	24 420
Imports	5 229	7 510	7 512	14 905	25 140	27 674	11 248	15 252	16 828	17 786	24 705	32 574
Services	738	614	1 249	733	1 842	1 785	260	654	1 426	57	150	-209
Transport	120	66	119	390	661	624	-176	76	102	576	605	..
Travel	875	657	1 320	890	1 242	1 122	503	658	1 288	-190	-200	..
Other services	-257	-109	-190	-547	-62	39	-67	-80	36	-329	-255	..
Non-classified transactions	–	–	–	–	–	–	1	25	–	–	7 754	7 153
Investment income	-124	-93	-53	-20	-106	-680	-1 446	-1 845	-1 454	-2 081	-628	-366
Transfers	459	645	823	126	579	390	909	1 127	995	1 916	6	224
Capital and financial account	584	902	975	3 371	8 226	4 081	3 254	7 011	-1 645	1 812	2 759	4 788
Capital transfers	–	–	–	–	–	–	–	–	156	–	–	86
Foreign direct investment	98	81	340	749	2 526	1 388	1 097	4 410	1 986	542	1 134	2 741
Portfolio investment	–	–	–	855	1 362	726	2 464	2 043	-861	-624	1 171	191
Long-term capital[g]	476	618	367	1 109	3 367	2 725	-1 267	-853	-1 914	-444	-1 239	-239
Assets	242	50	-22	419	52	-524	36	118	311	-11	29	-32
Liabilities	234	568	389	690	3 315	3 249	-1 303	-971	-2 225	-433	-1 268	-207
Short-term funds	10	203	241	659	971	-758	960	1 411	-1 012	2 338	1 693	2 009
Valuation effects[d]	–	–	–	–	–	–	–	–	–	1 108	748	-846
Errors and omissions	101	1 304	742	-950	595	-424	[h]	[h]	1864	-228	-27	480
Overall balance	788	495	588	2 372	7 458	-819	-657	4 531	-1 459	1 748	8 935	3 070
Reserves (net)	-789	-495	-588	-2 372	-7 458	820	656	-4 532	1 458	-1 748	-8 935	-3 070

(For source and notes see end of table.)

TABLE 3.6.18 (continued)

Balance of payments of selected transition economies, 1994-1996

(Million dollars)

	Romania			Slovakia			Slovenia			The FYR of Macedonia		
	1994	1995	1996	1994	1995	1996[a]	1994	1995	1996	1994	1995	1996
Current account balance	-428	-1 639	-2 336	712	646	-1 401	540	-36	47	-158	-222	-491
Trade balance	-411	-1 577	-2 130	105	24	-1 597	-338	-954	-853	-185	-220	-522
Exports	6 151	7 910	8 238	6 727	8 546	8 097	6 830	8 345	8 366	1 086	1 205	898
Imports	6 562	9 487	10 368	6 622	8 521	9 694	7 168	9 298	9 218	1 271	1 425	1 421
Services	-171	-190	-271	657	546	38	723	725	786	-155	-200	-155
Transport	-22	2	80*	377	315	242	68	69	76	7	-138	-101
Travel	-35	-107	-120*	285	302	183	590	666	769	-125	-8	-6
Other services	-114	-85	-231*	-5	-71	-386	66	-10	-59	-37	-54	-48
Investment income	-129	-241	-314	-120	-9	-23	108	147	89	-47	-29	-32
Transfers	283	369	378	69	85	181	47	45	24	229	228	218
Capital and financial account	952	847	2 935	130	974	1 572	97	377	552	97	307	445
Capital transfers	12	242	200	87	46	30	-4	-14	-5	30	2	–
Foreign direct investment	341	417	210	170	157	64	131	170	180	24	9	6
Portfolio investment	75	-21	–[i]	272	243	68	-33	-10	638	–	3	28
Long-term capital[g]	1 125	325	1852	518	406	495	-9	521	68	43	294	410
Assets	-3	2	50	182	69	495	-125	-96	-14	..	..	..
Liabilities	1 128	323	1802[i]	336	337	–	116	617	82	..	..	..
Short-term funds	-489	440	631	-917	39	915	12	-290	-329	..	..	..
Other funds (net)	-112	-556	42	–	83	–	–	–	–	–	–	–
Errors and omissions	94	535	-382	449	-41	–	5	-120	-12	102	15	39
Overall balance	618	-257	217	1 290	1 579	171	642	220	587	41	101	-8
Reserves (net)	-618	257	-217	-1 290	-1 579	-171	-642	-220	-587	-41	-101	8

	Baltic states			Estonia			Latvia			Lithuania		
		Jan.-Sept.			Jan.-Sept.			Jan.-Sept.			Jan.-Sept.	
	1995	1995	1996	1995	1995	1996	1995	1995	1996	1995	1995	1996
Current account balance	-827	-443	-908	-185	-108	-223	-27	37	-273	-614	-373	-412
Trade balance	-1 970	-1 293	-1 778	-692	-464	-678	-580	-380	-614	-698	-448	-486
Exports	5 935	4 255	4 927	1 862	1 333	1 446	1 368	988	1 104	2 706	1 935	2 377
Imports	7 905	5 548	6 705	2 554	1 797	2 124	1 947	1 368	1 718	3 404	2 383	2 863
Services	832	622	679	378	265	369	466	347	284	-13	10	25
Transport	645	498	506	152	116	130	498	375	379	-5	7	-3
Travel	232	151	207	266	174	276	-4	-3	-120	-29	-20	51
Other services	-46	-27	-35	-40	-24	-38	-28	-26	25	21	23	-23
Investment income	8	-1	-61	3	-3	3	19	16	-9	-13	-15	-54
Transfers	303	230	253	126	95	83	68	54	66	109	81	104
Capital and financial account	1 509	975	728	259	203	230	691	464	134	559	308	365
Capital transfers	-40	-42	1	-1	-1	-1	–	–	–	-39	-41	1
Foreign direct investment	518	375	290	202	165	44	245	162	161	72	48	86
Portfolio investment	-37	-42	-57	-16	-25	16	-37	-38	-168	16	21	96
Long-term capital[g]	470	264	438	74	52	64	52	40	142[g]	344	172	233
Assets	-26	-97	-18	-18	-18	-1	-3	-19	-21	-5	-60	4
Liabilities	496	361	457	92	70	65	55	59	163	349	232	229
Short-term funds	598	420	56	–	12	107	432	300	–	166	108	-51
Other funds (net)	–	–	–	–	–	–	–	–	–	–	–	–
Errors and omissions	-380	-335	357	32	-9	42	-699	-542	252	288	216	64
Overall balance	302	197	178	105	86	48	-35	-41	113	232	151	17
Reserves (net)	-302	-198	-178	-106	-87	-48	35	41	-113	-232	-151	-17

(For source and notes see end of table.)

TABLE 3.6.18 (concluded)

Balance of payments of selected transition economies, 1994-1996
(Million dollars)

	Belarus			Republic of Moldova			Russian Federation			Ukraine		
		Jan.-Sept.			Jan.-Sept.			Jan.-Sept.			Jan.-Sept.	
	1995	1995	1996	1995	1995	1996	1995	1995	1996	1995	1995	1996
Current account balance	-567	-365	-752	-115	-129	-134	9 305	8 609	10 243	-1 152	-753	-553
Trade balance	-777	-476	-1 046	-55	-64	-106	22 173	17 726	17 474	-2 702	-2 009	-2 998
Exports	4 689	3 355	3 858	739	470	517	81 540	59 171	63 310	14 244	10 115	11 497
Imports	5 466	3 831	4 904	794	534	624	59 367	41 445	45 836	16 946	12 124	14 495
Services	182	104	296	-82	-85	-30	-9 703	-7 097	-3 697	1 512	1 124	2 446
Transport	..	..	..	-88	-84	-38	495	356	444	..	..	..
Travel	..	..	..	-56	-4	-4	-7 287	-5 430	-2 580	..	..	..
Other services	..	..	..	6	3	13	-2 911	-2 023	-1 561	..	..	..
Investment income	-51	-42	-43	-11	-7	-13	-3 273	-2 033	-3 399	-434	-240	-391
Transfers	79	49	40	33	28	16	108	13	-135	472	372	390
Capital and financial account	455	217	341	146	104	84	8 946	7 982	-5 833	1 572	889	631
Capital transfers	7	7	47	–	–	-1	-347	-267	-715	6	6	–
Foreign direct investment	15	5	12	64	59	68	1 711	841	1 167	257	182	306
Portfolio investment	–	–	–	-1	–	20	-1 437	-1 499	2 333	4	1	57
Long-term capital[g]	81	60	57	1	29	-3	953	4 827	-1 221	1 305	700	395
Assets	–	–	–	-70	7	-65	-809	1 019	-12 778	-1 574	-1 609	..
Liabilities	–	–	–	71	22	61	1 762	3 808	11 557	2 879	2 309	..
Short-term funds	352	145	225	83	16	–	8 066	4 080	-7 397	..	..	-127
Other funds (net)	–	–	–	–	–	–	–	–	–	–	–	–
Errors and omissions	213	324	414	-19	3	27	-7 867	-9 457	-7 198	49	364	150
Overall balance	100	176	3	13	-22	-22	10 384	7 134	-2 788	469	499	228
Reserves (net)	-100	-176	-3	-13	22	22	-10 386	-7 137	2 787	-469	-499	-228

Source: National balance of payments statistics.

Note: Data for 1996 are preliminary.

[a] Full-year data except for Albania and Bulgaria which are for January-September and Slovakia which are for January-November.

[b] Excludes details of services for Poland.

[c] Non-classified current transactions of Hungary and Poland.

[d] Includes valuation effect of Poland.

[e] Official projections except for foreign direct investment, which is annualized on the basis of January-September data.

[f] Breaks in Polish series. See tables 3.6.1 and 3.6.2.

[g] Includes short-term funds for The FYR of Macedonia and Ukraine, and Latvia in 1996.

[h] Included in short-term funds.

[i] External bond issues are included in long-term capital.

Chapter 4

THE CENTRAL ASIAN ECONOMIES, 1991-1996[1]

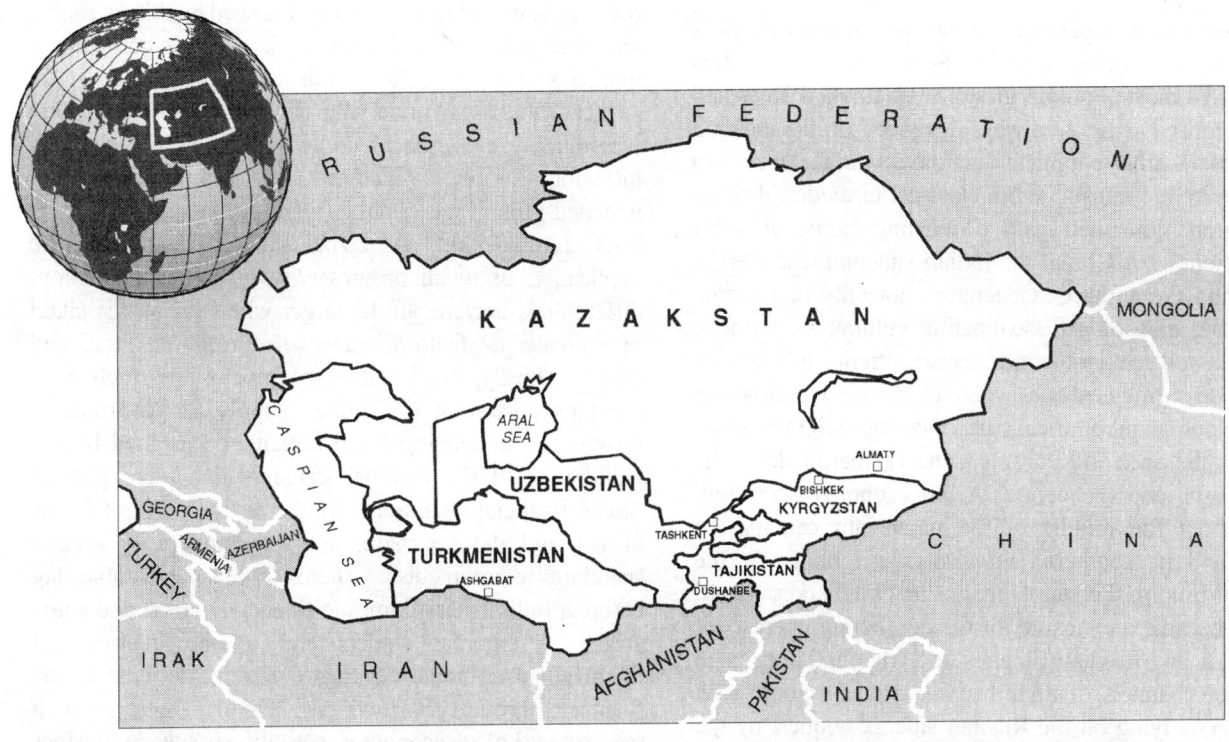

4.1 The impact of quitting a centripetal economy

In 1991, the 52 million people of five central Asian states secured independence from the collapsing USSR, but their livelihood remained heavily dependent on the ex-Soviet republics. The year 1996 was something of a turning point, for more than half their imports and their exports (51 per cent of each) were by then with partners other than CIS members. This diversification was especially notable for imports, just 39 per cent of which had come from outside the CIS in 1995; export destinations had been widening earlier, and 48 per cent were outside the CIS by 1995. The authorities in Moscow had directed the flows of funds between Union Republics and in the last year of central planning (1991) gross transfers from the Union exceeded total tax revenue in three republics (Kyrgyzstan, Tajikistan and Uzbekistan) and were large in Kazakstan and Turkmenistan. Losing such inflows before any mechanism was available for public borrowing, governments monetized their fiscal deficits. That coincided with the liberalization of most wholesale and retail prices after seven decades of arbitrary control: high, even hyper-, inflation resulted. By 1996, however, three countries (Kazakstan, Kyrgyzstan and Uzbekistan) had brought consumer price inflation down to an annual 30-50 per cent and four had reduced budget deficits to between 4 and 6 per cent of GDP. Finally, two states in 1996 had begun, on a GDP measure, the upturn from the recession which struck all republics when the trading areas of the USSR and CMEA evaporated and market demand replaced the

---

[446] This special study appears 40 years after the ECE secretariat published "Regional economic policy in the Soviet Union: the case of central Asia", UN/ECE, *Economic Bulletin for Europe*, Vol. 9, No. 3, November 1957, pp. 49-75. Geographically, that study followed the Soviet convention at the time, "Srednyaya Aziya" excluding Kazakstan, judged a separate Large Economic Zone, but from 1993 the five governments have considered themselves as constituting "Tsentral'naya Aziya". The 1957 study, the first by the United Nations devoted to the central Asian economies, was the result of a mission led by the first Executive Secretary, Gunnar Myrdal, who was later awarded the Nobel Prize for Economics. The other members of the mission were Ester Boserup, who later wrote the seminal work, *Woman's Role in Development* (London, Allen and Unwin, 1970), Mogens Boserup, subsequently Professor of Economics at Copenhagen University and Danish Commissioner for Greenland, and Michael Kaser, subsequently Professorial Fellow of St. Antony's College and Director of the Institute of Russian and East European Studies, University of Oxford, who drafted the present text as consultant to the ECE.

plan. This chapter surveys the five full years of independence of the central Asian states in the light of their economic inheritance and the policies they have subsequently pursued. Geography, ethnicity and political history (the present frontiers were set in 1924 with little subsequent change) justify their treatment as an identifiable group within the global economy.

In a long-run perspective transport infrastructure is the key to the development of ex-Soviet central Asia. It is the world's most populous group of landlocked states and its prosperity has hence always depended on the creation of routes to other suppliers and buyers or for access to cheap maritime transit. From classical to medieval times the region flourished as the meeting point of great civilizations – of China, the Indian subcontinent, Persia, Rome and Byzantium. Generators not only of material wealth but also of a disseminating culture, its farmers, craftsfolk and scholars worked astride arteries of Eurasian commerce during centuries when the maritime option was puny shipping and piratical seas. Oceanic freight routes – and later the Suez and Panama Canals – marginalized the Silk Roads, and the central Asian economy stagnated. From about the middle of the nineteenth century the region and its southern and eastern neighbours – the Ottoman Empire, Persia, Afghanistan, east Turkestan and Tibet – became a cynosure for the conflicting interests of the British and Russian Empires and what had once been profitable channels of trade became a buffer zone, with central Asia lying on the Russian side. Conquest by the Tsarist army, between the fortification of Akmolinsk (Akmola, the proposed new capital of Kazakstan) in 1830 and the annexation of eastern Tajikistan in 1895, placed the regional economy almost exclusively in the Russian sphere; three of the five capital cities – Almaty, Ashgabat and Bishkek – were created as Russian towns, a colonial city was established adjoining Tashkent[447] and one was set up in Soviet times – Dushanbe (Stalinabad). Slav immigration and the penetration of the Russian rail network – the Trans-Caspian and Orenburg-Tashkent lines (completed respectively in 1885 and 1900) – reinforced the ties, which Soviet central planning and east-west international confrontation rendered even tighter. Central Asia was politically and physically remote from the globalization that has marked the rest of the world since the Second World War.

Emergence into political independence confronted governments with economic problems of four kinds, each of which is considered in the following sections. First, the five economies were as open as, on a world ranking, their level of income would indicate, but because their trade and payments relations had been dominatingly operated from outside – by the planning and industrial agencies for the whole USSR in Moscow – their now autonomous administrators were unused to much more diverse relationships and to controlling the transmission of external changes to domestic activity. Taking place at the moment of wider nation building, it could be described as an exogenous management shock. Secondly, the decision of the reformist Russian government to decontrol most retail and wholesale prices on 2 January 1992 had to be followed within a rouble area that persisted until mid-1993. With a competitive market system scarcely established, a price-wage spiral (which had begun under the decontrols of perestroika) was quickly exacerbated under conditions of relatively inflexible supply. Income and wealth distribution widened after wage controls were discarded (save for a brief reimposition in Uzbekistan) and before the application of much progressivity in personal taxation; differentials became all the larger when account is taken of a greater de facto tolerance of corruption, fraud and crime. Thirdly, that inflationary impetus was fuelled by rapid expansion of the money supply as governments resorted to monetizing fiscal deficits, occasioned by the withdrawal of the external transfers that were part of Soviet financial operations, by the restructuring of forms of tax and the tax base, and by temporarily greater spending requirements. There was an inevitable lag before a rudimentary domestic financial market and some access to external lenders and to the international financial institutions began to enable deficits to be financed through borrowing. Finally, there was a recession shock: economic activity shrank as former "planned" relations – internal and external – were liquidated more rapidly than their replacement by market linkages: the mix and level of demand underwent considerable change.[448] Investment plummeted and the utilization of labour and of capacity significantly diminished. A considerable retro-emigration of Slavs has enhanced the ethnic homogeneity which can help nation building, but it has contributed to productivity decline until sufficient skilled replacements are trained.

The chapter then goes on to consider the institutional changes deployed to further the establishment of viable market practices in the region, notably new currencies, broad-scale privatization and the operational and legal infrastructure which transmits and responds to market generated signals such as interest and exchange rates and securities prices. It considers the new opportunities being taken to foster a diversified economy and to repair a damaged environment and misuse of natural resources. Export concentration, as a result of Soviet planning, on minerals and cotton has continued in independence because these – especially newly exploited hydrocarbons – offer current comparative trade advantage

---

[447] The Uzbek capital transferred from Samarkand in 1930.

[448] For a fuller consideration of demand modification in transition recession, see this *Survey*, chap. 3.2, complementing the supply-side analysis in UN/ECE, *Economic Survey of Europe in 1995-1996*, pp. 67-72.

and attract foreign capital. Support in this reorientation and in broader restructuring has been furnished by the United Nations family of agencies and others – especially the European Union, the European Bank for Reconstruction and Development (EBRD) and the Asian Development Bank (ADB) – and membership is sought of the trade-conducive World Trade Organization (WTO). But dangers loom of the "Dutch Disease" and over-reliance on a few price sensitive commodities. The chapter therefore concludes on the current and prospective problems facing the central Asian members of the United Nations after their first five years of statehood.

(i) The management shock

The new states inherited an open economy, which was nevertheless focused overwhelmingly on the Soviet republics and, to a much lesser extent, the then disappearing CMEA, comprising eastern Europe and Mongolia. The Soviet expansion of trade with industrial market economies and with politically selected developing countries brought by international détente had had little effect on central Asia, although exchanges with China and India were burgeoning.[449] In 1988, total trade (imports plus exports) as a percentage of GDP was 34 per cent in Kazakstan, 39 per cent in Turkmenistan and Uzbekistan, 42 per cent in Tajikistan and 45 per cent in Kyrgyzstan.[450] Such ratios are typical among "middle-income" economies – 42 per cent on average for those thus classed by the World Bank – but, save for Kazakstan, are higher than those for "low-income" countries – 35 per cent on World Bank statistics.[451] On per capita GNP (at official exchange rates) the Bank puts Tajikistan and Kyrgyzstan within the "lower income" group ($360 and $630, respectively), and Uzbekistan, Kazakstan and Turkmenistan within the "middle income" ($960, $1,160 and $2,900, respectively).[452] But between 86 and 89 per cent of that 1988 trade turnover was with Soviet republics.[453] The central Asian states were thus, as open economies, sensitive to exogenous shock, but the shock at the time of their independence was the stronger because, as table 4.1.1 shows, production and effective

TABLE 4.1.1

GDP growth in the central Asian economies, 1992-1996
(Annual percentage change)

	1992	1993	1994	1995	1996
CIS	-13.9	-9.8	-14.5	-5.7	-5.3
Kazakstan	-13.0	-12.9	-18.8	-8.9	0.5
Kyrgyzstan	-13.9	-15.5	-20.1	-5.4	6.0
Tajikistan	-31.0	-17.3	-12.7	-12.4	-17.0
Turkmenistan	35.7	10.0	-18.0	-16.0	–
Uzbekistan	-11.0	-2.3	-5.2	-1.2	2.0

Source: Appendix table B.1.

demand plummeted in the partner economies responsible for over four fifths of their external trade. For three countries, dependence on the other Soviet republics became even greater in 1991 than it had been in 1988. The former Soviet Union share of total trade (imports plus exports) was 96 per cent in Tajikistan, 91 per cent in Kyrgyzstan and 89 per cent in Kazakstan, and was still high in the other two – 85 per cent in Uzbekistan and 84 per cent in Turkmenistan. It happened also that import demand was at the time relatively weak in the few market economies with which central Asia had trade relations.

Domestic transmission of the external shock was rapid because it came at a time when the planning mechanism, which under other circumstances could have absorbed acute changes, was being dismantled, and market relationships – unfamiliar to government and enterprises alike – were hesitantly being applied. As a regulator, from the microeconomic to the macroeconomic scale, the state apparatus was in turmoil. First, even with central planning, it had not exercised control over all the enterprises and institutions on its territory: the Soviet lines of command ran straight to and from Moscow in the case of All-Union ministries, and authority was divided with Moscow in Union Republican ministries; at the sub-national level dual subordination was the rule, the duality being a ministry and the regional or local authority. Some budget financed institutions were national, others – including military and space agencies[454] – were financed and administered from Moscow. Secondly, it had had few rights over its own finances, and such as it had by 1991 were newly won in what under perestroika had been termed "the war of the laws". Most tax rates were set by the Ministry of Finance in Moscow – some imposts were paid direct to it – and the Union Republican and sub-national authorities had budgets which were balanced by the simple allowance to them of sufficient revenue to cover expenditure as established by centrally determined norms. Where such allotted revenue did not equal outlay,

---

[449] Cuban cane sugar displaced Kyrgyz sugar beet and was imported as material for the republic's refineries.

[450] IMF, Economic Review, *Common Issues and Interrepublic Relations in the Former USSR* (Washington, D.C.), April 1992, table 1.

[451] World Bank, *World Development Report 1996: From Plan to Market* (New York, Oxford University Press for the World Bank, 1996), GDP for 1994 from table 12, aggregated imports and exports of merchandise from table 15.

[452] The World Bank cautions that these estimates are preliminary and gives no figure for Turkmenistan (while ranking it between Estonia at $2,820 and Brazil at $2,970) in terms of GNP per head in 1994. World Bank, *World Development Report 1996: ...*, table 1.

[453] The bounds were Kazakstan and Tajikistan at 86 per cent and Turkmenistan at 89 per cent. IMF, op. cit.

[454] After independence Russia recognized the Baikonur Space Centre as Kazak state property but, after lengthy negotiations, leased it for a minimum of 20 years at an annual rental of $115 million, plus reimbursement of maintenance costs incurred by Kazakstan in 1992 and 1993. Russia is, however, establishing a new space centre in its Far East.

payment was made from central funds, generating the large transfers to which reference is made above. Thirdly, central Asian officials (and academics who could be called on for advice) had had far fewer opportunities to study economic theory or practice in developed market economies; and their universities, still less their government departments, had but rare professional visitors from abroad. There were exceptions, but even today there are cases of rivalry in the advice tendered governments by the western educated and the Soviet trained.[455] Administrative rules and conventions for seven decades had been those of a state bureaucracy operating a command economy and subordinate to the Communist "Partocracy".

(ii) The shock of price liberalization

The Soviet rouble circulated throughout central Asia and Russia until mid-1993 and in January 1992 there were no constraints on trade among them. Russia's decontrol of most wholesale and retail prices on 2 January 1992 – decided the preceding November – had to be followed in the other republics, and by the end of that month all central Asian governments had done so. Some limitations were retained on incomes (most of which had been state determined in the past), but they largely followed prices in a familiar spiral. The degree of liberalization (and the subsequent reimposition of some controls) varied somewhat, but the measure was a profound shock. Enough has been written on the failure of the Russian government to limit price liberalization to the expected "corrective inflation" for the explanations not to be reiterated here,[456] but two of the constituents must be noted as especially relevant to the central Asian inflations. Competitive supplies, which in normal circumstances would have been attracted by the rapid price upswing and following money incomes, were still rarer in central Asia than in Russia.[457] The contractual relationships between enterprises were abandoned and the republics' firms, nominally state owned but effectively in the hands of "insiders", exercised their monopoly power wherever they could; the low share of non-CIS trade made for few competitive imports (which in any event were still subject to Soviet style controls). Secondly, the central banks of the five states, hastily redesignated from Union Republican branches of the USSR Gosbank, could create rouble credits without hindrance, and even if they were to have exercised restraint, the lack or inadequacy of bankruptcy procedures or of branch bank monitoring and sanctions allowed the accumulation of vast inter-enterprise indebtedness and tax arrears.

TABLE 4.1.2

Consumer price inflation in the central Asian economies, 1992-1996
(Annual averages, percentage change)

	1992	1993	1994	1995	1996
Kazakstan	1 504	1 663	1 880	176	39
Kyrgyzstan	855	1 209	278	43	30
Tajikistan	822	2 885	350	682	422
Turkmenistan	770	1 631	2 714	1 005	500
Uzbekistan	415	1 232	1 550	316	50
Russian Federation	1 529	875	309	197	48

Source: Appendix table B.7; EBRD, *Transition Report 1996* (London), pp. 196-209.

As in Russia and all CIS members, the liberalization shock led to rapid inflation, but, as table 4.1.2 shows, to rates significantly less than in Russia. The exception was Kazakstan where inflation was virtually identical to the Russian, probably because retail interchange was considerable between southern Russia and the Slav majority zones of northern Kazakstan. By the imposition of price controls and a special purchase constraining scheme for consumer goods, Uzbekistan had the lowest inflation in the initial "shock" year,[458] but it was just temporarily hidden. In annual terms, consumer price inflation was in four digits for three years in Kazakstan and Turkmenistan, for two years in Uzbekistan, and for one year in Kyrgyzstan and Tajikistan.[459] Inflation was lowered to two digits only in 1995 – in Kyrgyzstan – and in 1996 in Kazakstan and Uzbekistan.

4.2 Macroeconomic instabilities

Until the establishment of separate currencies during 1993, rouble devaluation was a factor in these inflations, but a more potent force was monetization of fiscal deficits. As table 4.2.1 shows, fiscal deficits were as large as 17 to 31 per cent of GDP in the first full year of independence – Kazakstan was in single figures (7 per

---

[455] Exemplified in the advice to impose arbitrary exchange controls in Uzbekistan in November 1996, tendered by the President of the central bank, F. Mulajanov, with his entirely Soviet experience, and the contrary counsel offered by the western trained economist, R. Azimov, who chairs the National Bank for Foreign Economic Activities.

[456] UN/ECE, *Economic Survey of Europe in 1992-1993*, pp. 160-168.

[457] So disrupted were intra-CIS trading relationships that a Czech firm found it profitable to purchase motor cars from the Volga Automobile Plant (VAZ) in Russia and a mere 220 kilometres from the Kazak frontier, have them transported to the Czechoslovak frontier and reship them forthwith to Kazakstan. This particular commerce persisted until 1996.

[458] In addition to some rationing, coupons were issued with the wages and salaries of all state employees, cooperative members and social transferees equivalent to 70 per cent of the remuneration or benefit: these coupons had to be surrendered for purchases in state enterprises (except for newspapers, periodicals and books).

[459] The Tajik index does not cover regions (the majority of the territory) not ruled from Dushanbe, where some controls are enforced; it is likely that price increments have been higher elsewhere. At the time of the peace agreement of 23 December 1996, the Dushanbe and Kulyab regions were held by the government; Garm, north Komsomolabad and west Badakhshan were controlled by the Islamic and the democratic opposition; the rest of Badakhshan had its Ismaili separatists; and the remainder of the west and north was run by local clan chiefs. See map in *Central Asia Newsfile*, January 1997, p. 2.

TABLE 4.2.1

Budget disequilibrium in relation to GDP and general government expenditure in the central Asian economies, 1992-1996

	1992	1993	1994	1995	1996
Kazakstan					
Per cent of GDP	-7	-1	-7	-3	-3
Per cent of expenditure	-23	-4	-29	-10	-12
Kyrgyzstan					
Per cent of GDP	-17	-14	-8	-14	-6
Per cent of expenditure	-51	-46	-30	-59	-29
Tajikistan[a]					
Per cent of GDP	-31	-23	-5	-10	-7
Per cent of expenditure	-54	-86	-12	-66	-42
Turkmenistan[b]					
Per cent of GDP	13	-1	-1	-1	-1
Per cent of expenditure	31	-3	-9	-11	-3
Uzbekistan					
Per cent of GDP	-18	-10	-6	-4	-8
Per cent of expenditure	-37	-24	-19	-12	-20

Source: 1992: EBRD, *Transition Report 1996* (London), pp. 196-209; other years: IMF, European Department II Database.

[a] Tajikistan data refer to the state budget only.

[b] Turkmenistan data for both revenue and expenditure are incomplete.

cent) and there was a surplus on such outlays and revenue as the Turkmen government chose to consolidate in the central budget.[460] In 1993, in that part of war torn Tajikistan under Dushanbe rule the deficit was almost as large as central budget expenditure (and just on a quarter of the shrunken GDP) but by 1996 it was in the same range as Kyrgyzstan and Uzbekistan (respectively 6 and 8 per cent of GDP). Kazakstan had kept its deficit at 3 per cent of GDP for two years. Measured against the GDP share taken by government expenditure (table 4.2.2), however, only Kazakstan exhibited a magnitude then being experienced in eastern Europe – 20 per cent in Bulgaria, 19 per cent in Romania, 16 per cent in Hungary and 14 per cent in Slovakia.[461] Thus in their first years of running their own currency (Kazakstan and the uncertain data of Turkmenistan excepted) the central Asian states were in comparison with more developed transition states running bigger deficits to spend proportionately more.

In the early years of independence that relatively high expenditure could not be covered by public borrowing, due to the absence of any domestic capital market and the inevitable delay before the international financial institutions could provide credit. Soviet practice had drawn upon private savings only by commandeering funds from the amount due to depositors in the State Savings Bank; the new governments could at the time gain little from these banks when ownership passed to the republics, because the real value of deposits was soon greatly shrunk by rapid inflation. The republican groups of the USSR Savings Bank were renamed the People's Bank in Kazakstan and Uzbekistan, Elbank in Kyrgyzstan and Khatlon Bank in Tajikistan, but the title was retained in Turkmenistan. Each still remains wholly state owned.[462]

Under these limitations, the five governments could pursue measures along three lines to moderate the extent of monetizing their fiscal deficits. First, all but Kazakstan made heavy use of membership of the "rouble zone" to issue credits on the Central Bank of Russia; this had to stop in 1993, as is discussed in section 4.4(i) below. Secondly, they strove to increase revenue, but were slowed by the need to reconstitute the fiscal system and to restructure the institutions for tax collection. Thirdly, they made efforts to constrain central and local expenditure, but the coincident impact of recession evoked further subsidies to state enterprises and the payment of more social benefits. The requirement for such transfers has moderated, but some implicit subsidies remain through foreign exchange rationing (such as surrender quotas) and price and procurement controls.

(i) Policies to increase revenue

In analysing the "shock" to tax revenue, the indicator used in table 4.2.2 is its percentage of GDP. By 1996 tax revenue was about half its late Soviet proportion in Kazakstan, Kyrgyzstan and Tajikistan, but about the same in Turkmenistan and Uzbekistan. Over that period GDP had fallen sharply in Kazakstan and Kyrgyzstan (see appendix table B.1), but had been more closely maintained in Turkmenistan and Uzbekistan. Combined with the withdrawal of Union grants, Ministries of Finance were confronted with a profound real decline (receipts indexed by a GDP deflator) at a time when the profile of taxpayers and the composition of the tax base was in flux. First, large profits were generated in trade and currency dealing, little of which could initially be captured by taxation. As noted in section 4.4(i), a partial liberalization was undertaken of foreign exchange controls and of the quantitative restrictions imposed by central planning. "Shuttle traders"[463] import – on Kazak customs service estimates[464] – considerably more than they export and are doubtless financing the balance of their purchases from "flight

---

[460] IMF Staff Country Report No. 96/30, *Turkmenistan – Recent Economic Developments* (Washington, D.C.), April 1996, p. 15, notes, on introducing the Treasury system, that "the most pressing remaining issue is the need to bring the 29 self-supporting ministries into the budget".

[461] UN/ECE, *Economic Survey of Europe in 1995-1996*, table 3.7.1, 1994 data.

[462] The National Bank of the Kyrgyz Republic closed Elbank in March 1996 for thorough recapitalization.

[463] *Chelnoki* in the region's lingua franca, Russian, is from a weaver's shuttle, characterizing persons who earn a substantial livelihood by taking consumer goods in and out of the country, often to distant destinations (China, Singapore, India and Dubai are especially favoured).

[464] Government of the Republic of Kazakstan/Centre for Economic Reforms, *Kazakstan Economic Trends*, Second Quarter 1996, p. 115.

TABLE 4.2.2

Government revenue and expenditure in relation to GDP in the central Asian economies, 1989-1996

(Per cent of GDP)

	Tax revenue	Non-tax revenue	Grants	Expenditure
Kazakhstan				
1992	21.5	1.3	1.7	31.9
1993	15.6	4.2	–	23.5
1994	11.6	5.7	–	17.3
1995	12.9	3.9	–	20.2
1996	10.5	2.2	–	17.2
Kyrgyzstan				
1989	28.0	2.9	7.1	35.9
1990	26.3	1.3	10.9	38.3
1991	17.5	5.5	12.7	31.1
1992	11.2	1.6	–	26.2
1993	13.5	1.3	8.1	36.6
1994	13.9	3.9	2.5	28.6
1995	15.1	1.1	0.3	30.3
1996	13.5	1.3	1.4	23.7
Tajikistan				
1991	20.8	–	19.6	38.6
1992	25.7	0.9	..	57.8
1993	25.2	1.9	..	52.1
1994	42.2	2.3	..	55.0
1995[a]	14.7	0.4	..	24.3
1996	11.8	0.5	..	19.7
Turkmenistan				
1989	15.5	14.6	–	32.0
1990	14.1	29.4	–	42.3
1991	11.6	26.6	–	35.7
1992	36.1	8.9	–	30.9
1993	22.7	1.5	–	26.9
1994	6.8	3.6	–	11.9
1995	8.0	2.2	–	11.8
1996	14.0	1.0	–	15.7
Uzbekistan				
1989	24.5	2.3	11.8	..
1990	24.4	2.5	19.8	..
1991	19.3	11.1	19.4	32.4
1992	20.6	10.7	–	43.4
1993	28.5	7.5	–	37.1
1994	23.7	5.5	–	38.2
1995	28.3	1.8	–	38.1
1996	29.9	1.7	–	35.6

Source: IMF Economic Reviews, *Kazakhstan,* 1994, table 19; *Kyrgyz Republic,* 1993, table 13; *Tajikistan,* 1994, table 20; *Turkmenistan,* 1994, table 26; *Uzbekistan,* 1994, tables 25, 26 (Washington, D.C.); IMF Staff Country Reports, *Kazakstan – Recent Economic Developments,* March 1996, table 20; *Kyrgyz Republic – Recent Economic Developments,* September 1996, table 18; *Tajikistan – Recent Economic Developments,* June 1996, table 28; *Turkmenistan – Recent Economic Developments,* April 1996, table 21a; *Uzbekistan – Selected Issues and Statistical Appendix,* August 1996, table 4 (Washington, D.C.).

Note: Capital revenue is not included in the table: all shares reported by the IMF are less than 1 per cent except for Kazakhstan in 1993 (2.5 per cent). The coverage of data on Tajikistan is incomplete and non-comparable over time.

a Fiscal data are not comparable with previous years because cash and non-cash receipts were differently valued until the May 1995 currency reform.

funds": both the duties at the frontier and tax on the proceeds are under-collected. Secondly, the recession particularly hit state enterprises, which proved less flexible to changing market conditions and more sensitive to ruptures of longstanding contracts with state firms elsewhere in the former Soviet Union. Because of previous close – often personal – relations with finance and industrial ministries, their managements were able to negotiate the tolerance, even the remission, of tax arrears. Thirdly, it became imperative to adjust forms of taxation from transfers of bank balances from one state account to another to the imposts which are more cost effective and appropriate to a market system. A Soviet Ministry of Finance could levy turnover tax on a state enterprise, deduct its chosen share of profit, oversee the appropriation to investment or depreciation, deduct income tax at source (at a low non-progressive rate) and encash social security premia as a set proportion of the authorized wage bill. Ministries of the independent states had no such automatic rights over state enterprises when they were corporatized[465] or privatized. Finally, the expansion of the private sector – *ab initio* as property rights were defined and protected, and from privatization – multiplied the number of potentially taxable enterprises and taxpayers. "Small" privatization in all five states took precedence over that of medium and large firms, thereby substituting a myriad of taxpayers for a small number of groupings under the administrations of the Ministry of Local Economy, of municipal and rural authorities or under state industrial and transport enterprises. Inadequate legal and law enforcement infrastructures enabled an abnormal proportion of the new businesses to avoid tax payment or to conceal criminal activities.

Of the new range of taxes – the application of which involved a "learning curve" – value added tax (VAT) and personal income tax were the most pervasive. By 1995 VAT receipts constituted 33 per cent of total revenue (other than grants) in Kyrgyzstan and in Turkmenistan, 25 per cent in Tajikistan, 19 per cent in Uzbekistan and 18 per cent in Kazakstan; individual income tax was 11 per cent in Kyrgyzstan and 9 per cent in Uzbekistan, but only 6 per cent in Tajikistan and 3 per cent in Turkmenistan; consolidated returns for enterprise and individual income taxes in Kazakstan in 1995 were 34 per cent.[466] Kyrgyz VAT legislation was amended with effect from 1 July 1996 to introduce a modern invoice-based tax, based on the destination principle, to absorb the retail sales tax (inappropriately cumulated on VAT) and to conform to the requirements of the Customs Union with Belarus,

---

[465] That is, converted to joint-stock companies, wholly-owned by a state entity; state firms had already gained some financial autonomy under Soviet legislation after 1987.

[466] Sources are as for table 4.2.2, supplemented by the Government of the Republic of Kazakstan/Centre for Economic Reforms, *Kazakstan Economic Trends, Monthly Update,* May 1996, table 8.2, and IMF Staff Country Report No. 96/98, *Kyrgyz Republic – Recent Economic Developments* (Washington, D.C.), September 1996, pp. 18-19.

Kazakstan and Russia.[467] The Tajik and Turkmen collection of personal income tax was low due to numerous exemptions which international advisers are seeking to eliminate, or at least to moderate.

(ii) Policies to reduce expenditure

As Soviet Union republics, the central Asian states were remarkably homogeneous in the share of output commanded by direct state expenditure. Table 4.2.2 shows (except for Tajikistan) at least one year of 1989-1992 in which government expenditure was 31 or 32 per cent of GDP.[468] Government absorption of a shrinking gross product has been reduced in Kazakstan, Kyrgyzstan and probably Tajikistan (where comparability was interrupted by the treatment of cash and bank money in the 1995 currency reform)[469] and Turkmenistan (where many ministries are off-budget and considered self-financing).[470]

The Uzbek government on the other hand maintained, and even in the two initial years of independence raised, its take of national resource flows, which were themselves only slightly diminished from the previous level. This policy followed from the implementation of three of President Karimov's five Basic Principles for "Uzbekistan's Own Model for Transition to a Market Economy" – "the state guarantees economic transformation"; "social protection is a major function of the state"; and "the consistent and phased implementation of economic reforms".[471] The Uzbek government's declared preference for "gradualism" – partly impelled by demonstrations in 1992 against price liberalization and the reduction of student grants[472] – led to subsidization both of producers and consumers. Direct subsidies to state enterprises were quickly tapered down and – as privatization progressed – mostly replaced by subsidized credit on prolonged terms or by support from extrabudgetary funds. Consumer subsidies peaked at 7 per cent of GDP in 1992, were down to 4 per cent in 1993 and almost fully withdrawn during 1994. Social security payments, other than those (such as unemployment benefit) paid from a separate fund, steadily rose – from just under 3 to over 5 per cent of GDP between 1992 and 1995. Credits to farmers, however, remained the weakest link in the stabilization chain. The poor 1996 grain and cotton harvests and a decline in world prices for the latter evoked widespread calls for rural credits, to which the government acceded: reserve money increased by 50 per cent in October alone, both from this cause and from business generally following an ill-advised cut in the central bank rate in August.

Kazak social security spending through the budget declined as a proportion of GDP, but extrabudgetary funds, which took over pension and unemployment charges, ran a deficit (reaching some 2 per cent of GDP) which was compensated from the budget. Subsidies of all kinds took as much as 5.5 per cent of GDP in Kyrgyzstan in 1993, but direct consumer subsidies were abolished in 1995 and replaced by the so-called "unified cash benefit" paid to households whose per capita income falls below a determined level – payment is triggered when they have been three months below that level. By such targeting, the aggregate disbursement is expected to be much lower than was paid in subsidies.

Both producer and consumer subsidies were high in Turkmenistan after independence: the latter equalled 9.6 per cent of GDP in 1992, but the trimming to 1.1 per cent in 1995 was partly illusory, being achieved by shifting them to extrabudgetary agencies. Subsidies may in fact have increased that year because grain and livestock prices were substantially raised without compensatory adjustment to retail prices. The dependence of Turkmen households on these food subsidies is noted in section 4.3(v) in assessing poverty. Transport tariffs are heavily subsidized. But the other biggest subsidy is non-transparent, as the state owned gas and oil enterprises are required to supply citizens' households and small businesses without or at nominal charge.[473]

The extrabudgetary pension and social security funds run a deficit covered by the budget at about 1 per cent of GDP. In the light of the disturbed conditions since Tajik independence and the limited area over which the government has control, it is unsurprising that the deficit on social security funds has been as high as 5 per cent of GDP in 1994. Partly financed by the sequestration of some of the foreign receipts from aluminium and cotton exports, as described in section 4.4(i), consumer subsidies on flour and bread alone then amounted to a further 8 per cent of GDP. The subsidy on cheap or free fuel just noted came from the same foreign earnings seizure. In 1995, however, raising the controlled

---

[467] The IMF points out that unilateral change to the destination principle, while other Union participants remained on an origin principle, would have involved double taxation on imports from those partners. IMF, loc. cit., p. 18.

[468] Alternative series are published by the EBRD, notably exhibiting higher shares of GDP in Uzbekistan and Tajikistan (1991 only) and a variable relationship for the others; Kazakstan in its series gives 31.4 and 32.9 as percentages of GDP in 1990 and 1991. EBRD, *Transition Report 1996* (London), annex tables.

[469] As section 4.4(i) describes, cash exchanged at 100 to 1 and non-cash at 1,000 or 1,200 to 1.

[470] Their debts or carry-over funds do not revert to the Ministry of Finance at the end of the financial year (as under the Treasury system).

[471] As appendix table B.1 shows, Uzbek GDP by 1996 was still only 18 percentage points below the 1989 level; Turkmenistan was at that level but the other three states were between 44 and 67 points below.

[472] The President's reaction also reflected previous civil disturbances during which hundreds had died – in June 1989 in the Fergana Valley and in June 1990 near Osh.

[473] Resident Russians have the same benefit if they have taken out dual citizenship.

price of bread cut that particular subsidy to 5 per cent of GDP and some social transfers were curtailed. The funds paying unemployment benefit and pensions were also in deficit despite being in arrears to claimants. In 1996 a more precise targeting of benefits was initiated, with a view to contributing to the reduction of the overall budget deficit to 5.4 per cent of GDP; in the event, the deficit was 6.7 per cent of GDP.

All governments save that of Uzbekistan have reduced their consumption of a smaller national cake. In table 4.2.2 statistics from the IMF show that the 1996 proportion of expenditure to GDP was less than half the share exhibited in the last pre- (or immediate post-) independence year in Kazakstan, Tajikistan and Turkmenistan, was significantly lower in Kyrgyzstan and was about constant in Uzbekistan. As GDP has fallen least in Turkmenistan and Uzbekistan, the implication is that the relationship of these aggregates more closely approximates to those of the late Soviet period than it does in the other three countries.

(iii) Overhaul of the public finance and statistical agencies

Tax collection became inevitably more costly as the tax base widened from a determinate set of state enterprises and trade agencies to a universe embracing a diverse private sector, potentially all the gainfully employed and a ring of customs posts at land and air entry points. The inputs supplied included new legislation, the training of officials and physical structures. But it was of considerable importance to establish a self-standing agency of government to administer taxation and to separate tax-liable state entities from a "cosy relationship"[474] with the Ministry of Finance and other supervisors. The diffusion of computerization in tax administration and expenditure monitoring and management, where it had been notably weak, was assisted by such concentration of work. More fundamental to proper financial management was the abrogation of earmarked taxes in a general conversion to a Treasury system for budget financed departments and the establishment of self-funding executive agencies for such separable functions as health and unemployment insurance and social security. Such fiscal concentration is still underway, and has only just begun in Tajikistan, and is complemented by the establishment of taxation agencies separate from the direct hierarchy of the Ministry of Finance. Such agencies need staffing and of particular importance is the training and retraining of civil servants. With much still to do, the extirpation of corrupt practice, the tolerance of tax arrears and the transparency of public contracting remain difficult tasks.

The European Union (TACIS Programme), the IMF and the United Nations Development Programme (UNDP) have taken the lead in advising central Asian governments on statistical and cognate problems. Intensive use has been made of successive IMF country studies[475] – Economic Reviews (1993-1994) and Staff Country Reports (1995 onward). The IMF began sending specialized statistical missions to Ministries of Finance in early 1994. Other notable steps were taken in 1996. The European Expertise Service of the European Commission began publication jointly with the Centre for Economic Reforms of the Government of Kazakstan of a quarterly *Kazakstan Economic Trends*, with a *Monthly Update*. Partly staffed by the German Institute for Economic Research (DIW), it was modelled on the successful *Russian Economic Trends* (also with a *Monthly Update*) launched by the Russian Government as early as 1992 with EU finance and the collaboration of the London School of Economics. A corresponding *Uzbekistan Economic Trends* was started in 1997. The monitoring of societal as well as economic dimensions has been enhanced by the UNDP publication of a country series under the title *Human Development Report*. Among many studies commissioned by the European Union TACIS Programme, an example of a substantial survey in the socio-economic field is the three-volume *Social Policy and Enterprise Restructuring in Kazakstan*, produced in 1996.

The economic and business analysis of the central Asian states has been hitherto complicated by gaps in national and sub-national statistics and by sometimes conflicting data as either reported by domestic agencies or reproduced by international organizations. In the preparation of this chapter an effectively arbitrary choice has at points been exercised among differing statements of what appears to be the same statistical fact. The advice tendered by bodies noted in the preceding paragraph will doubtless remove many such difficulties in future, although some divergences will persist, due either to variant methodologies or to concealment by practitioners (such as tax avoidance, capital flight, illegal and corrupt transactions, trade in narcotics or arms, and so on).

4.3 The post-independence recession

(i) Aggregate decline and upturn

Thus far only one transition economy – Poland – has surpassed the pre-independence (1989) aggregate production level, although the eastern *Länder* of Germany, Slovakia and Slovenia have breasted 90 per cent (appendix table B.1). Neither GDP nor any other single value can calibrate progress in systemic change,

---

[474] The term is J. Kornai's to describe the mutual interaction of a state enterprise, its supervisory agency and the Ministry of Finance in a Soviet type system.

[475] Examples of these are noted in the sources to some of the tables and footnotes in this chapter.

but regaining a quantitative baseline and a majority share for the private sector are helpful indicators of the end of an abnormal period. There are many objections to GDP as a surrogate for welfare, that is for the outcomes of economic activity, but in the present context its estimation during a transition process may fail to measure a greater proportion of some "comprehensive" production aggregate than was the case in a planned economy. As already mentioned, newly created businesses (some tempted to escape taxation or the law) are difficult for transition governments, themselves undergoing change, to monitor. In the particular case of an output base such as the last planned economy year (1989), the mix of goods and services provided under repressed inflation, arbitrary pricing and enterprise production constraints must have differed greatly from that which would have operated under market conditions. The composition and relative valuation of demand and supply was profoundly affected by the government's predominance in procurement (of defence goods and services, farm produce and producer goods) and its control over foreign trade and payments — all of which segregated a "passive" form of money from that, in household and foreign hands, which had an "active" role. Many goods and services, such as finance, the media and advertising, were either unprovided or supplied in a form which rapidly diversified as political and economic restrictions were lifted.

Table 4.3.1 exhibits the available detail of the decline in economic activity of the post-communist states, that is, since 1991. On the official data, the level of real GDP in 1996 was almost one fifth below that of 1991 in Uzbekistan, well over a third in Kazakstan and Kyrgyzstan and two thirds in Tajikistan. Only Turkmenistan reports national income (real NMP) near the 1991 level,[476] but there too real gross output of the main sectors of production was still in deep recession. Production in mining and manufacturing has declined steeply in all countries but Uzbekistan, and in most faster than in agriculture. One reason is that trade with "traditional" partners in temperate farm produce has held up more than for industrial materials and products (Kazakstan, Kyrgyzstan and Turkmenistan) and "non-traditional" partners have been found for cotton

---

[476] The real NMP data of Turkmenistan are rather puzzling, possibly reflecting inappropriate deflation procedures. The relatively high level of the NMP index in 1996 stems from the substantial real NMP growth reported for 1992 and 1993 (see table 4.1.1), which the Turkmen statistical authority attributes to trading gains from fuel exports in intra-CIS trade and allocates to the value added of the industrial sector (growth of 132.1 per cent "at constant prices" in 1992). CIS Statistical Committee, *Ofitsial'naya statistika stran Sodruzhestva Nezavisimykh Gosudarstv* (Moscow), October 1996 (database on CD-Rom). An alternative series of NMP growth data used by the IMF and the international banks (e.g. IMF, *World Economic Outlook* (Washington, D.C.), October 1996, table A7) shows NMP declines of 5.3 and 10.0 per cent for the same years, but seems to be based on NMP data "exclusive of natural gas exports". IMF Economic Reviews, *Turkmenistan*, No. 3 (Washington, D.C.), March 1994, pp. 88-90).

TABLE 4.3.1

Production aggregates in the central Asian economies, 1996
(Index numbers, 1991=100)

	GDP^a	Industry	of which: Consumer goods	Agriculture
Kazakstan	56	49	32	60
Kyrgyzstan	58	36	28	68
Tajikistan	36	40	21	42
Turkmenistan	103	73	..	71
Uzbekistan	83	104	119	69

Source: For GDP and industry: appendix tables B.1 and B.4; for consumer goods and agriculture: CIS Statistical Committee, *Sodruzhestvo nezavisimykh gosudarstv (SNG) v 1996 g., Statisticheskii spravochnik* (Moscow), 1997, pp. 210, 223, 262, 275, 288.

a NMP for Turkmenistan.

(Uzbekistan and even war damaged Tajikistan). Another is the deliberate protection of the existing agrarian institutions, with only nominal change in their membership, in contrast to the upheaval of privatization and labour shedding in industry. Services of all kinds were little favoured by Soviet planners and their opening up to private entrepreneurship (such as trade, catering and banking) and (still inadequately) to foreign providers (notably telecommunications) has made this general branch the most dynamic of the transition period. Kazakstan and Uzbekistan have notably shared in that dynamism. Finally, trends in construction have been conflicting: on the one hand civilian investment and military and security construction have respectively shrunk and all but disappeared, but on the other hand the privatization of housing and the widening of income differentials have fostered residential building and repair.

The decline has been much steeper (1991 to 1996) in consumer goods production in three states (Kazakstan, Kyrgyzstan and Tajikistan) due to competition from imports. All post-communist countries, from the former German Democratic Republic to Mongolia, experienced a buying spree of western goods for which household demand had been so long pent up. But it seems to have been longer lasting in central Asia. An impression of the magnitude involved may be gained by comparing Kazakstan's 1995 imports of consumer goods of $423 million, or 25.8 billion tenge at the current exchange rate, with retail sales of 139.9 billion tenge, i.e. some 18 per cent before any mark-ups which could easily double or triple this share.[477] In Kazakstan's retail sales of textiles, the proportion of imports, much of which come from east and south Asia, comes close to three quarters.

---

[477] Imports (calculated from product groups) and retail sales from Kazakstan State Statistical Committee, *Kratkii statisticheskii ezhegodnik Kazakstana 1995*, pp. 46, 140-142; annual exchange rate (61.125 tenge to the dollar, averaging monthly rates). World Bank, *World Development Report 1996:* ..., table 1.

Unrecorded imports (see section 4.2(i)) would have increased the volume of competitive imports, and more generally the recession would be recorded as less deep and the start of recovery earlier if activities could be enumerated that currently fall outside "measured GDP". The strengthened statistical agencies described in section 4.2(iii) are engaged in the investigation of this penumbra, but measurement of the informal sector of production is still in its early stages. One estimate is that of the Kyrgyz Statistical Committee, which puts it at adding 19-20 per cent to measured GDP in 1994.[478]

The immediate and initial causes of the general downturn – the loss of virtually guaranteed demand in the former USSR and the rest of the CMEA, the collapse of military purchasing and the competition from imports – were gradually offset by entry into "non-traditional" markets, the stimulation of domestic production by small and medium privatization, and by the entry of foreign corporations as buyers at large privatizations or partners in joint ventures or as direct investors, mainly in energy (oil, gas and electricity in Kazakstan and oil in Turkmenistan). The issue was not only that of a collapse of the volume of orders and supplies, but also that the officials who had run the former trading network had little or no responsibility for the new relationships. Such an "institutional hiatus" took time to fill, as other administrators and business people took over, or as earlier contacts and payments facilities with firms in other CIS countries – especially Russia – were re-established.

Recovery, nevertheless, was delayed by the real appreciation of the new currencies (as demonstrated in section 4.4(i), except for Turkmenistan) and, at points, by officials obstructive of local business or big foreign companies. There were substantial turn arounds of industrial output in all but Tajikistan in 1996, following negative growth in 1995 in Kazakstan, Kyrgyzstan and Turkmenistan and constant output in Uzbekistan. GDP showed positive growth in 1996 in Kyrgyzstan (6 per cent), Uzbekistan (2 per cent) and – although by less than 1 per cent – Kazakstan; there was at least no further fall in Turkmenistan. Considerable foreign investment and a resumption of domestic capital formation were crucial to the upswing, as section 4.5(iv) points out.

(ii) Disinvestment and new capital formation

The proportion of investment under-reported must be less than that of unmeasured GDP and, as the magnitude of the latter is likely to have increased during transition, the share of investment in national accounting aggregates would have decreased faster than shown by the official estimates. Comparable series even of the latter are not, however, fully available. As presented by the EBRD in a major study of investment, its determinants and finance,[479] the diminution of the proportion of gross domestic investment in GDP between 1990 and 1994 was from 27 to 21 per cent in Kyrgyzstan, and from 32 to 22 per cent in Uzbekistan. Other estimates for Kyrgyzstan (almost certainly on a different definition) show investment as low as 12 per cent of GDP in 1994 (but 20 per cent in 1995 and 17 per cent in 1996), matching a fluctuation of savings from a nadir of 7 per cent of GDP in 1994 to 12 per cent of GDP in 1996.[480] The World Bank reports investment as high as 30 per cent of GDP in Uzbekistan, to which foreign inflow must have contributed. The infusion of All-Union funds into Kazakstan in 1990 was particularly high, putting the gross investment share as high as 48 per cent of GDP, and included early development of the Tengiz oilfield, such that even with much foreign investment (see section 4.5(iv)) 24 per cent for both 1994 and 1995 constitutes a fall.[481] Foreign expenditure on oil and gas exploitation was large – greater than in any other CIS country, including Russia (as section 4.5(iv) recounts) – but domestic capital spending would seem to have been smaller. Hostilities in Tajikistan must have virtually annihilated investment, and information on Turkmenistan is insufficient.

The reduced volume of capital formation must imply net disinvestment, taking place along two lines. One is the underutilization of depreciation allowances, which were a compulsory component of Soviet enterprise accounting (part for capital replacement, part for major repair), while enterprises faced declining demand and could take autonomous decisions, either as corporatized or as private entities. The other is total disuse – the "rust-belt" phenomenon – where factories and mines no longer appear viable at current market price relativities, or where the costs of conformity to modern environmental and health and safety criteria are too great.[482] The low level of investment of course hinders retooling, change of production profile and the installation of more efficient techniques. In the longer term such underinvestment may not prove to have been disadvantageous. First, some activities may be unrevivable on environmental grounds. Secondly, Soviet planners favoured capital intensive branches and located enterprises on Union-wide (albeit

---

[478] IMF Staff Country Report No. 96/98, *Kyrgyz Republic – Recent Economic Developments* (Washington, D.C.), September 1996, p. 70.

[479] EBRD, *Transition Report 1995* (London), chaps. 3-7; statistics from table 3.1.

[480] Graphics accompanying the text of President Akaev's address to the World Economic Forum, "Kyrgyzstan: signs of recovery", mimeo, Davos, February 1997.

[481] World Bank, *Statistical Handbook 1996: States of the Former USSR* (Washington, D.C.), p. 199 and *World Development Report 1996: ...*, table 13.

[482] Two large plants assessed by western companies for part purchase were calamitously below modern health and safety standards. BAT bought into and repaired the Tashkent tobacco enterprise, but no firm was interested in the Tursunzade aluminium refinery.

doubtful) criteria. Thirdly, the ratio of productive assets to GDP at the end of the Soviet period was at or above western levels. The ratio in Kyrgyzstan and Tajikistan was equal to that of the United Kingdom and above that of the USSR as a whole, while that of Turkmenistan and Uzbekistan was still higher.[483] The suitability of many of those assets under new cost and demand conditions must be doubtful, and in any event a lower capital to output ratio than that in a typical industrial economy may be acceptable if coupled with a higher labour to output ratio. Central Asia was the one large Soviet region in the second half of the 1980s in which the active labour force was continuing to increase. Elsewhere in the USSR the young cohorts entering working age were fewer than the sum of retirements and deaths in active ages. Central Asian enterprises could become more labour intensive to absorb the active population, as discussed in the next section.

Expectations for greater capital funding in the medium term depend throughout the region on the mobilization of domestic savings and the attraction of external finance. As section 4.4(iii) shows, central Asian banking systems need considerable strengthening and some present restrictions on foreign banking could be modified without prejudicing that development. In section 4.5(iv) the substantial entry of foreign investors, both direct and portfolio, is noted. Legislation in Kazakhstan, Kyrgyzstan and Uzbekistan has been made more attractive to them,[484] but the immediate prospect has been clouded by officials' mishandling of some tenders, and by the imposition of arbitrary restrictions on import payment and profit repatriation.

(iii) Labour supply and migration

A major constituent of latter-day Soviet economic failure was the inadequate provision of civilian investment and of technology and management to offset the aggregate reduction in the labour force, consequent upon a serious fall in the birth rate and a rise in mortality during working life; the share of young people declined and that of the elderly rose. The central Asian republics were exceptions to this trend and the active population continued to rise in the 1980s. Together with the Azeri population of Azerbaijan, the titular ethnic populations of central Asia were alone in the USSR in recording at the time of the 1989 census natural demographic rises exceeding 20 per 1000 (up to 36.8 for Tajiks in Tajikistan) – except for Turkmens, these increments were greater than at the 1979 census. The overpopulation problem was exacerbated in the two decades before independence by labour productivity growth in the five republics which was half the average in the rest of the USSR and was actually negative in Turkmenistan.[485] Because a high proportion of the republics' labour force was engaged in agriculture (41 per cent in Tajikistan, 37 per cent in Turkmenistan, 34 per cent in Uzbekistan, 32 per cent in Kyrgyzstan and 22 per cent in Kazakstan),[486] underemployment was "kept behind the farm gate" in collective and state farms. The economic viability of, and the political justification for, these and other rural establishments began to be questioned seriously by governments under perestroika.

The need for off-farm job creation is underscored by the prospect of persistent future additions to the working age population of central Asia. A mere 13 years hence, as table 4.3.2 shows on the United Nations "medium variant" projection, with some allowance for migration, the 15 to 64 age cohorts will increase against 1995 by 64 per cent in Tajikistan, 53 per cent in Uzbekistan, 52 per cent in Turkmenistan, 42 per cent in Kyrgyzstan and 21 per cent in Kazakstan. The variant perspective for Kazakstan is attributable to the similarity of the demographic characteristics of the Slav and some of the Kazak population to those of the Slav republics of the CIS. Kazakstan also differs from the other central Asian republics in not experiencing significant agrarian underemployment at current levels of technique.[487] The massive example of underemployment is Uzbekistan, where some research suggests that as many as 4 million may be only nominally or partially employed. A UNDP consultant, in reporting this, comments that "policies need to be aimed at making major reductions in the labour market overhang".[488] If the proportion of underemployment to the working age population in the other three republics were the same as in Uzbekistan, its absorption would involve some 3 million people. Adding the increments from 1995 to 2010 of the four republics' working age cohort of 11 million (table 4.3.2), yields a

---

[483] M. Kaser and S. Mehrotra, "The central Asian economies after independence", in R. Allison (ed.), *Challenges for the Former Soviet South* (Washington, D.C., Brookings Institution Press, 1996), p. 254, compile for 1990 ratios of 1.8 for Kyrgyzstan, 1.83 for Tajikistan, 1.98 for the USSR, 2.04 for Uzbekistan and 2.59 for Turkmenistan, against 1.8 for the United Kingdom in 1987; the inclusion of "non-productive assets", especially housing, would show lower central Asian ratios than in the United Kingdom.

[484] The EBRD secretariat assess the legal environment on a scale of three out of four (high) for Uzbekistan, one for Turkmenistan and two for the three other states. EBRD, *Transition Report 1996* (London), table 2.1.

[485] For NMP per worker in 1970-1990 see W. Easterly and S. Fischer, "The Soviet economic decline: historical and republican data", *NBER Working Paper*, No. 4735 (Cambridge, MA), 1995, pp. 1-56.

[486] Estimates for 1990 in World Bank, *World Development Report 1996: ...*, table 4; these shares were on average 2 percentage points lower than 10 years previously.

[487] This is attributable to retro-migration of Slavic settlers brought to grain farming in the northern regions during the "Virgin Lands Campaign" of the second half of the 1950s and to the predominance of large-scale mechanization in the arable areas.

[488] Internal report by A. Agafonoff, Chief Technical Adviser to the UNDP-funded Macroeconomic Policy and Training Project, Tashkent.

TABLE 4.3.2

Projections of working age^a and total population in the central Asian economies, 1995 and 2010
(Thousands and per cent)

	1995			2010		
	Working age	Total	Inactive (per cent)	Working age	Total	Inactive (per cent)
Kazakhstan	10 814	17 111	37	13 132	19 327	32
Kyrgyzstan	2 708	4 745	43	3 856	5 970	35
Tajikistan	3 209	6 101	47	5 255	8 881	41
Turkmenistan	2 309	4 099	44	3 503	5 463	36
Uzbekistan	12 707	22 843	44	19 440	30 703	37

Source: The Sex and Age Distribution of the World Populations. The 1994 Revision (United Nations publication, Sales No. E.95.XIII.2).

a Working age is represented by the 15 to 64 age group.

job creation requirement of perhaps 17 million; the same cohort would enlarge by more than 2 million in Kazakstan, implying at least 19 million to be found employment in the region as a whole by the twentieth anniversary of independence.

From that standpoint net emigration should be encouraged. But the skill mix of those who have already left suggests that the withdrawal should not be precipitate – nor on sociocultural grounds may it even be desirable. Between 1989 and 1995, 1.45 million Russians left the five republics, 4-5 per cent of the total population of Kazakstan and Kyrgyzstan and 1-2 per cent of the population of the other three republics. The proportion of Russians settled in these republics, as well as the proportion of those leaving, varied considerably: in Kazakstan, the Russian population was some 37 per cent of the total at the 1989 census, and 10-11 per cent of this group departed; in Kyrgyzstan, the population share was some 20 per cent, and one fifth of them emigrated. The Russian population was of smaller importance – 7-9 per cent – in the other countries; almost half departed in Tajikistan, and close to one fifth in Turkmenistan and Uzbekistan.[489] Civil war in Tajikistan patently accounted for the high departure rate there, while the option of dual nationality offered soon after independence may well be responsible for the lower rate in Turkmenistan.[490] That option is now available also in Tajikistan,[491] but has been ruled out in Kazakstan and Uzbekistan. Inter-ethnic tension played a role in decisions to migrate – including the reverse flow of central Asian nationalities from the Slav republics – but army demobilization and the steep rise in transport tariffs, rendering kith and kin contact more costly, were also relevant. Central Asia also lost population in external departures common to all other republics: 245,000 of German stock and 102,000 Jews left the former USSR between 1989 and 1995.[492] The outflow from central Asia has been slackening – partly due to the high unemployment in Russia and Ukraine – but it has withdrawn a significant layer of professional and technical personnel from mining, manufacturing and services. Departures from Tajikistan have included many professionals and technicians of the titular nationality, some as a gesture of political withdrawal which could be reversed if the peace process goes further.[493] In a certain offset, some of central Asian nationality occupying similar posts in the Slav republics and the Caucasus have repatriated themselves.

Education is a key requirement both to replace the skills lost by the net balances of migration and to contribute to the general raising of qualifications needed for the productive use of the underemployed. The other crucial input is capital and technology, by no means all of which can be generated domestically, as the preceding section, 4.3(ii), points out. Central Asian governments have been willing to squeeze proportionately more for education out of their constrained post-independence budgets. Thus the percentage share of education in total government expenditure rose from 16.6 in 1992 to 17.7 in 1994 in Kyrgyzstan, from 13.8 to 15.6 in 1995 in Turkmenistan and from 23.3 to 24.7 in 1994 in Uzbekistan; the 1994 shares were 13.4 in Kazakstan and 21.6 in Tajikistan.[494] To meet the rise in salaries (despite which many teachers and university staff have been tempted into the private sector) and the cost of educational supplies (including the publication of new textbooks appropriate to a non-Marxist and nationalist inspiration), the capital component of expenditure has been severely cut, and preschool education almost eliminated.[495] A UNESCO questionnaire to educational authorities in 1994 illustrated some of the problems.

A significant proportion of the school buildings (at least 10 to 15 per cent) are dilapidated and considered to

---

[489] A. Michugina and M. Rakhmaninova, "The national composition of migrants in the flows of population between Russia and foreign countries", *Voprosy statistiki*, No. 12 (Moscow), 1996, p. 45.

[490] For an annual breakdown by republic of Russian emigration to Russia, totalling 1,142,400 during 1990 to 1994, and survey results on reasons and intentions for departure see M. Brill Olcott, "Demographic upheavals in central Asia", *Orbis*, Vol. 40, No. 4, 1996, pp. 537-555.

[491] The Russian Duma ratified the agreement with Turkmenistan on 25 November 1994 and with Tajikistan on 12 December 1996.

[492] A. Michugina and M. Rakhmaninova, op. cit., p. 47.

[493] The dissident group, focused on the "National Renaissance of Tajikistan" in Moscow, have been refused participation in the peace brokered between the government of President Rakhmonov and S. Nuri, leader of the United Tajik Opposition. *Central Asia Newsfile*, February 1997, p. 1.

[494] I. Kitaev, "Challenge of realities: an overview of trends and developments in educational finance", in I. Kitaev (ed.), *Educational Finance in Central Asia and Mongolia* (Paris, UNESCO: International Institute for Educational Planning, 1996), pp. 73-74.

[495] The loss in provision has partly been offset by the establishment of private preschool facilities (many using staff earlier employed by the local authority). UNDP, *Kazakstan: Human Development Report 1996* (Almaty), p. 42.

be unsafe. Between half and two thirds of schools lack space and work on two or three shifts, affecting from 25 per cent (in Uzbekistan) to 35-40 per cent (in Turkmenistan, Tajikistan, Kyrgyzstan) and just under 60 per cent (in Kazakstan) of the total enrolments. Some schools have no canteens: only some of the pupils receive a hot meal (18 per cent in Tajikistan, about 50 per cent in Uzbekistan).[496]

Human capital investment remains high – in 1994 in Kazakstan tertiary graduates constituted 19.2 per cent of the population of normal graduation age, of which science graduates, however, were only 9.8 per cent.[497]

It is past investment in human resources that marks the central Asian states from other developing economies of a similar income level, that is, towards the top of the "low income countries" or among the "lower middle income countries". Thus the poorest of the five, Tajikistan, recorded 98 per cent adult literacy (99 among men, 97 among women) but is ranked in income terms between Zambia at 22 (males 14, females 29) and Benin at 63 (males 51, females 74); the richest on the income scale, Turkmenistan, showed the same literacy rates as Tajikistan[498] but, of its income coevals, Iran had 28 per cent literacy (22 males, 34 females) and Brazil a trivial 17 per cent (the same for each sex). The other states show virtually the same literacy rate (98 per cent in Kazakstan, 97 in Kyrgyzstan and Uzbekistan) and all retained children for virtually the full 10 years of compulsory education, although completion rates were and remain lower in rural than in urban areas. In the five republics in 1989 Soviet statistics showed 4.1 million people with specialized secondary and tertiary education in a total wage and salary employment of 10.1 million.[499]

(iv) The inactive population

The democratic pressure for jobs to be found for the additional 19 million (plus or minus migration) could be rendered somewhat less serious by the prospective reduced burden of inactive persons to be borne by the gainfully employed. Table 4.3.2 also indicates the percentage of the "inactive", those below the age of 15 and above 65, in the total projected population of the year 2010. In all countries their weight decreases. A mere calculation of age groupings takes no account of the withdrawal from – or non-entry into – gainful occupation of those of working age (notably of women for household work or childbearing and care) or of the deferment of entry by education.[500] The profile of the population by religion can be expected to show over the next few years more practising adherents of Islam. No central Asian country has abandoned a secular stance in favour of a state religion, but after decades, first of persecution and then of official marginalization, Muslim institutions are increasingly pervasive. Moral suasion for women to confine themselves to the household would reduce the number of jobs to be created, but would increase the dependency ratio. That ratio can also be altered by social security legislation: thus in 1995, in response to undercapitalization of the pension fund, Kazakstan raised the pensionable age. The number of those seeking work is further likely to be reduced by an increase in tertiary and specialized post secondary enrolments, to support industrial diversification, especially into technology intensive branches. A greater proportion of students adds to the cost of dependency. Finally, there is the migration issue. The more successful the central Asian economies vis-à-vis the Slav republics, the fewer potential migrants, and vice versa. It is similar for the legal and sociocultural environment insofar as it is attractive or repellent to continued Slav residence. Influences include the availability of bilingual education for the children of non-titular nationalities, the drive to differentiate a national identity from the common Soviet past (epitomized in plans to change the script of the titular language away from Cyrillic), and multinational participation in political parties and other links of a civil society. For the period chosen for these demographic projections the quantitative balance of any further net outflow cannot be hazarded.

The lower dependency ratios early next millennium to some extent relax the prospective resource claims of health care, on the grounds that children and the elderly require more input than the rest of the population. But the central Asian heritage is that improving medical and prophylactic provision has been a complement to educational investment. The quality of life and life expectancy, just like literacy, symbolize returns on past expenditure which place the five states well above the parameters that rule for market based developing economies. Thus in 1994 Tajikistan, despite its hostilities and supply difficulties, had a life expectancy at birth of 67 years, while its peers, ranked on income per head, had 47 years (Zambia) and 50 (Benin); at the richer end, the differential disappears, for Turkmenistan had 66 years and its near-income countries had 68 (Iran) and 67

---

[496] G. Skorov, "Synthesis of questionnaires and reports of central Asia countries", in I. Kitaev (ed.), *Assessment of Training Needs in Educational Planning and Management (with Special Reference to Central Asia)* (Paris, UNESCO: International Institute for Educational Planning, 1995), p. 27.

[497] UNDP, op. cit., p. 97.

[498] The standardization of curricula and of finance assured similar basic results throughout the USSR. Data from World Bank, *World Development Report 1996:* ..., tables 1 and 7.

[499] USSR Goskomstat, *Narodnoe khozyaistvo SSSR v 1989 g.* (Moscow), 1990, pp. 53 and 58.

[500] Table 4.3.2 is drawn from projections in five-year cohorts and the active population therefore starts at age 15. Compulsory schooling under Soviet law was from 7 to 17 but there is advocacy of beginning at 6, partly for educational reasons, but also because budget constrained local authorities have closed many preschool facilities.

(Brazil). Moreover, despite problems with the supply of medicaments, equipment and qualified staffing, the conventional indicator of general health, infantile mortality, has stayed low.[501] Morbidity from some diseases has risen, some of which may be due to the widening of real income differentials. The emphasis here is on "real" because Soviet practice assured access to health-care facilities for the employed and farm cooperative members at the workplace, in addition to an evenly spread network of local provision: better access was obtainable by personal influence or by payment of relatively modest honoraria. The other principal cause of recent rises in disease incidence is severe budget cuts: government spending on health care in 1995 in Kazakstan, for example, was 40 per cent less than in 1990 in real terms; morbidity from all diseases was 400 per 1000 of population in 1990, but 500 in 1994.[502] In some districts, mostly in Kazakstan, industrial and nuclear pollution has long exacerbated morbidity and accelerated mortality, while sanitation almost everywhere needs radical investment.

Partial replacement of free social programmes by insurance has been accepted by all governments. In Kyrgyzstan a Presidential Decree of November 1996 introduced obligatory medical insurance for citizens, and linked the scheme into a "one-stop service" including pensions and social security benefits.[503]

(v) Levels of living

Free, or highly subsidized, education and health care contributed to the perceived level of living of the later Soviet period, as in a more general way does unfettered choice of occupation. It just happens that the extension of free and compulsory secondary education to rural areas (1952) was soon followed by the abandonment of labour direction (1956), except for collective farmers whose occupational decontrol was delayed until 1976. It is certain that because many more of the services previously offered as a citizenship right, such as education and health care, will have to be paid for in future, the welfare effects will depend on household wealth and income distribution. But there is one Soviet right that can scarcely be guaranteed in the coming market economies – job security. The typical Soviet enterprise not only contributed to numerous social provisions, which are being dismantled under restructuring or privatization, but

---

[501] To cite the World Bank rank comparisons again, in 1994 Tajikistan had 41 deaths under one year of age per 1,000 live births, but Zambia had 108 and Benin had 96. At the richer end there was convergence: Turkmenistan recorded 46 per 1000, while Iran had 47 and Brazil 56.

[502] UNDP, op. cit., p. 36.

[503] The required computerization, under which each citizen will have a personal identification number for all schemes, is being supported by the World Bank, the European Union TACIS programme, the Soros Foundation Kyrgyzstan and the German Agency for Technical Assistance (GTZ).

TABLE 4.3.3

Registered unemployment and not gainfully occupied population of working age in the central Asian economies,[a] **1992, 1995 and 1996**

(Percentages of labour force and of those of working age)

	Registered unemployment		Not gainfully occupied	
	1995	1996	1992	1996
Kazakstan	2.1	4.1	11.2	12.3
Kyrgyzstan	3.0	4.5	10.0	15.0
Tajikistan	1.8	2.4	19.7	25.5
Turkmenistan	..	..	13.1	21.2
Uzbekistan	0.3	0.4	11.5	18.8

Source: Cols. (1) and (2): appendix table B.6; cols. (3) and (4): CIS Statistical Committee, *Report on the Socio-economic Situation in the CIS Countries in 1996* (Moscow), 1997.

[a] Students are counted as "gainfully occupied" for the purposes of this table.

also assured staff of lifetime employment. The present changes are rapidly eroding that guarantee, with resultant unemployment. As table 4.3.3 shows, Kazakstan and Kyrgyzstan already report significant unemployment; in the former, staff kept on the books of an enterprise and accorded a token wage (a form of "suppressed" unemployment, in ILO usage) added a further 2 per cent. Such suppressed unemployment exists elsewhere and an indication might be found in those of working age who are not gainfully employed, also shown in table 4.3.3. Women engaged in household duties doubtless form the bulk, and the stronger social influence of Islam in Tajikistan and Turkmenistan may account for the higher levels in those states. But it is the increase between 1992 and 1996 that is significant. Hostilities in Tajikistan and the closure of many non-farm enterprises account for much of the Tajik increment – which has qualitative effects because half the Russian labour force has emigrated – but rises of 5 percentage points in Kyrgyzstan, 7 points in Uzbekistan and 8 points in Turkmenistan illustrate the extent of redundancies and withdrawal from the labour market.

Measurement of the average change in real household income over the period so far affected by the transition recession is far from straightforward. Income and wealth distribution has certainly widened. In Soviet times some disparities in consumption were "hidden", because favoured persons had access to priority goods and services at the standard prices which would have been charged to the general public had a supply been made available. Some such differentiation persists, but goods and services are by and large purchasable on an open market and money income hence gives command over them. The new "hidden" disparities arise from the concealment of income and wealth, allowing misleading conclusions from such published statistics as are available. The smuggling of armaments to the various hostile parties in Afghanistan and Tajikistan and of narcotics from the poppy-growing tracts of the remoter

TABLE 4.3.4

Allocation of total household expenditure in the central Asian economies, 1991, 1993 and 1995-1996

(Per cent)

	Food	Alcoholic drink	Other goods	Services
Kazakstan				
1991	42.6	4.3	44.4	8.7
1993	52.0	5.0	35.7	7.3
1996[a]	56.7	2.2	26.5	14.6
Kyrgyzstan				
1991	42.9	4.2	44.2	8.7
1993	58.0	2.8	29.6	9.6
1996[a]	60.3	1.6	24.7	13.4
Tajikistan				
1991	49.1	1.7	42.1	7.1
1993[b]	70.3	2.2	22.5	5.0
Turkmenistan				
1991	47.2	2.6	43.6	6.6
1993[b]	52.1	3.9	39.2	4.8
Uzbekistan				
1991	49.3	3.2	39.6	7.9
1993	61.3	2.9	30.3	5.5
1995[c]	69.3	1.9	19.6	9.2

Source: CIS Statistical Committee, *SNG v 1996 g., Statisticheskii spravochnik* (Moscow), 1997, pp. 133-134.

[a] January-September.
[b] Latest available data.
[c] Provisional.

upland zones of the region generate large cash incomes, as do bribery and corruption and income earned abroad from assets bought with flight capital. Such estimates as can be found of measured income, show a widening of the spread. Two recent studies, one by the IMF, the other by the World Bank[504] can be combined to show a rise in income dispersion in Kazakstan and Kyrgyzstan between 1990 and 1993: respectively the Gini coefficient rises from 0.297 to 0.327 and from 0.308 to 0.353.

An indicator of the level of living may be found in the proportion of the household budget spent on food. Sample surveys are summarized in table 4.3.4, which show very large rises between 1991 and 1995 or 1996 for Kazakstan, Kyrgyzstan and Uzbekistan and up to 1993 for Tajikistan. The latter's extreme widening suggests that the maintenance of the bread subsidy in 1995, as noted in section 4.2(ii), could have kept many out of penury, while its reduction in Turkmenistan could have put more into poverty. Consumer subsidization in Turkmenistan, also known only to 1993, helped to keep the food share close to that of the Soviet period. As also observed in section 4.2(ii), services are heavily subsidized in Tajikistan and Turkmenistan and these are the two countries of central Asia where the share devoted to services has fallen. In the three other states, a certain opening to market forces has pushed up tariffs for these faster than consumer price inflation as a whole, but the proportionate household budget increase has been least in Uzbekistan and there was even an intervening fall. A rise in the share of household budgets spent on alcoholic drinks is sometimes associated with greater poverty and such a construction could perhaps be put on the rise in war torn Tajikistan. But there may be factors other than changes in living levels behind the falls in the share of alcohol in Kazakstan, Kyrgyzstan and Uzbekistan – the emigration of Slavs and greater adherence to Islamic abstention from alcohol among the autochthonous peoples.

4.4 Institutions for a market economy

(i) Establishing new currencies

The central Asian governments had no option on independence but to continue use of the Soviet rouble. The largest CIS economy outside Russia, Ukraine, broke away in 1992, but for the others a single rouble area was seen as a benefit to the maintenance (or restoration) of trade and payments relations within it. The EU's Maastricht Agreement was perceived as a model for the inheritance of a common currency which should not needlessly be dismantled:[505] a substantial core within the CIS – Armenia, Belarus, Kazakstan, Kyrgyzstan, Russia and Uzbekistan – in Bishkek in October 1992 laid plans for an Interstate Bank of the CIS. The CIS Heads of State meeting in Tashkent in May 1992 heard the representative of the IMF propose that the Central Bank of Russia "provide enough rouble currency for the central banks of the other states in the rouble area to satisfy the demands for currency in those states". The IMF pointed out that those other banks had a reciprocal responsibility to establish uniform central bank lending rates and reserve requirements and to constrain their own credit issue, but because the Interstate Bank was never set up, there was no enforcement of such parameters. Monetary destabilization proceeded at widely disparate rates. Measured by rates of consumer price inflation, changes had been remarkably uniform in 1992, as would be expected within a single currency area: the indexes deviated from the CIS average by little more than they had in 1991 (that is, prior to general price decontrol) – a coefficient of variation of 0.39 in 1992, against 0.35 in

---

[504] The IMF staff study shows 1990 data: P. Cornelius and B. Weder, "Economic transformation and income distribution: some evidence from the Baltic countries", IMF, *Staff Papers*, Vol. 43, No. 3 (Washington, D.C.), 1996, pp. 587-604; the Bank staff study shows 1993 data: K. Deininger and L. Squire, "A new data set measuring income inequality", *World Bank Economic Review*, Vol. 10, No. 3 (Washington, D.C.), 1996, pp. 565-591. They both cite Gini coefficients (the higher they are, the greater the inequality) but one gives the decile ratio, the other a quintile ratio.

[505] The Head of the European Commission's mission in Moscow, M. Emerson, voiced a widespread advocacy of such emulation in a press article, "The CIS on the Maastricht Road?", *Ekonomicheskaya gazeta*, December 1992.

1991. But in 1993 the spread was very much wider – the coefficient of variation was 1.17 (and stayed high in 1994, at 0.93).

The Kyrgyz authorities had already notified the other CIS governments early in 1993 that Kyrgyzstan would introduce its own currency, for it was already clear that the Russian Central Bank could not indefinitely tolerate credit issue outside its control. Both policies were applied in May of that year – change to the Kyrgyz som, and the Russian government's attempt to decelerate its own inflation with an exchange rate anchor. The IMF attached as conditions of a $1.5 billion Systemic Transformation Facility to Russia targets for a base rate and credit emission which would appreciate the real rate of exchange – the change in the rouble-dollar rate was to be half that of domestic inflation. In order to implement this, the Russian Central Bank had to participate in the control of the monetary issue of the CIS central banks using roubles. It emitted new Russian roubles in July 1993, quickly demonetizing Soviet roubles and allowing other CIS central banks to have these only upon its terms.[506] Tajikistan alone of the five accepted; as the Kyrgyz exchange of currency had been completed by 21 May, the Kazak, Turkmen and Uzbek authorities were thus committed to new monies, unless they accepted the stringent conditions of the Russian Central Bank.[507] To offset some of the integration potential lost by a common currency, CIS governments other than Ukraine signed an Economic Union Treaty in September. This had been accompanied in April 1993 by a draft agreement on free trade and, at the October meeting of the CIS, a memorandum on the main directions of integrational development.[508]

To promote confidence in, and gain competitiveness for, the new currencies, the opportunity was taken to reduce the number of digits which inflation had added to currency units[509] and to devalue in the case of the Kazak tenge and the Uzbek sum. But the Kyrgyz som maintained parity at the prevailing rate of the exchanged roubles and the Turkmen manat was actually made to appreciate. In Kyrgyzstan, 200 Soviet roubles were exchanged for 1 som, except that deposits in the state Savings Bank (Sberbank) were recalculated at 150 to the som as a gesture towards those whose balances had previously been decimated by inflation. The government (supported by an IMF Stand-by Arrangement of $38 million) opted for full liberalization: all controls on current and capital transactions were lifted and the som was freely floated. The rate began to be set on an interbank market through twice weekly foreign exchange auctions by the National Bank, and maintained within a narrow band to the dollar, mostly by varying the amounts of foreign exchange offered by the National Bank. Although there were periods of depreciation (early in 1994 and for much of 1995), the rate has shown a real appreciation (that is, with account taken of inflation) as table 4.4.1 shows.

Quite the contrary policy was followed in Turkmenistan, when the manat replaced the rouble in November 1993. Two digits were cut off by exchanging 500 roubles for 1 manat and Savings Bank deposits of less than 10,000 roubles were set at 62.5 roubles for 1 manat as a gesture to the poorest hit by earlier inflation. But limits were imposed on the amounts that could be exchanged: no household could exchange more than stood to its members' credit at the Savings Bank on 1 September plus subsequent wage payments, and no individual could change more than 30,000 roubles. Exchanges of business deposits were capped at 50 billion each plus 25 per cent of any higher balance. The remaining 75 per cent was frozen in state bonds for later redemption. A three-way multiple exchange rate was applied: the official rate which was used for all government transactions, a commercial rate for authorized private transactions and a special rate for gas transactions (which, however, operated only between April 1994 and February 1995). Due to limits on purchases of foreign currency ($1,000 per transaction) and a surrender requirement on certain sales, a parallel unofficial market persisted, at a severely devalued rate. For almost two years foreign currency could legally circulate side by side with the manat, but this was forbidden at the end of December 1995. Over the three years, nevertheless, the manat has in real terms depreciated to a third of its introductory value (table 4.4.1).

The Kazak tenge was also introduced in November 1993, although an IMF Stand-by Arrangement was not finalized until the following January (for $173 million). The conversion rate of 500 roubles to the tenge was applied to small sums in cash (up to 100,000 roubles per person, or roughly the average wage in October) and to personal bank deposits at 1 October. Larger sums in cash and later deposits were held in blocked accounts until

---

[506] Among accounts of the breakup of the rouble area, see A. Åslund, *How Russia Became a Market Economy* (Washington, D.C., Brookings Institution, 1995); B. Granville, *The Success of Russian Economic Reforms* (London, Royal Institute of International Affairs, 1995); and T. D. Willett et al. (eds.), *Establishing Monetary Stability in Emerging Market Economies* (Boulder, CO, Westview Press, 1995).

[507] Thus, Kazak negotiations with Russian Vice-President A. Shokhin broke down on 29 October 1993 over the conditions for remaining on the rouble, which included Russian custody of the Kazak gold and currency reserve (then $222 million and $501 million), and payment of interest on rouble credits until the financial structure was aligned with the Russian. On rejection of those terms, a decision to introduce the manat was announced on 3 November. *Central Asia Newsfile*, November 1993, p. 8.

[508] Azerbaijan, Georgia and Turkmenistan joined the Economic Union after the Summit. On these meetings see M. Webber, *CIS Integration Trends: Russia and the Former Soviet South* (London, Royal Institute of International Affairs, 1997), chap. 5.

[509] At the basic rates, the new units exchanged for 1,000 roubles in Uzbekistan (sum-coupon), 500 in Kazakstan (tenge) and Turkmenistan (manat) and for 200 roubles in Kyrgyzstan (som).

TABLE 4.4.1

Real exchange rates of new central Asian currencies against the dollar
(Index, month of introduction=1)

	Official exchange rate	Consumer price index	Real exchange rate
Kazakhstan November 1993-December 1996	15.60	34.96	2.04
Kyrgyzstan May 1993-December 1996	4.31	11.21	2.35
Tajikistan May 1995-December 1996	6.42	15.24	2.28
Turkmenistan November 1993-October 1996	2 473	1 016	0.38
Uzbekistan November 1993-November 1996	41.64	60.44	1.33

Source: IMF, European Department II Database.

their legitimacy as assets could be verified. Business balances were similarly capped and holdings above that cap were blocked for verification as to origin. All non-cash transactions had to be denominated in tenge and, though foreign exchange could be used as cash, penalties were announced for refusal to accept tenge. A dual exchange rate was continued – an official and a commercial (plus, of course, an unofficial) – and from the introduction of the new units, weekly auctions of foreign currency were held and an interbank market was established for the tenge. Surrender requirements for encashment at the official rate were, however, enlarged: they had been at only 10 per cent of foreign exchange sold until shortly before the changeover and then were at 30 per cent, but at the start of 1994 the requirement was raised to 50 per cent. For its first two months the tenge maintained its real value, depreciation and inflation keeping pace, but it briefly depreciated in real terms in early 1994, before regaining its original level and, as indicated in table 4.4.1, eventually showing a three-year real appreciation. In what might be termed an inverse "Tequila effect" (see section 4.5(iv)), the tenge continued to show real appreciation into March 1997.

Uzbekistan introduced a transitional currency, the sum-coupon, in November 1993 without IMF support or advice. An informed observer, comparing its introduction with that of the Kyrgyz som, wrote that the Uzbek failure was attributable to "the continued easy credit and monetary policies of the Central Bank".[510] Despite the continuance of a dual exchange rate and foreign exchange surrender quotas at the official rate, the sum-coupon depreciated and inflation accelerated: whereas the monthly retail price index compared with the same month of the preceding year had been 450 in April 1993, and

895 in October, it was 1,039 in November and 1,295 in December.[511] Discussions with the IMF began in February 1994 but ended in May without agreement, such that the sum-coupon was replaced on 1 July 1994 by the sum on the basis of the government's judgement and reserves alone. The government had confidence in its ability to defend the new currency, having the right of pre-emption over the country's considerable gold production (see section 4.5(i)), having in 1993 been freed (like all CIS countries except Russia) of any share of Soviet debt, and having run a visible surplus on merchandise trade in 1993.[512] To limit a wage spiral after such devaluation and expected price rises, a one-for-one tax was imposed on any monthly wage bill increment that exceeded 70 per cent of the enterprise's sales.[513] The government's confidence extended to relaxing some current account exchange controls and to allowing virtually full exchange of sum-coupons, apart from a short freeze on bank deposits (that is, in contrast to the constraints imposed by the other central Asian replacements of the previous year). Its confidence was misplaced, and within five months the sum was a mere fifth of its rate to the dollar. The devaluation took place at first cosmetically, by announcing a "commercial" rate from 1 August 1995 and the next month unifying the official rate down to it (there was a spread between the rates between August 1995 and April 1996, after which the Central Bank supported the two together). Negotiations with the IMF were resumed and agreement was reached in January 1995. A Systemic Transformation Facility of $74 million was accorded against a budget cutting the deficit from 6 to 3.5 per cent of GDP and other disinflationary measures. Neither the deficit nor the inflation targets were in the event met, but with IMF and other international support, exchange controls and tariffs were somewhat liberalized in July 1995, and the real exchange rate appreciated. Central bank auctions and an interbank market (paralleled by an unofficial market) brought a more ordered procedure, but various surrender quotas (for non-centralized exports, 15 per cent with CIS partners and 30 per cent with others) persisted. Over the three years from its introduction to October 1996 Uzbekistan's new currency slightly

---

[510] M. Abdoulkadyrov, "Monetary reform: a comparison of the Kyrgyz Republic and the Republic of Uzbekistan", *Comparative Economic Studies*, Vol. 37, No. 3, Autumn 1995, pp. 36-56.

[511] Export earnings from non-CIS partners had been subject to a 60 per cent surrender quota until May 1993, but recipients had all the money due to them. When it was redesignated a tax at 35 per cent, that amount was forfeit, and exporters had only 65 per cent of their money, but at the commercial rate. In January 1994, the surrender quota was reintroduced at 15 per cent and the tax modified to 15 per cent. IMF Economic Reviews, *Uzbekistan* (Washington, D.C.), March 1994, pp. 36-37, retail price index from table 14.

[512] Aggregating CIS with rest of the world trade, the surplus was wholly derived from CIS partners and in 1994 was too small to offset the continuing visible deficit with the rest. IMF Staff Country Report No. 96/73, *Uzbekistan – Selected Issues and Statistical Appendix* (Washington, D.C.), August 1996, tables 31 and 33.

[513] This was operative only to the end of 1994.

appreciated at its official rate, but by mid-year the unofficial rate (50 sum against the official 38 in August) had returned in real terms to its 1993 level. It was about then that the Central Bank realized that current income from cotton (about half total exports) would be much lower and that imports were running at about 50 per cent above 1995. The course advised by the National Bank for Foreign Economic Activities was devaluation – thereby removing the prime cause of the import boom – but instead the Central Bank sold reserves and delayed exchange of sum into foreign currency, even for major investing firms such as Daewoo and BAT. In October, continuing to misjudge the situation, the Central Bank of Uzbekistan retroactively cancelled all 1,400 licences to buy foreign exchange.[514] Reversing partial current account liberalization, the panic measures seemed to prejudice Uzbekistan's Article VIII commitment to the IMF (for full current account convertibility and non-discriminatory currency arrangements) during 1997. That commitment had been accepted by Kyrgyzstan in March 1995 and by Kazakstan in July 1996. The Uzbek government, furthermore, imposed licensing for all imports. The IMF suspended disbursement of its Facility in December and a return mission in February 1997 declined to recommend resumption. At the unofficial rate (130 sum to the dollar in February 1997), towards which the official rate must eventually tend, the real exchange rate has depreciated since its replacement of the sum-coupon by half its original dollar value.[515]

Tajikistan was the exception to the exodus from the rouble in 1993, accepting the new Russian rouble. Civil war both within the country and in neighbouring Afghanistan rendered that part of the country under the administration of President Rakhmonov economically dependent on Russia and Uzbekistan. Economic weakness, and especially monetary instability, was unfavourable to the establishment of an autonomous currency. A dual currency emerged – cash was supplied by the Central Bank of Russia on commercial terms, but non-cash bank money could be created by the National Bank of Tajikistan without any constraint; the government needed funds to protect itself militarily and could raise few taxes in the disturbed social environment of the territory it nominally controlled. One of its financing sources was the sequestration of foreign exchange from the few earners of such – cotton plantations and the one aluminium plant, which staggered from idleness to polluting production and back. The foreign receipts were put into the extrabudgetary State Foreign Exchange Fund, on the operation of which nothing was published, and some of the earnings passed to their owners at an unfavourable exchange rate. The government used the foreign exchange it thus seized to subsidize imports of grain and fuel, sales of which could be at low controlled prices to households, as noted in section 4.2(ii). When the government in Dushanbe was reshaped in December 1994, an opportunity seemed to open for wider economic reform and an independent currency. From then until the changeover, on 10 May 1995, a tight credit policy was pursued, although it was in part illusory because enterprises paid taxes in increasingly worthless bank money. Also, because cash was in short supply and many retail prices were controlled and subsidized, inflation was repressed: from 7,344 in 1993 and back to 2,132 in 1995, one version of the consumer price index rising by just 1 per cent in 1994.[516] Due to the wedges that had opened up between the various forms of money, the rates of conversion were differentiated. Most non-cash money was converted at 1,000 roubles to 1 new Tajik rouble, but household and enterprise bank deposits were further disfavoured at 1,200 to 1, except for the amount standing in credit to a household on 1 January 1993, which benefited from the same rate as cash, 100 to 1. Public service wages (which had artfully been substantially increased on the eve of the reform), the minimum wage and social security payments were converted at the 100 rate. For many reasons – in which a rapid expansion of bank credit and the government's failure to collect taxes or sufficiently moderate subsidies played a large part – the economy and the new currency did not stabilize. In the last quarter of 1995, the budget deficit on a cash-flow basis was nearly 16 per cent of GDP (26 per cent on an accruals basis): in the full year it was 10 per cent of GDP, double the 1994 proportion. In 1996, conditions began to return towards normality after the peace agreement, signed under United Nations mediation in Ashgabat in July, and seemingly more definitively settled (after a renewal of conflict) in December.[517] A sign of increasing confidence in a more settled Tajikistan was the first EBRD involvement, incidentally making the Bank active in all 26 of its countries of operation: on the day the peace agreement was signed it announced investment in two private sector banks, Orienbank and Tajikbank, to service the needs of small and medium enterprises (an $8.6 million equity and loan package). For the year, the budget deficit was no

---

[514] Accounts of the handling of the crisis include M. Kaser, *The Economies of Kazakstan and Uzbekistan* (London, Royal Institute of International Affairs, 1997), pp. 30, 40-42; *Central Asia Newsfile*, October-November 1996, pp. 6-7; and K. Bartholdy, *CS-First Boston, Emerging Economies Research – Europe*, 28 February 1997.

[515] The period covered in table 4.4.1 begins with the introduction of the sum-coupon and refers to the official exchange rate.

[516] IMF Staff Country Report No. 96/55, *Republic of Tajikistan – Recent Economic Developments* (Washington, D.C.), June 1996, table 21 (variant data from this chapter's table 4.1.2); its table 21 gives the wholesale price index at each end-year as 5,996 in 1993, 295 in 1994 and 628 in 1995. The above discussion draws on that report, one of the very few sources on recent Tajik monetary and fiscal instability.

[517] The Moscow agreement between President Rakhmonov and the leader of the Islamic Renaissance Party, S. Nuri, provided for a reconciliation council headed by the latter's nominee, a general amnesty and the exchange of prisoners.

worse than that of two other central Asian states, Kyrgyzstan and Uzbekistan, at about 7 per cent of GDP. Institutionally there were improvements. Whereas until the currency reform there had been virtually no foreign exchange market, an Interbank Currency Exchange was set up in May 1995 to conduct regular auctions (initially only in terms of Tajik roubles to the dollar). After an interruption (August 1995 to February 1996), weekly auctions were instituted and for a short period the Tajik rouble, allowed to float, actually appreciated.

In Kazakstan, Kyrgyzstan and Uzbekistan, to minimize the consequences of fiscal deficits upon inflation, the interbank foreign exchange market is paralleled by an interbank money market, upon which the government sells and buys its own securities. By the issue of treasury bills public sector borrowing became a formal part of budgeting. The appreciation of the real exchange rate contributed to the deceleration of inflation in the same three states – Uzbekistan until late 1996 – while a floating exchange rate plus the relaxation of exchange control – complete in Kyrgyzstan and on current account in Kazakstan – has attracted foreign investment. A more stable currency has been associated in Kazakstan with external borrowing in place of monetizing still significant fiscal deficits: a £200 million bond issue was successful on the London market in December 1996, the first such flotation of any central Asian government.

(ii) Privatization

Essential as is state enterprise divestment to the transformation of a former command economy, there is no ideal share which the private sector should bear in a market economy, save that it should be dominant. Central Asian governments have held on to state ownership of their relatively few large enterprises more than is the case in most transition states, although, like the others, they have encouraged a private sector, both *ab initio* and by property transfer, among small- and medium-scale businesses. The secretariat of the EBRD estimates that by mid-1996 about 50 per cent of GDP was being generated in the private sector in Kyrgyzstan, and 40 per cent in Kazakstan and Uzbekistan, but only 20 per cent in Tajikistan and Turkmenistan. Three countries were ranked 3 out of 4 points on their progress in large privatization (Kazakstan, Kyrgyzstan and Uzbekistan), and Kyrgyzstan was accorded 4 out of 4 for small privatization. The others scored low – in both categories 2 for Tajikistan and 1 for Turkmenistan.[518] Due to this variant development and to the disparity in size between the smaller and the larger firms being privatized, statistics on enterprise numbers, brought together in tables 4.4.2 to 4.4.7, do not carry much analysable information. The

TABLE 4.4.2

Number of enterprises privatized in Kazakstan

	Before 1994	During 1994	During 1995	1996 QI-QIII
Small enterprises	153	2 645	3 920	2 423
Large and medium enterprises auctioned	–	382	1 056	590
Farms	602	888	472	378
Case-by-case sales	–	3	2	26
Total	755	3 918	5 450	3 417

Source: Government of the Republic of Kazakstan/Centre for Economic Reforms, *Kazakstan Economic Trends*, Third Quarter 1996.

three states most advanced in the process show increasing numbers transferred out of state ownership and a fairly common pattern by branch. By the end of 1996, 60 per cent of all enterprises in Kyrgyzstan had been privatized, and more than 13,500 enterprises in Kazakstan. In all three, trade and catering, being typically small scale and locationally dispersed, are the most privatized and where a programme has been announced (in Kazakstan for the year 2000) will be almost exclusively privately run (table 4.4.3). Consumer services are similarly advanced in two, but the government of Uzbekistan, by mid-1996 at least, had kept most such establishments in its own hands, and its industrial enterprises still generated a majority – albeit a bare majority – of sales, as table 4.4.7(iii) shows. In Kazakstan, by April 1996, only 5 Kazak state enterprises out of 180 large firms (with more than 5,000 employees) had been transferred, but 70 per cent of small enterprises (under 200 employees) or 11,000 businesses, had gone over to private hands and, in between, 1,700 medium enterprises had been privatized.[519] Turkmenistan has made a little headway (table 4.4.6), but only at the small business level. Of the 200 large enterprises on its territory, just 90 are scheduled for privatization and by the end of 1996 only 4 had actually been sold, with another 20 being prepared for privatization. The statistics on Tajikistan (table 4.4.5) show a lot of small disposals, exceeding 1,800 at the end of 1996, constituting 40 per cent of the stock of small state firms (less than 20 employees), but only 10 per cent of all state owned enterprises; moreover, even in the privatized firms the state generally retains a substantial holding.[520]

Early inflation which much reduced personal savings excluded the public offering of shares on the United Kingdom's privatization model, even if it had been seriously contemplated by the new governments of central Asia. Just independent, they were wary of opening purchases indiscriminately to foreigners, although a few joint ventures with state enterprises were

---

[518] EBRD, *Transition Report 1996* (London), table 2.1.

[519] But of the 1,700 some 700 were still in majority state ownership because no auction bids had been made for their shares.

[520] IMF Staff Country Report No. 96/55, *Republic of Tajikistan – Recent Economic Developments* (Washington, D.C.), June 1996, p. 43.

TABLE 4.4.3

Privatization plans in Kazakstan, 1994-1995 and 2000
(Percentage of enterprises in each branch)

	1994	1995	2000 Slow variant	2000 Fast variant
Industry and mining	17	25	30-35	35-40
Agriculture	46	50	57-62	63-67
Trade and catering	50	65	80-85	85-90
Business services	11	28	60-65	65-70
Consumer services	51	63	75-85	85-90

Source: Y. Kalyuzhnova, "The privatization of property in the Republic of Kazakstan", University of Reading, Department of Economics, *Discussion Papers in Economics*, No. 319, August 1995. Data for 1995 are provisional and in agriculture no farm is included in which some share is state owned.

TABLE 4.4.4

Number of enterprises privatized in Kyrgyzstan, 1992-1996

	1992	1993	1994	1995	1996 QI-QIII
Industry, construction and transport	222	564	726	970	1 009
Trade, catering and consumer services	2 042	3 437	3 597	3 666	3 699
Agriculture	53	235	319	342	329
Other branches	..	192	482	865	1 543

Source: CIS Statistical Committee, *SNG v 1996 g., Statisticheskii spravochnik* (Moscow), 1997, p. 54, IMF Staff Country Report No. 96/98, *Kyrgyz Republic – Recent Economic Developments* (Washington, D.C.), September 1996, table 13.

on the cards. They inherited a framework law from the USSR (on destatisation and privatization of July 1991) which envisaged a voucher system through the allocation to citizens of dedicated savings accounts from which equity purchases could be made. Kazakstan and Kyrgyzstan followed that route with adaptations, and voucher schemes have been considered in the other states. No republic offered compensation or restitution to former owners, Soviet expropriation of non-farm businesses being too remote in time and too few to justify reopening the issue.[521] All central Asian governments have reserved certain enterprises for retention in the wholly-owned state sector, most extensively in Turkmenistan and Uzbekistan, and have favoured some management and employee participation in the medium and large enterprise sector. Small enterprises have been sold off, usually to staff, as well as public housing, almost all to occupiers.

Kazakstan issued to citizens two forms of "privatization cheque". One was for share purchase (the value of which to a rural dweller was 20 per cent greater than that given to an urban resident) which had to be used through an "investment fund"; by end-1996 there were 170 such "investment privatization funds". The other was intended for housing, but could be deposited in the investment funds for equity purchase. In addition to these voucher arrangements, small state enterprises were offered for auction and large enterprises were sold on a case-by-case basis, including many to foreign companies, either wholly or in joint ventures. The Kyrgyz procedure was based on the dedicated account opened for each citizen at the State Savings Bank. The sum credited, uniquely for the purchase of shares, was, beyond a general minimum, calculated for those in, or retired from, employment on a scale related to monthly earnings and years of service. "Insider" ownership was facilitated by a provision that staff were entitled to buy at a discount an earmarked proportion of their own firm's privatization equity; the proportion was larger the smaller the firm's asset/staff ratio (30 per cent for the smallest enterprises, 5 per cent for the biggest). The coupons drawn on these accounts could be used both for buying shares or housing. Sales other than by coupon were for cash; intermediation by investment funds (termed specialized privatization funds) was possible under license. Few of these were established however and they accounted for less than a third of the coupon transactions.

In Tajikistan staff buy-outs as cooperatives, by lease or by individuals proceeded quite rapidly in the small-scale sector (fewer than 50 employees), but it was not until May 1996, with peace in prospect, that the Dushanbe government introduced a programme for medium and large enterprises. Part of wage and salary arrears were converted for civil servants and state enterprise employees into "privatization cheques", applicable for buying shares in the enterprises (initially 280) listed for privatization. Turkmen legislation has twice envisaged the issuance of privatization vouchers – in February 1992 a law embodied the 1991 Soviet model of dedicated accounts for buying state equity, but was not applied, and in May 1994 a Presidential Decree envisaged the provision of vouchers, in the use of which some priority would be given to employees of the enterprise. While the scheme awaits implementation, many small and medium firms (as already noted) have been transferred by either auction to individuals or sale to cooperatives. No investment funds have as yet been created in either country.

Although hints were given in 1996 that a voucher scheme might be offered to citizens in Uzbekistan, privatization to date has been confined to the all but complete disposal of small businesses and housing. Medium and large enterprises were set for privatization by a Presidential Decree of March 1994 and the programme was accelerated by another decree of July 1996. In contrast to practice in the other four states, no divestment will leave a majority of equity in the hands of a combination of the state and of insiders. Joint ventures,

---

[521] A fuller description of privatization and land reform procedures is in M. Kaser, "Economic transition in six central Asian economies", *Central Asian Survey*, Vol. 16, No. 1, 1997, pp. 5-26.

TABLE 4.4.5

Number of enterprises privatized in Tajikistan, 1991-1996

	1991-1995	1996 Q1
Industry	73	1
Consumer services	997	27
Trade and catering	552	33
Agriculture	25	–
Construction	55	4
Transport	18	3
Other branches	110	6
Total	1 830	74

Source: State Committee for the Management of State Property as reported in *Asia-Plus Bulletin*, No. 3, 1996.

TABLE 4.4.6

Privatization in Turkmenistan, end-1995

	Scheduled for privatization	Privatized
Small enterprises	3 980	1 652
Medium enterprises	1 645	4
Large enterprises	90	1
Total	5 715	1 657

Source: IMF Staff Country Report No. 96/30, *Turkmenistan – Recent Economic Developments* (Washington, D.C.), April 1996, p. 11.

TABLE 4.4.7

Privatization in Uzbekistan, 1992-1995

	1993	1995
(i) Number of enterprises, end-year	40 166	64 900
of which:		
As joint-stock companies	22 946	26 100
As collectives or private	16 600	25 900
Leased	560	400

	1992	1993	1994	1995
(ii) Number of enterprises by branch				
Industry, construction and transport	11 900	1 900	3 700	4 900
Agriculture	20	80	500	900
Trade, catering and consumer services	19 700	51 300	57 500	62 600

	State	Private	Corporate	Joint venture	Other
(iii) Percentage of sales in first half of 1996					
Industry and mining[a]	49.8	..	..	..	..
Agriculture	4.9	68.4	5.0	–	21.7
Trade and catering	6.2	69.3	5.6	2.4	16.5
Consumer services	71.0	12.7	9.6	–	6.7
Aggregate net material product	56.2	16.8	17.1	2.2	7.7

Source: (i) *Ekonomika i zhizn'*, No. 45 (Moscow), November 1996; (ii) State Committee on Forecasting and Statistics, *Basic Results of Social and Economic Development of the Republic of Uzbekistan* (Tashkent), January-June 1996; (iii) CIS Statistical Committee, *SNG v 1996 g., Statisticheskii spravochnik* (Moscow), 1997.

[a] The source does not distinguish between non-state owners in industry and mining.

including majority holdings, with foreign companies have been especially prominent.

The ECE study of 40 years ago pointed out that the postwar Soviet policy of farm amalgamation had resulted in the agriculture of four central Asian states becoming concentrated in a mere 3,600 state and collective farms, with an average of 1,450 hectares sown area, against 14,000 farms before the Second World War, with an average of less than 400 hectares.[522] Agrarian reform, which in an optimal world might turn such vast collective and state farms coexisting with smallholdings of rural households into medium-size agribusinesses, has everywhere been slow and formal. On the face of it, Kazakstan's privatization, by harvest time 1996, of 93 per cent of all farms and 80 per cent of farm land shows substantial progress along that path, but ownership titles are still ill-defined and scant restructuring has taken place. The formal renunciation of collectivization was historically significant because of the coercion exercised by the Soviet authorities in the settlement of predominantly nomadic Kazaks in the early 1930s, in the course of which 1 million died and the nation became a minority within its own borders. The lost generation was replaced on state farms by deportees in Stalin's time (more into Kazakstan than any other Union Republic other than Russia) and by young Slav volunteers under Khrushchev's "Virgin Lands Campaign". Collective farms have been restructured as joint-stock companies, 20 per cent of the shares being available for purchase by outsiders, provided they have experience of farming and are Kazakstan citizens; employees of state farms (two thirds of farm land) have been given long leases. The shares and leases may be sold and inherited, but – household plots and built-on land excepted – land remains formally in state ownership. In Kyrgyzstan, individuals could have their separate leases (about 10 per cent of farm output came from that sector in 1995) but most collective and state farmers agreed to reinstate their entities as cooperatives; all these leases are for 99 years, but freehold ownership is to be enacted. While land is still exclusively leasehold, Uzbekistan has broken up two thirds of its collective farms into peasant *(dekhan)* smallholdings, the remaining farm land being run as cooperatives. Tajikistan's de facto division has brought mixed experiences. In that part of the territory under control from Dushanbe, local authorities are authorized to lease land, all of which is vested in the state, for up to 10 years, but less than 10 per cent of farm land has been so taken over; however, it produces a third of agricultural output. In eastern Badakhshan, peasant cooperatives

---

[522] UN/ECE, *Economic Bulletin for Europe*, Vol. 9, No. 3, November 1957, p. 55.

have been established with financial and advisory support from the Aga Khan Foundation, as aid to those of the Ismaili sect of Islam. The constitution of Turkmenistan recognizes private property in land, but implementation awaits enactment of a new Civil Code.

(iii) Market generating infrastructure

The estimates of the EBRD secretariat quoted at the start of section 4.4(ii) concern also the stage of transition to the market reached on four regulatory or institutional criteria – competition policy; banking reform and interest rate liberalization; securities markets and non-bank financial institutions; and extensiveness and effectiveness of legal rules on investment. Ranking is, as for the other measures, on a scale up from 1 to 4. In mid-1996 Kazakstan and Kyrgyzstan ranked 2 on all four measures, and Uzbekistan could be considered slightly more advanced with the same rank of 2 on three, but a higher rating, 3, on legal reform. Tajikistan and Turkmenistan are uniformly lower – all score the bottom 1, except that legal reform is given 2 in Tajikistan (and of course only for territory under Dushanbe control).

The command economy from which the central Asian states emerged had already been partly dismantled under perestroika, particularly with respect to foreign trade, compulsory sales to, and purchases from, specified state entities (such as farm procurement and industrial "state orders") and constraints on enterprise finance. But there was much more to do once the aim of the independent governments had become a full market system. The "gradualism" which all those decision makers patently preferred meant that more of those instruments of the past economic mechanism were retained than would have been under the "shock therapy" option. In particular, as already cited in section 4.2(ii), Presidents have reserved a definite role for the state in establishing what they judge to be an appropriate market structure for their economies.

Some such control by the state is a retention of administrative structures inherited from the command economy. Thus, Uzbekistan retains state procurement quotas for cotton (as did Tajikistan until 1996, and Kazakstan for grain until 1995) and "state orders" (the *goszakazy* of Soviet perestroika) for industrial products were kept in Tajikistan until 1996, and still operate in Turkmenistan. All but Kazakstan and Kyrgyzstan still channel much of their exports and some of their imports through state trading corporations. The state's regulatory function in support of a functioning market environment is a different matter and is important in view of the low scores estimated by the EBRD on "emerging market" conditions. Among the many mechanisms and practices which especially still require the government's encouragement are appropriate banking and other financial services and bankruptcy procedures.

It is characteristic of all five states that banking is dominated by state owned institutions; that discriminatory restrictions are made on foreign banks; and that both coexist with weak private sector domestic banks. The World Bank early undertook an advisory role in the transformation of Union Republican branches of Gosbank USSR into central banks, but of the commercial banks, many still state owned, the IMF's observation of Kazakstan applies generally: "the weak financial position of major, formally fully state-owned banks, as well as the fragile position of newly emerging private banks ... a large proportion of loans classified as uncollectable losses, although many of these were issued with government guarantees, insufficient capital to meet minimum international standards, poor or non-existent policies related to risk management, inadequate accounting and reporting systems, and excessive exposure to a few large borrowers".[523] In Kazakstan, reserve requirements were increased in April 1995 and the regulatory role of the National Bank was strengthened by laws implemented by a decree of September 1995; it has so far only authorized foreign banks as subsidiaries, joint ventures or representative office. Minimum reserve requirements have been applied in Kyrgyzstan to close a number of insolvent banks, the latest being Dramds Bank in October 1996; foreign banks are allowed to operate under license. Higher reserve requirements were set by the National Bank of Tajikistan in January 1996, but the commercial banks, largely state owned, remain prudentially very exposed; only minority foreign ownership is allowed. The Central Bank of Turkmenistan similarly supervises poorly effective commercial banks which – apart from two half foreign owned – are still state run or "pocket banks", in the Russian phrase, that is owned by enterprises to which they predominantly lend. The regulatory function of the Central Bank of Uzbekistan was enhanced in December 1995, but state ownership is still high in commercial banking, apart from "pocket banks" and three joint-venture banks.

The poor monitoring of debt by banks and the insufficiency of the threat of bankruptcy are two of the principal deficiencies that allow the accumulation of payment arrears; the active factors perpetuating the widespread mutual indebtedness are the readiness of governments to subsidize ailing enterprises and of managements and workers to tolerate overstaffing. The latter issues have been noted in section 4.2(ii), and evidence of the former may be seen in statistics of indebtedness in Kazakstan, where the aggregate net debts of the enterprise sector rose from 177 billion tenge in August 1995 to 444 billion tenge in August 1996 (of which wage arrears from 23 to 46 billion).[524]

---

[523] IMF Staff Country Report No. 96/22, *Kazakstan – Recent Economic Developments* (Washington, D.C.), March 1996, p. 38.

[524] Government of the Republic of Kazakstan/Centre for Economic Reforms, *Kazakstan Economic Trends*, Third Quarter 1996, p. 138, table

4.5 Diversification, competitiveness and sustainability

(i) Dependence on natural resources

The mineral wealth and the arable and grazing land of the region are not evenly distributed among the five republics. Overall the most diverse and the richest is Kazakstan with its many metalliferous ores,[525] and hydrocarbons; Turkmenistan in the mineral sphere relies almost wholly on hydrocarbons; while Uzbekistan has less hydrocarbon deposits, but a range of other minerals. Kyrgyzstan and Tajikistan are unimportant in oil and gas, but have some valuable minerals. Converted to tonnes of oil equivalent,[526] Turkmenistan has 2,610 million of gas (96 years' production), and 200 million of oil (40 years); Uzbekistan has 1,710 million of gas (42 years) and 500 million of oil (63 years); and Kazakstan has 1,620 million of gas (320 years) and 700 million of oil (34 years). These estimates do not include deposits under the Caspian Sea, the division of which among the riparian states has not yet been determined. Coal mining was greatly expanded in Kazakstan and Kyrgyzstan in the Soviet period, especially during the Second World War, when other branches of mining and metal fabricating were developed (mostly by prison labour) to offset losses in the occupied zones. Among exportable commodities are barite, beryllium, bismuth, chromite, copper, gold, iron, lead, manganese, silver, tantalum, titanium, tungsten, uranium and zinc in Kazakstan; antimony, arsenic, gold, gypsum, lead, mercury and uranium in Kyrgyzstan; bismuth, boron, columbium, copper, gold, lead, mercury, molybdenum, silver, tantalum, tungsten and zinc in Tajikistan; bromine, iodine, strontium and sulphur in Turkmenistan; and copper, fluorspar, gold, lead, molybdenum, tungsten, uranium and zinc in Uzbekistan.

Output of many commodities slackened after the breakup of the USSR, because effective demand fell,[527] because transport and contractual links were interrupted, because equipment deteriorated and repair and investment outlays were insufficient, and on environmental grounds. Oil extraction declined in four of the states and rose only in Uzbekistan, but is now increasing again in Kazakstan and Turkmenistan. The vast Tengiz field has been developed by a number of transnational oil corporations, including Chevron (the first to negotiate, well before independence), Mobil, Agip, Total, Shell, BP and Statoil; other Kazak fields are being exploited with Exxon, Texaco and Amoco. The Turkmenistan government has plans to quintuple extraction (to 500,000 barrels/day) by the year 2000, and had intended to keep hydrocarbon production in state hands – apart from an early concession to the Argentine company, Bridas, which it had sought unsuccessfully to withdraw. But the patent inefficiency of Turkmanneft led to a new oil and gas law which, while not ceding ownership, envisaged production sharing agreements and joint ventures. In January 1997, Petronas of Malaysia and Monument Oil of the United Kingdom signed a production sharing agreement for the exploitation of most of the western deposits. Both governments have opened their refining operations to foreign companies in preparation for the anticipated larger throughputs and greater domestic value added. The privatization of the Chimkent refinery[528] somewhat tarnished the reputation of the Kazak authorities, which disallowed the auction winner and accepted the bid of a Canadian company, Hurricane Hydrocarbons. Also in 1996 the Turkmen government signed a $470 million construction package to enhance the continuingly state owned Turkmenbashi oil refinery on the Caspian coast.

Turkmenistan's greatest export potential lies in its gas deposits, but there is as yet no sign that these are to be developed other than by the state. Extraction has declined since independence because Gazprom, the Russian production and pipeline monopoly which provides the sole evacuation route, has offered transit capacity only to Ukraine, where consumption has declined with recession and payment has fallen into serious arrears. Over the post-independence period Kazak gas extraction has also fallen, but is now beginning a rapid upturn. The present export potential is at Karachaganak, being operated by British Gas, Agip, Texaco and Gasprom, but there are substantial deposits elsewhere in the country, notably at Pridorozh adjoining the Kumkol oilfield.

Western company participation in metals mining has been substantial in gold, the first joint venture for which, as for oil, antedates independence. Newmont of Denver, Colorado, agreed a joint venture in August 1991 with the Uzbek Navoi Mining and Metallurgical Combine to extract gold in the Kyzyl Kum desert at Zarafshan, a large opencast operation which had become the largest single gold-mine in the USSR; from 55 tonnes in 1995, production is expected to reach 83 tonnes in the year

---

II.1; a trade union source notes that workers in the Kazak Lead Combine were unpaid for 9-12 months but managers paid themselves 6 months in advance to invest in a private venture. D. Rosenblum, *Kennan Institute Meeting Report*, March 1997.

[525] More than 3,000 occurrences of economically valuable minerals, of which over 800 are currently exploited. J. Dorian, *Mining in the CIS*, Financial Times Management Reports, 1993, p. 23.

[526] Estimates by British Petroleum, converted at 1 cubic metre of gas = 0.9 tonnes of oil, of reserves and production for 1995.

[527] Although some supply continued even to non-paying customers, for example, from Turkmenistan to Ukraine.

[528] Historically and at present Chimkent refines crude piped from Siberia, but it is close to the Kumkol deposits, soon to be opened up; a joint venture of two Russian oil companies (Lukoil and Yuzhneftegaz) envisaged use of both Russian oil, which they have, and of the Kazak crude for which they tendered.

2000. Lonrho of London entered a joint venture with the Navoi Combine to work its other big opencast facility, at Amantaytau, aiming at 15 tonnes by the millennium. Negotiations in 1996 were breaking new ground, as the Uzbek government offered a majority shareholding to two companies (Mitsui of Japan, 20 per cent, and Newmont, 40 per cent) in the Angren mine (160 tonnes of reserves). Kyrgyzstan's Kumtor gold deposits are seen as the world's eighth largest deposit (estimated at 516 tonnes) and are being exploited in a joint venture of the state owned Kyrgyzaltyn with Cameco of Canada,[529] with two smaller deposits being exploited with other western firms. Kazakstan and Tajikistan have, on the other hand, dealt badly with foreign participation in gold-mining, both on pre-emption and participation. The Kyrgyz and Uzbek governments require all gold to be delivered to its agencies, but pay a proportion (respectively 25 and 30 per cent) in convertible currency,[530] whereas the others pay only in domestic currency. The two principal Kazak mines have had mixed fortunes with western companies. Shares in the Bakyrchik (340 tonnes reserves) joint venture with the state owned Altynalmas were floated on the London Stock Exchange but had to be rescued by a Canadian investor, who more recently has negotiated with the government for a half-share with Altynalmas in a smaller, but still promising deposit (200 tonnes reserves) at Vasilkovskoe. The tendering process had previously been seriously flawed and has jaundiced foreign estimations of participation in Kazak mining. The Tajik government, owning majority shares through its Commission on Precious Metals and Gems, in joint ventures with Nelson Gold of Canada at Zeravshan (in the peaceful north of the country) and with Gold and Mineral Excavation of the United Kingdom at Devaz (in the former battle zone), demanded a veto over all gold sales: only a long operational shut-down by the minority owners (who have management control) forced the Commission – at least temporarily – to yield.

Foreign interests are also participating in the revival of other metal extraction and fabrication. Among these are, in Kazakstan, the chrome industry at Aktyubinsk, the copper combine at Zhezkagan, and the iron, steel and coke complex at Karaganda. The large Uzbek steel plant at Bekabad (mostly using imported iron ore) remains wholly in state ownership with contracts with German and Russian firms for reconstruction. No takers have been found for the Tajik aluminium plant at Tursunzade discussed among environmental problems in the following section.

As primary producers, enterprises in central Asia seek to diversify their clientele away from the Soviet period reliance on Russia and Ukraine. This is being assisted by new transport investments recently, and prospectively to be, undertaken. A railway now links the line between Turkmenistan and Uzbekistan to the Iranian network and that from Kazakstan into China has been completed and activated. With three major trunk lines already into Russia and thence to Europe or to the Northern Pacific, railway freight can now be carried to the Indian Ocean and to the China Sea. Two important highways are being upgraded to modern standards in Kyrgyzstan from Bishkek to Osh and in Turkmenistan from Ashgabat to Mary. For the crucially important oil and gas expansion, much more than the existing lines of Transneft and Gazprom are needed. Currently all Turkmen gas and nearly all Kazak oil is exported via Russian pipelines.[531] Turkmen gas exports by 1994 had fallen to a third of their pre-independence volume and Kazak gas exports had halved.

The potential now being opened up in the two countries for both hydrocarbons involves massive investment in pipelines, and many political and economic problems.[532] The Caspian Pipeline Consortium (CPC)[533] is to build a pipeline, operated by Transneft, to the Russian Black Sea port of Novorossisk which should be sufficient to carry all Tengiz and possibly Emba oil exports until the year 2015. The Kazakstan Pipeline Consortium plans to take other Kazak oil under the Caspian, with landfall near Baku, thereby avoiding Russia (because the transit of CPC across Chechnya causes political concern); it would there collect Azerbaijan crude, and pass through Georgia[534] for export either there (at Supsa on the Black Sea) or across Turkey to the Mediterranean at Ceyhan. The latter variant is one mode of reducing oil tanker transit of the Bosphorus, which Turkey limits on environmental grounds; the other mode is to put oil in tankers across the Black Sea, for a pipeline uptake over to the Adriatic (an option apparently superseding a project across Bulgaria and Greece to the Aegean). Both Kazak and Turkmen oil are to be supplied

---

[529] At 4,400 metres high in a remote part of the Tien Shan mountains, the mine has required much material infrastructure and personal endurance due to the rarefied oxygen in the atmosphere; 290 tonnes are planned to be extracted over 19 years. *Financial Times*, 9 December 1996.

[530] These arrangements were secured in Tashkent only after lengthy negotiations, due it seems more to bureaucracy than a matter of principle.

[531] A little Kazak oil is shipped by tanker across the Caspian; Chevron received special permission from the United States government to deliver to Iran, despite the United States embargo on United States firms. The reduced export of Kazak gas has so far been put through Gazprom's network.

[532] Among studies devoted to these issues are R. Forsythe, *The Politics of Oil in the Caucasus and Central Asia*, Adelphi Paper 300, Oxford University Press for the Institute for Strategic Studies, 1996; J. Roberts, *Caspian Pipelines* (London, Royal Institute of International Affairs, 1996); and G. McDonell, "The Euro-Asian corridor", in R. Allison (ed.), *Challenges for the Former Soviet South* (London, Royal Institute of International Affairs, 1996), pp. 307-352.

[533] As signed in April 1996, the CPC has a 24 per cent state holding by Russia, 19 per cent by Kazakstan and 7 per cent by Oman, with the remaining 50 per cent distributed among eight western, Russian and Kazak companies.

[534] Avoiding Armenia, with which Azerbaijan fought a war over the Nagorno-Karabakh region.

to northern Iran, against which Iran would supply on their account world markets from its Gulf outlets. A gas pipeline from Turkmenistan into Iran is to start throughput in 1997. An oil pipeline from Turkmen deposits across Afghanistan to an Indian Ocean deep water promontory in Pakistan is under negotiation. Projects for gas pipelines break more audacious ground. Unocal of the United States and Delta Oil of Saudi Arabia have preliminary agreements with Turkmenistan, Iran and Pakistan for exporting gas from the large Daulatabad deposit; this could be paralleled for much of its route by an oil pipeline from the neighbouring Chardzou field also to an outlet on the Indian Ocean. More visionary is a Japanese sponsored gas pipeline for taking Turkmen and Uzbek gas into China and Japan.

With hydrocarbon exports as secure as could be expected for the long term – subject always to the complex of political risks, domestic and international, that could be envisaged – the two countries which will be deriving large shares of their revenue (export earnings and taxes) therefrom have been reminded in a World Bank report of the perils of the "Dutch Disease".[535] In the case of the Netherlands, the 1960s windfall of abundant gas brought substantial tax revenue and appreciation of the guilder. Manufacturing became less competitive, but the redundancies occasioned could be supported by generous unemployment benefit. In the aftermath of the gas boom, industry had to be restored to competitiveness and jobs recreated. The possible lesson for Kazakstan (the currency of which has indeed appreciated) and Turkmenistan is that over-reliance on energy exports must be matched by industrial diversification. The same message may in part be relevant to Uzbekistan, as it liberates itself from cotton monoculture. In 1995, cotton still contributed about two thirds of Uzbek export earnings, but less than one third of Turkmen exports. As the following section observes, the environmental damage effected by excessive plantation of cotton is being modified. The cotton area had already been cut back after independence in Uzbekistan and the crop has been declining. The cotton harvest was severely hit by bad weather in 1996 and export earnings were also affected by a 15 per cent fall (in dollar terms) of the cotton price on world commodity markets, on which – following a virtual elimination of direct contractual sales to Russia – all sales are made. A mainstay of Kazak exports, grain, experienced a record low harvest in 1995, but gained higher prices. The unit-value divergence may be attributed to a long-run declining trend for cotton as stocks increase and crop yields rise,[536] and to the significant share of Kazakstan in the world grain trade.[537]

Industries are already established which could be redeveloped in a policy of diversification, but they had their origin in a centrally planned economy which assured them of outlets by and large irrespective of their production or transport costs. To nurture some of them into a competitive relationship with imports may require a period of protection when WTO membership is negotiated and as the two customs unions with some of the states are finalized. Foreign investment is effecting some of that modernization, especially among consumer goods industries which use local materials. Significant takeovers under privatization, or in joint ventures with state entities, include Philip Morris in the Kazak tobacco industry and BAT in that of Uzbekistan; Daewoo in fabrics and television electronics in Uzbekistan; RJR Nabisco in Kazak confectionery; and André of Switzerland in mineral water in Kyrgyzstan. Little such intervention is yet found in Tajikistan and Turkmenistan.

(ii) Environmental damage and repair

The "liquid assets", so to speak, of central Asia are oil and water. They are to be husbanded for national development more than in the past: the hydrocarbons have been under-exploited and water has been overexploited. A key issue in Turkmenistan and Uzbekistan has been the waste of water and its misuse to drain off noxious chemicals applied in farming (notably cotton plantations). As part of its mission the EBRD requires all those with whom it sponsors projects to assure compliance with an Environmental Action Plan. It is publishing an *Investor's Guidebook on Environmental Health and Safety* for each of the five states,[538] and is advising each government on a National Environmental Action Plan. The irrigation system requires upgrading: the water is wastefully used on cotton plantations and the Main Turkmen Canal, being long and unlined, loses much of its throughput by evaporation and leakage. There and in Uzbekistan over-application of pesticides and fertilizer during the Soviet targeting of gross production alone has polluted rivers and reduced yields on watered land.

So much of the region is desert that the water availability per capita classes four of the countries as in a state of "absolute stress", that is less than 500 cubic metres per year; even Kazakstan with its lakes and rivers has so much desert and semi-arid territory that it is classed as in a state of "relative scarcity". Table 4.5.1 shows countries with similar water ratios to the central Asian states.[539] Part of the solution must be specific charges for water, currently supplied free, and the other must be investments to rationalize consumption. Intergovernmental negotiations

---

[535] World Bank, *Kazakstan. The Transition to a Market Economy* (Washington, D.C.), 1993, pp. 30-31.

[536] *The World Cotton Complex*, Report by Rabobank International (summarized in *Financial Times*, 21 November 1996).

[537] Deputy Prime Minister Z. Karibzhanov, *Panorama* (Almaty), 18 October 1996.

[538] These are scheduled to appear during the course of 1997.

[539] The terminology and concordances are from A. Baer, *International Social Science Journal*, No. 148, June 1996, as cited in table 4.5.1.

TABLE 4.5.1

Annual water resources per person, average 1989-1990
(Cubic metres)

Country	Cubic metres	Country	Cubic metres
Kazakstan	762	Algeria	690
Kyrgyzstan	186	Bahrain	184
Tajikistan	156	..	..
Turkmenistan	198	Barbados	195
Uzbekistan	376	Israel	407

Source: USSR Goskomstat, *Narodnoe khozyaistvo SSSR v 1989 g.* (Moscow), 1990, pp. 17, 247; A. Baer, "Not enough water to go round?", *International Social Science Journal*, No. 148, June 1996, table 1.

particularly concern the shrinkage of the Aral Sea and the timing of upstream water release. The volume of the Aral Sea has diminished by three quarters since 1960 as the annual input of water from the Amu-Darya and the Syr-Darya was reduced from 60 cubic kilometres to 5 cubic kilometres, due to the increasing offtake for irrigating cotton plantations and the wasteful practices employed. Complementing the expansion of the irrigated area, bigger cotton harvests were derived from increased yields as pesticides and fertilizer were liberally applied. Water pollution from these chemicals, soil salination from inefficient canalization and airborne salt from the dessicated bed of the Aral Sea are among the environmental legacies of the Soviet campaign for "white gold". After many previous meetings by them, by officials and at international conferences, the five Presidents reached agreement in February 1997 to devote 0.3 per cent of their fiscal revenue to an International Fund for Saving the Aral Sea, to which the World Bank will contribute $380 million over four years. A trilateral agreement of the Presidents of Kazakstan and Uzbekistan with the President of Kyrgyzstan in May 1996 provides for scheduling the flow of the Syr-Darya. In Kyrgyzstan the upper waters of that river and its tributaries and the river Talas have electricity generating potential of an annual 140 billion kilowatt-hours, but at present only a tenth of that is generated; until more of that power is captured (by substantial investment), Kyrgyzstan seeks to retain water for hydroelectricity when the downstream states need it for irrigation. A bilateral arrangement on the Amu-Darya was made by the Turkmen and Uzbek Presidents in January 1996, but a headwaters agreement awaits a more settled Tajikistan.

Among the polluting legacies of Soviet plant construction, one of the most serious is the Tajik Aluminium Works at Tursunzade, the uncontrolled emissions from which cause fluorosis downwind of the plant and into Uzbekistan.[540] Noxious and persistent particulate discharge is particularly high from copper installations at Balkhash and ferrous metallurgy at Karaganda in Kazakstan, but mining uranium and antimony in Kyrgyzstan and the chemical warfare installation at Aralsk (on the Kazak side of the border with Uzbekistan) have caused deep concern. The environmental cleansing of the nuclear weapons test site near Semipalatinsk in Kazakstan cannot be completed in the foreseeable future.

The prospective cost of making good environmental damage and preventing its recurrence is a particular deterrent to foreign investors taking up offers of joint ventures or privatizations. They have (as in the example cited of the Uzbek tobacco industry) higher standards of their own, and as a foreign company would be more open to criticism than a domestic one.

(iii) New avenues of trade

Small economies in world terms, the five central Asian states formulate their policies on international trade in two familiar contexts. The first is to use trade to recover, and then develop, from the transition recession, a considerable factor in which was the cessation of trade with enterprises in the former USSR and other members of the CMEA. Because it would be suboptimal to revive all those exchanges,[541] some of the trade expansion must be (and already has been) with other partners. To benefit from present trends in economic globalization the indications would be generally to reduce barriers to trade and in particular to participate fully in the World Trade Organization. But two dangers emerge in such widening: at current terms of trade, reliance may become excessive on natural resource exports if rational use is not made of the profit arising from present comparative advantage – to avoid the "Dutch Disease" discussed in section 4.5(i) – while in the interests of resumed political relationships, economic integration may be unwisely promoted – the "trilateral" and "quadrilateral" customs unions of which some central Asian states are members. A second danger lies in the temptation to ignore the signals of competition and profitability coming from abroad which allow trade to be an instrument of marketization. Price distortions and branch or enterprise subsidies have persisted partly because of trade barriers. "Dismantling these controls reduces the distortions and promotes more efficient resource allocation – a potentially dramatic

---

[540] A UNICEF/WHO mission drew attention to this in the first months of independence, but the pollution has only ceased when lack of imports or capture by rebels stopped production.

[541] B. Kaminski, "Trade performance and access to OECD markets", in C. Michaelopoulos and D. Tarr (eds.), *Trade in the New Independent States* (Washington, D.C., The World Bank, 1994), displays gravity models of trade which justify as optimal merely 15 to 40 per cent of the trade which took place among the Union Republics of the former USSR. The IMF calls such trade "traditional" and trade with the rest of the world "non-traditional"; Russian commentaries often refer respectively to the "near abroad" and the "far abroad".

TABLE 4.5.2

Trade with CIS members and non-CIS countries, 1995
(Million dollars)

	CIS			Non-CIS		
	Imports from	Exports to	Balance with	Imports from	Exports to	Balance with
Kazakstan	2 609	2 631	22	1 172	2 343	1 171
Kyrgyzstan	353	269	-84	169	140	-29
Russian Federation	13 526	14 244	718	33 155	65 666	32 512
Tajikistan	478	252	-226	321	497	176
Turkmenistan	745	930	185	619	951	332
Uzbekistan	1 118	1 109	-9	1 630	1 712	82
Total CIS	30 924	29 326	-1 598	44 431	79 409	34 978

Source: Chapter 3, tables 3.5.7 and 3.5.9.

TABLE 4.5.3

Trade with CIS members and non-CIS countries, 1996
(Million dollars)

	CIS			Non-CIS		
	Imports from	Exports to	Balance with	Imports from	Exports to	Balance with
Kazakstan	3 028	3 565	537	1 241	2 787	1 546
Kyrgyzstan	495	405	-90	395	101	-294
Russian Federation	14 095	15 427	1 332	31 191	70 991	39 800
Tajikistan	384	334	-50	269	432	163
Turkmenistan	350	1 172	822	918	532	-386
Uzbekistan	1 455	847	-608	3 116	2 825	-291
Total CIS	31 978	31 620	-358	49 532	87 089	37 557

Source: CIS Statistical Committee, *SNG v 1996 g., Statisticheskii sbornik* (Moscow), 1997, p. 47.

process, especially for the smaller economies among the newly independent states."[542]

On the evidence of 1995 and 1996 (tables 4.5.2 and 4.5.3), there is a clear contrast between the two hydrocarbon exporters, Kazakstan and Turkmenistan, which ran visible surpluses (in 1996 the Turkmen surplus with CIS partners offset a deficit with the rest of the world), and Kyrgyzstan, which was in deficit in both years to CIS and to non-CIS partners. Both Kazak and Turkmen exports of energy (and metals from the former) are likely to expand in 1997, and the start of Kumtor gold exports and of hydroelectricity sales may turn Kyrgyz deficits around. Uzbekistan had been in slight deficit with CIS and in surplus with non-CIS partners in 1995 but was in serious deficit with each in 1996. The turn around shocked the central bank into severe foreign currency rationing: even before the reimposition of controls in the autumn (section 4.4(i)), long delays were being imposed on the release of currency for import payments and profit repatriation. Unless those measures are relaxed, trade volumes in 1997 will be much lower

TABLE 4.5.4

Exports to and imports from non-CIS partners, 1994-1996
(Per cent of GDP)

	Exports			Imports		
	1994	1995	1996	1994	1995	1996
Kazakstan	10.8	14.0	13.9	11.0	7.0	6.2
Kyrgyzstan	10.6	9.4	5.8	9.7	11.4	22.5
Tajikistan	41.0	82.5	41.7	40.8	53.3	25.9
Turkmenistan	13.2	31.7	17.7	20.8	20.6	30.6
Uzbekistan	14.8	16.9	20.3	18.5	16.0	22.3

Source: CIS Statistical Committee, *SNG v 1996 g., Statisticheskii spravochnik* (Moscow), 1997 and UN/ECE computations.

Note: Trade data were reported in dollars; GDP data reported in national currencies were converted to dollars at the current (average annual) exchange rate. The trade/GDP ratios are thus very sensitive to exchange rate behaviour. The large swings in the ratios in the face of rather small changes in trade volume, notably in Tajikistan, mainly reflect exchange rate volatility (the new Tajik currency was probably much undervalued at its introduction in 1995 and appreciated substantially in real terms in 1996); similar factors may have affected the ratios for other countries.

than in 1996. Tajikistan is, as in most other economic indicators, a special case. While civil war raged, it traded predominantly with the two states, Russia and Uzbekistan, which militarily supported the Dushanbe government (95 per cent of imports and 75 per cent of exports in 1994), but as hostilities wound down – the peace accords were reached in July and December 1996 – such reliance diminished: the two states accounted for 57 per cent of imports and 65 per cent of exports in 1996.[543]

As table 4.5.4 shows, three states have been enlarging the proportion of their exports to non-CIS members – the destinations of some 14 per cent of GDP for Kazakstan, 18 per cent for Turkmenistan and 20 per cent for Uzbekistan – but the 1996 share for Kyrgyzstan was not only small (6 per cent of GDP) but about half what it had been of a larger GDP in 1994.[544] Conditions in Tajikistan are still unstable. In 1995, 51 per cent of total Tajik exports was aluminium shipped to non-CIS customers, nearly all in Benelux; other than cotton (a third of all exports) and aluminium, Tajikistan had little else to sell. For the three states, continued growth of exports to western markets depends upon the usual determinants of trade direction, of which two merit special attention. One is the inflow of foreign direct (and to some extent portfolio) investment. Not only do the equipment and restructuring of enterprises part or fully purchased by external companies usually involve imports, often from the owner's country, but they also stimulate exports. A French analysis of French and United States

---

[542] M. Bruno, "Foreword", in C. Michaelopoulos and D. Tarr, *Trade Performance and Policy in the New Independent States* (Washington, D.C., The World Bank, 1996).

[543] CIS Statistical Committee, *Vneshneekonomicheskaya deyatel'nost' gosudarstv Sodruzhestva v 1994 g.*, p. 17; *Vneshneekonomicheskaya deyatel'nost' gosudarstv Sodruzhestva v 1995 g.*, p. 11; *Sodruzhestvo nezavisimykh gosudarstv v 1996 g.*, p. 101 (Moscow).

[544] These ratios of trade to GDP, however, are very sensitive to exchange rate movements. See note to table 4.5.4.

TABLE 4.5.5

Progress in trilateral and quadrilateral integration, 1994-1996
(Million dollars)

	1994	1995	1996
Imports of Kazakstan, Kyrgyzstan and Uzbekistan from each other	785	701	704
Exports of Kazakstan, Kyrgyzstan and Uzbekistan to each other	785	664	645
Imports of Belarus, Kazakstan, Kyrgyzstan and Russian Federation from each other	9 913	10 113	12 123
Exports of Belarus, Kazakstan, Kyrgyzstan and Russian Federation to each other	8 268	10 212	12 076

Source: CIS Statistical Committee, *Vneshneekonomicheskaya deyatel'nost' gosudarstv Sodruzhestva v 1994 g.*, p. 17; *Vneshneekonomicheskaya deyatel'nost' gosudarstv Sodruzhestva v 1995 g.*, p. 11; *SNG v 1996 g.*, p. 101 (Moscow).

TABLE 4.5.6

Comparison of partner trade returns, 1995
(Million dollars)

	National returns		Partner returns	
	Exports	Imports	Exports	Imports
Kazakstan/Germany	170.4	201.5	285.0	95.0
Kazakstan/Russian Federation ...	2 102.6	1 834.1	2 586.0	2 726.4
Kazakstan/United Kingdom	111.7	82.8	41.9	77.2
Uzbekistan/Germany	22.4	371.9	374.0	187.6
Uzbekistan/Russian Federation ...	549.2	719.6	824.0	888.7
Uzbekistan/United Kingdom	222.8	27.8	23.8	2.5

Source: Kazakstan State Statistical Committee, *Kratkii statisticheskii sbornik Kazakhstana 1995*, pp. 134-137; CIS Statistical Committee, *Vneshneekonomicheskaya deyatel'nost' gosudarstv Sodruzhestva v 1995 g.* (Moscow); United Nations Comtrade Database.

---

trade and direct investment with 40 partners over 1984-1994 shows that $1 of French investment abroad led to 80 cents in additional exports to, and 60 cents in imports from, the host country; $1 of United States investment abroad generated $2.50 in both exports and imports with the host country.[545] The other determinant is the balance between trade creation and trade diversion consequent upon the formation of the trilateral and quadrilateral customs unions. The period covered by table 4.5.5 is too short to show the effects of the trilateral agreement of 1994 forming a "united economic zone" of Kazakstan, Kyrgyzstan and Uzbekistan – in fact, mutual trade fell in value, and more in real terms, in 1995-1996. The quadrilateral customs union embracing Belarus, Kazakstan, Kyrgyzstan and Russia was established only in March 1996 and the outcomes are still to be seen. However, the rise in the value of exchanges shown for 1996 may suggest some re-establishment in different forms of previous inner Soviet goods flows, notably by Kazak-Russian "financial-industrial groups", which exist in ferrous and non-ferrous mining and metallurgy, heavy engineering and telecommunications. In the transition period before tariff harmonization, Kazakstan first considerably raised some tariffs to, then sharply reduced them below, the Russian level. Thus, in August 1995 the duty on motor cars was increased from 15 to 30-40 per cent and then cut to 2 per cent in April 1996 – a typical product the home production of which Russia protects, but which Kazakstan imports.[546] A problem of revenue sharing has arisen because most non-CIS imports enter Kazakstan and Kyrgyzstan through Belarus or Russia and the duty is collected at the first customs union frontier traversed.

Because Kazakstan and Russia share a common frontier, their respective trade returns should not differ by more than valuation margins.[547] The divergence in their reported mutual trade in table 4.5.6 is probably attributable to the differences of treatment by each customs authority of the unrecorded trade to which reference has already been made (section 4.2(i)) and of cross-border shopping. The large divergences of trade returns with remoter partners, however, raise other issues. Thus, ignoring the f.o.b./c.i.f. differential, Uzbekistan reported in 1995 as sales to Germany only one eighth of the value Germany recorded as importing from Uzbekistan, and contrariwise it reported selling to the United Kingdom no less than 89 *times* the value the United Kingdom wrote down as coming from Uzbekistan. But partner records of the return flows – that is, Uzbek imports – were identical with the mirrored exports. There seems less reporting divergence between imports of those same countries and Kazakstan's exports, but the gap was in fact wider for their exports compared with Kazakstan's imports. There is some ready explanation for Uzbek exports to the United Kingdom, in which its customs statistics record cotton sold to a United Kingdom broker (A. Meredith Jones and Co.) as going to the United Kingdom, whereas the broker disposes most of it to buyers in other countries. Such brokerage, which may apply to transactions with German firms, is significant for cotton, as the Russian mills no longer buy direct from Uzbekistan as they did when under Soviet planned deliveries.

A policy of enhancement of value added to commodities in which the central Asian states have a comparative advantage is everywhere to be promoted. Although some redirection of crude oil and natural gas away from export into refining and petrochemicals is

---

[545] L. Fontagné and M. Pajot, "The links between trade and FDI", *The CEPII Newsletter*, No. 6 (Paris), 1996, p. 5.

[546] On the 1995 tariffs in anticipation of the quadrilateral union, see Z. Usenova and S. Horton, "New customs union leads Kazakstan to liberalize import/export constraints", *Central Asian Monitor*, No. 4, 1995, pp. 30-36.

[547] The valuation differential arises between exports f.o.b. and imports c.i.f. (conventionally assessed at 10 per cent on exports) and the exchange rates into dollars used to convert the respective customs declarations (significant in periods of rapid inflation, although both the tenge and the rouble were appreciating against the dollar in 1995).

underway, the markets envisaged are in or near the region itself. In the case of Uzbekistan, the upgrading of the Fergana refinery is import saving, as its throughput changes from Russian to Uzbek crude oil. But investment in higher technology is export promoting, notably in allowing better ferrous and non-ferrous products to be sold, in replacement of ingot metals. The EBRD is assisting the Kasansay-Tekmen woollen plant to process Uzbek wool into higher quality textiles than the country has hitherto produced and to make fabrics, carpets and blankets for export; likewise, a Turkmen plant is being aided to weave high quality denim.

Trade, particularly that between neighbouring states, is intended to be stimulated by the creation of special economic zones – so far three in Kazakstan, four in Kyrgyzstan and seven in Turkmenistan. But a principal objective is to encourage foreign investors to establish manufacturing facilities therein, imports and exports being duty-free.

(iv) Foreign investment and external support

Beneficial participation in a global economic system is served by membership of a range of worldwide and regional institutions. Following entry into the United Nations in March 1992, all five states joined the international institutions in that "family", and have particularly received advice and assistance from the UNDP, the IMF and the World Bank. At end-January 1997, the IMF had lending arrangements with Kazakstan (Extended Fund Facility of $431 million), Kyrgyzstan (Enhanced Structural Adjustment Facility of $123 million) and Uzbekistan ($174 million);[548] a very small facility for Tajikistan had been exhausted and no arrangement had been made with Turkmenistan. All five states are members of two United Nations regional Commissions, the ECE and ESCAP, and of two intergovernmental regional banks, the EBRD and (except for Tajikistan and Turkmenistan) the Asian Development Bank.[549] The EBRD now has projects in all five states, among which are bank finance for small and medium private sector lending (Kazakstan, Tajikistan and Uzbekistan), gold production (Kyrgyzstan and Uzbekistan), port reconstruction (Kazakstan and Turkmenistan), telecommunications, an agribusiness company and electricity supply (Kyrgyzstan), textiles (Uzbekistan) and home appliance manufacture (Kazakstan).[550] The ADB supports a number of social infrastructure projects in Kazakstan and Kyrgyzstan, farm restructuring in the same two, with water management in Kazakstan, and rail and road development in Uzbekistan. A mission from the Islamic Development Bank (IDB) visited Kazakstan, Turkmenistan, Uzbekistan and Azerbaijan in September 1996 to examine further projects.[551] The central Asian states are at different stages in the process of obtaining membership of the World Trade Organization: the requests for membership of Kazakstan, Kyrgyzstan and Uzbekistan have been lodged and the first two are currently under consideration by accession working parties; but Tajikistan and Turkmenistan have yet to apply.[552] In seeking membership their MFN tariff levels and their customs union relationships will need to be negotiated.

Kazakstan and Kyrgyzstan have each created agencies, described as "one-stop shops" for foreign investors: in the words of the establishing Kazak decree of March 1997, the State Committee on Investments, headed by the First Deputy Prime Minister, Akhmetzhan Yasimov, is "the sole government body with the right to execute government policies to support direct investment".[553] These two states each offer duty-free import of equipment and guaranteed conversion of profits and repatriated capital, and in March 1997 Kazakstan enacted tax concessions paralleling those in Kyrgyzstan on the first five years' profit of a foreign enterprise. Tajikistan in 1995 offered improved protection for foreign investment and preferential terms for joint ventures in which more than 30 per cent is foreign owned. Turkmenistan also permits duty-free import of capital equipment and concessions on the taxation of dividends and profits. Uzbekistan allows the duty-free import of equipment, and licence-free exports, together with tax concessions for an initial two years. Portfolio investment in Kazakstan, Kyrgyzstan and Uzbekistan has been stimulated by a secondary market in privatization equity and by the establishment of a stock exchange. A further positive consideration for investors choosing between central Asia and, for example, Russia is personal security: lawlessness does not involve murder or kidnap in business disputes or competition, but administrators may range from the overzealous and the inefficient to rent-seekers and the corrupt.

---

[548] *IMF Survey* (Washington, D.C.), 24 February 1997, p. 57, converted at SDR=$0.71.

[549] The ADB has lent $230 million to Kazakstan for projects in agriculture, road rehabilitation and education.

[550] The $12.5 million equity and $25 million loan deal for two plants in Almaty (granted in May 1995) was to a joint venture of the (Kazak) Kramds National Joint Stock Corporation and the (Turkish) Simtel Industry and Trade Co., but may be prejudiced by the collapse of Kramds Bank in October 1996 with debts of $28.6 million.

[551] IDB projects in Turkmenistan include a 126 kilometre irrigation canal (Kazandjik to Kyzyl Atrek) to reduce water waste and return some to the Aral Sea, and in Kazakstan road modernization.

[552] The accession working party to consider Kyrgyzstan's application met for the first time on 10-11 March 1997, that for Kazakstan on 19-20 March 1997. The accession process for Uzbekistan will only begin when it completes the documentation required for the working party to start its work.

[553] As cited in *Financial Times*, 6 March 1997.

The currency crisis in Uzbekistan has made Kazakstan relatively more attractive to foreign investors, such that while the sum (in its unofficial exchange rate) has depreciated since delays in central bank settlements began and arbitrary controls were imposed, the tenge has appreciated (both on estimates of real change). An examination of 20 countries during 1959-1993 showed that "a currency crisis elsewhere in the world increases by about 8 per cent the probability that a country will be hit at the same time by a speculative currency attack"; one consideration is that "a crisis-induced currency devaluation in one country would put its major trading partners at a competitive disadvantage, thus applying downward pressure on their currencies".[554] As Kazakstan and Uzbekistan are largely non-competitive in current account transactions, enterprises in the former are unlikely to expect to be disadvantaged in that sphere by the latter's devaluation, but as FDI is deflected from the latter to the former (as seems to be the case), the increase in the demand for tenge rather than for sum appreciates the Kazak currency. The swing of investors' preference between the two countries seems initially (i.e. after privatization and sales to foreigners had started) to have been for Kazakstan in 1994, then to Uzbekistan, and now back to Kazakstan. Among the factors influencing investor ranking were the hydrocarbon and other capital-attractive options and earlier privatization in Kazakstan; opinion soured with Chevron's disaffection and tendering procedures for Vasilovskoe gold and Chimkent refining,[555] while Uzbekistan was attracting many major investors, such as BAT and Daewoo.[556] Kazakstan's foreign investment drive, following the Law of 30 December 1994, then offered a counterforce[557] and was fortified in October 1996 — by coincidence, just as the Uzbek crisis was becoming transparent — when IBCA (the Europe credit rating agency), and shortly afterwards Standard and Poor's and Moody's, accorded Kazakstan a rating (BB-),[558] but declined to rate Uzbekistan.

The branch profile of investible projects also militated in favour of Kazakstan, that is, in the potential for energy exports. Thus, Kazakstan and Turkmenistan exhibit high ratios of FDI to GNP — in 1995, 5 and 8 per cent respectively, and cumulated inflows from 1991 of $1.7 billion and $0.4 billion (table 4.5.7). The higher cumulated estimates compiled by the National Bank of

---

[554] B. Eichengreen, A. Rose and C. Wyplosz, "Contagious currency crises", *NBER Working Paper*, No. 5681 (Cambridge, MA), 1997.

[555] Headlines in the *Financial Times* were "Kazakstan gets black mark from EBRD", 9 August 1995, and "Kazakstan tries to woo disgruntled gold companies," 26 April 1996.

[556] By end-1995 Korean firms had invested $109 million in Uzbekistan. UN/ECE, *East/West Investment News*, Autumn 1996, p. 23.

[557] UN/ECE, *East/West Investment News*, Summer 1995, p. 2 and Autumn 1995, p. 6.

[558] *Financial Times*, 18 January 1997 (Moody's Ba3 is equivalent to the others' BB-).

TABLE 4.5.7

Net inflows of foreign direct investment in the central Asian economies, 1991-1996
(Million dollars)

	1994	1995	1996	1991-1995[a]	FDI/GDP[b]
Kazakstan	519	859	451[c]	1 706	5.1
Kyrgyzstan	45	191	..	246	12.7
Tajikistan	12	13	..	42	2.2
Turkmenistan	103	233	..	416	7.8
Uzbekistan	72	120	..	280	1.2
Total	751	1 416		2 690	4.4

Source: World Bank, *Statistical Handbook 1996. States of the Former USSR*, 1996 and *World Debt Tables*, Vol. 2, 1996 (Washington, D.C.); national balance of payments statistics; UN/ECE secretariat estimates.

[a] Cumulative FDI.
[b] FDI 1995/GDP 1995.
[c] January-June.

Kazakstan – $2,769 million by 1 January 1996 – were 74 per cent in oil and gas, and 24 per cent in manufacturing.[559] The divergence of estimates does not obscure the country ranking by value of capital inflow since independence — Kazakstan, Turkmenistan, Uzbekistan, Kyrgyzstan and Tajikistan.

Capital inflow as equity (including portfolio and joint venture) generates an outflow of repatriable profits, all governments having undertaken to assure their convertibility. Loan funding requires service in the denominated currency. It is obviously necessary for the host economy to gain enough future foreign earnings while maintaining an appropriate exchange rate. For Uzbekistan (Bukhara, Khiva and Samarkand) and Kyrgyzstan (mountain sports and Lake Issyk-Kul), tourist income should be substantial, but most of the foreign revenue must be from merchandise trade. For stability, those earnings should be sold domestically or should fructify in factor income from abroad, but hitherto a significant amount — as elsewhere in the CIS — has gone into capital flight. In 1994, the ratio of external indebtedness to export earnings (merchandise and services) was already high in Kyrgyzstan (130 per cent) and substantial in Kazakstan (88 per cent), but it was low in Uzbekistan (33) and Turkmenistan (17).[560] Over the two years 1994 and 1995, as the first and last columns of table 4.5.8 indicate, three states have more or less uniformly doubled their indebtedness, mostly with long maturities, while Tajikistan and Turkmenistan, which at end-1993 had engaged themselves very little in external finance, now have more significant external debt. Such stability also requires a suitable margin of reserves of foreign exchange and gold. Over the same two years

---

[559] UN/ECE, *East/West Investment News*, Winter 1996, pp. 11-21.

[560] World Bank, *World Development Report 1996:* ..., table 17.

TABLE 4.5.8

External indebtedness in the central Asian economies, 1993 and 1995
(Million dollars, end of year)

	1993 Total	1995 Multilateral debt	1995 Other long-term debt	1995 Short-term debt	1995 Total
Kazakstan	1 706	1 365	1 711	50	3 126
Kyrgyzstan	301	307	263	1	572
Tajikistan	42	48	232	–	280
Turkmenistan	10	58	224	–	282
Uzbekistan	829	404	1 121	162	1 689

Source: OECD, *External Debt Statistics 1995*, table 2, and *1996*, table 1 (Paris).

Kazak currency and gold reserves rose by about two thirds (and at latest report, stood at just under $2 billion) and Uzbek reserves increased by about one third (to a recent $1.5 billion). On the books of banks reporting to the Bank for International Settlements as at end-September 1996, Kyrgyz and Tajik entities had trivial, and Uzbek substantial, sums to their credit ($3 million, $19 million and $576 million, respectively); Turkmen entities showed trivial, and Kazak large, debits ($20 million and $576 million).[561]

4.6 Economic aims after five years of independent statehood

Each President has formulated an economic and social strategy for his country and each has a mandate by election or referendum to hold office until the eve of, or into, the next millennium. President Akar Akaev has promulgated a "Strategy of Social and Economic Development of the Kyrgyz Republic to the Year 2005".[562] It is posited on a general economic expansion – a doubling of real GDP between 1995 and 2005 (about 7 per cent annually) – and requires completion of land reform and of industrial restructuring, special concern for small and medium enterprises, the development of an appropriate financial infrastructure and the attraction of FDI. In the "Uzbek Model for Transition to a Market Economy",[563] President Islam Karimov has stressed welfare objectives in Paretian terms – the establishment of market reforms such that everyone gains without worsening the situation of others. In Kazakstan, President Nursultan Nazarbayev had a multi-ethnic state in mind in evoking "the development of democracy, property rights reform and the movement towards a full-fledged market

system" as "conducive to the rise of a nation state".[564] President Saparmurat Niyazov has colourfully evoked a future for Turkmenistan as "the Kuwait of the Karakum"[565] and has set his sights on strengthening relations with his southern neighbours – Iran,[566] Afghanistan[567] and the states of the Arabian Peninsula.[568] Finally, the Tajik parliament adopted in November 1995 a reform programme to the year 2000, which promises the liberalization of foreign trade (except for the retention of the aluminium and cotton state monopsony), further privatization and the achievement of an open and transparent economy with favourable conditions for foreign investment and promoting exports.

The combination for four of the states is of strong presidential authority with the encouragement in a market system of small- and medium-sized domestic entrepreneurship and foreign co-owners for larger firms and resource development. The unification and reconciliation still to come in Tajikistan is not merely dependent on national accord, but on events outside its frontiers. What are needed urgently are orderly and predictable political and economic structures within which both domestic and foreign investors can assess and make capital decisions in a competitive context

Misinvestment – in the sense of an inheritance of assets with comparative disadvantage – was proportionately less important in central Asia than in most other command economies, but their low per capita output makes investment for generating growth more urgent than early welfare enhancement. The welfare component of new capacity is mainly to be found in job creation, needed in the light of the relative and absolute growth of the population of working age (a prospect which distinguishes central Asia from many other post-communist states). It is in this aspect that some inconsistency may be perceived among the presidential objectives. At the low – but far from abysmal – levels of living extant, it was natural for governments to offer to protect existing standards across the board, and it is certainly right to inhibit any widening of wealth and income distribution that could seriously disrupt the establishment of a normal civil society. The little survey work that has been undertaken so far on income distribution suggests that in central Asia, as elsewhere in the CIS, the spread has widened considerably. If the

---

[561] BIS, *International Banking and Financial Market Developments* (Basle), February 1997, table 5A.

[562] Summarized in the President's address to the World Economic Forum, "Kyrgyzstan: signs of recovery", mimeo, Davos, February 1997.

[563] I. Karimov, *Building the Future* (Tashkent, Uzbekistan Publishers, 1993).

[564] N. Nazarbayev, *A Strategy for the Development of Kazakstan as a Nation State*, Kazakstan Embassy (Washington, D.C.), 1994.

[565] Cited in *Central Asia Newsfile*, November 1993, p. 6.

[566] Sealed by President Rafsanjani's state visit to Ashgabat in October 1993.

[567] Turkmenistan in mid-1993 opened a consulate-general in Mazar-e-Sharif, headquarters of General Dostum, whose opposition to the Taliban corresponds to that of Iran.

[568] Two dozen small luxury hotels on the outskirts of Ashgabat particularly serve wealthy visitors for falconry in the neighbouring desert.

experience in Russia[569] is repeated in this region, the recent slackening of inflation could moderate that dispersion, but active government measures, including the comprehensive monitoring of income for progressive taxation and restraint on criminal or rent-seeking activity, could also play a significant role.

It is, rather, against the maintenance of existing real incomes by social transfers that governments may have to be more cautious than hitherto. Enterprises confronted with declining demand or rising costs have made claims on all transition governments for subsidies or various non-transparent devices such as quantitative import restrictions, price controls and procurement procedures. The loss of job security adds to the demands on government to provide protection from some of the consequences of the new organization of economic activity. The "gradualism" which all five governments have embraced for the pace and sequence of transition measures can in retrospect be seen as having partly served as a cover (as in all CIS members) to delay the impact of market oriented restructuring. Where expressed in money flows, such subsidies, tax arrears and inter-enterprise and foreign indebtedness both weaken the discipline of market signals and hold back the attainment of fiscal stability. Where explicit and implicit instruments reflect policy makers' unwillingness to forsake state controls, the highest authorities need to assess the appropriate balance for their country between restriction and regulation. Insufficient state regulation is still patent in weak financial institutions and the facilitation of firms' entry (as implemented in competition policy) and exit (through bankruptcy and compulsory rehabilitation or capitalization). Each state is engaged in the building of an independent nation and the creation of a market based economy. The construction of a responsive administration with trustworthy public servants has a crucial role to play in that process; incompetent, bureaucratic or corrupt procedures can only apply a brake to economic reform.

The many criteria that can be applied to good government and economic stability have particular force in making resources available for investment. The need for capital from both domestic and external sources argues for policies to foster investment which all central Asian governments are formulating. Six considerations are particularly relevant to the type of capital formation that is being sought, although the profile of needs (and its financing) differs in each. *One* is the installation of more efficient technology in substitution for that to which Soviet factories or mines were constrained by slow domestic diffusion, by preference for Soviet or CMEA sourcing and by the CoCom embargo. Major re-equipment by foreign owners currently includes chrome mining and steel smelting in Kazakstan, and cigarette manufacture there and in Uzbekistan, but less is needed where the enterprise had enjoyed high Soviet technological priority, such as the Chkalov aircraft plant in Uzbekistan. A *second* imperative is the redirection of trade away from the "traditional" Soviet and CMEA state enterprise recipients towards globally competitive buyers or the production of goods to counter imports; both moves are made more difficult where there is exchange rate overvaluation. This is typified in the installation of foreign companies in local manufacturing, such as consumer electronics in Kazakstan, farm produce processing in Kyrgyzstan and buses and trailers in Uzbekistan. A *third* investment line must be the enhancement of value added to minerals and commodities in which the region has a natural advantage. The most general example is oil refining – major overhauls and additions to existing capacity are underway in Turkmenistan and Kazakstan – while the accord of international acceptance of ingot gold from a Kazak refinery ironed out a difficulty between the government and the mining companies. The weight of cotton in Uzbek exports, both as a bulky commodity and as constituting some half of annual earnings, pinpoints a deficiency in Soviet industrial policy to which the ECE secretariat drew attention 40 years ago.[570] Foreign investment is rectifying this with, for example, shirt and bedlinen production in Uzbekistan and high quality denim in Turkmenistan, and with another regional speciality, silk ribbon making, also in Uzbekistan. The three other calls for investment are common to all transition economies – the restructuring of enterprises after privatization; the foundation of new private businesses; and the improvement of infrastructure. In the small- and medium-sized enterprise sector, the potential of relatively cheap but satisfactorily educated labour and the national need for job creation suggest a choice of high labour-to-capital ratios. All five states have banks oriented to rural business, which encourage rural location and hence minimize migration to towns.

Reference is made above to the support by the ADB, the EBRD and the World Bank for infrastructure development, which so far is chiefly in air, rail and road transport, and by major project consortia for pipelines. The repair and prevention of environmental damage and the protection of precious water resources is more a matter of intergovernmental agreement and operation, but pollution from individual factories (Tajik aluminium is

---

[569] B. Granville and J. Shapiro with O. Dynnikova, "Less inflation, less poverty: first results for Russia", Royal Institute of International Affairs, *Discussion Paper,* No. 68, 1996, note that "with lower inflation starting in spring 1995, the exploding inequality which characterized the first years of the Russian Federation has been moderately reversed", p. 22.

[570] "With 85 per cent of the cotton production of the Soviet Union, central Asia accounted for only 4 per cent of the spinning and 5 per cent of the weaving of cotton. Similarly the bulk of central Asian wool and silk was processed in other regions of the Soviet Union". UN/ECE, *Economic Bulletin for Europe,* Vol. 9, No. 3, November 1957, p. 61.

the salient case) has to be dealt with by the owner, appropriate allowance being made in the price at privatization.

Foreign participation, as the selection just given shows, is already widespread and in three states adds significantly to investible funds.[571] It is especially valuable at this early phase of transition because of the weak mobilization of domestic savings. The incidence of bank failures, the limitations on foreign bank operations, and the inadequacy of client monitoring all require further attention by the regulatory authorities. Because governments have generally preferred to privatize their large enterprises into joint ventures or part-sale to foreign firms, there are few big domestic firms which have profits to plough back and to invest in diversification, or which command enough confidence to successfully float new issues on the national stock exchange (which in any event does not yet exist in Tajikistan and Turkmenistan). Governments have appreciated the value of foreign direct investment, which can bring not only finance, but technology, managerial methods and access to new markets; as noted in this chapter, they offer considerable incentives to such inflow. Kyrgyzstan and Uzbekistan offer certain tax rebates to foreign firms, but other governments may be wary of discriminating against domestic enterprises.

Incentives to attract foreign capital and the minimization of disincentives (such as non-transparent tendering, unfair practice and corruption) are important, but the chief criteria which transnational investors apply are profitability, the assurance of profit and capital repatriation and an orderly and predictable economic and political environment in which they can operate. As magnets to FDI the central Asian states are small players on a large stage: cumulated over 1992-1995, they attracted just 1 per cent of the capital that went into developing and transition countries during those years.[572] They are even smaller players on world portfolio markets. Governments need to be reminded that there are 45 substantial "investment quality" markets into which capital may be placed, and that so far only Kazakstan in the central Asian region qualifies for "speculative quality". From a safe backwater of Soviet and CMEA protectionism, the five countries are sailing into the wilder waters of global competition.

---

[571] The ratio of FDI to GDP in 1995 was higher in Kyrgyzstan (12.7 per cent) than in any other transition economy, including Hungary (10.1 per cent), the east European leader. Turkmenistan (7.8 per cent) is above the second-ranking east European group (Czech Republic, Estonia and Latvia, all with ratios of about 5.5 per cent), and above Kazakstan (5.1 per cent). Tajikistan and Uzbekistan rank among the larger group of east European countries which have not received much FDI (tables 4.5.7 and 3.6.15).

[572] World Bank, *World Debt Tables 1996*, Vol. 1 (Washington, D.C.).

STATISTICAL APPENDIX

For the user's convenience, as well as to lighten the text, the *Economic Survey of Europe* includes a set of appendix tables showing annual changes in main economic indicators over a longer period. The data are presented in two sections, following the structure of the text: *Appendix A* provides macroeconomic indicators for the market economies of western Europe and North America for 1982-1996, *Appendix B* does the same for the east European countries, the Baltic states and the Commonwealth of Independent States for 1980-1996.

Data for the transition economies are preliminary because re-estimated historical series are not yet available for all countries, and longer time series could in some instances be obtained only by splicing older data with the new statistics (as explained in the notes to the tables).

Data were compiled from international and national statistical sources. Regional aggregations are ECE secretariat estimates based, variously, on 1990-1992 weights. All figures for 1996 are preliminary estimates based on data available in the middle of March 1997.

APPENDIX TABLE A.1

Real GDP in the ECE market economies, 1982-1996
(Percentage change over preceding year)

	1982	1983	1984	1985	1986	1987	1988	1989	1990	1991	1992	1993	1994	1995	1996
Western Europe	1.0	1.8	2.5	2.6	3.0	3.2	4.1	3.4	3.3	1.5	1.2	-0.1	2.5	2.6	1.8
4 major countries	0.8	1.7	2.3	2.5	2.9	2.8	4.3	3.3	2.9	1.6	1.0	-0.5	2.9	2.3	1.4
France	2.5	0.7	1.3	1.9	2.5	2.3	4.5	4.3	2.5	0.8	1.2	-1.3	2.8	2.2	1.3
Germany[a]	-0.9	1.8	2.8	2.0	2.3	1.5	3.7	3.6	5.7	5.0	2.2	-1.1	2.9	1.9	1.4
Italy	0.2	1.0	2.7	2.6	2.9	3.1	4.1	2.9	2.1	1.2	0.7	-1.2	2.1	3.0	0.7
United Kingdom	1.7	3.7	2.3	3.8	4.3	4.8	5.0	2.2	0.4	-2.0	-0.5	2.1	3.8	2.5	2.1
17 smaller countries[b]	1.3	2.1	2.9	2.9	3.2	4.0	3.7	3.5	3.9	1.4	1.5	0.7	1.7	3.2	2.8
Austria	1.1	2.0	1.4	2.5	1.2	1.7	4.1	3.8	4.2	2.8	2.0	0.4	3.0	1.8	0.8
Belgium	1.7	0.1	2.2	0.7	1.6	2.1	4.9	3.4	3.7	1.6	1.7	-1.4	2.3	1.9	1.4
Cyprus	6.3	5.3	8.8	4.7	3.8	7.0	8.5	7.9	7.4	0.6	9.8	1.5	6.1	5.3	2.4
Denmark	3.0	2.5	4.4	4.3	3.6	0.3	1.2	0.6	1.4	1.3	0.2	1.5	4.4	2.8	2.2
Finland	3.2	2.7	3.0	3.3	2.4	4.1	4.9	5.7	–	-7.1	-3.6	-1.2	4.5	4.5	3.2
Greece	0.4	0.4	2.8	3.1	1.6	-0.5	4.5	3.5	-0.6	3.4	0.8	-0.5	1.5	2.0	2.2
Iceland	2.2	-2.2	4.1	3.3	6.3	8.5	-0.1	0.3	1.2	1.3	-3.3	0.9	3.5	2.1	5.4
Ireland	2.3	-0.2	4.4	3.1	-0.4	4.7	4.3	7.4	8.6	2.1	4.0	3.1	6.5	10.3	6.3
Israel	1.4	2.6	2.2	4.4	3.7	6.3	3.5	0.2	7.1	6.3	6.6	3.5	6.8	7.1	3.8
Luxembourg	1.1	3.0	6.2	2.9	4.8	2.9	5.7	6.7	3.2	3.1	1.9	–	4.2	3.7	2.4
Malta	2.3	-0.6	0.9	2.6	3.9	4.1	8.4	8.2	6.3	6.3	4.7	4.5	4.0	9.0	4.0
Netherlands	-1.4	1.4	3.2	2.6	2.7	1.2	2.6	4.7	4.1	2.3	2.0	0.8	3.4	2.1	2.7
Norway	0.3	4.6	5.7	5.3	4.2	2.0	-0.5	0.6	1.8	2.9	3.3	2.8	5.0	3.3	4.8
Portugal	2.1	-0.2	-1.9	2.8	4.1	5.5	5.8	5.7	4.3	2.1	1.1	-1.2	0.8	2.4	2.6
Spain	1.6	2.2	1.5	2.6	3.2	5.6	5.2	4.7	3.7	2.3	0.7	-1.2	2.1	2.8	2.3
Sweden	1.0	1.8	4.0	1.9	2.3	3.1	2.3	2.4	1.4	-1.1	-1.4	-2.2	3.3	3.6	1.5
Switzerland	-0.9	1.0	1.8	3.7	2.9	2.0	2.9	3.9	2.3	–	-0.3	-0.8	1.0	0.1	-0.7
Turkey	3.6	5.0	6.7	4.2	7.0	9.5	2.1	0.3	9.3	0.9	6.0	8.0	-5.5	7.3	7.5
North America	-2.3	4.0	6.8	3.8	3.0	3.0	3.9	3.3	1.2	-1.0	2.6	2.3	3.5	2.1	2.4
Canada	-3.2	3.2	6.3	4.8	3.3	4.2	5.0	2.4	-0.2	-1.8	0.8	2.2	4.1	2.3	1.5
United States	-2.2	4.0	6.8	3.7	3.0	2.9	3.8	3.4	1.3	-1.0	2.7	2.3	3.5	2.0	2.4
Total above	-0.7	2.9	4.6	3.2	3.0	3.1	4.0	3.3	2.2	0.2	1.9	1.1	3.0	2.3	2.1
Memorandum item:															
Japan	3.1	2.3	3.9	4.4	2.9	4.2	6.2	4.8	5.1	4.0	1.1	0.1	0.5	0.9	3.7
Total above, *including Japan*	-0.1	2.8	4.5	3.4	3.0	3.3	4.3	3.6	2.7	0.8	1.7	0.9	2.6	2.1	2.3

Source: National statistics.

Note: Growth rates of regional aggregates have been calculated as weighted averages of growth rates in individual countries. Weights were derived from 1991 GDP data converted from national currency units into dollars using purchasing power parities.

[a] West Germany 1982-1991.
[b] Excluding Israel.

APPENDIX TABLE A.2

Real private consumption expenditure in the ECE market economies, 1982-1996
(Percentage change over preceding year)

	1982	1983	1984	1985	1986	1987	1988	1989	1990	1991	1992	1993	1994	1995	1996
Western Europe	1.1	1.7	1.7	2.6	4.1	3.6	3.8	3.1	3.4	2.3	1.6	0.2	1.3	2.0	2.2
4 major countries	1.0	1.8	1.7	2.6	4.4	3.9	4.2	3.1	3.0	2.3	1.4	0.1	1.6	1.8	1.8
France	3.5	0.9	1.1	2.4	3.9	2.9	3.3	3.1	2.7	1.4	1.3	0.2	1.4	1.8	2.3
Germany[a]	-1.3	1.5	1.8	1.7	3.5	3.4	2.7	2.8	5.4	5.7	2.8	0.3	1.0	1.8	1.3
Italy	1.2	0.7	2.0	3.0	3.7	4.2	4.2	3.5	2.5	2.7	1.1	-2.4	1.5	1.7	0.7
United Kingdom	1.0	4.5	2.0	3.8	6.8	5.3	7.5	3.2	0.6	-2.2	-0.1	2.5	2.6	1.9	3.0
17 smaller countries[b]	1.4	1.4	1.7	2.7	3.6	3.1	2.8	3.1	4.1	2.3	1.8	0.5	0.9	2.5	3.1
Austria	1.2	5.0	-0.1	2.4	1.8	3.1	3.6	3.5	3.3	2.9	2.8	0.7	2.5	1.9	1.2
Belgium	1.4	-1.4	0.9	2.0	2.0	2.5	2.9	3.9	2.7	2.9	2.3	-0.8	1.3	1.2	1.4
Cyprus	11.5	7.3	4.5	5.9	1.7	5.4	10.5	6.9	9.0	9.9	3.2	-5.8	4.8	11.5	2.5
Denmark	1.4	2.6	3.4	5.0	5.7	-1.5	-1.0	-0.4	–	1.2	1.9	2.4	6.6	2.3	2.4
Finland	4.9	3.1	3.1	3.7	4.0	5.2	5.1	4.3	–	-3.6	-4.9	-2.9	1.9	3.7	3.3
Greece	3.9	0.3	1.7	3.9	0.7	1.2	3.6	5.6	2.1	2.3	1.8	0.2	1.5	1.6	1.9
Iceland	4.9	-5.6	3.7	4.2	6.9	16.2	-3.8	-4.2	0.5	4.1	-4.4	-4.5	1.8	4.6	6.8
Ireland	-7.1	0.9	2.0	4.6	2.0	3.3	4.4	6.2	1.3	2.3	4.4	1.7	6.6	3.9	6.3
Israel	8.2	8.7	-6.9	1.3	16.4	8.9	4.5	0.4	5.6	7.2	7.8	7.3	9.2	7.3	6.0
Luxembourg	0.4	0.5	1.4	2.7	3.4	5.0	3.9	3.9	4.0	6.5	1.7	-0.9	2.4	2.4	2.1
Malta	2.9	1.4	4.0	5.0	1.4	0.5	9.0	9.2	3.8	3.8	4.3	0.8	2.3	11.2	9.4
Netherlands	-1.2	0.9	1.0	2.4	2.6	2.7	0.8	3.5	4.2	3.1	2.5	1.0	1.8	1.8	3.0
Norway	1.8	1.5	2.7	9.9	5.6	-1.0	-2.8	-0.6	0.7	1.4	2.2	2.2	4.1	2.6	4.7
Portugal	2.4	-1.4	-2.9	0.7	5.6	6.0	5.7	3.6	6.9	4.8	3.7	0.4	0.2	1.7	2.0
Spain	-0.1	0.3	-0.2	3.5	3.3	5.8	4.9	5.7	3.6	2.9	2.2	-2.2	0.9	1.5	1.8
Sweden	0.6	-2.4	1.4	2.7	4.9	4.8	2.6	1.4	-0.4	0.9	-1.4	-3.1	1.8	0.8	1.5
Switzerland	–	1.7	1.6	1.4	2.8	2.1	2.1	2.2	1.5	1.5	-0.2	-0.6	0.9	0.7	0.2
Turkey	6.9	6.7	8.1	-0.6	5.8	-0.3	1.2	-1.0	13.1	1.9	3.3	8.4	-5.3	7.6	9.6
North America	0.9	5.0	5.1	4.7	4.0	3.2	4.0	2.4	1.6	-0.7	2.7	2.7	3.1	2.3	2.5
Canada	-2.6	3.4	4.6	5.2	4.4	4.4	4.5	3.4	1.0	-1.6	1.3	1.6	2.9	1.4	2.4
United States	1.2	5.2	5.2	4.7	4.0	3.1	3.9	2.3	1.7	-0.6	2.8	2.8	3.1	2.3	2.5
Total above	1.0	3.4	3.4	3.7	4.1	3.4	3.9	2.8	2.5	0.8	2.1	1.5	2.2	2.1	2.3
Memorandum item:															
Japan	4.4	3.3	2.6	3.3	3.5	4.2	5.3	4.8	4.4	2.5	2.1	1.2	1.8	1.7	2.9
Total above, *including Japan*	1.5	3.4	3.3	3.6	4.0	3.5	4.1	3.1	2.8	1.0	2.1	1.4	2.1	2.1	2.4

Source: National statistics.
Note: See appendix table A.1.
[a] West Germany 1982-1991.
[b] Excluding Israel.

APPENDIX TABLE A.3

Real general government consumption expenditure in the ECE market economies, 1982-1996
(Percentage change over preceding year)

	1982	1983	1984	1985	1986	1987	1988	1989	1990	1991	1992	1993	1994	1995	1996
Western Europe	1.2	2.6	1.6	3.0	2.8	3.0	2.2	1.2	2.8	2.2	2.3	1.2	0.7	1.3	1.4
4 major countries	1.4	1.8	1.8	2.0	2.2	2.2	2.3	0.1	2.0	1.7	2.4	0.9	1.0	1.1	1.2
France	3.8	2.1	1.2	2.3	1.7	2.8	3.4	0.4	2.1	2.8	3.4	3.5	1.1	0.9	1.5
Germany[a]	-0.9	0.2	2.5	2.1	2.5	1.5	2.1	-1.6	2.2	0.3	4.3	–	1.3	2.0	2.4
Italy	2.6	3.4	2.3	3.4	2.6	3.4	2.8	0.8	1.2	1.6	1.0	0.5	–	-0.5	-0.4
United Kingdom	0.9	2.1	0.8	-0.1	1.6	1.0	0.7	1.4	2.5	2.6	-0.1	-0.1	1.7	1.5	0.8
17 smaller countries[b]	0.9	4.3	1.4	4.9	4.1	4.5	2.0	3.3	4.2	3.3	2.1	1.8	0.1	1.7	1.9
Austria	2.3	2.2	0.2	1.9	1.7	0.4	0.3	0.8	1.2	2.6	2.2	3.1	2.2	2.1	0.5
Belgium	-1.3	0.2	0.4	2.5	1.8	0.3	-0.9	-1.1	-0.4	2.2	0.1	1.4	1.4	1.0	1.8
Cyprus	3.6	6.1	3.8	3.7	3.6	5.8	10.5	1.9	16.8	4.5	13.8	-14.3	4.1	2.9	9.0
Denmark	3.1	–	-0.4	2.5	0.5	2.5	0.9	-0.6	-0.4	-0.1	0.4	3.0	1.4	0.8	1.0
Finland	3.3	3.7	2.7	4.5	3.1	4.3	2.3	2.3	3.8	2.5	-2.2	-5.3	-0.3	1.7	2.9
Greece	2.3	2.7	3.0	3.2	-0.8	0.9	5.7	5.2	2.2	2.0	0.9	0.9	1.1	0.6	3.6
Iceland	5.9	4.7	0.6	6.5	7.3	6.5	4.7	3.0	4.4	3.1	-0.8	2.3	3.7	2.0	2.8
Ireland	3.2	-0.4	-0.7	1.8	2.6	-4.8	-5.0	-0.9	5.8	3.0	2.4	1.1	3.4	3.0	3.0
Israel	-6.5	-4.6	5.8	3.6	-9.8	14.1	-2.6	-8.5	6.6	4.6	-0.3	5.4	0.3	0.6	4.2
Luxembourg	1.5	1.9	2.2	2.0	3.1	2.7	3.8	1.9	3.2	3.8	3.5	1.8	2.0	2.2	2.0
Malta	6.0	-1.1	-2.5	5.6	4.4	9.1	6.0	12.7	5.7	10.9	8.9	6.0	6.4	7.5	5.9
Netherlands	0.7	1.2	-0.6	1.3	3.6	2.9	1.4	1.5	1.6	1.5	1.7	1.5	0.2	0.6	0.6
Norway	3.9	4.6	2.4	3.3	2.2	4.0	0.5	2.2	4.1	3.9	5.5	2.3	0.7	0.2	1.6
Portugal	3.7	3.8	0.2	6.4	7.2	3.8	8.0	4.4	5.7	3.0	1.4	–	1.4	1.8	1.8
Spain	5.3	3.9	2.4	5.5	5.4	8.9	4.0	8.3	6.6	5.6	4.0	2.4	-0.3	1.3	0.9
Sweden	1.0	0.8	2.3	2.4	1.4	1.1	0.7	2.1	2.6	2.8	–	0.2	-0.7	-1.0	-1.8
Switzerland	1.1	3.9	1.2	3.3	3.7	1.8	4.3	4.1	4.7	1.5	-0.1	-1.2	1.3	-0.1	-0.5
Turkey	-10.6	16.6	1.9	14.1	9.2	9.4	-1.1	0.8	8.0	4.5	3.8	5.4	-3.5	6.7	7.6
North America	1.6	2.0	1.6	4.7	4.3	2.1	2.2	2.8	2.4	1.1	–	–	–	-0.4	0.3
Canada	2.4	1.4	1.2	3.2	1.6	1.7	4.1	4.0	3.2	2.7	1.0	0.5	-1.7	-0.7	-1.8
United States	1.5	2.1	1.6	4.8	4.6	2.2	2.0	2.7	2.3	1.0	-0.1	–	0.2	-0.3	0.5
Total above	1.4	2.3	1.6	3.8	3.6	2.6	2.2	2.0	2.6	1.7	1.1	0.6	0.4	0.5	0.8
Memorandum item:															
Japan	2.9	2.5	2.3	0.3	5.1	1.6	2.3	2.0	1.5	2.0	2.0	2.4	2.2	2.0	2.3
Total above, *including Japan*	1.6	2.4	1.7	3.3	3.8	2.4	2.2	2.0	2.4	1.7	1.3	0.9	0.6	0.7	1.1

Source: National statistics.

Note: See appendix table A.1.

[a] West Germany 1982-1991.

[b] Excluding Israel.

APPENDIX TABLE A.4

Real gross domestic fixed capital formation in the ECE market economies, 1982-1996

(Percentage change over preceding year)

	1982	1983	1984	1985	1986	1987	1988	1989	1990	1991	1992	1993	1994	1995	1996
Western Europe	-1.6	0.6	1.4	2.8	4.6	7.4	8.0	6.7	4.2	-0.6	-0.8	-5.0	1.6	3.9	1.8
4 major countries	-1.9	1.0	2.2	1.7	3.2	5.2	8.3	6.1	3.4	-0.1	-0.2	-6.2	2.3	2.4	0.1
France	-1.4	-3.6	-2.6	3.2	4.5	4.8	9.6	7.9	2.8	–	-2.8	-6.7	1.3	2.6	-0.6
Germany[a]	-5.4	3.1	0.1	-0.5	3.3	1.8	4.4	6.3	8.5	5.8	3.5	-5.6	4.2	1.5	-0.8
Italy	-4.7	-0.6	3.6	0.6	2.2	5.0	6.9	4.3	3.8	0.6	-1.7	-12.8	0.2	5.9	1.2
United Kingdom	5.4	5.0	8.9	4.2	2.6	10.3	13.9	6.0	-3.5	-9.5	-1.5	0.6	2.9	-0.1	1.0
17 smaller countries[b]	-1.1	-0.2	-0.1	4.9	7.3	11.6	7.4	7.8	5.8	-1.4	-1.9	-2.7	0.3	6.7	5.4
Austria	-8.2	-0.6	2.1	5.0	3.7	3.1	6.0	6.2	5.7	6.3	1.7	-1.6	6.7	2.3	1.4
Belgium	-1.6	-4.2	1.9	0.7	4.5	5.6	15.7	11.6	12.0	-5.0	1.8	-5.0	0.3	3.0	2.2
Cyprus	3.8	-2.2	21.5	-7.7	-7.1	5.0	10.6	20.0	-2.8	-1.6	16.2	-12.8	-2.7	1.8	2.4
Denmark	7.1	1.9	12.9	12.6	17.1	-3.8	-6.6	1.0	-1.7	-5.7	-4.2	-4.7	3.0	10.2	5.6
Finland	5.1	3.7	-2.1	2.2	-0.4	4.9	9.8	14.8	-4.1	-20.3	-16.9	-19.2	0.2	8.5	5.5
Greece	-1.9	-1.3	-5.7	5.2	-6.2	-5.1	8.9	7.1	8.4	-4.5	0.5	-2.7	0.5	5.8	9.2
Iceland	0.1	-12.7	9.3	1.0	-1.5	18.8	-0.2	-7.9	3.0	2.0	-11.3	-11.4	-1.1	2.2	21.4
Ireland	-3.4	-9.3	-2.5	-7.7	-2.8	-5.7	-1.6	13.5	12.8	-7.4	-2.5	-2.1	8.7	10.1	10.3
Israel	5.8	12.8	-11.9	-7.7	0.1	14.4	-1.6	-3.5	23.0	38.8	7.2	1.9	12.7	8.2	8.1
Luxembourg	-0.5	-11.8	0.1	-9.5	31.2	14.7	14.1	8.9	2.5	9.8	-2.1	3.9	-14.9	3.5	5.9
Malta	14.7	15.8	-7.0	-4.0	-8.7	30.7	6.1	1.0	17.9	–	-0.2	11.1	6.8	21.2	-2.4
Netherlands	-4.1	2.1	5.4	6.8	6.9	0.9	4.5	4.9	1.6	0.2	0.6	-2.8	1.6	6.7	5.3
Norway	-11.0	5.8	10.9	-13.9	23.9	-2.1	1.6	-8.8	-11.9	-1.3	-3.3	4.3	6.9	4.5	3.1
Portugal	2.3	-7.1	-17.4	-3.5	10.9	16.8	11.2	4.3	6.8	2.4	5.4	-4.8	3.9	4.6	5.8
Spain	2.1	-2.4	-6.9	6.1	9.9	14.0	13.9	13.6	6.6	1.6	-4.4	-10.6	1.8	8.2	1.0
Sweden	-0.5	1.9	6.9	6.3	1.0	7.9	6.0	11.7	1.3	-8.9	-10.8	-17.2	1.9	10.9	4.7
Switzerland	2.6	4.1	4.1	5.3	7.9	7.4	6.9	5.8	2.6	-2.5	-5.0	-2.5	7.2	2.3	-0.1
Turkey	-5.7	2.6	0.9	11.5	8.4	45.1	-1.0	2.2	15.9	1.2	4.3	24.9	-15.9	8.3	18.2
North America	-8.2	6.3	14.5	6.3	2.3	1.2	2.2	2.3	-1.5	-6.3	4.7	4.7	7.7	4.8	6.1
Canada	-11.0	-0.7	2.1	9.5	6.2	10.8	10.3	6.1	-3.5	-2.9	-1.5	0.6	5.9	-0.1	6.3
United States	-8.0	7.0	15.6	6.1	1.9	0.4	1.5	2.0	-1.4	-6.6	5.2	5.1	7.9	5.2	6.1
Total above	-4.9	3.5	7.9	4.6	3.4	4.3	5.1	4.5	1.3	-3.4	1.9	-0.2	4.7	4.3	4.0
Memorandum item:															
Japan	-0.2	-1.1	4.3	5.0	4.8	9.1	11.5	8.2	8.5	3.3	-1.5	-2.0	-1.0	0.9	9.0
Total above, *including Japan*	-4.2	2.7	7.4	4.6	3.6	5.1	6.1	5.1	2.5	-2.4	1.4	-0.4	3.8	3.8	4.8

Source: National statistics.

Note: See appendix table A.1.

[a] West Germany 1982-1991.

[b] Excluding Israel.

APPENDIX TABLE A.5

Real total domestic demand in the ECE market economies, 1982-1996
(Percentage change over preceding year)

	1982	1983	1984	1985	1986	1987	1988	1989	1990	1991	1992	1993	1994	1995	1996
Western Europe	0.9	1.5	2.0	2.5	4.1	4.1	4.5	3.6	3.4	1.3	1.1	-1.1	1.9	2.7	1.7
4 major countries	0.9	1.7	2.1	2.2	3.9	3.7	5.0	3.1	2.7	1.4	1.2	-1.6	2.6	2.0	1.0
France	3.5	-0.7	0.4	2.5	4.5	3.3	4.7	3.9	2.8	0.6	0.2	-2.2	3.0	2.0	1.0
Germany[a]	-2.2	2.4	1.9	1.0	3.3	2.4	3.6	2.9	5.2	4.9	2.8	-1.3	2.8	2.1	0.8
Italy	0.6	0.2	3.5	2.8	3.0	4.2	4.4	2.8	2.5	1.9	0.8	-4.5	1.6	2.3	0.2
United Kingdom	2.4	5.1	2.8	2.9	4.9	5.3	7.9	2.9	-0.6	-3.1	0.2	2.0	2.9	1.5	2.0
17 smaller countries[b]	0.9	1.0	1.8	3.1	4.6	4.9	3.6	4.5	4.7	1.0	1.1	-0.1	0.6	4.2	3.0
Austria	-1.4	2.9	2.6	2.2	1.8	2.6	4.3	3.2	4.0	3.2	2.3	0.7	4.5	2.8	0.8
Belgium	0.8	-2.3	2.2	0.5	2.6	3.4	4.5	4.4	3.6	1.3	2.0	-1.5	1.4	1.6	1.6
Cyprus	7.9	4.9	9.3	2.2	-1.5	8.4	11.3	9.7	6.2	5.2	9.3	-10.1	7.0	9.1	3.5
Denmark	3.5	1.4	5.1	5.4	6.1	-2.2	-1.2	0.5	-1.0	-0.4	-0.1	0.8	5.8	4.6	1.1
Finland	4.5	2.6	2.2	3.3	2.2	5.2	6.6	7.6	-0.8	-9.7	-6.4	-6.4	3.8	3.8	2.8
Greece	1.7	0.4	0.2	5.2	-0.9	-1.0	6.6	4.0	1.6	2.8	-0.2	0.4	1.1	3.4	3.7
Iceland	5.0	-8.1	6.0	2.9	4.8	15.6	-0.9	-4.3	1.5	5.2	-5.3	-4.1	1.4	4.3	7.8
Ireland	-2.4	-2.2	1.1	1.2	1.2	1.2	1.6	8.8	6.3	0.2	–	0.7	5.9	6.0	6.7
Luxembourg	4.2	4.8	-3.1	-0.6	6.4	8.3	2.6	-2.5	9.1	12.6	5.4	6.0	7.0	6.5	5.7
Israel	1.0	-0.3	2.3	0.6	7.0	4.3	6.2	5.3	4.9	7.7	0.8	-0.3	5.9	2.7	4.0
Malta	10.5	-3.5	0.8	3.5	–	3.5	11.5	8.3	7.6	4.9	0.1	4.8	4.1	13.4	7.1
Netherlands	-0.9	1.5	1.8	3.3	3.9	1.4	1.8	4.7	3.5	1.9	1.6	-1.1	2.9	2.2	2.4
Norway	2.0	1.1	6.0	4.7	8.1	-1.4	-3.1	-2.4	-0.7	0.5	1.5	3.2	4.3	3.7	2.5
Portugal	2.2	-5.6	-6.2	1.5	8.4	10.5	7.9	3.7	6.7	4.3	4.3	-0.9	1.5	2.6	2.8
Spain	1.5	0.5	-1.0	3.4	5.4	8.1	7.0	7.8	4.8	2.9	1.0	-4.3	1.2	3.1	1.7
Sweden	0.3	-1.3	3.3	4.3	2.6	4.1	3.0	3.9	0.8	-2.1	-1.8	-5.2	2.6	2.3	–
Switzerland	-0.8	2.3	2.2	2.5	5.6	3.6	2.8	4.1	2.3	-0.5	-3.3	-1.8	3.5	1.9	-0.5
Turkey	1.4	5.8	6.4	3.2	7.0	8.9	-1.3	1.5	14.6	-0.9	5.2	13.5	-12.3	13.3	11.6
North America	-1.7	5.2	8.1	4.1	3.3	2.9	3.1	2.8	0.8	-1.7	2.7	2.8	3.9	1.9	2.4
Canada	-5.4	3.6	4.7	5.3	4.2	5.3	5.5	4.3	-0.5	-1.2	0.4	2.0	3.1	1.0	1.6
United States	-1.4	5.3	8.4	4.0	3.2	2.7	2.9	2.7	0.9	-1.7	2.9	2.9	4.0	2.0	2.5
Total above	-0.4	3.3	5.0	3.3	3.7	3.5	3.8	3.2	2.1	-0.2	1.9	0.9	2.9	2.3	2.1
Memorandum item:															
Japan	2.8	1.7	3.2	3.8	3.9	5.1	7.4	5.6	5.2	2.9	0.4	0.1	0.8	1.7	4.6
Total above, *including Japan*	0.1	3.0	4.8	3.4	3.7	3.7	4.4	3.6	2.6	0.3	1.7	0.8	2.6	2.2	2.5

Source: National statistics.
Note: See appendix table A.1.
[a] West Germany 1982-1991.
[b] Excluding Israel.

APPENDIX TABLE A.6

Real exports of goods and services in the ECE market economies, 1982-1996
(Percentage change over preceding year)

	1982	1983	1984	1985	1986	1987	1988	1989	1990	1991	1992	1993	1994	1995	1996
Western Europe	2.4	3.6	8.8	4.3	1.2	5.0	6.1	7.4	6.3	4.3	4.0	2.3	9.3	7.5	4.9
4 major countries	0.4	1.6	7.6	4.8	1.0	3.2	5.0	8.7	7.4	4.3	3.1	1.1	8.3	7.7	3.7
France	-1.7	3.7	7.0	1.9	-1.4	3.1	8.1	10.2	5.4	4.1	4.9	-0.4	6.0	6.0	3.6
Germany[a]	3.9	-0.8	8.2	7.6	-0.6	0.4	5.5	10.2	11.0	10.9	-0.3	-4.9	8.0	5.9	4.9
Italy	-2.4	2.3	8.5	3.2	2.5	4.7	5.4	8.8	7.0	0.5	5.0	9.1	10.5	11.6	-0.3
United Kingdom	0.8	1.8	6.5	6.0	4.5	5.8	0.5	4.7	5.0	-0.7	4.1	3.5	9.2	8.0	6.3
17 smaller countries[b]	6.1	7.5	11.1	3.2	1.7	8.3	8.1	5.0	4.4	4.4	5.9	4.5	11.3	7.0	7.3
Austria	2.7	3.2	6.1	6.9	-2.7	2.4	9.0	10.3	8.5	5.8	1.2	-1.6	5.2	5.4	3.7
Belgium	1.5	2.7	5.2	1.2	5.5	6.1	8.5	7.5	3.8	3.1	3.6	1.8	9.2	5.0	4.4
Cyprus	7.9	8.2	15.1	1.1	-1.7	13.7	13.5	16.8	6.7	-7.3	18.7	-1.5	7.4	4.5	4.6
Denmark	2.5	4.9	3.5	5.0	–	5.1	7.8	4.2	6.9	7.7	1.4	-1.6	7.9	3.7	3.3
Finland	-1.1	2.0	5.0	1.1	1.2	2.7	3.7	1.3	1.4	-6.6	10.0	16.7	13.3	8.2	3.8
Greece	-7.2	8.0	16.9	1.3	14.0	16.0	9.0	4.5	1.5	15.6	8.5	-0.7	7.7	1.8	1.7
Iceland	-8.9	11.0	2.4	11.0	5.9	3.3	-3.6	2.9	–	-5.8	-1.7	6.6	9.8	-2.3	5.2
Ireland	5.5	10.5	16.6	6.6	3.1	13.7	8.9	10.3	8.9	5.3	13.5	9.7	13.6	17.0	10.0
Israel	-3.5	1.7	13.6	10.0	5.9	10.4	-1.5	3.9	2.0	-2.0	14.9	10.7	12.4	10.9	3.7
Luxembourg	-0.3	5.3	18.0	9.5	3.2	6.5	7.5	6.9	2.6	3.6	1.3	-2.4	4.4	4.4	2.8
Malta	-13.8	-1.8	4.0	7.4	7.0	12.6	6.1	10.7	13.3	7.5	9.7	5.3	7.1	4.7	-5.7
Netherlands	–	3.5	7.4	5.4	1.8	3.6	9.0	6.7	5.3	4.7	2.9	1.5	6.7	6.9	3.9
Norway	-0.1	7.6	8.2	6.9	1.6	1.2	5.5	10.7	8.6	6.1	5.2	3.2	8.2	3.8	8.2
Portugal	4.7	13.6	11.6	6.7	6.8	10.6	7.9	13.3	10.5	0.5	6.1	-5.1	10.7	11.4	9.6
Spain	5.0	10.0	11.7	2.7	1.9	6.3	5.1	3.0	3.2	7.9	7.4	8.5	16.7	8.2	10.3
Sweden	5.7	9.9	6.9	1.2	3.4	4.2	2.8	3.1	1.6	-2.3	2.3	7.6	14.0	12.6	5.6
Switzerland	-2.9	1.1	6.3	8.3	0.4	1.7	5.8	5.0	3.0	-0.7	3.4	1.6	3.4	3.0	1.9
Turkey	34.0	13.1	25.4	-1.9	-5.1	26.4	18.4	-0.3	3.1	3.1	11.0	7.7	15.2	6.7	15.0
North America	-6.7	-1.9	9.0	3.0	7.1	10.4	15.4	10.8	8.1	5.9	6.7	3.5	8.7	9.2	6.3
Canada	-2.2	6.4	17.7	6.0	4.5	3.5	9.5	0.8	4.1	1.4	7.6	10.4	14.7	12.0	4.5
United States	-7.1	-2.6	8.3	2.7	7.4	11.0	15.9	11.7	8.5	6.3	6.6	2.9	8.2	8.9	6.5
Total above	-2.2	0.9	8.9	3.6	4.2	7.7	10.7	9.1	7.2	5.1	5.4	2.9	9.0	8.3	5.6
Memorandum item:															
Japan	0.9	4.8	14.8	5.4	-5.7	-0.5	5.9	9.1	6.9	5.4	4.9	1.3	4.5	5.0	2.1
Total above, including Japan	-1.7	1.5	9.8	3.9	2.6	6.4	10.0	9.1	7.2	5.2	5.3	2.7	8.3	7.8	5.1

Source: National statistics.

Note: See appendix table A.1. Data on a national accounts basis.

[a] West Germany 1982-1991.

[b] Excluding Israel.

APPENDIX TABLE A.7

Real imports of goods and services in the ECE market economies, 1982-1996

(Percentage change over preceding year)

	1982	1983	1984	1985	1986	1987	1988	1989	1990	1991	1992	1993	1994	1995	1996
Western Europe	1.9	1.7	6.6	4.2	5.4	8.8	7.8	8.5	7.5	3.5	3.8	-1.6	6.6	7.8	4.5
4 major countries	1.3	0.9	7.2	4.0	4.7	7.0	8.0	7.9	6.6	3.9	3.4	-3.9	7.2	6.4	2.6
France	2.6	-2.7	2.7	4.5	7.1	7.7	8.6	8.1	6.1	3.0	1.2	-3.5	6.7	5.3	2.2
Germany[a]	-1.1	1.4	5.2	4.5	2.7	4.2	5.1	8.3	10.3	11.6	2.0	-5.7	7.6	6.4	2.6
Italy	-0.3	-1.4	12.3	3.9	2.9	9.1	6.8	7.6	8.0	3.4	4.6	-8.1	8.9	9.6	-1.7
United Kingdom	4.9	6.6	9.9	2.6	6.9	7.8	12.6	7.4	0.5	-5.2	6.6	3.0	5.4	4.4	7.8
17 smaller countries[b]	3.0	3.4	5.3	4.7	6.5	12.2	7.4	9.7	9.1	2.8	4.5	3.0	5.5	10.7	8.1
Austria	-3.3	5.5	9.9	6.2	-1.2	4.7	9.4	8.5	7.8	6.4	1.8	-0.7	8.2	7.3	3.4
Belgium	–	-1.4	5.6	0.8	7.4	8.5	8.1	9.1	3.8	2.8	4.0	1.8	8.3	4.8	4.6
Cyprus	12.8	5.5	16.4	-4.5	-11.0	13.5	13.4	20.4	6.0	2.0	18.2	-18.3	8.6	11.2	6.3
Denmark	3.8	1.8	5.5	8.1	6.8	-2.0	1.5	4.5	1.2	4.1	0.8	-3.9	12.3	8.2	1.0
Finland	2.2	3.1	1.6	6.4	2.6	9.2	11.1	8.9	-0.6	-11.7	1.1	0.8	12.8	6.9	4.9
Greece	7.0	6.6	0.2	12.8	3.8	16.6	8.0	9.2	11.3	10.6	2.3	2.3	4.1	6.2	6.5
Iceland	-0.6	-9.7	9.2	9.4	1.0	23.3	-4.6	-10.3	1.0	5.5	-7.8	-8.6	4.1	3.8	12.8
Ireland	-3.1	4.7	9.9	3.2	5.6	6.2	4.9	12.7	5.5	2.3	7.9	7.2	14.3	12.3	11.5
Israel	3.6	6.8	-1.1	-0.9	9.2	19.0	-2.9	-4.9	9.1	15.5	9.0	14.2	10.9	8.4	7.6
Luxembourg	-0.3	1.2	13.9	7.0	6.1	7.8	8.5	6.1	4.3	8.1	0.2	-0.1	3.8	3.3	4.5
Malta	-2.0	-7.2	3.9	9.2	0.1	12.3	11.1	11.1	15.7	5.4	3.0	5.9	7.5	10.3	-1.7
Netherlands	1.1	3.9	5.0	6.6	3.5	4.2	7.6	6.7	4.2	4.1	2.1	-2.1	6.5	7.7	3.6
Norway	3.7	–	9.5	5.9	9.9	-7.3	-1.7	2.2	2.5	0.2	0.7	4.4	6.9	5.1	2.5
Portugal	3.9	-6.1	-4.4	1.4	16.9	20.2	16.5	7.9	13.7	5.4	11.1	-3.2	8.5	8.6	7.6
Spain	4.8	-0.3	-1.8	7.9	14.4	20.1	14.4	17.3	7.8	9.0	6.9	-5.2	11.4	8.8	7.5
Sweden	3.4	0.8	5.4	7.8	4.4	7.1	4.9	7.4	0.7	-4.9	1.1	-2.5	13.1	10.3	3.5
Switzerland	-2.6	4.4	7.1	5.1	7.1	5.5	5.3	5.4	2.9	-1.7	-3.8	-0.8	8.9	6.6	1.9
Turkey	8.3	16.9	19.7	-6.6	-3.5	23.0	-4.5	6.9	33.1	-5.3	10.9	35.8	-21.9	30.0	29.6
North America	-2.4	12.3	23.7	6.7	8.4	6.2	4.7	4.1	3.7	-0.3	7.4	9.1	11.9	8.1	6.3
Canada	-15.2	9.0	17.1	8.7	7.6	7.0	13.8	6.3	2.0	3.3	5.6	8.8	11.5	8.7	5.1
United States	-1.3	12.6	24.3	6.5	8.4	6.1	3.9	3.9	3.9	-0.7	7.5	9.1	12.0	8.0	6.4
Total above	-0.3	7.0	15.1	5.4	6.9	7.5	6.3	6.3	5.6	1.6	5.6	3.8	9.3	7.9	5.4
Memorandum item:															
Japan	-2.5	-3.0	10.4	-1.4	2.4	7.8	18.7	17.6	8.6	-4.1	-0.4	2.7	9.0	13.5	11.4
Total above, *including Japan*	-0.6	5.5	14.4	4.4	6.2	7.5	8.2	8.1	6.1	0.7	4.6	3.6	9.2	8.8	6.3

Source: National statistics.

Note: See appendix table A.1. Data on a national accounts basis.

[a] West Germany 1982-1991.

[b] Excluding Israel.

APPENDIX TABLE A.8

Industrial output in the ECE market economies, 1982-1996
(Percentage change over preceding year)

	1982	1983	1984	1985	1986	1987	1988	1989	1990	1991	1992	1993	1994	1995	1996
Western Europe	-1.0	1.2	2.7	3.6	2.6	2.5	4.4	4.1	2.4	-0.2	-0.9	-2.6	4.4	3.9	0.6
4 major countries	-1.6	0.2	1.9	3.6	2.2	1.9	4.8	3.9	1.9	-0.2	-1.4	-3.5	4.6	3.0	–
France	-0.7	-0.7	0.3	1.9	0.7	1.2	4.6	3.7	1.5	-1.3	-1.1	-3.8	3.8	1.8	0.4
Germany[a]	-3.3	0.6	3.0	4.9	1.8	0.5	3.5	5.0	5.2	2.9	-2.3	-7.4	3.8	2.1	0.2
Italy	-3.0	-2.4	3.3	1.3	4.1	2.6	7.0	3.9	-0.7	-0.9	-1.3	-2.1	6.3	6.1	-1.8
United Kingdom	2.0	3.6	–	5.6	2.2	4.1	4.7	2.1	-0.3	-3.6	–	2.1	5.0	2.5	1.1
14 smaller countries	0.3	3.5	4.6	3.7	3.5	4.1	3.4	4.5	3.4	-0.2	0.2	-0.6	4.0	6.0	2.2
Austria	-0.9	1.1	5.2	4.6	1.1	1.0	4.4	5.9	7.4	1.6	-1.1	-2.0	4.0	5.4	..
Belgium	–	1.9	2.6	2.4	0.8	2.1	5.9	3.4	3.7	-2.0	–	-5.2	1.8	4.2	0.7
Denmark	..	..	..	..	..	..	..	..	..	..	..	..	..	..	..
Finland	0.9	3.2	4.6	4.4	1.6	4.5	4.4	2.4	0.4	-9.7	2.3	5.2	11.4	7.5	3.9
Greece	0.9	-0.3	2.3	4.2	-1.0	-1.5	5.1	1.8	-2.3	-1.4	-1.2	-2.1	0.9	2.3	–
Ireland	-0.7	7.9	10.0	3.3	2.2	8.9	10.7	11.6	4.7	3.3	9.1	5.6	11.9	18.8	7.9
Luxembourg	..	..	..	..	..	..	..	..	..	0.1	-0.8	-2.5	5.9	0.8	-3.1
Netherlands	-3.9	1.8	5.0	4.8	0.2	1.1	0.1	5.1	2.4	1.7	-0.2	-1.2	4.2	2.1	3.3
Norway	-0.3	9.1	8.2	3.1	3.2	6.6	3.0	9.6	1.7	2.1	6.4	3.2	7.0	6.0	5.2
Portugal	7.8	3.6	2.5	0.5	7.3	4.4	3.7	6.8	9.1	–	-2.3	-2.6	-0.2	4.6	0.7
Spain	-1.1	2.7	0.8	2.0	3.0	4.6	3.1	4.5	–	-0.7	-2.8	-4.7	7.3	4.7	-0.7
Sweden	-1.7	4.0	7.1	3.0	0.4	2.5	1.3	3.7	1.1	-5.1	-1.5	-0.2	11.3	9.9	2.9
Switzerland	-4.9	–	2.6	6.3	3.5	1.1	7.9	1.0	3.1	–	-1.0	-2.0	5.2	2.0	-1.3
Turkey	8.2	8.1	11.0	5.9	11.8	10.4	1.6	3.6	9.4	2.7	5.0	8.0	-6.2	12.2	5.6
North America	-5.8	4.0	9.2	2.0	1.0	4.7	4.5	1.7	-0.5	-2.2	3.1	3.5	5.2	3.3	2.7
Canada	-9.7	6.4	12.2	5.6	-0.7	4.9	5.3	-0.2	-3.3	-4.2	1.1	4.4	7.0	3.4	1.6
United States	-5.4	3.7	9.0	1.6	1.1	4.7	4.5	1.8	-0.2	-2.0	3.3	3.4	5.0	3.3	2.8
Total above	-3.2	2.5	5.7	2.9	1.9	3.5	4.5	3.0	1.1	-1.1	0.9	0.1	4.8	3.6	1.5
Memorandum item:															
Japan	0.3	3.2	9.4	3.6	-0.1	3.4	11.0	4.8	4.3	1.9	-5.8	-4.2	1.2	3.3	2.6
Total above, including Japan	-2.5	2.6	6.4	3.1	1.5	3.5	5.8	3.4	1.7	-0.5	-0.4	-0.7	4.1	3.6	1.7

Source: National statistics; OECD, *Main Economic Indicators* (Paris), various issues.

Note: Growth rates of regional aggregates have been calculated as weighted averages of growth rates in individual countries. Weights were derived from GDP originating in industry in 1990. Data were converted from national currency units into dollars using purchasing power parities.

[a] West Germany 1982-1991.

APPENDIX TABLE A.9

Total employment in the ECE market economies, 1982-1996
(Percentage change over preceding year)

	1982	1983	1984	1985	1986	1987	1988	1989	1990	1991	1992	1993	1994	1995*	1996*
Western Europe	-0.4	-0.2	0.4	0.9	1.1	1.3	1.6	1.7	1.6	0.2	-1.1	-1.6	–	0.9	0.3
4 major countries	-0.6	-0.6	0.6	0.6	0.7	0.9	1.5	1.5	1.5	–	-1.5	-1.8	-0.4	0.3	-0.3
France	0.3	-0.1	-0.9	-0.3	0.4	0.3	0.9	1.3	1.0	0.1	-0.7	-1.2	-0.1	1.2	-0.5
Germany[a]	-1.2	-1.4	0.2	0.7	1.4	0.7	0.8	1.5	3.0	2.5	-1.8	-1.7	-0.7	-0.3	-1.1
Italy[b]	0.6	0.6	0.4	0.9	0.8	0.4	0.9	0.2	0.9	0.8	-1.0	-2.9	-1.5	-0.4	0.1
United Kingdom[c]	-1.8	-1.2	2.7	1.2	0.1	2.3	3.4	3.0	1.0	-3.2	-2.1	-1.5	0.6	1.1	0.5
17 smaller countries[d]	-0.2	0.3	0.1	1.2	1.7	1.8	1.6	2.0	1.6	0.5	-0.5	-1.2	0.6	1.7	1.2
Austria	-1.2	-0.8	0.1	0.2	0.4	–	0.6	1.4	1.9	2.0	1.6	–	0.2	-0.4	-0.7
Belgium[c]	-1.3	-1.0	-0.2	0.6	0.6	0.5	1.5	1.6	1.4	0.1	-0.4	-1.1	-0.7	0.4	-0.1
Cyprus	1.7	2.5	3.8	3.7	1.1	3.1	4.7	3.9	2.8	0.6	4.5	0.1	1.8	2.2	..
Denmark	0.5	0.3	1.7	2.5	2.6	0.8	-0.6	-0.6	-1.0	-1.5	-0.6	-1.0	-0.6	1.6	1.2
Finland	1.1	0.3	0.4	0.1	-0.4	0.5	0.6	0.7	-0.5	-5.2	-7.0	-6.5	-1.1	1.5	1.1
Greece	-0.8	1.1	0.4	1.0	0.4	-0.1	1.6	0.4	1.3	-2.3	1.5	0.9	1.9	0.9	0.8
Iceland[b]	3.1	0.4	1.4	3.6	3.2	5.7	-3.0	-1.4	-0.9	-0.1	-1.4	-0.8	1.2	0.9	1.7
Ireland[e]	–	-1.9	-1.9	-2.2	0.2	-0.1	1.0	-0.1	3.3	0.7	0.4	0.6	3.1	4.8	3.6
Israel	1.4	3.2	1.5	0.7	1.4	2.6	3.5	0.5	2.1	6.1	4.2	6.1	6.9	5.2	2.5
Luxembourg	-0.3	-0.3	0.6	1.4	2.6	2.8	3.1	2.7	4.1	4.1	2.5	1.8	2.3	2.5	2.3
Malta[f]	-4.1	0.1	0.6	1.3	2.1	5.9	2.5	0.9	0.8	2.5	1.0	0.5	1.2	4.2	2.6
Netherlands[b]	-2.5	-1.8	0.1	1.9	2.1	1.7	1.6	1.9	2.3	1.3	1.0	-0.1	0.1	1.9	2.0
Norway	0.1	-0.3	0.6	2.7	3.0	2.1	-0.1	-2.8	-0.8	-0.8	-0.3	0.2	1.2	2.1	2.7
Portugal	0.2	4.7	-1.0	-0.5	2.7	2.3	2.3	1.8	0.8	3.0	0.9	-2.0	-0.1	-0.6	0.6
Spain	-1.1	-0.5	-2.7	1.1	2.2	3.1	2.9	4.0	2.6	0.2	-1.9	-4.3	-0.9	2.7	2.8
Sweden	-0.2	0.2	0.8	1.0	0.6	0.8	1.4	1.5	0.9	-1.5	-4.4	-5.2	-1.0	1.5	-0.5
Switzerland	0.5	–	1.0	2.0	2.3	2.5	2.6	2.7	3.2	1.2	-1.6	-0.6	-0.2	0.2	–
Turkey[g]	1.1	1.0	1.6	1.7	1.9	2.3	1.5	2.6	1.7	1.7	0.8	0.9	2.5	2.5	2.4
North America	-1.8	1.1	4.7	2.4	1.9	2.9	2.9	2.3	0.8	-1.1	-0.2	1.8	2.3	1.6	1.4
Canada	-3.2	0.6	2.7	3.0	3.0	2.7	3.2	2.1	0.6	-1.9	-0.6	1.3	2.1	1.6	1.3
United States[b]	-1.7	1.2	4.9	2.3	1.7	2.9	2.8	2.4	0.8	-1.0	-0.2	1.8	2.3	1.6	1.4
Total above	-1.0	0.3	2.1	1.5	1.4	2.0	2.1	2.0	1.2	-0.3	-0.7	-0.2	1.0	1.2	0.8
Memorandum item:															
Japan	0.8	1.5	0.3	0.6	0.5	0.4	1.2	1.5	1.7	2.0	1.1	0.4	0.1	0.1	0.6
Total above, including Japan	-0.7	0.5	1.8	1.3	1.2	1.7	2.0	1.9	1.3	0.1	-0.4	-0.1	0.8	1.0	0.7

Source: National statistics; OECD, *National Accounts Detailed Tables, Vol. II*, various issues and *OECD Economic Outlook,* December 1996 (Paris); UN/ECE secretariat estimates.

Note: National accounts statistics, where available; otherwise annual labour force surveys. Unless otherwise indicated, the data refer to the number of persons employed i.e. no adjustment is made for part-time workers.

[a] West Germany 1982-1991.
[b] Full-time equivalent data.
[c] June of each year.
[d] Excluding Israel.
[e] April of each year.
[f] End of year.
[g] Civilian employment.

APPENDIX TABLE A.10

Standardized unemployment rates[a] in the ECE market economies, 1982-1996

(Per cent of total labour force)

	1982	1983	1984	1985	1986	1987	1988	1989	1990	1991	1992	1993	1994	1995	1996*
Western Europe	8.0	8.9	9.3	9.5	9.4	9.3	8.7	8.1	7.6	7.9	8.7	10.3	10.6	10.2	10.3
4 major countries	7.9	8.7	9.0	9.3	9.3	9.2	8.6	7.9	7.3	7.7	8.4	9.8	10.2	9.8	10.1
France	7.7	8.1	9.7	10.1	10.2	10.4	9.8	9.3	9.0	9.5	10.4	11.7	12.3	11.6	12.3
Germany[b]	5.9	7.7	7.1	7.1	6.4	6.2	6.2	5.6	4.8	4.2	4.6	7.9	8.4	8.2	9.0
Italy	6.8	7.7	8.1	8.4	9.2	9.9	10.0	10.0	9.1	8.8	9.0	10.3	11.4	11.9	12.0
United Kingdom	11.1	11.1	11.1	11.5	11.5	10.6	8.7	7.3	7.1	8.8	10.1	10.5	9.6	8.8	8.2
17 smaller countries[c]	8.1	9.2	9.8	9.8	9.7	9.5	9.0	8.4	7.9	8.3	9.3	10.9	11.4	10.8	10.6
Austria	3.1	3.7	3.8	3.6	3.1	3.8	3.6	3.1	3.2	3.5	3.6	4.2	3.6	3.8	4.1
Belgium	10.1	11.1	11.1	10.4	10.3	10.0	8.9	7.5	6.7	6.6	7.3	8.9	10.0	9.9	9.8
Cyprus	2.8	3.3	3.3	3.3	3.7	3.4	2.8	2.3	1.8	3.0	1.8	2.7	2.7	2.6	3.0
Denmark	9.8	10.4	10.1	9.0	7.8	7.8	6.1	7.4	7.7	8.5	9.2	10.1	8.2	7.1	6.0
Finland	5.3	5.4	6.2	6.3	6.8	5.2	4.6	3.5	3.5	7.6	13.0	17.6	17.9	16.6	15.7
Greece	5.8	7.9	8.1	7.8	7.4	7.4	7.7	7.5	7.0	7.7	8.7	9.7	9.6	10.0	10.1
Iceland	0.7	1.0	1.3	0.9	0.6	0.5	0.6	1.7	1.8	1.5	3.0	4.4	4.8	5.0	4.5
Ireland	11.6	14.0	15.5	16.9	16.8	16.6	16.1	14.7	13.4	14.8	15.4	15.6	14.3	12.4	12.3
Israel	5.0	4.5	5.9	6.7	7.1	6.1	6.4	8.9	9.6	10.6	11.2	10.0	7.8	6.9	6.6
Luxembourg	3.0	3.5	3.1	2.9	2.6	2.5	2.0	1.8	1.7	1.7	2.1	2.7	3.2	2.9	3.1
Malta[d]	8.6	8.5	8.6	8.1	6.9	4.4	4.0	3.7	3.8	3.6	4.0	4.5	4.0	3.6	3.7
Netherlands	8.2	9.7	9.3	8.3	8.3	8.0	7.5	6.9	6.2	5.8	5.6	6.6	7.1	7.0	6.6
Norway	2.7	3.5	3.2	2.6	2.0	2.1	3.3	5.0	5.3	5.6	6.0	6.1	5.5	5.0	4.8
Portugal	7.5	7.8	8.5	8.7	8.4	6.9	5.5	4.9	4.6	4.0	4.2	5.7	7.0	7.3	7.3
Spain	15.3	17.5	20.3	21.7	21.2	20.6	19.5	17.2	16.2	16.4	18.5	22.8	24.1	22.9	22.3
Sweden	3.5	3.9	3.4	3.0	2.8	2.3	1.9	1.6	1.8	3.3	5.8	9.5	9.8	9.2	10.0
Switzerland	0.4	0.9	1.1	0.9	0.7	0.7	0.6	0.5	0.5	1.0	2.5	4.5	4.7	4.2	4.7
Turkey	7.0	7.7	7.6	7.1	7.9	8.3	8.4	8.6	8.0	7.9	8.0	7.7	8.1	7.5	7.2
North America	9.8	9.8	7.9	7.5	7.3	6.5	5.7	5.5	5.8	7.1	7.9	7.3	6.5	6.0	5.8
Canada	11.0	11.9	11.3	10.5	9.6	8.8	7.8	7.5	8.1	10.4	11.3	11.2	10.4	9.5	9.7
United States	9.7	9.6	7.5	7.2	7.0	6.2	5.5	5.3	5.6	6.8	7.5	6.9	6.1	5.6	5.4
Total above	8.8	9.3	8.7	8.6	8.5	8.1	7.5	7.0	6.8	7.6	8.4	9.0	8.9	8.4	8.4
Memorandum item:															
Japan	2.4	2.7	2.7	2.6	2.8	2.8	2.5	2.3	2.1	2.1	2.2	2.5	2.9	3.1	3.4
Total above, *including Japan*	7.7	8.2	7.7	7.6	7.6	7.2	6.6	6.2	6.1	6.7	7.4	8.0	7.9	7.6	7.6

Source: National statistics; OECD, *Quarterly Labour Force Statistics,* No. 4, 1996, *Main Economic Indicators*, various issues and *OECD Economic Outlook*, December 1996 (Paris); UN/ECE secretariat estimates.

Note: Comparisons with previous years are limited due to changes in methodology in Austria (1995), Denmark (1988), Finland (1984), Germany (1984 and 1993), Israel (1986), Norway (1996) and the United States (1990 and 1994).

[a] Adjusted to achieve comparability between countries except for Austria (1982-1994), Cyprus, Denmark (1982-1987), Greece, Iceland, Israel, Malta, Switzerland and Turkey.

[b] West Germany 1982-1992.

[c] Excluding Israel.

[d] End of year.

APPENDIX TABLE A.11

Consumer prices in the ECE market economies, 1982-1996

(Percentage change over previous year)

	1982	1983	1984	1985	1986	1987	1988	1989	1990	1991	1992	1993	1994	1995	1996*
Western Europe	10.3	8.1	6.8	5.9	3.4	3.2	3.6	5.0	5.5	5.0	4.3	3.5	2.9	3.0	2.4
4 major countries	10.1	7.8	6.1	5.5	2.7	2.8	3.3	4.9	5.1	4.6	4.2	3.2	2.7	2.9	2.4
France	11.8	9.6	7.4	5.8	2.7	3.1	2.8	3.6	3.3	3.2	2.4	2.1	1.7	1.7	2.0
Germany[a]	5.2	3.3	2.4	2.0	-0.1	0.2	1.3	2.8	2.7	3.6	5.1	4.5	2.7	1.8	1.5
Italy	16.5	14.7	10.8	9.2	5.9	4.7	5.0	6.3	6.3	6.4	5.4	4.2	3.9	5.4	3.9
United Kingdom	8.6	4.6	5.0	6.1	3.4	4.1	4.9	7.8	9.5	5.9	3.7	1.6	2.5	3.4	2.4
17 smaller countries[b]	10.7	9.0	8.3	6.8	5.2	4.0	4.2	5.3	6.2	5.8	4.5	4.0	3.3	3.0	2.3
Austria	5.4	3.3	5.6	3.2	1.7	1.4	2.0	2.5	3.2	3.4	4.0	3.7	3.0	2.2	1.9
Belgium	8.2	7.7	6.3	4.9	1.3	1.6	1.3	3.0	3.4	3.2	2.4	2.8	2.4	1.5	2.0
Cyprus	6.5	5.1	6.0	5.1	1.2	2.8	3.4	3.8	4.5	5.0	6.5	4.9	4.7	2.6	2.8
Denmark	10.1	6.9	6.3	4.7	3.6	4.0	4.6	4.8	2.7	2.4	2.1	1.3	2.0	2.1	2.0
Finland	9.6	8.3	7.1	5.9	2.9	4.1	5.1	6.6	6.2	4.3	3.0	2.1	1.1	1.0	0.5
Greece	21.0	20.2	18.5	19.3	23.0	16.4	13.5	13.7	20.5	19.5	15.8	14.4	10.9	9.3	8.5
Iceland	51.0	84.3	29.2	32.6	22.7	18.8	25.8	20.8	15.4	6.8	3.7	4.1	1.5	1.7	2.2
Ireland	17.1	10.5	8.6	5.4	3.9	3.2	2.1	4.0	3.4	3.2	3.0	1.5	2.4	2.5	1.6
Israel	120.3	145.7	373.8	304.6	48.1	19.9	16.3	20.2	17.2	19.0	12.0	11.0	12.3	10.1	11.3
Luxembourg	9.3	8.7	5.6	4.1	0.3	-0.1	1.5	3.3	3.7	3.1	3.2	3.6	2.2	2.0	1.4
Malta	5.8	-0.9	-0.4	-0.3	2.0	0.5	0.9	0.9	3.0	2.5	1.6	4.1	4.1	4.0	2.5
Netherlands	5.7	2.7	3.2	2.3	0.3	-0.8	0.7	1.1	2.5	3.1	3.2	2.6	2.7	2.0	2.1
Norway	11.3	8.4	6.2	5.7	7.2	8.7	6.7	4.6	4.1	3.4	2.3	2.3	1.4	2.4	1.3
Portugal	22.4	25.5	29.3	19.3	11.7	9.4	9.6	12.7	13.4	11.4	8.9	6.5	5.2	4.1	3.0
Spain	14.4	12.2	11.3	8.8	8.7	5.3	4.8	6.9	6.7	5.9	5.9	4.6	4.8	4.6	3.6
Sweden	8.6	9.0	8.0	7.4	4.2	4.2	5.8	6.5	10.5	9.3	2.2	4.7	2.2	2.5	0.5
Switzerland	5.6	3.0	3.0	3.4	0.7	1.5	1.8	3.2	5.4	5.9	4.0	3.3	0.9	1.8	0.8
Turkey	27.1	31.4	48.4	45.0	34.6	38.9	73.7	63.3	60.3	66.0	70.1	66.1	106.3	88.0	79.8
North America	6.6	3.4	4.3	3.6	2.1	3.7	4.1	4.8	5.3	4.3	2.9	2.9	2.4	2.8	2.8
Canada	10.9	5.7	4.4	3.9	4.2	4.4	4.0	5.0	4.8	5.6	1.5	1.8	0.2	2.1	1.6
United States	6.2	3.2	4.3	3.6	1.9	3.6	4.1	4.8	5.4	4.2	3.0	3.0	2.6	2.8	3.0
Total above	8.5	5.8	5.6	4.8	2.8	3.4	3.8	4.9	5.4	4.7	3.6	3.2	2.6	2.9	2.6
Memorandum item:															
Japan	2.8	1.8	2.3	2.0	0.6	0.1	0.7	2.2	3.1	3.3	1.7	1.3	0.7	-0.1	0.1
Total above, *including Japan*	7.5	5.1	5.0	4.3	2.4	2.9	3.3	4.4	5.0	4.4	3.3	2.9	2.3	2.4	2.2

Source: National statistics.

Note: Regional aggregates exclude Israel and Turkey.

[a] West Germany 1982-1991.

[b] Excluding Israel.

APPENDIX TABLE B.1

Real GDP/NMP in the transition countries, 1980, 1983-1996
(Indices, 1989=100)

	1980	1983	1984	1985	1986	1987	1988	1989	1990	1991	1992	1993	1994	1995	1996
Eastern Europe	89.2	89.1	92.9	94.6	97.6	99.3	100.7	100.0	92.5	82.0	77.7	77.2	80.3	84.7	88.1
Albania	79.4	87.7	86.6	88.2	93.1	92.4	91.0	100.0	90.0	65.1	60.4	66.2	72.4	78.6	..
Bosnia and Herzegovina	..	..	..	..	..	..	..	..	..	..	..	..	..	..	..
Bulgaria	76.2	84.6	87.5	89.9	93.6	99.3	101.9	100.0	90.9	80.2	74.4	73.3	74.6	76.2	68.5*
Croatia[a]	99.0	97.6	99.7	99.8	102.6	102.5	101.6	100.0	92.5	74.2	65.9	65.4	65.8	66.9	69.8*
Czech Republic	..	..	..	91.3	93.2	93.7	95.7	100.0	98.8	84.7	79.2	78.5	80.6	84.4	88.1
Hungary	86.2	91.9	94.4	94.1	95.5	99.4	99.3	100.0	96.5	85.0	82.4	81.9	84.3	85.5	86.0*
Poland	91.1	82.4	87.1	90.3	94.1	95.9	99.8	100.0	88.4	82.2	84.4	87.5	92.1	98.5	104.5
Romania	88.5	97.7	103.5	103.4	105.8	106.7	106.2	100.0	94.4	82.2	75.0	76.1	79.1	84.7	88.2
Slovakia	..	..	..	91.0	94.8	97.1	99.0	100.0	97.5	83.3	77.9	75.0	78.7	84.0	89.8
Slovenia	98.9	98.0	99.9	100.9	104.1	103.5	100.5	100.0	91.9	84.5	79.9	82.1	86.5	89.9	93.0
The FYR of Macedonia[a]	93.3	93.7	96.7	96.0	102.7	101.4	98.1	100.0	89.8	78.9	68.4	58.7	54.5	52.9	53.8
Yugoslavia[a]	95.6	96.9	98.4	98.7	101.4	100.2	98.8	100.0	92.1	81.4	58.7	40.6	41.7	44.2	46.1*
Baltic states	68.5	79.0	82.2	82.0	86.1	88.6	95.7	100.0	95.9	84.8	59.3	47.6	47.5	48.2	49.7
Estonia	74.5	83.2	85.5	85.7	88.2	89.2	93.8	100.0	91.9	82.7	71.0	64.9	63.2	65.0	67.0*
Latvia	70.1	79.0	82.8	82.6	86.4	87.7	93.1	100.0	102.7	92.0	59.9	51.0	51.3	50.5	51.7*
Lithuania	64.7	77.0	80.1	79.8	84.9	88.9	98.4	100.0	93.1	80.9	53.4	37.2	37.5	38.7	40.1*
CIS[b c]	77.4	86.3	88.8	90.2	92.3	93.8	97.9	100.0	96.9	91.2	78.5	70.9	60.6	57.1	54.1
Armenia	73.5	85.2	90.5	95.9	97.7	94.5	92.2	100.0	94.5	86.2	50.2	45.7	48.2	51.5	53.6
Azerbaijan	79.6	91.0	95.8	98.8	100.6	105.1	109.7	100.0	88.3	87.7	67.9	52.2	41.9	36.9	37.3
Belarus	65.7	79.0	83.0	85.2	88.9	91.3	92.4	100.0	98.0	96.8	87.5	78.2	68.3	61.5	63.4
Georgia	79.1	89.8	95.5	99.8	98.7	96.8	103.6	100.0	84.9	67.8	40.5	24.5	17.2	17.6	19.5
Kazakstan	87.0	89.7	89.6	90.8	92.3	92.1	100.1	100.0	99.0	87.4	76.0	66.2	53.8	49.0	49.2
Kyrgyzstan	69.1	80.8	84.0	82.9	83.6	84.7	95.6	100.0	104.8	96.6	83.2	70.3	56.2	53.2	56.3
Republic of Moldova	72.1	87.5	90.4	82.6	89.2	90.3	91.9	100.0	98.6	81.3	57.6	57.0	39.2	38.0	35.0
Russian Federation	78.1	86.5	89.0	90.7	92.9	94.2	98.4	100.0	97.0	92.2	78.8	71.9	62.8	60.2	56.6
Tajikistan	80.8	87.5	89.8	91.8	95.0	93.9	106.9	100.0	100.2	91.7	63.3	52.4	45.7	40.0	33.2
Turkmenistan[c]	80.3	87.6	87.9	89.3	93.3	97.1	107.5	100.0	101.8	97.1	131.7	144.9	118.8	99.8	99.8
Ukraine	75.0	85.5	88.4	88.7	90.0	93.4	95.2	100.0	96.4	88.0	79.3	68.0	52.5	46.3	41.6
Uzbekistan	76.0	86.8	84.9	88.1	88.0	88.4	97.0	100.0	99.3	98.9	88.0	85.9	81.5	80.5	82.1
Total above	80.5	86.9	89.7	91.2	93.6	95.2	98.6	100.0	95.6	88.4	77.7	72.0	65.9	64.9	63.8
Memorandum items:															
CETE-5	89.2	85.7	89.5	92.0	95.0	97.1	99.7	100.0	92.3	83.3	82.5	84.0	87.7	92.8	97.3
SETE-7	89.2	94.9	98.5	99.0	102.0	102.9	102.5	100.0	92.8	79.9	69.5	65.9	67.8	71.3	72.6
Czechoslovakia[b]	84.9	86.9	90.0	92.7	95.1	97.0	99.3	100.0	98.5	84.4	78.8	..	..	..	..
Yugoslavia (SFR)[a]	97.7	98.4	100.1	99.5	103.0	101.0	99.4	100.0	92.4	..	..	..	..	..	..
Former Soviet Union[b]	77.0	86.0	88.5	89.9	92.0	93.6	97.8	100.0	96.9	90.9	..	..	..	..	..
Ex-GDR Länder	..	..	..	..	..	..	..	100.0	84.5	68.2	73.6	80.1	88.0	92.7	94.6

Source: UN/ECE Common Database, derived from national and CIS statistics (IMF and World Bank data for Albania).

Note: Data for the east European countries are based on a GDP measure, except where otherwise mentioned. For the countries of the former Soviet Union, NMP data for 1980-1990 were chain-linked to GDP data from 1990. Country indices were aggregated with previous year weights (viz., previous year's GDP at 1992 prices).

[a] Gross material product (1980-1989 for Croatia).
[b] Sum of individual country data for former members.
[c] Net material product for 1980-1990 (all years in the case of Turkmenistan).

APPENDIX TABLE B.2

Real total consumption expenditure in the transition countries, 1980, 1983-1996
(Indices, 1989=100 or earliest year available thereafter)

	1980	1983	1984	1985	1986	1987	1988	1989	1990	1991	1992	1993	1994	1995	1996
Bulgaria	..	..	..	..	..	..	..	100.0	100.6	92.3	89.4	86.2	82.3	79.9	..
Czech Republic	..	..	..	..	..	91.3	93.1	100.0	105.0	80.8	88.2	89.9	92.7	94.2	98.1*
Hungary	92.2	96.1	97.3	99.2	101.5	104.9	102.0	100.0	97.3	92.3	92.9	97.9	95.7	89.9	87.6*
Poland	108.0	97.3	101.9	105.1	109.3	111.8	114.7	100.0	88.3	94.9	98.2	103.3	107.3	111.7	119.7
Romania	83.9	83.1	86.0	85.4	85.8	88.7	90.6	100.0	108.9	96.0	90.6	91.7	95.2	108.8	113.4
Slovakia	..	..	..	81.8	85.4	89.2	92.1	100.0	103.3	76.9	75.5	74.3	71.7	73.8	82.7
Slovenia	..	..	..	..	..	..	..	..	100.0	91.6	88.8	98.9	102.8	108.9	..
Estonia	..	..	..	..	..	..	..	..	..	..	100.0	94.7	98.8	..	..
Latvia	..	..	..	..	..	..	..	..	100.0	76.9	50.0	47.5	48.5	47.9	48.6
Belarus	..	..	..	..	..	..	..	..	100.0	93.4	83.8	80.7	72.0	60.1	..
Kyrgyzstan	..	..	..	..	..	..	..	..	100.0	83.5	72.8	64.3	51.4	43.8	..
Republic of Moldova	..	..	..	..	..	..	..	..	..	100.0	67.7	66.6	38.5	..	..
Russian Federation	..	..	..	..	..	..	..	..	100.0	93.9	89.0	88.1	87.0	83.6	..
Ukraine	..	..	..	..	..	..	..	..	100.0	94.4	88.6	71.2	64.1	..	..

Source: UN/ECE Common Database, derived from national and CIS statistics.

APPENDIX TABLE B.3

Real gross fixed capital formation in the transition countries, 1980, 1983-1996
(Indices, 1989=100 or earliest year available thereafter)

	1980	1983	1984	1985	1986	1987	1988	1989	1990	1991	1992	1993	1994	1995	1996
Bulgaria	..	..	..	..	..	..	..	100.0	100.0	80.1	74.3	61.3	61.9	67.4	..
Czech Republic	..	..	..	..	..	93.4	99.4	100.0	97.9	80.5	87.7	81.0	95.0	108.4	127.8*
Hungary	114.7	104.6	100.5	94.6	100.8	110.7	100.6	100.0	92.9	83.2	81.0	82.7	93.0	89.0	81.0*
Poland[a]	124.6	98.5	104.9	111.4	116.4	116.5	126.5	100.0	75.2	60.1	52.3	60.0	65.5	83.5	101.6*
Romania	163.7	149.9	158.9	161.5	163.3	161.0	157.6	100.0	64.4	44.0	48.9	53.0	63.9	69.4	73.1
Slovakia	..	..	..	..	..	..	..	..	100.0	74.8	71.5	68.5	65.0	68.7	91.7
Slovenia	..	..	..	..	..	..	..	..	100.0	88.5	76.2	85.3	96.0	116.1	..
Estonia	..	..	..	..	..	..	..	..	..	..	100.0	110.0	121.1	..	..
Latvia	..	..	..	..	..	..	..	..	100.0	36.1	25.7	21.6	21.8	24.6	26.0
Belarus	..	..	..	..	..	..	..	..	100.0	104.4	85.5	72.3	59.9	43.7	..
Kyrgyzstan	..	..	..	..	..	..	..	..	100.0	88.7	62.8	49.1	34.9	59.6	..
Republic of Moldova	..	..	..	..	..	..	..	..	..	100.0	69.1	40.6	9.5	..	..
Russian Federation	..	..	..	..	..	..	..	..	100.0	84.5	49.4	36.7	27.1	24.1	..
Ukraine	..	..	..	..	..	..	..	..	100.0	81.6	69.4	48.3	28.5	..	..

Source: UN/ECE Common Database, derived from national and CIS statistics.

[a] Gross capital formation (including stock change).

APPENDIX TABLE B.4

Real gross industrial output in the transition countries, 1980, 1983-1996
(Indices, 1989=100)

	1980	1983	1984	1985	1986	1987	1988	1989	1990	1991	1992	1993	1994	1995	1996
Eastern Europe	83.1	84.7	89.0	91.6	95.5	98.0	100.6	100.0	83.9	68.4	61.8	60.2	64.2	69.1	73.9
Albania	77.0	86.3	88.5	87.5	91.9	93.3	95.2	100.0	86.7	50.4	35.2	31.7	25.8	23.9	..
Bosnia and Herzegovina	106.0	102.1	104.1	104.1	104.1	101.1	98.1	100.0	88.7	74.3	..	..	..	..	..
Bulgaria	71.3	82.1	85.5	88.2	92.4	98.0	101.1	100.0	83.2	64.7	54.4	48.5	52.6	55.5	54.9
Croatia	88.7	87.8	93.1	95.1	99.4	102.0	100.6	100.0	88.7	63.4	54.2	51.0	49.6	49.7	51.3
Czech Republic	81.5	86.5	89.0	92.0	94.5	96.5	98.5	100.0	96.5	73.0	67.2	63.6	65.0	70.9	75.8
Hungary	92.9	98.7	101.4	102.1	104.0	106.4	105.3	100.0	90.7	74.1	66.9	69.6	76.2	79.7	82.3
Poland	86.3	80.2	84.4	88.5	92.3	95.5	100.5	100.0	75.8	66.8	69.4	73.8	82.6	90.4	98.6
Romania	76.9	83.5	89.1	90.5	96.8	99.2	101.9	100.0	81.9	63.3	49.4	50.1	51.7	56.6	62.1
Slovakia	76.7	82.5	87.4	91.6	95.2	98.6	100.8	100.0	95.5	78.7	67.6	60.5	63.2	68.4	70.1
Slovenia	90.3	97.6	99.7	101.0	102.7	101.6	98.9	100.0	89.5	78.4	68.0	66.2	70.4	71.8	72.5
The FYR of Macedonia	72.1	79.5	85.9	88.6	95.0	97.3	95.6	100.0	89.3	73.8	62.4	53.7	48.0	43.3	44.7
Yugoslavia	80.0	84.8	91.2	93.6	96.8	97.6	98.4	100.0	87.2	71.9	56.5	35.4	35.8	37.2	39.7*
Baltic states	72.6	79.2	82.7	85.5	89.4	93.1	96.8	100.0	99.0	95.7	64.5	43.4	36.6	36.9	37.5
Estonia	78.5	85.0	87.9	90.2	93.5	96.3	99.3	100.0	100.0	92.8	60.1	42.9	42.0	42.8	43.3
Latvia	72.5	79.8	83.5	86.6	89.9	93.7	97.1	100.0	100.7	100.0	65.4	44.3	41.3	38.7	38.9
Lithuania	70.0	76.3	79.8	82.6	87.2	91.2	95.6	100.0	97.4	94.0	65.8	43.1	31.0	33.0	33.9
CIS[a]	73.0	80.9	84.1	87.2	91.0	94.4	98.1	100.0	99.9	93.6	79.1	69.9	54.8	51.7	49.9
Armenia	76.3	89.3	95.4	100.7	105.3	110.3	109.1	100.0	92.5	85.4	44.2	39.7	41.8	42.8	43.2
Azerbaijan	76.1	89.9	94.1	94.5	92.7	96.1	99.3	100.0	93.7	98.2	74.9	69.7	53.9	44.6	41.6
Belarus	61.1	70.3	74.5	79.0	84.3	89.9	95.6	100.0	102.1	101.1	91.6	82.4	68.3	60.3	62.3
Georgia	70.6	80.4	86.1	91.8	93.9	96.2	99.3	100.0	94.3	73.0	39.6	29.1	17.5	15.8	17.0
Kazakstan	72.4	79.8	82.8	85.8	90.2	94.1	97.6	100.0	99.2	98.3	84.8	72.2	51.9	47.7	47.8
Kyrgyzstan	66.7	76.1	80.8	84.2	87.8	89.0	95.1	100.0	99.4	99.1	72.9	54.5	39.2	32.2	35.7
Republic of Moldova	68.7	83.3	87.3	85.1	87.3	91.6	94.6	100.0	103.2	91.8	66.9	67.1	48.5	46.6	42.6
Russian Federation	74.4	81.9	84.9	87.8	91.8	95.0	98.6	100.0	99.9	91.9	75.3	64.7	51.2	49.5	47.0
Tajikistan	72.9	81.6	84.7	87.1	88.7	93.1	98.2	100.0	101.2	97.6	74.0	68.2	50.9	48.3	38.7
Turkmenistan	75.4	80.7	84.2	86.0	90.1	92.9	96.9	100.0	103.2	108.1	92.0	95.7	72.1	67.5	79.5
Ukraine	72.6	79.8	83.5	86.4	90.0	93.4	97.3	100.0	99.9	95.1	89.0	81.9	59.5	52.4	49.7
Uzbekistan	68.5	78.7	80.4	86.0	91.2	93.4	96.5	100.0	101.8	103.3	96.4	99.9	101.5	101.6	107.6
Total above	75.7	81.8	85.3	88.3	92.2	95.3	98.7	100.0	95.7	87.1	74.1	66.5	56.8	55.8	55.8
Memorandum items:															
CETE-5	85.8	84.5	88.2	91.5	94.8	97.4	100.6	100.0	83.4	70.3	68.6	70.1	76.7	83.1	89.2
SETE-7	79.3	85.0	90.1	91.8	96.6	98.9	100.5	100.0	84.5	65.7	52.2	46.0	46.5	49.2	52.3
Czechoslovakia	80.5	85.5	88.8	91.9	94.8	97.2	99.2	100.0	96.5	75.2	67.6	..	..	..	..
Yugoslavia (SFR)[a]	86.9	89.8	94.6	96.4	99.3	99.7	98.7	100.0	88.3	..	..	..	..	..	..
Former Soviet Union	73.0	80.8	84.0	87.1	91.0	94.4	98.1	100.0	99.9	93.7	..	..	..	..	..
Ex-GDR Länder	75.2	82.7	85.7	89.5	92.5	94.7	97.7	100.0	72.7	37.0	34.8	36.8	42.0	44.4	46.3

Source: UN/ECE Common Database, derived from national and CIS statistics (IMF and World Bank data for Albania).

Note: Data for former Czechoslovakia and SFR of Yugoslavia for 1980 to the breakup obtained as sum of individual country data for former members. For the countries of the former Soviet Union, data for 1980-1990 were chain-linked to national or CIS data from 1990. Country indices were aggregated with previous year weights (viz., previous year's GDP at 1992 prices).

[a] Generated from components.

APPENDIX TABLE B.5

Total employment in the transition countries, 1980, 1983-1995
(Indices, 1989=100)

	1980	1983	1984	1985	1986	1987	1988	1989	1990	1991	1992	1993	1994	1995
Eastern Europe[a]	94.5	94.7	95.2	96.1	96.8	97.3	97.3	100.0	97.1	91.5	86.8	84.3	84.4	84.0
Albania	..	..	..	..	..	..	..	100.0	99.2	97.5	76.0	72.7	80.7	78.7
Bulgaria	100.0	102.1	102.1	102.2	102.5	102.8	102.4	100.0	93.9	81.6	75.0	73.8	74.3	75.9
Croatia	87.4	92.4	93.8	95.9	98.6	100.6	100.4	100.0	96.9	88.5	77.9	76.5	74.8	73.9
Czech Republic	95.3	95.0	95.7	97.5	98.5	98.9	99.4	100.0	99.1	93.6	91.2	89.7	90.4	92.8
Hungary[b]	104.2	102.6	102.2	102.0	102.1	101.4	100.6	100.0	96.9	87.7	79.5	75.6	73.9	72.6
Poland	101.9	99.7	100.0	100.8	101.1	100.8	100.1	100.0	95.8	90.1	86.3	84.3	85.1	86.7
Romania[b]	94.6	95.5	95.9	96.7	97.5	97.9	98.7	100.0	99.0	98.5	95.5	91.9	91.5	86.7
Slovakia[b]	90.9	93.1	94.5	96.2	97.5	99.0	100.0	100.0	98.2	85.9	86.8	84.6	83.7	85.4
Slovenia[c]	84.0	86.3	87.6	88.9	90.4	101.9	101.3	100.0	96.1	88.7	82.8	81.0	79.5	79.3
The FYR of Macedonia	79.1	88.2	90.4	92.7	96.0	100.5	100.1	100.0	98.5	95.6	91.2	86.2	81.7	73.9
Yugoslavia	83.4	89.4	91.4	93.7	96.5	99.0	99.8	100.0	97.0	94.1	90.9	88.3	86.5	85.3
Baltic states	..	..	..	98.6	99.4	100.0	100.2	100.0	98.4	99.6	96.1	89.6	85.9	84.5
Estonia	..	..	..	99.7	100.9	100.7	100.7	100.0	97.9	100.0	93.5	82.4	81.4	80.2
Latvia	..	..	..	99.1	99.8	100.4	100.4	100.0	101.1	99.3	95.6	88.5	85.6	84.5
Lithuania	..	..	..	97.7	98.5	99.5	99.8	100.0	97.3	99.7	97.5	93.4	88.0	86.4
CIS	..	..	..	98.0	98.6	99.1	99.5	100.0	100.1	99.1	96.7	94.3	91.3	90.1
Armenia	..	..	..	97.2	98.6	99.4	101.4	100.0	102.4	105.0	99.1	96.9	93.5	92.7
Azerbaijan	..	..	..	89.8	91.8	94.5	94.9	100.0	100.0	104.7	100.4	100.2	98.0	97.5
Belarus	..	..	..	98.4	98.8	99.0	99.5	100.0	99.1	96.6	94.0	92.8	90.3	84.7
Georgia	..	..	..	100.9	102.6	102.9	103.2	100.0	102.4	93.3	73.6	66.5	64.9	64.1
Kazakstan	..	..	..	95.6	96.5	97.7	99.1	100.0	101.3	103.3	101.4	92.8	88.1	87.7
Kyrgyzstan	..	..	..	92.8	94.9	97.9	98.7	100.0	100.5	100.9	105.6	96.6	94.6	94.4
Republic of Moldova[d]	..	..	..	99.5	99.6	99.7	98.9	100.0	99.0	99.0	98.0	80.7	80.4	80.0
Russian Federation	..	..	..	99.2	99.5	99.7	99.9	100.0	99.7	97.7	95.4	93.8	90.6	87.9
Tajikistan	..	..	..	89.4	91.2	94.5	96.9	100.0	103.1	104.8	101.5	98.5	98.1	98.6
Turkmenistan	..	..	..	89.7	91.7	94.3	96.8	100.0	103.4	105.3	105.4	110.1	111.6	112.1
Ukraine	..	..	..	100.7	100.5	100.4	99.9	100.0	99.4	98.3	96.3	94.1	90.6	93.3
Uzbekistan	..	..	..	86.8	88.9	93.2	95.9	100.0	104.2	109.2	108.5	108.3	106.9	107.2
Memorandum items:														
CETE-5	99.7	98.5	98.8	99.7	100.2	100.5	100.1	100.0	96.7	89.9	85.9	83.6	83.8	84.9
SETE-7[a]	87.0	89.3	89.9	90.8	91.9	92.8	93.2	100.0	97.6	93.7	88.1	85.2	85.2	82.5
Former Soviet Union	..	..	..	98.0	98.6	99.2	99.5	100.0	100.1	99.1	96.6	94.1	91.1	89.9

Source: UN/ECE Common Database, derived from national and CIS statistics (IMF and World Bank data for Albania).

[a] Albania excluded until 1989.

[b] End of year.

[c] Self-employed excluded until 1986.

[d] Excluding Transnistria since 1993.

APPENDIX TABLE B.6

Registered unemployment in the transition countries, 1990, 1992-1996

(Thousands and per cent of labour force, end of period)

	Thousands						Per cent of labour force					
	1990	1992	1993	1994	1995	1996	1990	1992	1993	1994	1995	1996
Eastern Europe	2 773	6 768	7 443	7 190	6 583	6 114	..	12.4	14.0	13.6	12.5	11.8
Albania	150	394	301	262	171	158	9.5	27.0	22.0	18.0	13.1	12.1
Bosnia and Herzegovina	..	..	..	..	..	..	..	..	..	..	..	..
Bulgaria	72	577	626	488	424	479	1.8	15.6	16.4	12.8	11.1	12.5
Croatia	196	261	243	248	249	269	..	17.8	16.6	17.3	17.6	15.9
Czech Republic	39	135	185	167	153	186	0.7	2.6	3.5	3.2	2.9	3.5
Hungary	101	663	632	520	496	478	1.7	12.3	12.1	10.9	10.4	10.5
Poland	1 126	2 509	2 890	2 838	2 629	2 360	6.5	14.3	16.4	16.0	14.9	13.6
Romania	150	929	1 165	1 224	998	658	1.3	8.2	10.4	10.9	9.5	6.3
Slovakia	40	260	368	372	330	330	1.6	10.4	14.4	14.8	13.1	12.8
Slovenia	55	118	137	124	127	124	..	13.3	15.5	14.2	14.5	14.4
The FYR of Macedonia[a]	156	173	177	196	229	245	..	26.8	30.3	33.2	37.2	39.8
Yugoslavia[a]	688	749	718	751	777	827	..	24.6	24.0	23.9	24.7	26.1
Baltic states	..	115	176	197	245	237	..	2.1	4.5	5.3	6.6	6.4
Estonia[b]	..	15	34	35	34	37	..	1.6	5.0	5.1	5.0	5.6
Latvia	..	34	77	84	83	91	..	2.3	5.8	6.5	6.6	7.2
Lithuania	..	66	65	78	128	109	..	3.6	3.4	4.5	7.3	6.2
CIS	..	3 837	4 524	6 011	7 185	8 027	..	2.7	3.6	4.4	5.4	6.4
Armenia	..	56	103	92	132	159	..	3.5	6.3	6.0	8.1	9.7
Azerbaijan	..	6	20	24	28	32	..	0.2	0.7	0.9	1.1	1.1
Belarus	..	24	66	101	131	183	..	0.5	1.3	2.1	2.7	4.0
Georgia	..	19	40	76	61	58	..	0.3	2.0	3.8	3.4	3.2
Kazakstan	..	34	41	70	140	282	..	0.4	0.6	1.0	2.1	4.1
Kyrgyzstan	..	2	3	13	50	77	..	0.1	0.2	0.8	3.0	4.5
Republic of Moldova	..	15	14	21	25	23	..	0.7	0.7	1.0	1.4	1.5
Russian Federation[c]	..	3 594	4 120	5 478	6 431	6 788	..	4.7	5.5	7.5	8.9	9.3
Tajikistan	..	7	22	32	35	46	..	0.4	1.1	1.8	1.8	2.4
Turkmenistan	..	..	..	..	..	..	..	..	..	..	..	..
Ukraine	..	71	84	82	127	351	..	0.3	0.4	0.3	0.6	1.5
Uzbekistan	..	9	13	22	25	28	..	0.1	0.2	0.3	0.3	0.4
Total above	..	10 720	12 143	13 398	14 013	14 378	..	5.4	6.6	6.9	7.5	7.9
Memorandum items:												
CETE-5	1 361	3 685	4 212	4 020	3 735	3 478	..	11.3	13.3	12.8	12.0	11.3
SETE-7	1 412	3 083	3 230	3 170	2 848	2 636	..	14.2	15.1	14.6	13.3	12.5
Russian Federation[d]	..	578	836	1 637	2 327	2 506	..	0.8	1.1	2.1	3.2	3.4

Source: National statistics and direct communications from national statistical offices to UN/ECE secretariat.

[a] The data reported on employment cover only the social sector in agriculture, hence unemployment rates are biased upwards. In Yugoslavia, according to the labour force survey, the unemployment rate was 13.2 per cent in May 1996, instead of the officially reported rate of some 26 per cent. Also, in The FYR of Macedonia, the unemployment rate after inclusion of an estimate for the significant underrecording of the private sector in the labour force data, was some 24 per cent in December 1995 instead of the officially reported rate of 37 per cent.

[b] Job seekers.

[c] Based on Russian Federation Goskomstat's monthly estimates according to the ILO definition, i.e. including all persons not having employment but actively seeking work.

[d] Registered unemployment.

APPENDIX TABLE B.7

Consumer prices in the transition countries, 1989-1996
(Annual average, percentage change over preceding year)

	1989	1990	1991	1992	1993	1994	1995	1996
Albania	..	..	35.5	193.1	85.0	21.5	8.0	..
Bosnia and Herzegovina[a]	36.8	594.0	116.2	64 218.3	38 825.1	553.5	-12.1	-21.2
Bulgaria	9.2[a]	50.6[ab]	338.5	91.3	72.9	96.2	62.1	123.1
Croatia[a]	1 200.0	609.5	123.0	663.6	1 516.6	97.5	2.0	3.6
Czech Republic	1.4	9.7	56.6	11.1	20.6	10.0	9.1	8.9
Hungary	17.0	28.9	35.0	23.0	22.6	19.1	28.5	23.6
Poland	264.3	585.8	70.3	45.3	36.9	33.2	28.1	19.8
Romania	1.1	5.1	170.2	210.7	256.2	137.1	32.2	38.8
Slovakia	1.3	10.4	61.2	10.2	23.1	13.4	10.0	6.0
Slovenia	1 285.3[a]	549.7	117.7	201.3	31.8	19.8	12.6	9.7
The FYR of Macedonia[a]	1 246.0	608.4	114.9	1 505.5	353.1	121.0	16.9	4.1
Yugoslavia	1 265.0[a]	580.0[a]	122.0[a]	8 880.1	2.2E+14	7.9E+10[c]	71.8[d]	90.5
Estonia	4.0	18.0	202.0	1 078.2	89.6	47.9	28.9	23.1
Latvia[a]	5.2	10.9	172.2	951.2	109.1	35.7	25.0	17.7
Lithuania	2.2	9.1	216.4	1 020.5	410.1	72.0	39.5	24.7
Armenia	0.8	6.9	140.0	728.7	3 731.8	4 964.0	175.5	18.7
Azerbaijan	0.5	6.1	111.5	912.6	1 129.7	1 663.9	411.5	19.8
Belarus	1.7	5.5	98.6	971.2	1 190.9	2 219.6	709.3	52.7
Georgia	0.9	4.2	78.7	1 176.9	4 084.9	22 470.0	177.6	39.0
Kazakstan	1.8	5.6	78.7	1 504.3	1 662.7	1 879.5	175.9	39.1
Kyrgyzstan	1.6	5.5	113.9	854.6	1 208.7	278.1	42.9	30.3
Republic of Moldova	1.1	5.7	114.4	1 308.0	1 751.0	486.4	29.9	23.5
Russian Federation	2.5	5.3	100.0	1 528.7	875.0	309.0	197.4	47.8
Tajikistan	1.7	5.9	112.9	822.0	2 884.8	350.3	682.1	422.4
Turkmenistan	2.5	5.7	112.4	769.9	1 630.5	2 714.0	..	..
Ukraine	2.0	5.4	94.0	1 209.6	4 734.9	891.2	376.7	80.2
Uzbekistan	0.7	5.8	97.3	414.5	1 231.8	1 550.0	315.5	..

Source: UN/ECE Common Database, derived from national statistics.

Note: Retail prices for CIS countries, consumer prices elsewhere except where otherwise mentioned.

[a] Retail prices.

[b] December over May 1990.

[c] December over February 1994.

[d] February-December over the corresponding period in 1994.

APPENDIX TABLE B.8

Producer price indices in the transition countries, 1989-1996

(Annual average, percentage change over preceding year)

	1989	1990	1991	1992	1993	1994	1995	1996
Albania	..	..	..	..	..	..	..	..
Bosnia and Herzegovina	..	..	129.5	70 374.7	10 967.6	1 184.8	68.7	-4.8
Bulgaria	2.8	14.7	296.4	55.9	26.9	59.1	48.9	129.7
Croatia	1 346.7	455.3	146.3	826.0	1 510.4	77.7	0.8	1.3
Czech Republic	0.1	2.5	70.4	10.8	9.3	5.4	7.8	4.9
Hungary	15.4	22.0	32.6	12.3	14.1	12.3	28.5	22.3
Poland	212.8	622.4	48.1	28.5	32.6	31.0	26.0	13.2
Romania	..	26.9	220.1	184.8	165.0	140.7	35.3	50.0
Slovakia	..	..	68.8	5.3	17.2	10.0	9.1	4.0
Slovenia	1 413.3	390.4	124.8	215.7	23.3	17.8	12.4	6.7
The FYR of Macedonia	1 357.0	394.0	112.0	2 198.2	258.3	88.7	4.7	—
Yugoslavia	1 301.0	468.0	124.0	8 993.4	1.4E+13	7.9E+10	75.7	89.1
Estonia	..	..	..	..	75.2	36.2	25.6	14.7
Latvia	..	..	..	2 554.1	117.1	17.0	12.0	13.8
Lithuania	..	..	148.2	1 517.4	397.6	44.7	28.8	16.5
Armenia	..	..	120.0	947.0	892.0	4 394.4	187.8	36.7
Azerbaijan	..	..	135.0	1 303.0	1 040.7	3 971.6	1 340.1	70.6
Belarus	..	..	..	..	..	3 362.1	538.6	37.6
Georgia	..	..	..	..	..	..	..	..
Kazakstan	..	..	193.0	2 369.0	1 342.0	2 952.2	173.6	24.7
Kyrgyzstan	..	..	160.0	1 664.0	831.0	223.6	34.0	44.7
Republic of Moldova	..	..	130.0	1 210.9	1 078.5	711.7	52.2	30.2
Russian Federation	..	..	138.0	1 949.0	887.0	340.0	237.6	50.7
Tajikistan	..	..	163.0	1 323.0	1 080.0	665.5	351.7	341.9
Turkmenistan	..	..	211.0	999.0	1 610.0	911.0	..	..
Ukraine	..	..	122.0	2 491.7	4 698.3	901.2	450.8	54.6
Uzbekistan	..	..	147.0	1 296.0	1 119.0	2 162.6	792.5	..

Source: UN/ECE Common Database, derived from national statistics.

APPENDIX TABLE B.9

Nominal gross wages in industry in the transition countries, 1990-1996

(Annual average, percentage change over preceding year)

	1990	1991	1992	1993	1994	1995	1996
Albania	..	..	..	69.5	34.5	29.3	..
Bosnia and Herzegovina	..	..	..	..	..	620.5	293.4
Bulgaria	21.9	175.4	193.0	13.2	61.4	59.8	105.6
Croatia	481.5	68.1	314.4	1 478.8	25.9	45.7	11.7
Czech Republic	3.1	20.5	21.0	20.0	15.7	18.1	17.7
Hungary	22.4	25.3	26.0	25.8	20.8	21.2	21.4
Poland	369.0	67.0	64.4	40.8	41.2	31.9	26.3
Romania	9.7	125.0	173.5	216.3	130.3	54.2	53.4
Slovakia	3.4	15.8	19.8	18.3	17.5	15.2	14.7
Slovenia	361.4	68.4	195.7	43.8	27.1	17.0	13.1
The FYR of Macedonia	300.0	100.0	925.0	353.5	103.7	10.4	2.7
Yugoslavia	431.0	110.0	4 527.0	-62.1	212.3	109.1	82.9
Estonia	23.0	136.4	-37.0	93.3	72.2	36.3	19.5
Latvia	..	103.9	599.0	111.5	63.1	28.1	20.3
Lithuania	..	..	..	-96.5	64.6	39.4	29.4
Armenia	..	32.0	316.0	875.5	-84.7	214.0	48.8
Azerbaijan	..	71.0	694.0	705.0	581.3	298.4	44.8
Belarus	..	103.0	822.0	1 110.4	58.3	668.3	60.5
Georgia	..	29.0	413.0	19.0	..	..	..
Kazakstan	..	69.0	949.0	1 304.2	1 235.2	186.7	36.2
Kyrgyzstan	..	65.0	592.0	644.2	185.8	60.1	26.9
Republic of Moldova	..	83.0	641.0	770.7	268.6	30.8	33.0
Russian Federation	..	90.0	1 009.0	904.2	270.0	125.6	48.0
Tajikistan	..	80.0	446.0	649.7	141.4	109.5	265.5
Turkmenistan	..	90.0	664.0	1 750.0	878.0	639.5	763.7
Ukraine	..	91.0	1 253.0	2 431.0	747.0	425.1	20.6
Uzbekistan	..	68.0	605.0	1 147.2	811.8	289.2	99.1

Source: UN/ECE Common Database, derived from national statistics.

Note: Indices for 1990-1992 derived from annual average nominal values in national currency reported or published by the countries; for CIS countries, indices published by CIS Statistical Committee. From 1993 onwards, indices derived from monthly nominal values in national currency reported to ECE secretariat, except Albania (all years), Baltic countries (1990-1994) and, in 1993, Croatia, Georgia, Turkmenistan, Ukraine.

APPENDIX TABLE B.10

Merchandise exports of the transition countries, 1980, 1984-1996
(Billion dollars)

	1980	1984	1985	1986	1987	1988	1989	1990	1991	1992	1993	1994	1995	1996
Eastern Europe	56.367	54.957	55.020	60.117	63.550	65.020	63.850	61.733	57.096	59.457	61.330	71.425	95.204	96.087
Albania	0.320	0.250	0.242	0.230	0.230	0.230	0.302	0.231	0.101	0.072	0.123	0.139	0.202	0.211
Bulgaria	7.160	7.234	7.389	7.599	7.841	7.554	6.651	5.232	3.288	3.908	3.693	3.994	5.362	4.500*
Czechoslovakia	10.475	10.001	10.552	12.160	12.355	12.381	11.988	10.728	11.319	12.475	..	..	..	30.739
Czech Republic	..	..	..	..	..	..	..	..	..	8.767	13.205	14.252	21.647	21.918
Slovakia	..	..	..	..	..	..	..	..	..	3.708	5.445	6.714	8.585	8.821
Hungary	8.609	8.617	8.472	9.170	9.584	9.999	9.673	9.731	10.226	10.681	8.921	10.738	12.905	13.113
Poland	13.071	10.897	11.100	13.130	14.095	14.573	14.665	18.291	14.912	13.187	14.202	17.272	22.898	24.350
Romania	9.217	9.466	8.376	8.159	8.580	8.971	8.076	4.570	4.266	4.363	4.892	6.151	7.910	7.660
Yugoslavia (SFR)	7.514	8.492	8.889	9.669	10.866	11.311	12.496	12.950	..	..	..	..	..	15.514
Bosnia and Herzegovina	..	..	..	..	..	1.550	2.100	1.850	..	..	..	..	..	..
Croatia	..	..	..	..	..	2.300	2.600	4.020	3.310	4.353	3.709	4.250	4.645	4.511
Slovenia	1.836	2.004	2.111	2.567	2.757	3.278	3.408	4.118	3.874	6.681	6.083	6.828	8.316	8.306
The FYR of Macedonia	..	..	..	..	..	..	0.654	1.113	1.095	1.199	1.055	1.086	1.204	0.855*
Yugoslavia	..	3.560	3.810	3.974	4.063	4.298	4.461	4.651	4.704	2.539	..	..	1.531	1.842
Baltic states	..	..	..	..	..	..	..	..	..	2.139	4.197	4.324	5.844	6.798
Estonia	..	..	..	..	..	..	..	..	..	0.444	0.802	1.305	1.835	2.076
Latvia	..	..	..	..	..	..	..	..	..	0.843	1.401	0.988	1.304	1.443
Lithuania	..	..	..	..	..	..	..	..	..	0.852	1.994	2.031	2.705	3.280
CIS	..	..	..	..	..	..	..	..	..	51.594	52.552	62.559	77.472	84.901
Armenia	..	..	..	..	..	..	..	..	..	0.026	0.030	0.058	0.104	0.160
Azerbaijan	..	..	..	..	..	..	..	..	..	0.754	0.351	0.363	0.329	0.329
Belarus	..	..	..	..	..	..	..	..	..	1.061	0.758	1.031	1.776	1.801
Georgia	..	..	..	..	..	..	..	..	..	0.068	0.069	0.039	0.058	0.063
Kazakstan	..	..	..	..	..	..	..	..	..	1.398	1.501	1.357	2.343	2.787
Kyrgyzstan	..	..	..	..	..	..	..	..	..	0.077	0.112	0.117	0.140	0.101
Republic of Moldova	..	..	..	..	..	..	..	..	..	0.157	0.178	0.160	0.279	0.252
Russian Federation	..	..	..	..	..	..	..	..	..	42.391	44.297	53.001	63.729	68.803
Tajikistan	..	..	..	..	..	..	..	..	..	0.111	0.263	0.320	0.497	0.432
Turkmenistan	..	..	..	..	..	..	..	..	..	0.908	1.049	0.494	0.951	0.532
Ukraine	..	..	..	..	..	..	..	..	..	3.774	3.222	4.653	5.554	6.816
Uzbekistan	..	..	..	..	..	..	..	..	..	0.869	0.721	0.966	1.712	2.825
Former Soviet Union	..	57.942	62.361	57.317	60.043	63.406	62.016	62.286	59.056	46.660	..	..	..	..
Total	114.310	117.318	112.337	120.161	126.957	127.035	126.136	120.788	103.756	113.190	118.078	138.308	178.520	187.785

Source: UN/ECE secretariat, based on national statistical publications and direct communications from national statistical offices.

Note: Data exclude intra-CIS trade, but now include the "new trade" among members of other recently dissolved federal states: former Czechoslovakia (from 1993), SFR Yugoslavia (from 1992), and the trade of the Baltic states with the former USSR (from 1992). Data excluding the "new trade" were shown in earlier issues of this publication. Changes in the method of recording trade are reflected from 1995 in data for the Czech Republic (inclusion OPT transactions, etc), Latvia (imports registered c.i.f.) and Lithuania; see box 3.1.1 in UN/ECE, *Economic Bulletin for Europe*, Vol. 48 (1996), p. 42.

All trade values for the years 1991-1996 are expressed in dollars at prevailing market exchange rates. For earlier years, values reported in national currencies were adjusted by the UN/ECE secretariat to remove distortions stemming from mutually inconsistent national rouble/dollar cross-rates in the valuation of the then important intra-CMEA trade flows. For details on the revaluation, see the note to table 2.1.4 and the discussion in box 2.1.1 in UN/ECE, *Economic Bulletin for Europe*, Vol. 43 (1991).

APPENDIX TABLE B.11

Merchandise imports of the transition countries, 1980, 1984-1996
(Billion dollars)

	1980	1984	1985	1986	1987	1988	1989	1990	1991	1992	1993	1994	1995	1996
Eastern Europe	65.443	50.830	53.875	59.532	62.033	60.158	61.185	63.408	61.496	68.311	74.340	83.729	116.640	129.977
Albania	0.320	0.280	0.278	0.230	0.230	0.280	0.385	0.381	0.409	0.524	0.421	0.549	0.650	0.910
Bulgaria	6.321	6.832	7.568	8.679	8.222	8.131	7.325	5.584	2.587	4.453	4.722	4.194	5.665	4.300*
Czechoslovakia	10.619	9.529	10.216	10.277	12.503	12.180	11.772	11.808	10.962	14.257	..	..	..	38.749
Czech Republic	..	..	..	..	..	..	..	..	..	10.368	12.859	14.968	25.262	27.824
Slovakia	..	..	..	..	..	..	..	..	..	3.889	6.543	6.705	8.917	10.925
Hungary	9.188	8.129	8.183	9.594	9.859	9.372	8.863	8.797	11.449	11.123	12.648	14.623	15.406	16.169
Poland	14.705	9.722	10.430	12.315	12.686	12.987	12.941	12.619	15.531	16.141	18.758	21.596	29.079	36.940
Romania	11.061	6.430	6.700	6.411	6.355	5.361	5.834	6.889	5.793	6.260	6.522	7.109	10.278	9.970
Yugoslavia (SFR)	13.229	9.909	10.500	12.026	12.178	11.847	14.064	17.330	..	..	..	..	..	22.939
Bosnia and Herzegovina	..	..	..	..	..	1.300	1.850	1.750	..	..	..	..	..	..
Croatia	..	..	..	..	..	2.900	3.750	5.133	3.811	4.346	4.166	5.198	7.508	7.790
Slovenia	2.463	1.928	2.075	2.740	2.722	2.914	3.216	4.727	4.131	6.141	6.501	7.304	9.492	9.397
The FYR of Macedonia	..	..	..	..	..	..	0.934	1.531	1.274	1.206	1.199	1.484	1.719	1.650*
Yugoslavia	..	4.428	4.659	4.753	4.851	4.915	5.383	6.701	5.548	3.859	..	..	2.665	4.102
Baltic states	..	..	..	..	..	..	..	..	..	1.802	4.101	5.251	8.005	9.922
Estonia	..	..	..	..	..	..	..	..	..	0.406	0.896	1.659	2.538	3.198
Latvia	..	..	..	..	..	..	..	..	..	0.794	0.961	1.240	1.818	2.320
Lithuania	..	..	..	..	..	..	..	..	..	0.602	2.244	2.352	3.649	4.405
CIS	..	..	..	..	..	..	..	..	..	42.364	33.354	36.742	44.432	49.697
Armenia	..	..	..	..	..	..	..	..	..	0.050	0.086	0.188	0.340	0.548
Azerbaijan	..	..	..	..	..	..	..	..	..	0.333	0.241	0.292	0.440	0.632
Belarus	..	..	..	..	..	..	..	..	..	0.751	0.779	0.975	1.887	2.331
Georgia	..	..	..	..	..	..	..	..	..	0.227	0.167	0.059	0.225	0.355
Kazakstan	..	..	..	..	..	..	..	..	..	0.469	0.494	1.384	1.172	1.241
Kyrgyzstan	..	..	..	..	..	..	..	..	..	0.071	0.112	0.108	0.169	0.395
Republic of Moldova	..	..	..	..	..	..	..	..	..	0.170	0.184	0.183	0.271	0.412
Russian Federation	..	..	..	..	..	..	..	..	..	36.984	26.807	28.344	33.155	31.351
Tajikistan	..	..	..	..	..	..	..	..	..	0.132	0.374	0.318	0.321	0.269
Turkmenistan	..	..	..	..	..	..	..	..	..	0.030	0.501	0.782	0.619	0.918
Ukraine	..	..	..	..	..	..	..	..	..	2.219	2.651	2.907	4.203	8.124
Uzbekistan	..	..	..	..	..	..	..	..	..	0.929	0.958	1.202	1.630	3.116
Former Soviet Union	52.218	53.947	54.763	55.016	53.794	58.044	64.983	64.963	45.405	..	..	..	..	..
Total	117.661	104.777	108.638	114.548	115.827	118.202	126.168	128.371	106.901	112.477	111.795	125.722	169.077	189.596

Source: UN/ECE secretariat, based on national statistical publications and direct communications from national statistical offices.

Note: See appendix table B.10.

APPENDIX TABLE B.12

Balance of merchandise trade of the transition countries, 1980, 1984-1996
(Billion dollars)

	1980	1984	1985	1986	1987	1988	1989	1990	1991	1992	1993	1994	1995	1996
Eastern Europe	-9.076	4.127	1.146	0.585	1.517	4.861	2.665	-1.675	-4.400	-8.854	-13.010	-12.304	-21.436	-33.890
Albania	–	-0.030	-0.036	–	–	-0.050	-0.083	-0.150	-0.308	-0.452	-0.298	-0.410	-0.448	-0.699
Bulgaria	0.839	0.402	-0.179	-1.080	-0.381	-0.577	-0.674	-0.352	0.701	-0.545	-1.028	-0.200	-0.303	0.200*
Czechoslovakia	-0.144	0.472	0.336	1.882	-0.147	0.201	0.216	-1.080	0.356	-1.783	..	..	..	-8.010
Czech Republic	..	..	..	..	..	..	..	..	..	..	-1.601	0.346	-0.716	-3.615	-5.906
Slovakia	..	..	..	..	..	..	..	..	..	..	-0.182	-1.098	0.010	-0.332	-2.104
Hungary	-0.579	0.488	0.289	-0.423	-0.276	0.627	0.810	0.934	-1.223	-0.442	-3.727	-3.885	-2.501	-3.055
Poland	-1.634	1.175	0.670	0.815	1.409	1.586	1.724	5.672	-0.619	-2.955	-4.555	-4.324	-6.182	-12.590
Romania	-1.844	3.037	1.676	1.748	2.225	3.610	2.242	-2.320	-1.528	-1.897	-1.630	-0.958	-2.368	-2.311
Yugoslavia (SFR)	-5.715	-1.416	-1.611	-2.357	-1.312	-0.536	-1.568	-4.380	..	..	..	..	..	-7.425
Bosnia and Herzegovina	..	..	..	..	..	0.250	0.250	0.100	..	..	..	..	..	..
Croatia	..	..	..	..	..	-0.600	-1.150	-1.113	-0.501	0.007	-0.457	-0.948	-2.863	-3.279
Slovenia	-0.626	0.076	0.035	-0.173	0.035	0.365	0.192	-0.609	-0.257	0.540	-0.418	-0.476	-1.176	-1.091
The FYR of Macedonia	..	..	..	..	..	..	-0.280	-0.418	-0.179	-0.007	-0.144	-0.398	-0.515	-0.795*
Yugoslavia	..	-0.868	-0.849	-0.779	-0.788	-0.617	-0.922	-2.050	-0.844	-1.320	..	..	-1.134	-2.260
Baltic states	..	..	..	..	..	..	..	..	..	0.337	0.096	-0.927	-2.161	-3.124
Estonia	..	..	..	..	..	..	..	..	..	0.038	-0.094	-0.353	-0.703	-1.122
Latvia	..	..	..	..	..	..	..	..	..	0.049	0.440	-0.252	-0.514	-0.877
Lithuania	..	..	..	..	..	..	..	..	..	0.250	-0.250	-0.322	-0.944	-1.125
CIS	..	..	..	..	..	..	..	..	..	9.229	19.198	25.817	33.039	35.203
Armenia	..	..	..	..	..	..	..	..	..	-0.024	-0.056	-0.130	-0.236	-0.388
Azerbaijan	..	..	..	..	..	..	..	..	..	0.422	0.110	0.071	-0.111	-0.303
Belarus	..	..	..	..	..	..	..	..	..	0.310	-0.021	0.056	-0.111	-0.530
Georgia	..	..	..	..	..	..	..	..	..	-0.159	-0.098	-0.020	-0.167	-0.292
Kazakstan	..	..	..	..	..	..	..	..	..	0.930	1.007	-0.027	1.171	1.546
Kyrgyzstan	..	..	..	..	..	..	..	..	..	0.006	–	0.009	-0.029	-0.294
Republic of Moldova	..	..	..	..	..	..	..	..	..	-0.014	-0.006	-0.023	0.008	-0.160
Russian Federation	..	..	..	..	..	..	..	..	..	5.407	17.491	24.657	30.573	37.446
Tajikistan	..	..	..	..	..	..	..	..	..	-0.021	-0.111	0.002	0.176	0.163
Turkmenistan	..	..	..	..	..	..	..	..	..	0.879	0.548	-0.288	0.332	-0.386
Ukraine	..	..	..	..	..	..	..	..	..	1.555	0.571	1.746	1.351	-1.308
Uzbekistan	..	..	..	..	..	..	..	..	..	-0.060	-0.237	-0.236	0.082	-0.291
Former Soviet Union	5.724	8.414	2.554	5.027	9.612	3.972	-2.697	-5.907	1.255	..	..	..	..	..
Total	-3.351	12.541	3.700	5.613	11.130	8.833	-0.032	-7.583	-3.145	0.712	6.283	12.586	11.380	-1.811

Source: UN/ECE secretariat, based on national statistical publications and direct communications from national statistical offices.

Note: See appendix table B.10.

APPENDIX TABLE B.13

Merchandise trade of the transition countries by direction, 1980, 1984-1996
(Shares in total trade, per cent)

	1980	1984	1985	1986	1987	1988	1989	1990	1991	1992	1993	1994	1995	1996
Eastern Europe, *to and from:*														
Exports														
World	100.0	100.0	100.0	100.0	100.0	100.0	100.0	100.0	100.0	100.0	100.0	100.0	100.0	100.0
Transition economies	50.0	47.5	50.0	55.5	53.0	49.3	46.6	41.1	30.0	24.2	30.5	27.3	26.5	27.6
Former Soviet Union	27.1	27.6	28.5	31.0	29.7	27.2	25.5	22.3	17.9	12.3	9.2	8.6	8.3	9.0
Eastern Europe	18.6	15.7	16.2	18.6	17.8	16.7	15.8	12.7	7.7	6.9	16.7	15.1	14.8	15.3
Other socialist countries[a]	4.3	4.2	5.4	6.0	5.6	5.4	5.3	6.1	4.4	5.0	4.6	3.6	3.4	3.3
Developed market economies	35.7	36.3	35.9	32.3	35.2	38.7	42.6	49.5	59.8	63.2	58.0	62.5	64.4	63.6
Developing economies	14.3	16.2	14.1	12.2	11.8	11.9	10.8	9.4	10.2	12.7	11.5	10.2	9.1	8.8
Imports														
World	100.0	100.0	100.0	100.0	100.0	100.0	100.0	100.0	100.0	100.0	100.0	100.0	100.0	100.0
Transition economies	47.5	53.5	55.0	58.6	54.7	49.3	45.7	36.8	29.3	25.8	30.1	27.2	26.5	25.8
Former Soviet Union	26.8	31.6	32.2	34.5	30.7	26.3	23.5	18.3	20.2	18.0	16.1	13.9	12.9	12.5
Eastern Europe	17.1	17.7	17.6	18.7	18.4	17.6	16.4	12.5	6.3	5.6	11.9	11.1	11.2	10.8
Other socialist countries[a]	3.6	4.3	5.3	5.4	5.6	5.4	5.8	6.0	2.8	2.3	2.1	2.1	2.3	2.5
Developed market economies	38.7	33.7	32.9	32.3	36.1	41.1	44.0	53.3	58.3	64.1	61.6	65.0	65.8	66.2
Developing economies	13.8	12.7	12.0	9.1	9.2	9.5	10.3	9.9	12.4	10.1	8.3	7.8	7.8	8.0
Former Soviet Union/Russian Federation, *to and from:*														
Exports														
World	100.0	100.0	100.0	100.0	100.0	100.0	100.0	100.0	100.0	100.0	100.0	100.0	100.0	100.0
Transition economies	39.5	36.5	41.3	46.4	40.4	35.7	32.4	25.9	30.0	30.7	26.3	23.7	26.2	26.1
Eastern Europe[b]	30.7	28.1	31.6	36.5	31.4	27.5	24.4	18.8	22.9[b]	17.7	17.0[b]	14.1	16.2	13.0
Other socialist countries[a]	8.8	8.4	9.7	9.9	9.0	8.1	8.1	7.1	7.0	13.0	9.3	9.7	10.0	13.1
Developed market economies	42.2	42.0	38.7	31.0	35.3	38.9	41.8	49.5	56.5	57.9	59.7	62.0	59.9	58.0
Developing economies	18.3	21.5	20.0	22.6	24.3	25.4	25.8	24.6	13.5	11.3	14.0	14.3	13.9	16.0
Imports														
World	100.0	100.0	100.0	100.0	100.0	100.0	100.0	100.0	100.0	100.0	100.0	100.0	100.0	100.0
Transition economies	38.6	38.4	41.3	46.5	45.3	38.5	32.9	29.4	31.5	23.7	22.1	18.8	22.2	18.1
Eastern Europe[b]	31.1	30.6	32.1	37.0	36.9	31.3	26.4	23.2	24.5[b]	12.1	10.7[b]	11.7	15.5	8.9
Other socialist countries[a]	7.4	7.8	9.1	9.5	8.4	7.3	6.5	6.2	6.9	11.6	11.4	7.1	6.7	9.1
Developed market economies	46.4	44.5	42.1	40.9	40.7	46.3	50.1	52.9	58.1	62.4	60.6	70.3	69.6	67.3
Developing economies	15.0	17.1	16.7	12.6	13.9	15.2	17.1	17.7	10.4	13.9	17.3	10.9	8.2	14.6

Source: UN/ECE Common Database, derived from national statistics.

Note: Data for 1980-1990 refer to the east European CMEA countries (Bulgaria, Czechoslovakia, German Democratic Republic, Hungary, Poland and Romania) and to the former Soviet Union. Trade data in national currencies were revalued at consistent rouble-dollar cross-rates (see the note to appendix table B.10). For 1991-1996, eastern Europe covers Bulgaria, former Czechoslovakia (from 1993, Czech Republic and Slovakia including their new mutual trade, which is responsible for the substantial shifts in trade shares in that year), Hungary, Poland and Romania, and the second panel reflects trade of the Russian Federation only.

[a] China, Cuba, Democratic People's Republic of Korea, Mongolia, Viet Nam, former SFR of Yugoslavia.

[b] Former CMEA countries.

APPENDIX TABLE B.14

Exchange rates of the transition countries, 1980, 1984-1996
(Annual averages, national currency units per dollar)

	Unit[a]	1980	1984	1985	1986	1987	1988	1989	1990	1991	1992	1993	1994	1995	1996
Albania	lek	..	..	..	..	..	..	..	8.90	24.20	75.03	102.06	94.62	93.14	104.33
Bulgaria	leva	0.86	1.01	1.03	0.94	0.87	0.83	0.84	0.79	17.45	23.42	27.85	54.13	67.08	177.88
Czechoslovakia	koruna	5.37	6.64	18.51	16.20	14.79	14.37	15.06	18.56	29.56	28.30	..	..	..	..
Czech Republic	koruna	..	..	..	..	..	..	..	..	..	..	29.15	28.79	26.54	27.14
Slovakia	koruna	..	..	..	..	..	..	..	..	..	..	30.80	31.93	29.71	30.68
Hungary	forint	32.64	48.04	50.12	45.83	46.97	50.41	59.07	63.21	74.73	78.98	91.91	105.11	125.69	152.65
Poland	zloty[b]	3.05	113.63	147.17	175.28	265.08	430.64	1 439	9 500	10 576	13 627	18 136	22 723	2.42	2.70
Romania	leu	4.47	21.28	17.50	16.80	16.00	16.00	16.00	22.43	71.84	307.98	760.12	1 654	2 033	3 085
Yugoslavia (SFR)	dinar[c]	24.64	152.82	270.16	379.22	737.00	2'523	28 760	11.32	19.64	..	..	..	..	..
Bosnia and Herzegovina	..	..	..	..	..	..	..	..	..	..	..	..	..	..	..
Croatia	kuna[d]	..	..	..	..	..	..	..	..	18.80	264.30	3 578	6.00	5.23	5.43
Slovenia	tolar	..	..	..	..	..	..	..	..	27.57	81.29	113.24	128.81	118.52	135.37
The FYR of Macedonia	denar[e]	..	..	..	..	..	..	..	..	19.69	508.07	23.26	43.25	38.05	39.92
Yugoslavia	dinar[f]	..	..	..	..	..	..	..	..	19.64	508.07	..	..	..	..
Estonia	kroon[g]	..	..	..	..	..	..	..	..	..	12.11	13.22	12.98	11.46	12.03
Latvia	lats[h]	..	..	..	..	..	..	..	..	..	..	0.67	0.56	0.53	0.55
Lithuania	litas[i]	..	..	..	..	..	..	..	..	..	..	4.37	3.98	4.00	4.00
Armenia	dram	..	..	..	..	..	..	..	..	..	..	8.66	288.35	405.93	413.47
Azerbaijan	manat	..	..	..	..	..	..	..	..	..	..	..	1 169	4 417	4 296
Belarus	rouble[j]	..	..	..	..	..	..	..	..	..	..	2 177	4 017	11 538	13 472
Georgia	lari[k]	..	..	..	..	..	..	..	..	..	..	..	1.10	1.29	1.26
Kazakstan	tenge	..	..	..	..	..	..	..	..	..	..	..	35.54	60.95	67.30
Kyrgyzstan	som	..	..	..	..	..	..	..	..	..	..	..	10.86	10.83	13.08
Republic of Moldova	leu	..	..	..	..	..	..	..	..	..	..	..	4.07	4.50	4.60
Russian Federation	rouble[l]	0.65	0.82	0.84	0.70	0.63	0.61	0.63	0.59	1.74	192.75	927.46	2 204	4 559	5 121
Tajikistan	rouble	..	..	..	..	..	..	..	..	..	..	..	..	107.59	292.89
Turkmenistan	manat	..	..	..	..	..	..	..	..	..	..	..	19.50	110.42	3 509
Ukraine	hryvnia[m]	..	..	..	..	..	..	..	..	..	..	4 796	31 700	147 314	1.83
Uzbekistan	sum[n]	..	..	..	..	..	..	..	..	..	..	932.15	9.96	29.81	40.15
Memorandum items:															
Ex-GDR Länder	mark[o]	3.30	3.64	8.14	8.14	8.14	8.14	8.14	8.14	1.66	1.56	1.65	1.62	1.43	1.50

Source: UN/ECE Common Database, derived from national, IMF and CIS statistics. Annual averages are unweighted arithmetic averages of monthly values. Change or redenomination of currency is indicated by a vertical slash.

Note: Under the central planning system with its state foreign trade monopoly, exchange rates served primarily statistical and accounting purposes (notably the conversion of foreign trade values for statistics expressed in domestic currency), without direct impact on domestic price formation. Market-based exchange rates and a meaningful link to domestic currency values emerged only with the transformations from 1989 onward. The official exchange rates of the earlier period are therefore not suitable for the conversion to dollars of macroeconomic and other data of these countries expressed in domestic currency. These strictures should be kept in mind in the interpretation and use of the data for the 1980s shown above.

[a] Currency unit of the last period shown. For prior periods, see footnotes.

[b] The zloty was redenominated at 1:10,000 from 1 January 1995.

[c] The dinar was redenominated 1:10,000 from 1 January 1990.

[d] The kuna replaced a Croat dinar on 3 May 1994 at 1:1,000; the 1994 average is shown in kuna terms.

[e] The denar (which had replaced the Yugoslav dinar 1:1 on 26 April 1992) was redenominated 1:100 on 1 May 1993; the 1993 average is shown in terms of that unit.

[f] The dinar was further redenominated on 1 July 1992 (1:10), 1 October 1993 (1:1 million), 1 January 1994 (1:1 trillion) and 24 January 1994 (1:13 million). Average annual exchange rates not available after 1992.

[g] The kroon replaced the Soviet rouble in June 1992 with a peg to the deutsche mark (8:1); the average shown for 1992 refers to June-December.

[h] The lats replaced an earlier Latvian rouble at 1:200 on 18 October 1993; the 1993 average is shown in lats terms.

[i] The litas replaced the earlier talonas at 1:100 on 1 June 1993; the 1993 average is shown in litas terms.

[j] The Belarus rouble was redenominated 1:10 on 10 August 1994; the 1994 average here assumes this applied to the entire year. Annual averages were computed from end-of-period monthly rates.

[k] The lari replaced a lari-kupon on 25 September 1995; the annual average for 1994 is shown in million lari-kupon, and that for 1995 in lari.

[l] 1980-1991: Soviet rouble/dollar rate used in the conversion of foreign trade data for statistical purposes.

[m] The hrynvia replaced the former karbovanets on 2 September 1996 at 1:100,000; the average for 1996 is shown in hrynvia terms.

[n] Sum-kupon in 1993.

[o] GDR mark through 1990, deutsche mark thereafter.

APPENDIX TABLE B.15

Current account balances of the transition countries, 1990-1996
(Million dollars)

	1990	1991	1992	1993	1994	1995	1996[a]
Eastern Europe (revised)[b]	..	..	..	..	..	-1 389	-12 935
Eastern Europe	-3 631	-3 068	-712	-8 175	-4 203	-9 143	-20 088
Albania	-118	-168	-51	15	-43	-15	-85
Bulgaria	-1 710	-77	-360	-1 098	-25	-26	-34
Croatia	..	-589	329	104	103	-1 711	-1 129
Czech Republic	-338	1 143	-305	115	-50	-1 362	-4 476
Hungary	127	267	324	-3 455	-3 911	-2 480	-1 678
Poland (revised)	..	..	..	..	..	5 455	-1 352
Poland	716	-1 359	-269	-2 329	-944	-2 299	-8 505
Romania	-1 650	-1 369	-1 460	-1 174	-428	-1 639	-2 336
Slovakia	-767	-786	173	-559	712	646	-1 401
Slovenia	518	129	926	192	540	-36	47
The FYR of Macedonia	-409	-259	-19	15	-158	-222	-491
Baltic states	..	..	681	374	-67	-827	-908
Estonia	..	..	153	40	-178	-185	-223
Latvia	..	..	207	417	201	-27	-273
Lithuania	..	..	322	-84	-90	-614	-412
CIS	-6 300	2 500	-5 700	1 780	9 629	7 471	8 804
Belarus	..	..	..	0	-506	-567	-752
Republic of Moldova	..	..	..	-155	-82	-115	-134
Russian Federation[c]	-6 300	2 500	-5 700	2 700	11 378	9 305	10 243
Ukraine	..	..	..	-765	-1 161	-1 152	-553

Source: National balance of payments statistics.

[a] January-September except for Slovakia (January-November) and Croatia, Czech Republic, Hungary, Poland, Romania, Slovenia and The FYR of Macedonia which are full-year figures.

[b] Incorporating revision of Poland's current account data.

[c] 1990-1993 excluding transactions with Baltic states and CIS.

APPENDIX TABLE B.16

Inflows of foreign direct investment in the transition countries, 1990-1996
(Million dollars)

	1990	1991	1992	1993	1994	1995	1996[a]
Eastern Europe	449	2 326	3 120	3 999	3 469	9 152	7 171
Albania	..	..	20	58	53	70	76
Bulgaria	4	56	42	40	105	90	82
Croatia	..	..	16	74	98	81	340
Czech Republic	120	511	1 004	568	862	2 562	1 428
Hungary	311	1 459	1 471	2 339	1 146	4 453	1 983
Poland	10	117	284	580	542	1 134	2 741
Romania	-18	37	73	94	341	419	210
Slovakia	18	82	100	134	170	157	119
Slovenia	4	65	111	113	128	176	186
The FYR of Macedonia	..	..	..	..	24	9	6
Baltic states	..	..	101	241	536	457	443
Estonia	..	..	58	160	225	205	104
Latvia	..	..	43	49	279	180	224
Lithuania	..	..	..	31	31	73	115
CIS	-400	-100	700	1 098	817	2 363	2 436
Belarus	..	..	..	..	9	15	16
Republic of Moldova	..	..	..	..	12	64	91
Russian Federation	-400	-100	700	900	637	2 017	1 907
Ukraine	..	..	..	198	159	267	421
Total above	49	2 226	3 921	5 339	4 822	11 972	10 050
Asian CIS	..	..	150	376	783	1 596	884
Armenia	..	..	..	..	3	19	..
Azerbaijan	..	..	..	..	22	155	..
Georgia	..	..	2	2	8	6	..
Kazakstan	..	..	100	228	519	859	884
Kyrgyzstan	..	..	..	10	45	191	..
Tajikistan	..	..	8	9	12	13	..
Turkmenistan	..	..	..	79	103	233	..
Uzbekistan	..	..	40	48	72	120	..
Total	49	2 226	4 071	5 715	5 605	13 567	10 934

Source: National balance of payments statistics, cash basis. For the Asian CIS (except Kazakstan), The World Bank, *Statistical Yearbook 1996, States of the Former USSR* (Washington, D.C.), 1996.

[a] Annualized on basis of January-September data, except for the Czech Republic, Hungary, Poland, Romania, Slovenia and The FYR of Macedonia for which full-year data are available.

OTHER RECENT PUBLICATIONS FROM
UNITED NATIONS ECONOMIC COMMISSION FOR EUROPE
DIVISION OF ECONOMIC ANALYSIS

- *Economic Bulletin for Europe,* Vol. 48 (1996), Sales No. E.96.II.E.29

Review of economic developments in the ECE region with special emphasis on international trade and payments of the transition economies of eastern Europe, the Baltic states and the CIS. Special topics discussed include:

- Enlarging the European Union to the transition economies
- The foreign trade of the transition economies
- The re-emergence of trade among the east European and Baltic countries
- External financial developments (including FDI) in the transition economies

- International Migration in Central and Eastern Europe and the Commonwealth of Independent States, *Economic Studies No. 8*, Sales No. GV.E.96.0.22

An analysis of international migration within and from central and eastern Europe and the CIS. An overview of the migration flows and issues is followed by 11 country studies plus one on the countries of the former Soviet Union.

- Fertility and Family Surveys in Countries of the ECE Region. Standard Country Report: Norway, *Economic Studies No. 10a*, Sales No. GV.E.96.0.32

This is the first of a series of country monographs, based on national sample surveys, on long-term changes in fertility and family structures in Europe (east and west) and North America. The research programme is coordinated by ECE's Population Activities Unit, an integral part of the Economic Analysis Division, and involves the participation of national experts and population centres as well as national statistical offices.

* * * * * * * *

To obtain copies of these publications contact:

Publications des Nations Unies
Section de Vente et Marketing
Organisation des Nations Unies
CH-1211 Genève 10
Suisse

Tele: (4122) 917 2612 / 917 2606 / 917 2613
Fax: (4122) 917 0027
E-mail: unpubli@unog.ch

United Nations Publications
2 United Nations Plaza
Room DC2-853
New York, NY 10017
USA

Tele: (1212) 963 8302 / (1800) 253 9646
Fax: (1212) 963 3489
E-mail: publications@un.org